Accession no.
36236858

The Role of Women in the History of Geology

The Geological Society of London
Books Editorial Committee

Chief Editor

BOB PANKHURST (UK)

Society Books Editors

JOHN GREGORY (UK)
JIM GRIFFITHS (UK)
JOHN HOWE (UK)
PHIL LEAT (UK)
NICK ROBINS (UK)
JONATHAN TURNER (UK)

Society Books Advisors

MIKE BROWN (USA)
ERIC BUFFETAUT (FRANCE)
RETO GIERÉ (GERMANY)
JON GLUYAS (UK)
DOUG STEAD (CANADA)
RANDELL STEPHENSON (NETHERLANDS)

Geological Society books refereeing procedures

The Society makes every effort to ensure that the scientific and production quality of its books matches that of its journals. Since 1997, all book proposals have been refereed by specialist reviewers as well as by the Society's Books Editorial Committee. If the referees identify weaknesses in the proposal, these must be addressed before the proposal is accepted.

Once the book is accepted, the Society Book Editors ensure that the volume editors follow strict guidelines on refereeing and quality control. We insist that individual papers can only be accepted after satisfactory review by two independent referees. The questions on the review forms are similar to those for *Journal of the Geological Society*. The referees' forms and comments must be available to the Society's Book Editors on request.

Although many of the books result from meetings, the editors are expected to commission papers that were not presented at the meeting to ensure that the book provides a balanced coverage of the subject. Being accepted for presentation at the meeting does not guarantee inclusion in the book.

More information about submitting a proposal and producing a book for the Society can be found on its web site: www.geolsoc.org.uk.

It is recommended that reference to all or part of this book should be made in one of the following ways:

BUREK, C. V. & HIGGS, B. (ed.) 2007. *The Role of Women in the History of Geology*. Geological Society, London, Special Publications, **281**.

FALCON-LANG, H. J. & MILLER R. F. 2007. Marie Stopes and the Fern Ledges of Saint John, New Brunswick. *In*: BUREK, C. V. & HIGGS, B. (eds) *The Role of Women in the History of Geology*. Geological Society, London, Special Publications, **281**, 227–245.

GEOLOGICAL SOCIETY SPECIAL PUBLICATION NO. 281

The Role of Women in the History of Geology

EDITED BY

C. V. BUREK
University of Chester, UK

and

B. HIGGS
University College Cork, Ireland

2007
Published by
The Geological Society
London

THE GEOLOGICAL SOCIETY

The Geological Society of London (GSL) was founded in 1807. It is the oldest national geological society in the world and the largest in Europe. It was incorporated under Royal Charter in 1825 and is Registered Charity 210161.

The Society is the UK national learned and professional society for geology with a worldwide Fellowship (FGS) of over 9000. The Society has the power to confer Chartered status on suitably qualified Fellows, and about 2000 of the Fellowship carry the title (CGeol). Chartered Geologists may also obtain the equivalent European title, European Geologist (EurGeol). One fifth of the Society's fellowship resides outside the UK. To find out more about the Society, log on to www.geolsoc.org.uk.

The Geological Society Publishing House (Bath, UK) produces the Society's international journals and books, and acts as European distributor for selected publications of the American Association of Petroleum Geologists (AAPG), the Indonesian Petroleum Association (IPA), the Geological Society of America (GSA), the Society for Sedimentary Geology (SEPM) and the Geologists' Association (GA). Joint marketing agreements ensure that GSL Fellows may purchase these societies' publications at a discount. The Society's online bookshop (accessible from www.geolsoc.org.uk) offers secure book purchasing with your credit or debit card.

To find out about joining the Society and benefiting from substantial discounts on publications of GSL and other societies worldwide, consult www.geolsoc.org.uk, or contact the Fellowship Department at: The Geological Society, Burlington House, Piccadilly, London W1J 0BG: Tel. +44 (0)20 7434 9944; Fax +44 (0)20 7439 8975; E-mail: enquiries@geolsoc.org.uk.

For information about the Society's meetings, consult *Events* on www.geolsoc.org.uk. To find out more about the Society's Corporate Affiliates Scheme, write to enquiries@geolsoc.org.uk.

Published by The Geological Society from:
The Geological Society Publishing House, Unit 7, Brassmill Enterprise Centre, Brassmill Lane, Bath BA1 3JN, UK

(*Orders*: Tel. +44 (0)1225 445046, Fax +44 (0)1225 442836)
Online bookshop: www.geolsoc.org.uk/bookshop

The publishers make no representation, express or implied, with regard to the accuracy of the information contained in this book and cannot accept any legal responsibility for any errors or omissions that may be made.

© The Geological Society of London 2007. All rights reserved. No reproduction, copy or transmission of this publication may be made without written permission. No paragraph of this publication may be reproduced, copied or transmitted save with the provisions of the Copyright Licensing Agency, 90 Tottenham Court Road, London W1P 9HE. Users registered with the Copyright Clearance Center, 27 Congress Street, Salem, MA 01970, USA: the item-fee code for this publication is 0305-8719/07/$15.00.

British Library Cataloguing in Publication Data

A catalogue record for this book is available from the British Library.

ISBN 978-1-86239-227-4

Typeset by Techset Composition Ltd, Salisbury, UK

Printed by Cromwell Press, Trowbridge, UK

Distributors

North America
For trade and institutional orders:
The Geological Society, c/o AIDC, 82 Winter Sport Lane, Williston, VT 05495, USA
Orders: Tel +1 800-972-9892
            Fax +1 802-864-7626
            E-mail gsl.orders@aidcvt.com

For individual and corporate orders:
AAPG Bookstore, PO Box 979, Tulsa, OK 74101-0979, USA
Orders: Tel +1 918-584-2555
            Fax +1 918-560-2652
            E-mail bookstore@aapg.org
            Website http://bookstore.aapg.org

India
Affiliated East-West Press Private Ltd, Marketing Division, G-1/16 Ansari Road, Darya Ganj, New Delhi 110 002, India
Orders: Tel +91 11 2327-9113/2326-4180
            Fax +91 11 2326-0538
            E-mail affiliat@vsnl.com

Contents

Preface	vii
BUREK, C. V. & HIGGS, B. The role of women in the history of geology: an introduction	1
BUREK, C. V. The role of women in geological higher education – Bedford College, London (Catherine Raisin) and Newnham College, Cambridge, UK	9
CREESE, M. R. S. Fossil hunters, a cave explorer and a rock analyst: notes on some early women contributors to geology	39
FRASER, H. E. & CLEAL, C. J. The contribution of British women to Carboniferous palaeobotany during the first half of the 20th century	51
HART, J. K. The role of women in British Quaternary science	83
WYSE JACKSON, P. N. & SPENCER JONES, M. E. The quiet workforce: the various roles of women in geological and natural history museums during the early to mid-1900s	97
BUREK, C. V. & KÖLBL-EBERT, M. The historical problems of travel for women undertaking geological fieldwork	115
CLARY, R. M. & WANDERSEE, J. H. Great expectations: Florence Bascom (1842–1945) and the education of early US women geologists	123
HIGGS, B. & WYSE JACKSON, P. N. The role of women in the history of geological studies in Ireland	137
KÖLBL-EBERT, M. The role of British and German women in early 19th-century geology: a comparative assessment	155
TURNER, S. Invincible but mostly invisible: Australian women's contribution to geology and palaeontology	165
BUREK, C. V. & MALPAS, J. A. Rediscovering and conserving the Lower Palaeozoic 'treasures' of Ethel Woods (neé Skeat) and Margaret Crosfield in northeast Wales	203
FALCON-LANG, H. J. & MILLER, R. F. Marie Stopes and the Fern Ledges of Saint John, New Brunswick	227
LAMING, S. & LAMING, D. Etheldred Benett (1776–1845): the first woman geologist?	247
MATHER, J. D. & CAMPBELL, I. Grace Anne Milne (Lady Prestwich): more than an amanuensis?	251
MORGAN, N. Anne Phillips: John Phillips's geological companion	265
ORR, M. Keeping it in the family: the extraordinary case of Cuvier's daughters	277
ROBINSON, E. The influential Muriel Arber: a personal reflection	287
SHINDLER, K. A knowledge unique: the life of the pioneering explorer and palaeontologist, Dorothea Bate (1878–1951)	295

WACHTLER, M. & BUREK, C. V. Maria Matilda Ogilvie Gordon (1864–1939): Scottish researcher in the Alps   305

WILLIAMS, T. P. T. The role of Annie Greenly in the elucidation of the geology of Anglesey   319

WYATT, A. R. Nancy Kirk: turning the world of graptolites upside down   325

Index   335

Preface

The seeds of this book lie back in 1998, during discussions which took place in the middle of the Atlantic Ocean at the Fundación Canaria Orotava de Historia de La Ciencia in La Oratava, Tenerife. At this time a European project, Penelope, under the Comenius initiative, was taking place (led by Agustin Isidro, coordinator of the project in Tenerife), to include the history of science within the secondary school European National Curricula (Cartwright & Burek 2000). Three countries – Spain, the United Kingdom and France (specifically Tenerife, Chester and Rennes, respectively) – had come together to develop an educational course to emphasise the importance of the history of a science subject as a context for that subject and as a possible recruitment agent. This resulted in a conference at Fundación Canaria Orotava de Historia de La Ciencia, La Orotava, España, on 22–28 July 2001, entitled 'The Role of the History of Sciences in Secondary Education' (González 2002).

It was hoped that this exposure might encourage students to study science by introducing it through a historical context. Serendipity played a part here as each country had one female scientist attached to it.

During these discussions it became apparent that, although mathematics, biology, geology, physics and the history and philosophy of science were to be represented, the role that women played did not feature. After discussion, it was decided to run a workshop supported by two short lectures on the work of famous female mathematicians and geologists. Anne Boyé of the University of Nantes would illustrate the former and Cynthia Burek the latter. Anne's job was made easier as a booklet aimed at high-school teachers on famous female mathematicians, such as Sophie Germaine, Maria Agnesi and Emmy Noether, had just been produced in France (Boyé 2002).

However, female geologists were another matter. Few scientific papers had been published and what Burek thought was to be a simple task turned out to be 4 years of hard research. Several researchers, such as Hugh Torrens (1995), Martina Kölbl-Ebert (1997) and Mary and Tom Creese (1994), with some others represented here, had touched on the subject and produced individual papers on Mary Anning, Mary Buckland and Etheldred Benett in the United Kingdom, Alice White in Canada (Sarjeant 1984) and Florence Bascom in the United States (Arnold 1977). However, other more general material, such as that produced in *The Scientific Lady – A Social History of Woman's Scientific Interests Between 1520–1918* by Phillips (1990) and, more recently, *Women and Science* (Le May 2005) barely mentioned female geologists. Indeed *Hypatia's Heritage* by Alic (1986), subtitled *A History of Women in Science from Antiquity to the Late Nineteenth Century*, has less than two pages out of 190 on female geologists. *Women of Science: Righting the Record*, edited by Kass-Simon & Farnes (1990), did contain a small chapter on 'Women in geology' by Aldrich, but it concentrated mainly on women in the United States. She did, however, cover some of the roles they adopted as illustrators and workers in the State surveys and in the early petroleum industry. One paragraph is devoted to British women collectors and authors in the 19th century. Work undertaken by Paula Gould (1998) at Cambridge University mentioned some female geologists, but only in passing.

During the next 6 years, some relevant papers were published and history of science books grasped the public's imagination such as *The Map that Changed the World* by Simon Winchester (2001), *The Dating Game* by Cherry Lewis (2000) and *The Dinosaur Hunters* by Dorothy Cadbury (2000). The role women played sometimes surfaced but it was clear that there was room for more research and certainly more exposure. Thus the conference entitled 'The role of women in the history of geology' was born in November 2005.

The context of this conference was to raise not only the work of female geologists but also the role that women played in the development of the science of geology. Individual papers on female geologists had been accepted for the conference but early on it was decided that, if this was the case, we would concentrate on those no longer able to tell their own story. The publishing house of the Geological Society of London then offered to publish a book based on the proceedings of the conference with other contributions not presented at the time. They recognized this as the first of its kind and we have to thank Angharad Hills and the reviewers of the original proposal for their foresight in this.

I must thank both my fellow convenors of the conference, Professor John Mather of the University of London and Dr Bettie Higgs of University College Cork, for their willingness to discuss, organize and take forward this vision, born like a volcano erupting out of the middle of the ocean from its hidden depths, to be seen by all.

I would also like to publicly acknowledge the History of Geology Group (HOGG), the Geological

Society of London, and the UK Resource Centre for Women in Science, Engineering and Technology (WiSET) for their active support and financial contributions, and those men and women back in Tenerife (especially Agustin Isidro), who did not laugh but actively supported the idea of women as role models in science.

We hope that this book will result in raising the profile of women geologists and the fundamental role they have played in the history of the science. This should then encourage other women to take up the baton rather than the hammer in today's world and advance geological knowledge. We would like to take as a conceptual analogy the subtitle of Kölbl-Ebert's paper in *Earth Science History* (2002): 'A conglomerate with a female matrix'. This is the way forward.

References

ALDRICH, M. L. 1990. Women in geology. *In*: KASS-SIMON, G. & FARNES, P. (eds) *Women of Science: Righting the Record*. Indiana University Press, Bloomington, 42–71.

ALIC, M. 1986. *Hypatia's Heritage: A History of Women in Science from Antiquity to the Late Nineteenth Century*. The Women's Press Ltd, London, 230 pp.

ARNOLD, L. B. 1977. American women in geology: a historical perspective. *Geology*, **5**, 493–494.

BOYÉ, A. 2002. Les femmes dans les mathematiques. *In*: GONZÁLEZ, M. H. (ed.) *Proyecto Penélope: The Role of the History of Science in Secondary Education*. Proceedings of a conference held in Fundación Canaria Orotava de Historia de La Ciencia, La Orotava, Tenerife, España, 22–28 July 2001. Fundación Canaria Orotava de Historia de la Ciencia, La Orotava, 206–213.

BUREK, C. V. 2002. Where are the women in science? A case study using women in the history of geology to develop a European curriculum. *In*: GONZÁLEZ, M. H. (ed.) *Proyecto Penélope: The Role of the History of Science in Secondary Education*. Proceedings of a conference held in Fundación Canaria Orotava de Historia de La Ciencia, La Orotava, Tenerife, España, 22–28 July 2001. Fundación Canaria Orotava de Historia de La Ciencia, La Orotava, 214–222.

CADBURY, D. 2000. *The Dinosaur Hunters: A True Story of Scientific Rivalry and the Discovery of the Prehistoric World*. Fourth Estate, London.

CARTWRIGHT, J. & BUREK, C. V. 2000. *The Penelope Project: International Perspectives on Teaching History of Science at Secondary Level*. World Wide Web Address: http://www.ucl.ac.uk/sts/bshs/papers/cartw6-6.doc.

CREESE, M. R. S. & CREESE, T. M. 1994. British women who contributed to research in the geological sciences in the nineteenth century. *British Journal of History of Science*, **27**, 23–54.

GONZÁLEZ, M. H. (ed.) 2002. *Proyecto Penélope: The Role of the History of Science in Secondary Education*. Fundación Canaria Orotava de Historia de la Ciencia, Tenerife, España, Proyecto Sócrates – Comenius legal deposit TF-168/2002, 241 pp.

GOULD, P. 1997. Feminity and physical science in Britain, 1870–1914. PhD thesis, University of Cambridge, Cambridge.

KÖLBL-EBERT, M. 1997. Mary Buckland née Morland 1797–1857. *Earth Science History*, **16**, 39–43.

KASS-SIMON, G. & FARNES, P. (eds) *Women of Science: Righting the Record*. Indiana University Press, Bloomington.

KÖLBL-EBERT, M. 2002. British geology in the early nineteenth century: a conglomerate with a female matrix: *Earth Science History*, **21**, 3–25.

LE MAY, S. 2005. *Women and Science: Social Impact and Interaction*. ABC Clio Science and Society Series, California, 409 pp.

LEWIS, C. 2000. *The Dating Game: One Man's Search for the Age of the Earth*. Cambridge University Press, Cambridge, 253 pp.

PHILLIPS, P. 1990. *The Scientific Lady: A Social History of Woman's Scientific Interests Between 1520–1918*. Weidenfeld & Nicolson, London, 279 pp.

SARJAENT, W. A. S. 1984. Women in Geology: 3. Alice Wilson. *Open Earth*, **22**, 10–11.

TORRENS, H. S. 1995. Mary Anning (1797–1847) of Lyme: 'The greatest fossilist the world ever knew'. *British Journal for the History of Science*, **28**, 257–284.

WINCHESTER, S. 2001. *The Map that Changed the World: William Smith and the Birth of Modern Geology*. Viking, London.

Professor Cynthia Veronica Burek
and
Dr Bettie Higgs

The role of women in the history and development of geology: an introduction

C. V. BUREK[1] & B. HIGGS[2]

[1]*Centre for Science Communications, Department of Biological Sciences, University of Chester, Parkgate Road, Chester CH1 4BJ, UK*
(e-mail: c.burek@chester.ac.uk)

[2]*Department of Geology, University College Cork, Cork, Ireland (e-mail: b.higgs@ucc.ie)*

A conference held at the London Geological Society, Burlington House, London, on 28 November 2005, hosting over 70 participants, was the first to deal solely with the role that women played in the history and development of the science of geology. Sixteen papers were read and there were two poster presentations. Prior to this, there have been individual articles written and papers published on the historical role of women in the history of the geosciences, but this collection of conference papers is, surprisingly, the first time a book has been published bringing the evidence together and giving an overview and a selection of detailed case histories.

In carrying out this project the authors ask: 'Can we really analyse the situation for women in the geosciences today without knowing what happened in the past?' Consequently, the collection of papers in this book mainly deals with the late 1700s to early 1900s, but also offers some links to the present day. It attempts to evaluate the contribution of women, and their changing roles, in the development of geology as a science. This undertaking has allowed a number of themes and common issues to emerge and be identified, which will be drawn out and discussed in this introduction. This work suggests that, in relation to our question, the past is the key to the present.

It is interesting to note that it is not only women who have researched these case histories; there are valuable contributions from respected male colleagues. The mix includes review papers referring not only to the development of geology in Great Britain, but also in other European countries, Australia, and North America. There are papers that look at a particular role, such as women as museum curators, or at a particular issue, such as travel for women during field studies. There are also several papers that focus on the contribution of a particular individual. The conference was publicized using the image of Etheldred Benett (1776–1845), and so an introduction to this early pioneer, originally submitted as a poster, is included in this book.

The book is not intended to be an exhaustive study of all women who played a role, as the work of well-known individuals, such as Mary Anning (1799–1847), (Torrens 1995; Tickell 1996; Burek 2001, 2002, 2004) are documented in detail elsewhere. There are other important histories still waiting to be uncovered, including some very influential women, such as Professor Janet Watson (1923–85), first female President of the Geological Society of London, Dr Doris Reynolds (1899–1985), who developed new ideas on the origin of granites, and Marie Tharp (1920–2006), whose work led directly to the first map of the Atlantic seabed and helped forward ideas on seafloor spreading.

The book is necessary because both historians and scientists have neglected the topic to a certain extent. Historians have sometimes omitted to mention a male geologist's female research assistant, or intellectual wife, sister or daughter, beyond stating that: 'she was following the fashion' or 'she was unusual for her time'. Modern scientists have considered these stories and the gender issue unimportant to their present-day specialized research. However, the picture emerging suggests that these stories are important to the present day. It seems that the treatment of women in the past has left a legacy that has not entirely been overcome in the 21st century. In this book we want to bring these issues to a wider audience, to highlight this legacy, and to ask: 'What *has* changed?'.

This book crosses the divide between science and the humanities. It is important to note that many contributors are writing outside of their normal discipline to document the role of women in the history of geology. They are palaeontologists, geoconservationists, geophysicists and hydrogeologists, to name but a few. Why would they do this? It is because there has been a gap in our understanding of the role that women have played, and because the research has uncovered fascinating stories. Those who began this research some time ago are becoming more proficient and expert in this interdisciplinary work. The book has benefited from a wide range

From: BUREK, C. V. & HIGGS, B. (eds) *The Role of Women in the History of Geology*. Geological Society, London, Special Publications, **281**, 1–8.
DOI: 10.1144/SP281.1　0305-8719/07/$15.00 © The Geological Society of London 2007.

of reviewers, including geologists, historians and educationalists. A variety of styles in this book is evident and is a result of the interdisciplinary nature of the project. During the review process the strict science historians, when reviewing, were quite critical of the style of some of the contributions, while disciplinary geologists were highly complimentary of the same papers and vice versa!

Characteristics of women who played a role

The early female scientists had a number of common characteristics. They were often born into influential families, for example Grace Milne who was 'born in 1832...eldest child of Louis Falconer and sister of the eminent botanist and palaeontologist, Hugh Falconer' (**Mather** 2007). This position in society allowed some women to work voluntarily, for no pay and usually no status, because they had a private income or were supported by a man. It was common for male scientists to have women assistants, and the well-known male geologists of the time encouraged women to do some of the time-consuming work of writing and illustrating. These women often went unacknowledged and become lost to history. For example, Mary Morland (1797–1857) illustrated some of George Cuvier's work before she became Mrs William Buckland. **Mary Orr** (2007) also writes of Cuvier's 'extraordinarily enlightened encouragement and employment of his daughter, and perhaps even more importantly his step-daughter, in his collaborative projects...as his "research assistants"'. Their stories have not been told before.

A number of women benefited from their associations and wrote geological articles under their own names. These were often for wider audiences, for example, in the case of Muriel Agnes Arber (1913–2004), it was her aim 'to present Earth history...in terms that the general public could understand' (**Robinson** 2007). Some of these geological articles were particularly for women and children. For example, Grace Milne (1832–99) wrote six chapters in *Every Girls Magazine*, and for women she wrote three articles in *Good Words* on the forces of nature that shaped the Earth (**Mather** 2007). This formed part of the whole genre of literature for women and children at this time (often written by women) and was in the form of conversations between a parent and child or between a teacher and pupil. For example, Arabella Buckley (1840–1929), Charles Lyell's long-time secretary, took to writing this kind of literature after Lyell's death in 1875 (**Burek** 2007). Her book entitled *The Fairy-land of Science* (Buckley 1879) is based on a series of lectures she gave to children and contains chapters on 'The history of a piece of coal' and 'The two great sculptors – water and ice'. It can be argued that, as Lyell's secretary, she was exposed to new geological ideas and, as a woman herself, she was well placed to pass on this knowledge to other women.

Several biographies of men were written after their death by women to whom they were related. These biographies show the depth of understanding the authors had of the geological material. Most telling was that some women were capable of understanding the importance of publishing their husband's work posthumously, such as Hugh Falconer's wife Lydia. Indeed William Buckland would not have finished the Bridgewater treatise without his wife Mary. However, in the 19th and 20th centuries, it was common for women to publish their own scientific work anonymously, or under the name of a male relation. **Wyse Jackson & Spencer Jones** (2007) in researching women as curators, report that:

Women typically were employed in the major national or university museums as preparators, illustrators or assistants and this trend continued until the 1930s. These women received little academic credit for their research as it was frequently incorporated into the publications of the men for whom they worked.

In the past, women with geological ambitions had to be pioneers (even during the last 40 years), sometimes 'ploughing a lonely [and marginalized] furrow' (**Watchler & Burek** 2007). However, in bringing together the histories of individual women, it can be seen that they were supported by a network of other women and male supporters. For example, Archibald Geikie supported Marie Ogilvie Gordon (1864–1939) (**Watchler & Burek** 2007), and Professors McKenny and Lapworth supported Ethel Wood (1871–1945) and other members of Newnham and Girton Colleges in Cambridge (**Burek** 2007). Several of the women highlighted in this book knew each other, either as contemporaries or role models. For example, Grace Milne met Mary Somerville on her travels, and later helped in the establishment of Somerville Hall (**Mather** 2007). Gertrude Elles (1872–1960) and Ethel Wood were collaborators with Margaret Crosfield (1859–1952) and Ethel Skeat (1865–1939) (**Burek & Malpas** 2007), and Catherine Raisin (1855–1945) acted as a role model for Doris Reynolds, as well as other early geologists (**Burek** 2007).

Influence of society

Women had to have a pioneering spirit because society as a whole did not support the ambitions of women to follow their geological interests until the late 20th century. The learned scientific societies were slow to accept women. It seems that those in power held real and deep-seated prejudices. For example, Murchison, who was president of the Geological Society, has private journals that show his low regard for women (**Mather** 2007; **Burek & Malpas** 2007). Women were regarded by some as weak and frivolous, reducing the seriousness of any study if allowed to participate (McEwan 1998). Thus, they were barred from scientific societies, such as the Royal Society and the Geological Society of London (until they were forced to accept women by law in 1919), due to the perception that women had a lack of intellectual rigour and understanding necessary to engage in serious science. They were also discouraged from fieldwork.

[Geologists and] geographers were not just men of science, but men of action. Women were therefore doubly excluded, first by the refusal to admit them to the 'rational sphere' of science, learning and public debate; but also by assumptions about their (lack of) ability to meet the physical challenge that fieldbased work involved. (Bracken & Mawdsley 2004)

Brickhouse (2001) offers insights into these attitudes towards women. Enlightenment epistemology was significant in that it persuaded society that dualisms, such as masculine and feminine, existed. The perceived masculine characteristics, such as objectivity and reason, were seen as superior and were aligned with science. In this way science was culturally defined in opposition to women. This work can help us to understand the attitudes of male scientists at the time and also allow us to gauge whether positive change has occurred.

To its credit, the Geologists' Association (GA), based in London, stands out from the other societies because it devoted itself to the needs of the amateur. From its foundation in 1858, women members had equal rights with men within the society (**Burek** 2007). In 2006, GA female members numbered about 439, out of 1650, or roughly 25%, of the members. This is higher than most other geological societies and associations except the Open University Geological Society (c. 50%). The Geological Society of London stands at an average of 17% female fellows, but this ranges from 46% (aged 21–25 years) to 7% (aged 51–55 years) in 2006 (Boning 2007) and demonstrates the slow leakage that occurs as women progress through their geological career. It is interesting to note that the Geological Society is taking this very seriously and is looking to set up a pilot project on the mentoring of younger fellows by their older colleagues (Boning 2007).

It is clear from the contributions in this book that access to education played a crucial part in the advancement of women's roles in the geological community. Women had to struggle for decades to change male opinion. The availability of schooling, and then tertiary level education, enabled the gender imbalance to begin to be redressed. In this book, the evolution of two female colleges of higher education, Bedford College, London, and Newnham College, Cambridge, both offering geological education within science, is outlined and evaluated (**Burek** 2007). Here role models existed in the forms of Dr Catherine Raisin, based at Bedford College, London, and Dr Gertrude Elles at Newnham College, Cambridge. In this struggle for education, women had their male opponents, such as the well-known mineralogist and educationalist Sir Robert Kane, but they also had their supporters, such as the professors of geology, Samuel Haughton and McKenny Hughes.

Several papers highlight the fact that, even when women gained a job, there were other barriers to overcome.

Anna Birchall Hastings was appointed to the British Museum of Natural History, but upon her marriage was required to relinquish her post, even though she had married a museum palaeontologist – he remained in his post while she became a volunteer. (**Wyse Jackson & Spencer Jones** 2007)

Compare this with Dorothea Bate (1878–1951), who never received payment for her work as an employee at the museum and never married (**Shindler** 2007). She was still producing research work into her 70s as was Maria Ogilvie Gordon (**Wachtler & Burek** 2007).

In the 1960s in Europe, female undergraduate students were told there were no professional jobs for women in geology and that teaching in a school was the only outlet (**Kölbl Ebert** 2007). Certainly geological mapping was considered unsuitable for women. So it must have come as a surprise to some when the first female field geologist, Audrey Jackson (**Higgs & Wyse Jackson** 2007), was appointed by the British Geological Survey in 1969. Interestingly, even into the 1970s, women postgraduate researchers had to use covert measures to gain access to coal mines for their sampling. Typically a student would write only their initials on a letter requesting a site visit (Langley pers. comm.). The mine manager, when responding positively, would assume the request was from a male student. Excuses then followed and varied from 'We do not have shower facilities for females' to 'It is unlucky to allow women down the mine'.

So, historically, barriers to employment as a field geologist existed and, until 1975, women who married had to resign from the Geological Survey and other Civil Service roles. Archives show that women typically worked for 2 or 3 years before disappearing from the records (**Burek** 2007; **Higgs & Wyse Jackson** 2007). The young male geologists of 30 years ago are now in senior positions of power, sitting on committees and selection panels. Do undergraduate women see academia as a no-go area for themselves, perpetuating the situation? Whitelegg & Smidt (2004) believe that, when there are no female role models, women have a dislike for the culture and develop pessimistic views about their own career potential.

The situation today

As Professor Dervilla Donnelly, an influential physicist at the Dublin Institute for Advanced Studies (DIAS) said, at the launch of a Women in Science, Engineering and Technology (WiSET) initiative at University College Cork, Ireland, in September 2006: 'Women are no longer discriminated against in science'. Is she correct (Higgs *et al.* 2005)? When asked why the present-day numbers do not tally, she admitted that there was a legacy to overcome. Members of the audience, in response, said that, even today, women continue to have challenges that their male counterparts do not have. The responsibility of bringing up the next generation may be shared to an extent, but the burden still weighs heavier on women. For example, in Ireland, only two women have succeeded in having professional geological careers in academia while also being mothers.

What is this legacy?

The legacy of the model that saw science as a masculine pursuit has given us a deficiency of female role models in the geological sciences. Most lecturers and leaders of geological field trips are male. The structures that still exist today were set up by men, for men.

The masculine culture of departments was universally recognised; there is a sense in which women have to become honorary men in order to survive as students. (Ward 1992)

Some literature suggests that having female role models could make progress in academia easier. The Women into Science and Engineering (WISE) initiative saw this as important and included positive role models to encourage female students into science. The GETSET women (Get Science Engineering & Technology) initiative used participation in 'empowering residential courses' to engage women in science (Whitelegg & Smidt 2004). In running the conference (The Role of Women in the History of Geology) we were fortunate to receive a grant from the UK Resource Centre for WiSET, based at Bradford University, as it was an innovative project.

It is clear that some changes are being made today to allow for equality of opportunity. However, Whitelegg & Smidt (2004) believe that unwelcoming behaviours still exist, both cultural and societal, and although they may be small and seemingly unnoticeable, yet cumulatively they are significant informal barriers. Indeed, some women will see their opportunities as limited and will not engage in science.

When women do engage in science do they get what they need? Should they be treated differently? There has been very little work done on gender inclusivity in the field-based sciences in higher education. Is the experience any different for female and male students? Some evidence, which suggests that it is, comes from the description of a field activity by Dixon (1999) in the United Kingdom. His work suggests that females favour collaborative and group work. Bracken & Mawdsley (2004), however, writing specifically about physical geography fieldwork, mention the diverse nature of fieldwork and seek to stress that fieldwork must not continue to be coded and perceived as a masculine area (see also **Hart** 2007). As fieldwork moves to more local domains, detailed field research may be more attractive to women scientists. Fieldwork problems in the past are covered by **Burek & Kölbl-Ebert** (2007).

It is good to note that, in awarding prizes for undergraduate fieldwork in some universities over the past 10 years, women have succeeded as often as men. However, encouraging women into academia is still problematic. In the geology departments of the higher education institutions in Ireland, for example, only two female full-time permanent staff members have been appointed, highlighting the deficiency in role models.

A report by the UK Earth, Marine and Environmental Science cost centre for the Higher Education Statistical Agency (HESA) in 1994/5 listed 4 female professors out of 160 male while in 2003/4 it had increased to 25 female professors out of a total of 325 male (HESA 1995, 2004), an increase from 2.5% to 7% of the total. In British universities, role models include the 20 professors who are researching and working in the broad area of earth science (see Table 1 & Fig. 1).

These 20 female professors are from 16 different universities, with Royal Holloway, University of London and University of Leeds, topping the list with three entries. In order to find these statistics, 37 university departmental websites were searched using the Geological Society of London listings. It is interesting to note that the majority of professors are researching in the field of geophysics, physical geography/Quaternary earth science and environmental issues. These last two perhaps reflect the growing concern with issues of climate change and the well-being of the planet in general. This is a very different picture from a century ago, when the majority of work undertaken by females at all levels was in the palaeontological area. Figure 1 shows the distribution of professors and their disciplines.

Table 1. *Female professors in Earth sciences January 2007 (including relevant archaeological applications) in the UK (listed alphabetically)*

Professor	University	Chair or area of research
Burek, Cynthia	Chester	Geoconservation
Burgess, Jacqueline	East Anglia	Environmental Geography
Collinson, Margaret	Royal Holloway, London	Plant Palaeobiology
Das, S.	Oxford	Earth Sciences seismology
Downes, Hilary	Birkbeck/University College, London	Geochemistry
Ebinger, Cynthia	Royal Holloway, London	Structural
Edwards, Dianne	Cardiff	Palaeobotany
Foulger, Gillian	Durham	Geophysics
Fowler, Mary	Royal Holloway, London	Geophysics
Francis, Jane	Leeds	Palaeoclimateology
Frostick, Lynne	Hull	Physical Geography
Hart, Jane	Southampton	Physical Geography
Heywood, Karen	East Anglia	Oceanography
Kneale, Pauline	Leeds	Applied Hydrology & learning
Petts, Judith	Birmingham	Environmental Risk Management
Plant, Jane	Imperial College, London	Applied Geochemistry
Ragnarsdottir, Vala	Bristol	Environmental Sustainability
Whaler, Kathryn	Edinburgh	Geophysics
Wilson, B. Marjorie	Leeds	Igneous Petrogenesis
Wintle, Ann	Aberystwyth	Quaternary geoscientist

So, acquiring positive role models is still problematic. For the most part, in the 'power situation', females are still the students; males are the teachers and supervisors. In 1997 nearly 75% of women geoscientists were under 40 years old and so few role models were available (Geological Society 1997). As Whitelegg & Smidt (2004) point out, there are still many factors combining to discourage women from progressing to academia. One exception is the Open University, United Kingdom, where a large number of women are employed to tutor Earth sciences, including field-based courses. This may initially seem to be very positive. However, these positions are typically part-time and temporary.

The lessons of the past presented in this book (**Burek** 2007; **Higgs & Wyse Jackson** 2007; **Kölbl-Ebert** 2007) show that most male geologists do not do U-turns and change their minds about the role of women in geology. The geological debate on gender focused on women as 'the problem', when it could have been focusing on teaching and societal attitudes as the problem. If women have different educational needs, and these needs are ignored, then equal treatment may fail to deliver equal

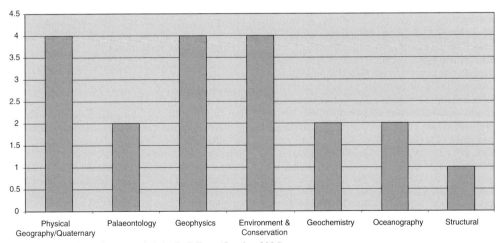

Fig. 1. UK female professors and their disciplines (October 2006).

outcome. This is an important realization and message for the future; 'equal treatment' does not necessarily mean 'equality of opportunity'.

Brickhouse (2001) documents a call to change unreflective 'masculine curricula' into empowering curricula, with an increase in collaborative group work; for changes to the traditional teacher/student hierarchy; and for new pedagogies which try to give more consideration to students' ideas and students' needs. In addition, Brickhouse (2001) recommends mentors to encourage female students to consider new territories. With appropriate scaffolding, linking students to the outside world, new identities can be built. Current practice in higher education may be failing to help with this finding identity, due to lack of awareness, and stereotypes may dominate. Using contexts with which female students identify, and which allow them to see a path they would like to follow, is more likely to motivate them to progress in science (Whitelegg & Edwards 2001). So, we must not simply add women to science, but change the structure of science to make it accessible to women who would like to pursue a career in science. This is discussed in detail by Schiebinger (1999).

A survey carried out by Burek & Higgs (2004), asking people to name ten female scientists, revealed that the public perception and knowledge of women's contributions in science is poor. This on-going study is part of a project on 'Public understanding of women in the history of science' and has tested the hypothesis that women were influential in the history of science, but have been forgotten. The countries sampled were the United Kingdom, Ireland, France, Spain and Germany. The authors found that people cannot name ten female scientists from any historical age, or nationality, even though they are there. From a total sample of 500 respondents, typically only one or two female scientists were named, with Marie Curie being the most well known and listed by 72% of participants across Europe; 28% of respondents listed her alone. The most common female geologist named in the British Isles was Mary Anning, 'the dinosaur woman'. She was listed by 10% of respondents.

On the positive side, there has been an increase in the number of women taking places in consultancy companies that engage in fieldwork. This is partly because the number of commercial geoscience companies has increased in recent years. It is the expansion of job opportunities in the geosciences during the past two decades that has allowed an increase in the number of female (as well as male) professional geologists. This is seen particularly in environmental geology, hydrogeology and geophysics and, during the last decade, in the petroleum, mining and quarrying industries. These women are mapping out paths and providing role models in industry and government bodies, and lifting the aspirations of current female students. Today there is a lack of young people entering the geoscience professions (Geological Society 1997). Perhaps by providing role models for half of the population, we can address this shortfall.

It was reported in a talk given by Annette Williams in 2004, at the UK All-Party Parliamentary Group for Earth Science Conference on improving the effectiveness of education resources for earth science and industry, that women are still hugely under-represented in science, engineering and technology, including the earth sciences. This is despite outperforming their male counterparts at GCSE and 'A' level examinations in key subjects. Thus women represented an untapped resource of creativity and innovation. The Science, Engineering and Technology (SET) unit at the University of Bradford has established an expert-women's database (GETSET) and is working to raise the profile of women in SET in the media. They hope this will make an impact on neutralizing gender stereotypes and overcoming cultural barriers. Women lost to industry or not using their qualifications in the fullest sense are classified as 'potential returners' to STEM (Science, Technology, Engineering and Maths). Many of these women may be working on the 'edges of science', for example as part-time workers at heritage centres, in middle management, in environmental education centres, etc. The question is 'Why do they choose to leave the industry?'. The SET unit is working with companies to look at retention, work-place issues and attitudes to women in the workplace. Whatever the answers, industry and some geological surveys have been ahead of the game, employing more women than the academic institutions. There are still some university geology departments that are into their third century of existence and have yet to appoint a female to a full-time permanent academic post (**Higgs & Wyse Jackson** 2007).

Conclusions

Our understanding of the present-day contribution of women in the geological sciences depends on an understanding of their past roles and the restrictions of travel, study and work placed upon them. Despite the difficulties of prejudice and adversity, especially within Victorian society, women have contributed significantly to the development and history of geology through the various roles they have played. These roles include researchers, teachers, illustrators, collectors and practitioners. Their contributions have often been lost, or buried under the wealth of new knowledge and publications emanating from the leading men within the field. The understanding of the societal context in which they (both men and women) operated is paramount, and the subsequent conclusions drawn, must be evidence based. For this reason, the overviews and case histories presented at the November 2005 conference represent many months of researching this evidence in archives, museums, private collections, libraries, and even in churchyards, as well as by personal communications and reflections. Some of the evidence is documented in previous books, which dealt with broader histories. If geologists did not venture outside of their normal specialized disciplines to research and write, there is a danger that some of this evidence would be lost. If non-geologists do not see the value of their research within geological science, the evidence would be lost too (**Orr** 2007). The bringing together of this evidence has allowed new connections to be made and will provide a valuable base on which future work will build.

It is hoped that both geoscientists and historians will expand on this work so that more stories are told and the effect of societal restrictions is not forgotten. Histories such as these help us to understand the present, to plan for the future, and to build a diverse, multi-talented and equitable community of scientists.

References

BONING, R. 2007. Narrowing the gender gap. *Geoscientist*, **17**, 4–5.

BRACKEN, L. & MAWDSLEY, E. 2004. 'Muddy glee': rounding out the picture of women and physical geography fieldwork. *Area*, **36**, 280–286.

BRICKHOUSE, N. W. 2001. Embodying science: a feminist perspective on learning. *Journal of Research in Science Teaching*, **38**, 282–295.

BUCKLEY, A. 1879. *The Fairy-land of Science*. Facsimile 2006, Yesterday's Classics, Chapel Hill, North Carolina, 228 pp.

BUREK, C. V. 2001. Where are the women in geology? *Geology Today*, **17**, 110–114.

BUREK, C. V. 2002. Where are the women in science? A case study using women in the history of geology to develop a European curriculum. *In*: GONZALES, M. H. (ed.) *Proyecto Penelope – The Role of the History of Science in Secondary Education.* Fundación Canaria Orotava de Historia de La Ciencia, La Orotava, Tenerife, Spain, 214–221.

BUREK, C. V. 2004. Anning Mary (1799–1845). *In*: LIGHTMAN, B. (ed.) *The Dictionary of Nineteenth-Century British Scientists.* Thoemmes Continuum, Bristol, **1**, 51–53.

BUREK, C. V. 2007. The role of women in geological higher education – Bedford College, London (Catherine Raisin) and Newnham College, Cambridge, UK. *In*: BUREK, C. V. & HIGGS, B. (eds) *The Role of Women in the History of Geology.* Geological Society, London, Special Publications, **281**, 9–38.

BUREK, C. V. & HIGGS, B. 2004. *European Perception of Women Scientists*, Unpublished Report.

BUREK, C. V. & KÖLBL-EBERT, M. 2007. The historical problems of travel for women undertaking geological fieldwork. *In*: BUREK, C. V. & HIGGS, B. (eds) *The Role of Women in the History of Geology.* Geological Society, London, Special Publications, **281**, 115–122.

BUREK, C. V. & MALPAS, J. A. 2007. Rediscovering and conserving the Lower Palaeozoic 'treasures' of Ethel Woods (nee Skeat) and Margaret Crosfield in northeast Wales. *In*: BUREK, C. V. & HIGGS, B. (eds) *The Role of Women in the History of Geology.* Geological Society, London, Special Publications, **281**, 203–226.

DIXON, C. 1999. Detailed underground mapping in a mine: learning to think in three Dimensions. *In*: KING, H. (ed.) *A Staff Resource Book to Support Earth Sciences Learning & Teaching in Higher Education,* Section 5, 22–28. Earth Science Staff Development Project. World Wide Web Address: http://www.gees.ac.uk/essd/resource.htm

GEOLOGICAL SOCIETY. 1997. Working party on 'Women and the Geological Society'. *In*: *Women and the Geological Society.* Supplement to the report of the working group, Geological Society, London, 65 pp.

HART, J. K. 2007. The role of women in British Quaternary science. *In*: BUREK, C. V. & HIGGS, B. (eds) *The Role of Women in the History of Geology.* Geological Society, London, Special Publications, **281**, 83–95.

Higher Education Statistical Agency. 1995. *HESA Staff Record 1995.* World Wide Web Address: www.setwomenstats.org.uk.

Higher Education Statistical Agency. 2004. *HESA Staff Record 2003–4.* World Wide Web Address: www.setwomenstats.org.uk.

HIGGS, B., BUREK, C. V. & WYSE JACKSON, P. N. 2005. Is there gender bias in the geological sciences in Ireland? *Irish Journal of Earth Sciences*, **23**, 132–133.

HIGGS, B. & WYSE JACKSON, P. N. 2007. The role of women in the history of geological studies in Ireland. *In*: BUREK, C. V. & HIGGS, B. (eds) *The Role of Women in the History of Geology.* Geological Society, London, Special Publications, **281**, 137–153.

KÖLBL-EBERT, M. 2007. The role of British and German women in early 19th century geology—a comparative assessment. *In*: BUREK, C. V. & HIGGS, B. (eds) *The Role of Women in the History of Geology.* Geological Society, London, Special Publications, **281**, 155–163.

MATHER, J. D. & CAMPBELL, I. 2007. Grace Anne Milne (Lady Prestwich): more than an amanuensis? *In*: BUREK, C. V. & HIGGS, B. (eds) *The Role of Women in the History of Geology.* Geological Society, London, Special Publications, **281**, 251–264.

MCEWAN, C. 1998. Gender, Science and physical geography in nineteenth century Britain. *Area*, **30**, 215–223.

ORR, M. 2007. Keeping it in the family: the extraordinary case of Cuvier's daughters. *In*: BUREK, C. V. & HIGGS, B. (eds) *The Role of Women in the History of Geology.* Geological Society, London, Special Publications, **281**, 277–286.

ROBINSON, E. 2007. The influential Muriel Arber: a personal reflection. *In*: BUREK, C. V. & HIGGS, B. (eds) *The Role of Women in the History of Geology*. Geological Society, London, Special Publications, **281**, 287–294.

SCHIEBINGER, L. 1999. *Has Feminism Changed Science?* Harvard University Press, Cambridge, Massachusetts, 252.

SHINDLER, K. 2007. A knowledge unique: the life of the pioneering explorer and palaeontologist, Dorothea Bate (1878–1951). *In*: BUREK, C. V. & HIGGS, B. (eds) *The Role of Women in the History of Geology*. Geological Society, London, Special Publications, **281**, 295–303.

TICKELL, C. 1996. *Mary Anning of Lyme Regis*. Philpott Museum, Lyme Regis, 32 pp.

TORRENS, H. 1995. Mary Anning (1799–1847) of Lyme; the greatest fossilist the world ever knew. *British Journal for the History of Science*, **28**, 257–284.

WARD, R. 1992. Women and geology at UK universities: 1985–1990. *Geology Today*, **8**, 185–188.

WACHTLER, M. & BUREK, C. V. 2007. Maria Matilda Ogilvie Gordon (1864–1939): a Scottish researcher in the Alps. *In*: BUREK, C. V. & HIGGS, B. (eds) *The Role of Women in the History of Geology*. Geological Society, London, Special Publications, **281**, 305–317.

WHITELEGG, E. & EDWARDS, C. 2001. Beyond the laboratory – learning physics using real-life contexts. *In*: DUIT, R. (ed) *Research in Science Education – Past, Present and Future. Netherlands*. Kluwer Academic Publishers, Dordrecht, 337–342.

WHITELEGG, E. & SMIDT, S. 2004. Block 4: inclusivity and diversity in science learning. *In*: *Contemporary Issues in Science Learning*, Open University, Milton Keynes, SEH806.

WILLIAMS, A. 2004. *UK Resource Centre for Women in SET*. Conference on improving the effectiveness of education resources for earth science and industry. *The All-Party Parliamentary Group for Earth Sciences & The Earth Science Education Forum for England & Wales, London*, 38–39.

WYSE JACKSON, P. N. & SPENCER JONES, M. E. 2007. The quiet workforce: the various roles of women in geological and natural history museums during the early to mid-1990s. *In*: BUREK, C. V. & HIGGS, B. (eds) *The Role of Women in the History of Geology*. Geological Society, London, Special Publications, **281**, 97–113.

The role of women in geological higher education – Bedford College, London (Catherine Raisin) and Newnham College, Cambridge, UK

C. V. BUREK

*Centre for Science Communication, Department of Biological Sciences,
University of Chester, Parkgate Road, Chester, CH1 4BJ, UK
(e-mail: c.burek@chester.ac.uk)*

Abstract: This paper explores the place of geology within science education and the part women have played in geological higher education through history. The context is set firstly by exploring the informal role women have played in education in general and secondly by examining in detail the positions they held after 1870, when female higher education was put on to a more formal footing. To illustrate this, the evolution of two female colleges of higher education, Bedford College, London, and Newnham College, Cambridge, both offering geological education within science, are evaluated within a wider educational context. Finally, the cases of Dr Catherine Raisin, who was based at Bedford College, and Dr Gertrude Elles, based at Newnham College as role models are highlighted within this wider framework. This is supported by examining the student experience offered at these institutions through the formal laboratory and lecture provision and also informally by societies, specifically the Sedgwick Club in Cambridge and the Natural History Society at Bedford College, which were both supported strongly by the aforementioned female role models. Some other key figures, both students and staff, are also examined and an analysis of student destinations from both institutions during the late 19th and early 20th centuries are tabulated. The place of women firstly within science education and then specifically within geological education forms the context of this paper. The conclusions develop a portrait of early geological role-model women.

Informal female educators

Women have always played a major informal part in family education, especially of their own offspring, and also in some cultures in a wider extended family network. There is a Fanti (Ghanaian) proverb: 'If you educate a man, you educate an individual, but if you educate a woman, you educate a family'. This has been expanded and modified several times and was quoted by UNESCO on the occasion of its 50th anniversary as:

When you educate a man, you educate an individual.
When you educate a woman, you educate a family, a nation.
(L'Oreal 2006)

We might also add 'And a family passes on what it learns to the next generation'.

This well-known saying is attributed to the scholar and clergyman, Dr James Emmanuel Kwegyir-Aggrey (1875–1927), one of the 20th century's greatest educators and one of Africa's most eminent men. Kwegyir-Aggrey probably used this proverb to convince African parents who were more willing to allow their sons to attend missionary schools than their daughters. Like many sayings, this one makes its point by unqualified exaggeration to capture our attention. The message here is that, once we know the value of education for men in society, we should allow women to have equal access to it. This applies equally to scientific education and was reiterated in 1894 by Sir Henry Roscoe in his inaugural speech at Bedford College, London:

If men are all the better for scientific training, why not women? A refined woman could not become unrefined by a knowledge of the truths of nature. (Busk 1894)

It is now necessary to put science education and women into a historical context in order to establish the place of women as geological educators and possible role models in higher education.

Science education through history – a female perspective

Looking at science in its entirety we see a change occurring in the 16th and 17th centuries from what is termed 'Old Science' to the 'New Science' of Bacon and Descartes. The world of learning moved from looking at the Earth and nature as a living organism to the Earth as a machine (Sheffield 2004). This shift in thinking had profound implications for women's role as contributors to formulating, advancing and securing scientific theory and the transfer of information.

Thus while Old Science recognized, respected and encouraged women's knowledge on reproduction and the natural world treating them as equals, the New Science, with its reductionist rather than holistic approach, cut women out of the loop. They were no longer the producers or recipients of technical books on domestic issues ranging from brewing beer to silkworm production or the more familiar midwifery skills. The New Science required active participation, experimentation and control over nature. It required breaking up knowledge to understand it and this mechanical philosophy effectively removed most women from study, which society perceived as a masculine pursuit. Women were regarded as weak and frivolous, reducing the seriousness of any study if allowed to participate. Thus they were barred from scientific societies, such as the Royal Society, because of their perceived lack of intellectual rigour and understanding necessary to engage in serious science. This had tremendous implications for women. They were no longer involved in scientific experimentation or research at any level, except occasionally as family or wife assistants (Kölbl-Ebert 2002).

During the 17th and much of the 18th centuries male children were educated to become responsible and independent individuals. Latin, for example, which was considered an importance entry qualification for university education right up to the 1960s, was not regarded as a suitable subject for females and therefore was not normally taught to girls. Thus access to higher education was severely limited if not effectively denied. Society generally thought that girls and boys should be given different types of education, as their future needs were different. Girls were educated to be dependent, taught French as a communication skill, and basic reading, music and needlework to enhance domestic skills as their expectations were to become wives and mothers. Of course there were exceptions, Anne Conway (1631–79), Maria Sibylla Merian (1647–1717) and Margaret Cavendish (1623–73) to name a few, but most upper and middle class girls were bored and ignorant (Sheffield 2004). The working classes had no time to be bored! However, as we journey through the 17th, 18th and 19th centuries, we see a gradual move from a perception of science in the United Kingdom being wholly masculine to a realization that a certain type of domesticated science for women was permissible.

Mode of education

Throughout the 18th century some people would question the validity of not educating young women and it was eventually recognized that science could promote domestic harmony, as portrayed in James Ferguson's (1710–76) *Young Gentlemen and Lady's Astronomy* (Ferguson 1748). By the late 18th century science, especially botany, was taught through drawing, collecting and writing. This particular science education was seen to support gardening, herbalism, modelling, walking, floral arrangements and would be taught by mothers, governesses or artists (often male). This could also include an appreciation and understanding of landscapes. Thus by the mid-19th century, botany had become a feminized science, so much so that, by 1887, J. F. A. Adams was asking 'Is Botany a Suitable Study for Young Men?' (Sheffield 2004).

Generally through the late 18th and 19th centuries, Middle- and Upper-class women were starting to obtain a science education through books written especially for them, such as John Bonnycastle's *Introduction to Astronomy* as a series of letters (Bonnycastle 1796) or John Lindley's (1799–1865) *Ladies Botany* (Lindley 1865). In 1889, Charles Kingsley published one of the first earth science books for children, *Madam How and Lady Why*, which introduces boys to geology through conversations about earthquakes, volcanoes, coral reefs, and so on. In the preface he starts 'My dear boys, when I was about your age, there were no such children's books as there are now'. It encouraged them to wonder about the distinctive features of the landscape and how they came to be that way. Some books were actually written by women for women, for example Jane Marcet's (1769–1858) *Conversations on Chymistry* (Marcet 1809) or Miss Arabella Buckley's (1840–1929) '*The Fairy-Land of Science*' and other books on nature and elementary science (Buckley 1879). The latter were recommended to Bedford College student teachers in an article by Miss Henrietta Busk, dated 1895 and entitled 'Geography as a school subject'. Arabella Buckley was Sir Charles Lyell's secretary for several years and therefore exposed to current geological thinking. These books were often conversations or letters between either a mother and her children or between teachers and pupils. These forms of communication were considered much easier for females to understand. This whole genre of literature was known as 'polite science' (Sheffield 2004).

So on the whole, women were students of science and teachers to their children. They were thus consumers not producers of science. This informality was to change in the 1870s as female science education became more formal.

The beginning of higher education for women

During the second half of the 19th century some higher education institutions for women were being founded. These were partly underpinned in an educational sense by two private schools which were spearheading female education at secondary level. They were established during the 1850s along enlightened lines and included North London Collegiate School and Cheltenham's Ladies College. Then, between 1871 and 1878, 16 'Girls Public Day School Trust' high schools were opened and universities started to open their doors to girls and allow them to take local exams. Bedford College had already been in existence since 1849 and had initially found this lack of high-quality female students a distinct challenge. The relationship between women's colleges in Cambridge and leading girl's secondary schools is dealt with in detail by Creese (1991).

Cambridge. It was in 1864 that Cambridge University open exams became available to women and, in partial response, two colleges were established: Girton College and Newnham College.

The former was started in 1869 by Emily Davies and the latter by Professor Henry Sidgwick and a small group of advanced thinkers. They brought Anne Clough (the sister of the poet Arthur Clough) from Liverpool to be the head in 1871 and five girls became students. When Professor Henry Sidgwick was asked why Miss Clough was appointed he stated: 'My desire for her co-operation was partly on account of her long devotion to the improvement of the education of women' (Sidgwick 1892).

The ethos of these two colleges was different. Girton insisted that the women were treated exactly the same as men. Students followed the same curriculum and did the same exams. There was no compromise. Newnham College, under Miss Clough, did allow some flexibility (Holt 1897). While Cambridge University allowed female students to attend lectures from 1870, they were not allowed to take formal exams until 1875.

By 1879 Somerville College and Lady Margaret Hall had been established in Oxford to cater for their female students. All these colleges were essentially residential. This differed substantially from other institutions set up in the capital or indeed civic colleges, which were aimed at a more local population.

Degrees were not awarded to women at Newnham until 1921. In 1926 women were allowed to become professors, and female students became eligible for studentships and prizes. In fact Newnham College did not formally become part of the University of Cambridge until 1948.

London: Bedford College. It must be remembered that the University of London was an examining and degree-giving body and did not teach, so students undertaking the exams were taught elsewhere.

In 1878, University College London allowed women into its lectures. However Bedford College had already been in existence for some time before this.

Bedford College has its beginnings in 1849 when a Ladies College in Bedford Square was inaugurated by Mrs Elizabeth Reid, a widow after only 13 months of marriage. She was a Unitarian who was driven by high moral values, to improve society by educating mothers. 'If they could but see and feel as I do, that we shall never have better men till men have better mothers' (letter from Mrs Reid to Henry Crabb Robinson 1858, Quoted in Tuke 1939). It was not learning for learning's sake. This was to come later. Two of Bedford College's early students had literary connections. Mary Ann Evans, alias George Eliot, arrived in 1851 for one term to study maths and Charles Dickens's daughter Catherine arrived in 1852 to attend lectures. So elements of an enlightened London society supported the move to educate its young ladies.

Mrs Reid's trustees for the college were Erasmus Darwin, older brother of the future famous Charles, Thomas Farrer, and Hensleigh Wedgwood, Erasmus's cousin and later chair of the general Committee. Erasmus was chair for 20 years (1850–69) and a member of the college until his death in 1891. A detailed history of the early difficult evolution of Bedford College is given in Tuke (1939). By 1869, the same year that Newnham College was established, Bedford College was on a firmer footing, with a new constitution, and the college was incorporated as an association under Board of Trade regulations and a public company not trading for profit. By the late 1870s the College effectively had to decide whether it was half school/half college or whether it should align itself with the university. It opted for the latter and, in 1877, entrance exams were established. The following year classes for matriculation and degree examinations for the University of London were offered.

During the first 45 years of Bedford College, numbers rose and fell, only really reaching stability when funding from the government became secure in 1894. By this time the staff numbered 17, with three assistants in Greek, Latin and combined mathematics and physics. All the assistants were female but only two of the lecturers (times have not changed greatly). Table 1 shows the range of subjects taught at this time.

All the staff were part time before 1894 and often taught elsewhere as well. Catherine Raisin, for example, in her letter of acceptance of the vice principal position in 1898 states:

With regard to outside work, I may say that I should not be willing myself to take any additional teaching . . . I have for years refused

Table 1. *Bedford College: subjects taught in 1894*

Biology (Animal)	History (Ancient)	Latin
Biology (Vegetable)	History (Modern)	English
Chemistry	Greek	French
Physics	Mathematics	Art
Geography & Geology	Mental & Moral Science	German
Pianoforte		

and given up all such (with the one exception which I mentioned) since I believed it was right for the college to keep for my own work any time which I could spare from my lectures and laboratory. (Raisin 1898)

So from 1894 onwards many positions were full time.

This situation changed dramatically by the time Sir Henry Roscoe (from Owen's College, Manchester) gave his inaugural lecture in 1894. Bedford College had just been awarded its first grant from the government, as announced by Dr Russell, then Chairman of the Council:

£700 a year had been awarded to Bedford College and this was the only Women's College that had received such recognition. (Busk 1894)

Sir Roscoe was very complimentary in his lecture and was loudly applauded. His concluding remarks show the improved status of the college in the eyes of the world (of men)

The fact that Bedford College was taking the lead in the matter of women's higher education.

This was further amplified in the reply by Professor Rücker:

The success of this Institution depends not only on the work done within it, but also on the opinions of the world.... We want the world to recognise what we are doing and although we are doing the very best we can *with the means we possess* [my italics], we mean to agitate for better means, and we hope some day to obtain better buildings in which to house the college. (Busk 1894)

These statements certainly show that Bedford College had increased its standing in society compared to 20 years earlier. In 1900, Bedford College became part of the University of London and this had a further tremendous impact on the public perception of the college.

Admission and awarding of degrees

In 1879 (Anon. 1888b), the University of London allowed female students to take their exams. It is interesting to note that the presentation of awards, whether degrees, diplomas, scholarships or prizes, were received by students at Burlington House from the Chancellor of the University of London (Fig. 1). The University Council report of 1878/9 states that 11 women in total took exams. By 1888, 145 women in total had taken a BA (47 from Bedford) and 19 a BSc (11 from Bedford). There were also 7 MAs: 4 from Bedford by this date (Anon. 1888b). However, on 9 May 1888 only one student from Bedford College received a BSc from Earl Grenville, Chancellor: a certain Miss Semmens (Anon. 1888a). In 1893, Burlington House again served as the venue for receipt of degrees but in this year there were no science students, only 10 for BAs and 2 BA students for postgraduate teaching diplomas (Anon 1893).

By 1897, seven students from Bedford College alone passed the intermediate science exam and three passed the BSc. (BHW 1897). A year later, in 1898, nine passed the intermediate exam and two the BSc. However it must be recognized that, initially, both Bedford and Royal Holloway Colleges, the two female colleges serving London, were looked down upon by the other London colleges and instruction of women was often left to individual professors, such as Huxley. However, Bedford College did offer evening classes for English, Latin and mathematics in preparation for the Cambridge local examinations as well as for the London University examinations.

In comparison, Victoria University, founded in 1851 as Owens's College (later Manchester University), the first of the civic colleges following the pattern of University College London (Roderick & Stephens 1972), did not admit female students until 1883 (Whitworth 2000), the same year that the Cavendish Laboratories in Cambridge opened their doors to women. Cambridge and Oxford did not award degrees to women until 1948 and 1920 respectively. However, women were not admitted to full university status until 1948 in Cambridge and 1959 in Oxford. The women were allowed into exams and to attend lectures but, because of male student outcry, they could not be publicly recognized.

As British universities did not award degrees to women until 1876 (Kölbl-Ebert 2001), it was quite common during the late 19th and early the 20th centuries for women to travel across the Irish Sea to Trinity College, Dublin, to receive their degrees (Higgs & Wyse-Jackson 2007). They were known as the 'Boat People' and included the geologists Gertrude Elles and Ethel Skeat from Newnham College, Cambridge. It is interesting to note that the issue of female attendance in lectures was something Professor McKenny Hughes from the Cambridge geology department was specifically

Fig. 1. Women receiving University of London Degrees.

asked in 1892 by Eleanor Mildred Sidgwick, the Principal of Newnham College.

> An effort is being made to get the classes of Trinity College, Dublin opened to women, and ... I have been asked to collect information as to the working of mixed classes of men and women here, which those who are promoting this movement in Dublin may lay before the authorities. (Sidgwick 1892)

And then towards the end of her letter:

> We have I think very few such classes in Cambridge, but it stuck me that the work done in your geological excursions and museum is something of the same kind and I am therefore especially desirous to be able to report your opinion on the subject.

This is interesting considering Dublin was prepared to give women degrees. A week later he answers the questions after admitting to consulting his colleagues in the Woodwardian Museum. They show the feeling of the geology department towards mixed teaching in the early 1890s.This is discussed in detail by Higgs & Wyse Jackson (2007).

Professor McKenny Hughes was renowned for his tolerance of ladies in geological fieldwork and is often pictured in the field with both male and female students, accompanied by his wife as chaperone (Fig. 2). If we look at Professor Hughes's excursion albums from both 1892 and later, in 1913, the percentage of women participants remained at about 33%.

The medical profession was the first discipline to allow women to take higher education courses and become 'professionalized'; this occurred during the late 1860s and early 1870s in Edinburgh, followed later by England during the 1870s and 1880s.

In this respect, it is interesting to note that Elizabeth Garrett Anderson, the first female, British doctor (MD, University of Paris, 1869) applied to become a physiology lecturer at Bedford College in 1867 once she had qualified as an apothecary in 1865, but was turned down because physiology was not considered a suitable subject for women. This did not change until it was offered as a course in 1882 15 years later.

Science teaching and laboratories

By the late 1880s, science was no longer taught and demonstrated by a lecturer at the front of the class,

Fig. 2. Field trip with Professor McKenny Hughes in 1896.

as Buckland had done. It was now necessary for the students to do experiments themselves. The first laboratories for female students were built for the ladies of Girton and Newnham College in Cambridge: the Balfour Biological Laboratories, which were in use between 1884 and 1914 (Sheffield 2004). However, women were not admitted to work as undergraduates in the Cavendish Laboratory for Physics until 1882 (Sheffield 2004) or the University Chemical Laboratories until 1910, 'much to the annoyance of the "lab boys", who evidently expected us to give them more trouble than the men' (Ball 1988).

In London by 1886 there were recommendations from a special subcommittee set up by Bedford College that new laboratories be built. Thus, in 1888, it was proposed to build a physics as well as a chemistry laboratory in the new wing, the biological laboratory (Fig. 3) being retained on the present premises (Anon. 1888a). There was no mention of geology. The Shaen Wing (named after Mr William Shaen, the lawyer and a supporter of female education from the beginning) for teaching physics and chemistry was ready by Easter 1890, and three laboratories were built and opened by the Empress Frederick of Prussia in that year. When the Empress Frederick and her daughter arrived to formally open the new Shaen wing in 1891, the *Bedford College London Magazine* recorded that:

The Empress Frederick came to see for herself what a working Ladies' College was like: she looked closely into everything and carried away the calendar of the college with its rules and regulations, which she said she intended carefully to read. (Story-Maskelyne 1891)

Thus Bedford College was considered more vocational and technical.

The importance of these science laboratories for women within higher education in London is highlighted in a quote from 1890 in *Nature*, which stated:

Bedford College in York Place, Baker St, which was one of the earliest institutions devoted to the higher education of women is taking a leading part in providing facilities for their instruction in science. Founded long before Oxford and Cambridge condescended to the 'weaker sex', it is the result of the work of enthusiasts who would not admit the possibility of defeat. It has had to struggle not only against the inevitable difficulties due to its early foundation but against the apathy of London. The colleges for women at Oxford and Cambridge share in the picturesque surroundings of those old homes of learning. Bedford College has had no such advantages. London Institutions are regarded as either Imperial or Parochial as too large or too small to interest its citizens. (Anon. 1890)

Fig. 3. (a) Biology laboratory at Bedford College; (b) Student working in the laboratory at Bedford College.

By 1894, Dr Russell was able to announce 'the Technical Board of the County Council [London] had this year granted £500 for the further expansion of the laboratories' (Busk 1894) and Sir Roscoe commented in his inaugural speech that he was:

... glad to hear that £500 was going to be spent on the laboratories, as science was no good without laboratories, and laboratories were no good without adequate fitting and apparatus. Classicists and mathematicians do very well with a black board and a piece of chalk, but the chemist, physicist, biologist, geologist and botanist must have more. (Busk 1894)

The support of royalty for this initiative started to change public opinion and, on 5 July 1913, Queen Mary herself came to open the new buildings and laboratories (Fig. 4).

Higher education in geology for women in England at the end of the 19th century

Having now set the scientific educational context, it is necessary to examine geology, in particular in higher education in England.

Formal female geological study at this time was difficult and the options were limited. Problems included fieldwork, religion, and socially accepted standards and etiquette. However there were two locations where it was possible for women to study geology: Newnham College, Cambridge, and Bedford College, London, but it must be remembered that they were started 20 years apart in different social climates and for different reasons. The clientele of these colleges was also different, with Newnham aimed at the higher classes of society. Bedford College, which, like its forerunner University College which was founded in 1826 and modelled on the University of Berlin (Roderick & Stephens 1972), was designed to provide higher education for those members of the middle classes unable to gain entry into Oxford or Cambridge Universities and to provide the study of subjects not available at the older universities, which concentrated on the classics and theology.

Both Bedford and Newnham Colleges produced a good crop of female geologists during the last years of Queen Victoria's reign and into the beginning of the 20th century. These students then went out to act as role models and to teach other girls the rewards of learning about science, including geology. For comparison, the University of Liverpool did not teach geology except as evening classes until a chair of geology was created at the beginning of World War 1 (Roderick & Stephens 1972).

While both Newnham and Bedford Colleges were residential and offered accommodation for female students, this facility did not initially allow

Fig. 4. Queen Mary's visit to Bedford College in 1913.

them to study as part of the wider universities of London and Cambridge. It was deemed unsuitable for female students to move around unchaperoned.

We did not take much part in the life of the University, and we suffered many restrictions. We were asked always to wear gloves in the town (and of course hats) we must not ride a bicycle in the main streets, nor take a boat out on the river in the day time unless accompanied by a chaperone who must be either a married woman or one of the College Dons. (Crowther 1896)

However, Bedford College in London had a higher percentage of day students and many of their students were local. Figure 5 shows one of the study bedrooms that the girls would have lived in, often isolated from the other members of the university and college. For some these were lonely times.

Geology at Newnham College, Cambridge University

Geology was not taught as a separate subject until 1877. So students undertaking geology were permitted into the Tripos examinations informally

Fig. 5. Study bedroom in the hall of residence, Garden Square, opened in 1883.

and only by the favour of private examiners. In 1881, Newnham students were formally allowed into the Tripos exams and were then awarded a certificate to say they had done so. The students at Newnham was overjoyed and it is recorded that there was much jubilation when this happened as the votes were 398 for and 32 against (Andrews 1881). By 1885, 80% of Newnham students were sitting for the exams (Tuke 1939).

The first strong influence on successful female participation in geology at Cambridge in the 1890s was the presence of Dr Marr and Professor McKenny Hughes, who were both liberal in their attitude to female participation at all levels (Hughes 1892; Creese & Creese 2006). They encouraged females to take up geology as part of their studies.

Dear Dr Marr would come stamping into the lecture room, his gown streaming behind him and cut and slice the country up with gusto. Glaciers streamed slowly down, rivers meandered or were beheaded, such 'noxious spots' as the Fens spread themselves, and only the ancient hills of his beloved Lakeland and Scotland remained to make habitable spots for decent folk. All the same we all panted joyfully after him, filing notebook after notebook with more or less legible writing. (Ball 1988)

The curriculum at this time included microscopy, mapwork, fieldwork, palaeontology and petrography and it is interesting to look at some of Professor McKenny Hughes's remarks on teaching women in these classes.

I have had more experience in the arrangement of classes adaptation of Museum and laboratories and regulation of even excursions for mixed classes than any of those who have spoken and I maintain that no difficulties have arisen from the presence of women. We have to subdivide classes because they have become too large or because we have not a sufficient number of microscopes or of window space to work at. We have recognised the convenience of dividing the classes into elementary and advanced but not into classes of men and classes of women.

These are transcribed from his notes of his discussion for admission of women into the university in 1887 (McKenny Hughes 1887).

The first student to include geology in her studies was Dilys Davies, who started as a student in 1876. She left in 1878 to become a teacher at the progressive North London Collegiate College just after Catherine Raisin had left. In 1889 she moved to Bangor and served on the education committee before becoming a governor of Bangor University College. She died in 1932 after a lifetime of service to education (see Fig. 15).

The second was Margaret Chorley Crosfield, who arrived in 1878 but left after a year due to poor health. However she returned 10 years later with special permission to just study geology. Her contribution to geological research is documented elsewhere in this volume (Burek & Malpas 2007).

The third was Margaret Isabella Gardiner, a student from 1882 to 1888, who became a teacher at Withington and then Aldeburgh School. She died in 1944. However, the main crop of female geologists occurred in 1891 with what I have termed the 'Newnham Quartet of palaeontologists'. These were Margaret Crosfield, Ethel Skeat, Ethel Wood and Gertrude Elles. Each contributed significantly but in a different way to the advancement of geological knowledge, and each also acted as a role model in her own way: as lecturer, researcher, teacher, participant or recorder. Margaret Crosfield's written legacy and Ethel Skeat's scientific contribution has been discussed elsewhere (Burek & Malpas 2007) while the work of Elles and Wood is known to any student of graptolite palaeoecology or indeed palaeontology in general (Elles & Wood 1901–1918). While Margaret and the two Ethels left Newnham College, Gertrude Elles remained associated with Newnham for over 40 years. However she did also temporarily demonstrate for Catherine Raisin at Bedford College in 1903 (See Table 11).

Ethel Wood (Fig. 6) moved to Birmingham and for 10 years (1896–1906) was research secretary to Professor Lapworth. However, she married Gilbert Shakespear, a fellow student from St John's Cambridge, in 1906 and had to give up her research job on marriage, as was usual at this time. She remained an associate of Newnham College from 1905 to 1920. In 1920, she was made a Dame of

Fig. 6. Ethel Wood.

the British Empire not for her geological work but for her contribution to the War Pensions Commission, of which she was the founder. Her friends and colleagues tried to persuade her to return to research after World War 1 but her social work did not allow it.

Ethel Skeat was a research scholar for 2 years and then moved to study in Munich under Zittel. On 8 March 1896, she was the first woman to be admitted as a guest to scientific lectures at Munich University, after a petition was placed before the senate by Zittel. Skeat had been recommended to him by colleagues from Cambridge (Kölhl-Ebert pers comm. 2004). In 1898 she moved to Penarth, South Wales, to teach and then, 5 years later, to Queen's School, Chester, where she is recorded as an excellent teacher and an inspiration to the girls she taught (Burek & Malpas 2007).

The second crop of students, in the form of Pauline Baguley, Beatrice Brailsford, Hilda Sharpe and Igerna Sollas, entered Newnham just as the first quartet left. Of these, only Igerna remained at Newnham, not as a geologist but as a zoology lecturer (Fig. 7). The other three left to become teachers initially in Bradford and Wallasey. Their training and subsequent teaching inspired other girls to undertake geology as a subject.

In 1900 Helen Drew and Ida Slater arrived to study geology as part of their degrees (Fig. 8). Helen Drew became a teacher, travelling to Bristol, London, Bradford and Newark, and died while still in post at the age of 46 years. Ida Slater, after undertaking research, became a demonstrator for Catherine Raisin at Bedford for 2 years from 1910.

Thus we have built up a picture of cohorts of girls arriving to study geology in Newnham and then leaving upon successful graduation, with the majority entering one of the few careers available to them: teaching science. This story continues until the end of World War 1 when other avenues started opening up for female geologists.

The strong influence of Dr Marr and Professor McKenny Hughes on female geological education in the United Kingdom is commented on by Creese & Creese (2006) but this is also augmented by a second influence: informal networking of like-minded individuals.

Sedgwick Club

The second influence was enhanced by the Sedgwick Club which, from 24 January 1896, allowed women to participate in their activities (Sedgwick Club 1896). As this is at odds with

Fig. 7. Staff of Newnham College (1907). Gertrude Elles sitting on grass second from left; Igerna Sollas sitting on grass far right; Mrs Sidgwick sitting on chair in centre dressed all in black.

Fig. 8. Freshers in 1900 at Newnham College. Helen Drew standing second from left Ida Slater front row far right.

much common practice at the time it is worth noting in detail. The minutes of 21 January 1896 read:

Mr Brend [of Sidney College] then gave notice that at the next meeting he would propose the following amendment of the Rules of the Club whereby ladies should become eligible to be admitted as members of the club. *Rule 3* to insert the words 'of lady members' so that the rule should read 'that the Club consist of ordinary members who shall be resident in the University of lady members and of honorary members ... etc'.

It is followed by:

Mr Brend then asked the President whether he might propose ladies as members of the club so that their election might be proceeded with at the next meeting if the requisite amendment of the rules were carried. The president said in answer that the names of ladies wishing to become members of the club might be given to the secretary and that in the event of their becoming elligible [sic] for membership at the next meeting, their names might then be taken as having been proposed at the previous meeting.

On 24 January 1896 Mr Brend was invited to propose the resolution and speak to it.

Mr Brend then said that the club existed with the object of promoting the study of geology and that as there were now ladies working at the Woodwardian Museum who had shown good ability in the pursuit of that science he thought that the club would benefit by admitting them as members.

There was criticism from one member but that was dismissed and

Mr Cunningham-Cray in seconding said that as a matter of common justice whatever advantages the Club offered should be open to students of the University whether men or women on exactly the same footing.

A slight amendment was then put forward as Girton and Newnham College were not actually members of the university so it read:

Rule 3 should read 'That the Club consist of ordinary members who shall be resident members (1) of the University or (2) of Girton and Newnham Colleges; and of honorary members, ... etc.'

The amendment was seconded and after discussion 'carried unanimously'. Rule 4 also had to be changed to read:

That the number of ordinary members shall not exceed twelve undergraduates and six ladies.

Again this was carried unanimously.

Thus the ladies of Cambridge entered into the Sedgwick Club, which was obviously more liberal than the university itself. This allowed additional networking, practical fieldwork and additional lectures to become available to females, which enhanced their student experience. While they were limited in numbers by the rules of the Club, this mixing was still something not readily available to other female members of the university community.

By May 1896, only 5 months after first being admitted, the full contingent of females had been reached.

Gertrude Elles went on to become secretary and then president in 1902, while another Newnham student Ida Slater became secretary in 1903.

Louisa Jebb was an interesting member because she was an agricultural student and became the first female to take the university agricultural diploma. She then ran her father's farm, becoming an authority on smallholdings. She created the Women's National Service conference, becoming president in 1916. It was from this the Women's Land Army developed and, for this, she was awarded the OBE in 1918.

The admittance of ladies into the Sedgwick Club allowed the students at the female colleges, especially Newnham, to participate socially in geological activities. It is interesting to note that Gilbert Shakespear from Trinity is photographed as a member of the Sedgwick Club in 1898 (Fig. 9); he was to become Ethel Wood's husband in 1906. The records and minutes of the Sedgwick Club make fascinating reading in this regard and tell of some of the activities undertaken by the students.

Thus in the last decade of Queen Victoria's reign (the 1890s), Newnham College students were favourably influenced by both the liberal nature of the geology lecturers and the male students in the

Fig. 9. Members of the Sedgwick Club, Cambridge, 1898. Gilbert Shakesphere is seated at the end of the second row on the right.

guise of Professor McKenny Hughes and Dr Marr and the networking allowed by the Sedgwick Club.

Thus geology at Newnham College started slowly but then went from strength to strength. As we enter the 20th century, Gertrude Elles played a key role in this development.

Gertrude Elles (1872–1960)

Gertrude Lilian Elles was born in Wimbledon, Surrey, near London, on 8 October 1872, of Scottish extraction. At the age of 19 years she went up to Cambridge University, studying for the Natural Science Tripos, gaining a first-class honours degree in 1895 and becoming the first female to be awarded a Cambridge University readership 30 years later. She participated fully in college life and was an ardent member of the hockey team in 1891 (Fig. 10) (Cockburn 1987). She never married but spent the majority of her life in Cambridge at Newnham College and was recognized as an excellent and enthusiastic teacher. One student from 1908–11 recalls:

The energy, kindliness and enthusiasm of Miss Elles made her perhaps the most inspiring influence in those glorious years, which culminated in an unforgettable Sedgwick Club trip to Church Stretton. (Ball 1988)

Another student, somewhat later (1920), comments on the dress code.

Our dons in those days wore a variety of suits and dresses, but their skirts were always long. Miss Elles I can see in brown tweeds. (Fig. 11); Halliday 1988

By 1926 another student comments on hats:

And when women began to lecture in 1928 they lectured in hats. We didn't wear gowns, and Newnham and Girton decided between them I think with the leadership of Girton, that the appropriate wear was a hat. So the Newnham Fellows had lecturing hats. Not too large, turned up in the brim, just neat and appropriate. The Don (Dr Elles) had a marvellous sort of russet-coloured felt hat that she plonked on her head; it had a brim which swept across her nose and the crown was almost non-existent. (Grimshaw 1988)

She was a very stimulating teacher, and not only of members of the College – she was in great demand. She taught geological mapping in the Sedgwick: everybody went to that. She was marvellously clear and very, very fierce.

By 1926 she was very deaf too and by 1930 she was Vice Principal:

being deaf she talked in a very quiet voice, but it was well worth the effort to hear her stories of the supernatural. (Ingledew 1988)

From these few student quotes it is easy to demonstrate how Gertrude Elles became a role model teacher.

However, her fame was made not in the field of teaching but in research. Her contribution to the study and classification of graptolites has not been surpassed to date. She spent 12 years compiling a treatise on British graptolites (with her colleague and friend Ethel Wood (later Dame Ethel Shakespear) under the guidance of Professor Lapworth at Birmingham University. This became a seminal work on the evolution and taxonomy of graptolites, and Gertrude and Ethel are inextricably linked with graptolite research (Elles & Wood 1901–18). Gertrude Elles's work on the genera of graptolites

Fig. 10. Gertrude Elles (front row left) in the Newnham College hockey team, 1891.

Fig. 11. Professor McKenny Hughes (centre) on a field trip with students, Malverns, 1913.

from North Wales, the Skiddaw Slates of the Lake District, and the Wenlock Shales of the Welsh borders eventually led to her receiving the prestigious Lyell Fund from the Geological Society of London. She was not able to receive it in person because women at that time (1900) were prohibited from attending meetings. She was one of the first people to look at not individual specimens of fossils but at the concept of communities of organisms. In 1919, she became one of the first women to become a Fellow of the Geological Society of London and the same year she received the Murchison medal from the Society in recognition of her work.

However, she was also generous in her help of others and readily gave advice to her research associates and the Cambridge old girl's network (see Burek & Malpas 2007).

Her work also concentrated on stratigraphy and she published over ten papers on Lower Palaeozoic stratigraphy (Elles & Wood 1895; Elles & Slater 1906; Elles 1909, 1922a, b).

Among her other accolades she received the MBE for her work with the British Red Cross in the United Kingdom during World War 1. She was an active worker with them for many years. She was also elected president of the British Association in 1923.

As she approached old age, she became even more deaf and eccentric in her ways, surrounded by fossils, rocks and papers. Her idiosyncrasies

Table 2. *Sedgwick Club: first six female members (compiled from the Sedgwick Museum Archives)*

Date	Name	College	Proposed by	College	Seconded by	College
24 Jan. 1896	Gertrude Elles	Newnham	Mr Brend	Sidney	Mr Cunningham-Cray	Clare
	Ethel Wood	Newnham	Mr Brend	Sidney	Mr Kitson	Trinity
	Louisa Jebb	Newnham	Mr Brend	Sidney	Mr Cunningham-Cray	Clare
3 Mar. 1896	Ethel Skeat	Newnham	Miss Elles	Newnham	Miss Jebb	Newnham
5 May 1896	Miss Hugon	Newnham	Miss Jebb	Newnham	Mr Brend	Sidney
	Miss Phillips	Newnham	Mr Brend	Sidney	Mr Pentecost	Christ

were remembered by many for years (Hancock pers. comm. 2000). However, she always considered fieldwork to be the key to a good geologist and an invaluable aid for a good understanding of palaeontology and stratigraphy. Eventually her love of her Scottish homeland, called her back permanently (she had always spent considerable time there fishing and researching metamorphic rocks) and Gertrude Elles died in Scotland in 1960 at the age of 88 years. Throughout her life, she had made a significant contribution to both the status of women in science, especially the field of Earth sciences, where she acted as an important role model, and to the understanding of graptolite palaeoecology and their recognition as stratigraphic zone fossils.

Geology at Bedford College, London

The subjects of geology, geography and botany had an intertwined history at Bedford College.

Geography was taught as a separate subject from the beginning in 1849 but had a chequered history (Table 3) and had no separate department until 1920. The place of physical geography within Bedford College is definitely a case of feast and famine. Its link with geology is positively shown, and it was Catherine Raisin, head of geology (Fig. 12), who argued for a separate geography department after World War 1. However, this situation was not that dissimilar to other higher education institutions, as no geography degree was offered anywhere until 1917, when the universities of Liverpool and Aberystwyth established them.

Geology was not taught as a separate subject until 1885 (see Table 1 for a list of subjects taught at Bedford in 1894), when a department was established.

The first woman graduated in geology in 1884 and the second in 1888. By 1897, geology was becoming increasingly popular and while

> of late years no one here has taken geology for the BSc but now the subject has come into favour again, and the students are most energetic over the work and the journeys to South Kensington Museum. (BHW 1897)

So, in the beginning, geology had its ups and downs but it was nevertheless taught for 100 years in Bedford College until its amalgamation with Royal Holloway College in 1985.

The Bedford College geology department had six leaders, either as head of department or as professors throughout its history. These are listed in Table 4, together with their contribution to the development of geology as a subject in the college. Until Catherine Raisin the position was part time. A close link existed between geology and biology as is reflected in Catherine Raisin running both departments from 1890 to 1898 and the shared laboratory facilities. The laboratory facilities were improved after 1897 and not before time (see Morton Sumner bequest). The laboratories had little equipment, few specimens or maps, nothing resembling the museum facilities available to Newnham students. Nevertheless, the congested state of the lecture rooms in Nos 8 and 9 York Place became so bad that it was necessary to move the training department next door to No. 10, along with 'the Botany and Geology so that Miss Raisin will have two laboratory-lecture rooms' (Anon. 1896).

There were two major turning points for the successful teaching of geology at Bedford College:

- Mrs Morton Sumner's bequest;
- New labs for old in 1897 and 1913.

Table 3. *Bedford College: position in physical geography (1849–1947)*

Years	Name	Notes	Number of students
1849–50	Rev. James Booth FRS	Taught scientific geography and astronomy	1st term: 3 2nd term: 12 3rd term: 18
1850–51	Rev. Thomas Wilson	Taught English, astronomy and Geography	
1851–54	Alexander Bain	Evening class taught by an educated weaver	
1854–66	Gottfried Kinkel	Professor of fine art who also taught geography	
1866–74	Professor William Hughes	Part time from King's College	
1882–85	E. G. Ravenstein	Cartographer and history of exploration	
1885–1920	Grenville Cole & Catherine Raisin	Geography marginalized (no degree in geography) and taught within geology department. Value seen during expansion of the British Empire and during World War 1.	
1920–47	Blanche Hosgood	Shadowy figure who published little but whose name is still attached to a prize, courtesy of Catherine Raisin's bequest	

Fig. 12. The Principal, art professor and science staff of Bedford College, 1903. Catherine Raisin standing far right.

Morton Sumner bequest

The Morton Sumner bequest was fundamentally important to the success of geology at Bedford College as it secured a named geology teaching position and allowed a dedicated curator and professor to be employed. From 1907 this was Catherine Raisin who, for many years, was only granted a yearly renewal of her contract, despite her requests for longer term contracts for planning purposes (Burek 2003b).

The terms of the bequest from Mrs Morton Sumner were given in the 1901 College magazine to the staff and student body of the College as:

£4000 and a large number of books, pictures, bookcases and geological specimens of great value. Of the sum of £4000, £2000 is to be invested and the income used for the payment of a salary to a curator of the books, pictures etc, while the other £2000 is to be expended at the discretion of the Council.

This is reflected in the salaries documented around this time.

Table 6 gives an idea of the salaries expected and the increase in salaries as the college gained both more students and a better public perception after University status in 1900.

Mrs Morton Sumner had presented books and specimens to the department before. For example, in 1894 she had presented

a cabinet of minerals, a cabinet of rocks with a catalogue as well as for Prestwick's Geology, Nicholson's Palaeontology, Teall's British Petrography and the Story of our Planet by Professor Bonney. Miss Raisin was the means of obtaining so large a donation from Mrs. Morton Sumner as she called on her and submitted to her lists of books which would be valuable in the library. (Jones 1894)

Indeed it can be surmised that Catherine Raisin was also responsible for the donation by Professor

Table 4. *Bedford College: leaders of the geology department (1885–1985)*

Years	Name	Full or part time	Contribution to department
1885–86	Mary Foster	Part time	Established the department
1886–90	Grenville Cole	Part time	Left to a full-time appointment at Trinity College, Dublin
1890–1920	Catherine Raisin	Full time	Established research and teaching and separated geography
1921–56	Leonard Hawkes	Full time	Introduced annual field trips
1956–77	Basil King	Full time	Provided an analytical approach to geochemistry and structural geology
1977–85	Alec Smith	Full time	Oversaw the amalgamation with Royal Holloway College

Table 5. *Bedford College: leaders of the botany department (1872–1948)*

Years	Name	Title
1872–91	A. W. Bennett	Head
1886–1908	Catherine Raisin	Lecturer then head
1908–16	Ethel Thomas	Head
1919–48	William Neilson Jones	Head

Compiled from Bedford College Archives.

Bonney of 'some microscopic slices of rocks' in 1898 (Anon. 1898).

However, Mrs Morton Sumner's bequest ensured that not only the teaching continued but the curatorial side of geology was covered too.

Until this year, Bedford College has been entirely without anything which could reasonably be styled a museum, but the generous bequest of the late Mrs Morton Sumner, has suddenly raised us from extreme poverty in this respect to comparative affluence. (Dale 1901)

The extent of the bequest must have seemed enormous to those studying geology. From bare walls and cupboards they now had a 'small Japanese cabinet containing slides' as well as three cabinets containing specimens of rocks.

In 1901 Maggie Dale encouraged others to come and see the delights:

In various cabinets and cases lining the walls of the North Lecture Room is to be seen a large and beautiful collection of mineralogical and crystallographical specimens.

She goes on to say:

Among the finest specimens which are displayed to great advantage in a handsome hexagonal crystal case, we notice in particular a splendid opal, an emerald, a diamond and other precious stones still embedded in their native ores. In the small cabinet next to this are gold and silver ores, a gold nugget, and two drawers full of small and delicate mineralogical treasures.

It is small wonder that the study of geology increased with such a magnificent set of specimens.

However, this situation changed. An important letter from Catherine Raisin, to Miss Tuke, dated 24 March 1916, when redundancy seemed

Table 6. *Bedford College: geology and botany salaries (1895 to 1907)*

Date	Amount
1895–96	£198
1899–1900	£405
1906–07	£447

imminent, gives her thoughts and opinions on the situation, as well as outlining her work on the curatorial side. It is worth quoting parts in full. Perhaps it is also worth remembering that, at this point, Catherine Raisin is 61 years old and has a lifetime of experience of Bedford College. Her prime concern is the teaching collection and the geological specimens held by the college.

I have to ask the Council if they will consider the arrangements which may be made in the future for the care of geological specimens. Doubtless modifications cannot take place until the war is over, but possibly a decision as to the general arrangements could be reached. It seems to me that certain provisions of Mrs Morton Sumner's bequest might be applied with more advantage to the college. The will states that the objects were bequeathed 'in order to form a collection or museum'. A museum always demands a curator and the special fund bequeathed by Mrs Morton Sumner for a custodian would provide this. As Head of Geology department, I have carried out, as far as time allowed, the duties of curator of the geological specimens. But only a small part of the work has been accomplished. And I cannot presume that any future holder of the post will give as much time as I have done. Yet I have not prepared for the general museum collection more than an extremely partial and incomplete catalogue, nor can I keep pace with alterations, which are necessary. I have succeeded in getting numbered and listed some parts of the collections for the ordinary Elementary and Final pass courses, but the collections for the advanced work are not catalogued, nor any of the special series of specimens.

She then goes on to list the duties of the curator in detail, including dusting and preparing catalogues. This attention to detail was present in all her work, whether academic, administrative or research.

New laboratories for geology

If the first turning point was the legacy from Mrs Morton Sumner, the second was the establishment in 1897 of new laboratory rooms for geology and botany. A quote from the *Bedford College London magazine* shows the relief felt by the geology and botany students.

The new rooms are light and airy looking on to the gymnasium on the East and to Baker Street on the west ... The laboratory has been provided with new 'Swift' microscopes which are greatly appreciated now that the initial difficulties as to getting them out of and into their cases have been overcome. (HBS 1897)

A quote for 1898 from the Bedford College Magazine further reiterates the gratitude felt.

The hearts as well as the backs of the Geology & Botany students were gladdened this term by new tables and chairs in their lecture room replacing the old desks. This room has also now been fitted with some new cabinets of drawers and cases which are beginning to be occupied with specimens.

They also now increased the number of geological maps. For the 1906/7 session a part-time demonstrator was appointed, Miss Gibson, who was shared between the botany and geology departments at £180 per annum, and a junior demonstrator at £50. By the following year Miss Gibson had moved to botany and two lecturers were appointed to geology: Miss E. N. Thomas and Miss Agnes Robertson. Miss Gibson resigned in June 1907 to take up a full-time teaching position. She was replaced by Miss Thomas. By November 1907 Catherine Raisin had decided that to run two departments was not beneficial to the college and she handed in her resignation for the botany position. The council responded and passed a resolution to the effect that:

Miss Raisin's resignation of the lectureship in Botany be accepted, and the thanks of the Council be conveyed to Miss Raisin for her admirable conduct of the department during the past twenty-one years.

They then offered her the Morton Sumner lectureship as a two-thirds appointment with a guaranteed income of £283.6.8d per session from Easter 1908. By March 1909 however she again writes to the Council:

The department of geology is almost the only science in the college without any assistance... Help is much required for the arranging and labelling of specimens, all of which has at present, to be done by myself.

She asks for a demonstrator for one half day a week at a fee not exceeding £5 (Raisin 1909). This is granted.

In 1910 there was a further call for demonstrators to help with the running of the department and the laboratories. The department had 14 applications. They decided to appoint two: Miss Ida Slater from Newnham College, Cambridge (Fig. 8), and Miss Bowen-Colthurst from Ireland. Their wages were £30 part time, and £100 full time. This was the same as the lecturers only 20 years earlier (Table 6). The importance of the position, or more likely the lack of other opportunities, is shown in the range of applications they had from Newnham College, Oxford and Dublin, University College and Durham.

Later, in 1913, a new building housing a geology laboratory, map room, museum and research rooms came into being. This led to an increase in the provision of demonstrators in the department. The person that oversaw the move and educational improvement brought about by these developments was Catherine Raisin. The problem of laboratory assistance was obviously contentious but vital for the work to be completed. The curriculum at that time is shown by Irene Lowe's timetable in 1914 (Table 7). She taught for 14 hours: 2 hours of lectures and 12 hours of practical, for which she was paid £200.

Natural History Society

The growth in the number of science students at Bedford College during the 1890s led to the establishment of the Natural History Society. While this could not and did not fulfil the same role as the Sedgwick Club in Cambridge, which was intercollegiate, it did allow the students to meet socially and enjoy additional benefits, such as field excursions and lectures. The students themselves had lobbied for such a society.

This wish at last resulted in a general meeting of students, held on Monday November 21st when Miss C.A. Raisin, DSc. was unanimously voted into the chair ... It was said that such a Society would supply a long-felt want and that it was very desirable that students should be encouraged to take an interest in subjects outside their college curriculum, to observe the facts of Nature for themselves, and gather knowledge directly from them. They should take an interest in the progress of science of the present day.

Miss Raisin pointed out that such a Society would help students to realise the unity of knowledge, and add to their interest in Nature. But its greatest value would depend on the bringing forward of *bits of real knowledge*, not got up entirely from books, but based on observations of Nature, made in the field and open air. This side of scientific work is in great danger of neglect in these days of minute investigation in the laboratory, though not by the most eminent scientists – such as Darwin, Wallace and Müller. (Quoted from the records of the Secretary, Eleanor Pearse)

How true this statement is today, when fieldwork is on the decline and the bigger picture is often lost along with the holistic approach to study.

The success of the Natural Science Society, as it became known, is evident in the events which it ran. Catherine Raisin gave a presidential address, quoted in full in the *Bedford College London Magazine*, on 'Natural sciences *versus* physical sciences'. Here she explores the benefits teaching natural sciences have over the physical sciences:

Imagining a country walk over chalk hills and through woodlands as an illustration, the President suggested some of the many observations in plants and animals and stones which might be made. Then hypotheses on interesting problems would arise, and careful experiments be made to establish or refute them. (Raisin 1907)

She then gives examples of how this could happen in the classroom.

Thus the Natural Sciences should be of use in schools in directing attention to all the various facts of nature, but also in yielding a training as severe and as exact as that supplied by the Physical Sciences.

She is training the teachers of tomorrow to look at the value of field observations for school children, something we tend to neglect today.

The outside fieldtrips were also valuable learning experiences.

The work of this Society belongs more to vacation than to term time. Unfortunately the weather during the Christmas vacation was such that even the most enthusiastic and omnivorous collectors could hardly carry out their noble resolutions. However, enough specimens have been forthcoming this term and we hope for better luck at Easter.

After each meeting, specimens of natural materials were shown, including rock and fossil samples. These small extracts from the society sections of the *Bedford College London Magazine* show how active and influential this society was in helping the teaching of all the natural sciences including geology. Catherine Raisin's name comes up continuously when reading the minutes of the Natural History Society which shows the influence that she exerted.

Catherine Raisin (1855–1946)

Catherine Raisin played a crucial role in the development of geological education at Bedford College from 1886 to 1920, as can be seen from the foregoing sections. She grew to become a significant role model and so it is necessary to examine the context of her background and upbringing.

Catherine Alice Raisin was born on 24 April 1855 in Camden New Town, in the county of Middlesex, just outside London, the youngest child and only daughter of Daniel Francis Raisin and Sarah Catherine Woodgate. Her mother, unusually at that time, was 45 years old and her father was employed at the Inner Temple as a pannierman (Burek 2003b). Catherine was educated at the progressive and pioneering North London Collegiate School, where she stayed on as a teacher until she was 20 years old.

From an early age, Catherine had an interest in geology, a debt she owed to Sir Charles Lyell 'whose *Principles of Geology* was one of the earliest books to arouse my enthusiasm' (Raisin 1893a). At the age of 18 years, she attended classes at University College London and studied geology and mineralogy, the latter the subject that was to become her own research area.

By 1878, London University opened its doors more widely to women (Burek 2002) and Catherine Raisin attended, emerging in 1879 with a pass in the intermediate exam. She returned the following year to complete her degree, studying under Professor T. G. Bonney and attending Thomas Huxley's zoological lectures at the Royal School of Mines. Her interest in female education was highlighted

Table 7. *Irene Lowe's teaching timetable in 1914 (taken from the staff notes in the Bedford College archive)*

TIME TABLE OF COURSES OF INSTRUCTION
Given by Miss Irene Helen Lowe
At Bedford College for Women (University of London)
(State in each case whether lecture or practical work)

Hours at which lectures are held	Monday	Tuesday	Wednesday	Thursday	Friday
11	Stratigraphy practical for 1st and 2nd yr pass students	Igneous Mineralogy practical 1st and 2nd yr students	Igneous Petrology Practical 1st and 2nd yr students	Stratigraphy practical for 1st and 2nd yr pass students	
12	Stratigraphy practical for 1st and 2nd yr pass students	Igneous Mineralogy practical 1st and 2nd yr students	Igneous Petrology Practical 1st and 2nd yr students	Stratigraphy practical for 1st and 2nd yr pass students	
2		Crystallography & Mineralogy Practical Intermediate	Crystallography & Mineralogy Lecture Intermediate		
3	Crystallography & Mineralogy Lecture Intermediate		Crystallography & Mineralogy Practical Intermediate		
4	Crystallography & Mineralogy Practical Intermediate		Crystallography & Mineralogy Practical Intermediate		

during this time when she set up the Somerville Club, a discussion group for women in 1880, which soon had over 1000 members in London (Edgell 1945). Catherine was just 25 years old. It was after this that her belief in equality of education became stronger. In 1884, she received her BSc honours in geology/zoology as top University College geology graduate. At the age of 38 years she received the Lyell Fund from the Geological Society of London and in 1899, aged 43 years a DSc, becoming only the second woman ever to achieve this.

At Bedford College she was the first full-time geology Head of Department, established teaching and research in geology, and was instrumental in the formation of a separate geography department (Page & Smith 2001; Burek 2003b). She cites the War as giving an impetus to the study of geography as well as the introduction of geography into the teaching curriculum and offers financial help.

I will be willing to contribute £50 a year for the three years to a guarantee Fund if that could be raised to cover the cost if increased fees were not obtained in sufficient amount. (Raisin 1916)

The increase in the number of students (Fig. 13) reflects the growing status of Bedford College during the time of Catherine Raisin. While obviously not all the students were in geology, the science department grew substantially, as is shown by the following quotes:

... as the large number of students makes the room almost too crowded for complete comfort. In fact if all the science classes continue to increase in size as they have done this session [1897], we shall soon overwhelm the art students in numbers, as we have already done in general activity. (HBS 1897)

In 1915:

The committee has also before them statistics ... of the numbers in the Department of Geology as compared with other science subjects and with the same subject in other Colleges in the University. In the latter respect the department compares favourably with those of the other Colleges. (Bedford Committee 1915)

Thus her future as a role model was ensured.

Figure 13 shows the number of students at Bedford College. The bold lines indicate the time Catherine Raisin was there. This shows a steady increase from about 120 to well over 500 by the time she retired in 1920.

She was to become a role model as a teacher, as a student, as an administrator and as a researcher.

Role model as an academic. Catherine Raisin's whole academic career was spent at Bedford College, London, where she become the first female head of department for both the geology and botany departments, a taxing administrative burden which impinged on both her teaching and her research (Burek 2003b). She is listed in the 'Present staff of Bedford College 1894' [lecturer in both Biology (Vegetable) – Botany and Geography and Geology]. It is a testament to her flexibility that, under geology and physical geography, she is listed in the Bedford Calendar as teaching both First and Second Classes (Levels):

The First Class lectures cover

- The building up of rocks and the sculpturing of the Earth's surface.
- Description and classification of rocks and their materials.
- Some account of the successive Geological Periods, including the modifications in their physical geography, and in the life forms which characterise them.

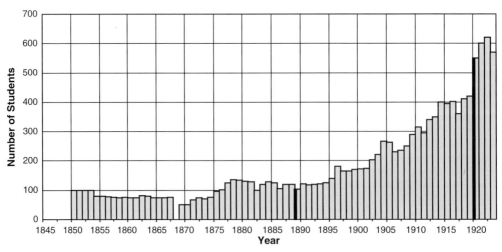

Fig. 13. Bedford College: number of students (1849–1923; after Tuke 1939). Heavy black line represents Catherine Raisin was present at Bedford College.

The First Class practical work covers

- Mineralogy and Petrology – The principal types of minerals and rocks are examined in hand specimen, and in a large series of sections for the microscope.
- Stratigraphy – is illustrated by geological maps of different districts, from which sections are drawn showing the succession of the strata.
- Palaeontology – A collection of the most important fossils is available for Student's work and these type specimens are examined and described.

Some visits to collections at the Geological Museums, and some field excursions will be arranged if possible during the session. Papers will be occasionally given for practice in answering questions.

These classes will meet the requirements of the BSc Examinations. (Bedford College 1894).

This curriculum will appear familiar to many geologists of today.

The Second Class contained

More detailed study of Petrology, British Stratigraphy and Palaeontology. Some account of foreign geology and the growth of continents.

The majority of this was taught by Catherine Raisin with occasional help and was a very full teaching load, which led Catherine to comment 'I am actually teaching nine half days in the week' (Raisin 1909).

However, she was admired not only for her academic research but also her pioneering attitude to female education. In a letter to Miss Calkin on her resignation as Vice Principal in 1901 she states:

The true ideal of a University standard must be kept in view. This will be for women (as in the older universities it has been for men) to encourage the desire to search for knowledge for its own sake. (Raisin 1901*a*)

She fought hard for her female students and colleagues to win their rightful place in higher education as either researchers or lecturers. Her generosity was legendary. She sometimes paid wages out of her own pocket and set up funds as academic prizes. This culminated in legacies at her death in 1946 of:

- a £300 annual prize to be awarded by the geography department of Bedford College, which she had helped financially and personally to set up.
- a £500 bequest to be used for the benefit of non-smoking students.
- £200 to the "Headmistresses Association the interest to be applied for an annual lecture on the duty of thrift".
- £300 to the National Society of Non-Smokers a controversial legacy that made the headlines of the *London Times*.

Her own career contained many Milestones, which are shown in Table 8.

Catherine Raisin died from cancer aged 90 years, after a pioneering and role-model life showing what determined women can achieve in a male-dominated profession.

In order to evaluate her status as a role model several different types of categories will be examined.

Role model for student life: work hard and play hard. Catherine Raisin played an active role in college life, as shown from the *Bedford College London Magazine* (HBS 1897).

In 1891:

Miss Raisin is President of the Debating Society for the second time.

In 1893:

Miss Raisin is now giving a very interesting course of lectures on the 'Ancient Volcanoes and Glaciers of North Wales' illustrated by means of the lime-light lantern.

In 1897:

Students are indeed grateful to Miss Raisin for the consideration that she has shown them in the arrangements of the timetable and especially for her sympathy with the devotees of hockey – the more so as this necessitates two nine o'clock demonstrations and a nocturnal lecture.

Also, in 1900:

The Natural History Society continues we are happy to say in [an] increasingly flourishing state. I think we should all fully recognise that the happy state of affaires is almost entirely due to the energy and kindness of the President Miss Raisin, without whom the Society would soon I am afraid languish and die. (Anon. 1900)

And in 1901, very grateful response to a popular lecturer when she resigned her Vice Presidency:

All have appreciated the kindly and active part which Miss Raisin has taken in the social life of the college. Her loss will be most keenly felt by the residents, for they know her best and will miss her repeated kindness. We hope it will not be only the geology and botany students who will see much of Miss Raisin in the future. (Anon. 1901)

Role model as an administrator. Her role as an administrator was recognized early. In 1898 Catherine was offered the post of Vice Principal, which she accepted in a letter dated 11 June 1898.

I have to thank the Council for the trust which they are willing to repose on me. I appreciate the responsibility which I am undertaking. (Raisin 1898)

However, Catherine started to feel that it detracted from her teaching and research and she tendered her resignation in 1901. But by that time she had made her mark.

Above average height with a striking profile, she looked what she was: a strong personality. Neither colleague nor students could lightly disregard her opinion on college matters. Her earnestness

Table 8. *Catherine Raisin's 'milestones'*

Dates	Achievements
1875	The first woman to study geology at University College
1890	The first woman to become Professor of a science department in Bedford College and head of a geology department
1893	The first woman to receive the Lyell Fund
1898	The first woman to become Vice Principal of a College
1898	The second woman to receive a DSc. from London University
1907	The first women to be appointed Morton Sumner lecturer and curator
June 1919	The ninth woman to become a Fellow of the Geological Society

and sincerity were always beyond question, even when one disagreed with her point of view. A quote from a colleague. (Edgell 1945)

She had respect and deference from her peers.

Dr Raisin's complete devotion of her time and energy was outstanding. Her attention to details of departmental organisation and her interest in the work and careers of students in the Geology department were unsurpassed while the firmness of her convictions and her outspoken expression of them made her throughout the many years of her connection with Bedford College a great force in College affairs. (Bedford Council, reported in Crook 2001)

By 1920 she had been appointed an examiner for the special intermediate exam for the University of London in geology along with one of her former students and subsequent demonstrator, Miss Irene Lowe.

Role model as a teacher. Catherine Raisin was a model teacher. She was affectionately known as 'the Raisin' or more surreptitiously 'The Sultana' (Hawkes 1946). This concern for her students and her teaching ability is shown in many different ways. In 1890 there were no petrological microscopes but by 1898 she had amassed 1100 fossils, 118 mineral specimens, a new laboratory and a £500 grant for equipment from the London County Council.

In 1915, in a letter to Miss Tuke, she states:

I have felt on recent occasions that I have never lectured better in my life.

Also in that year she offers:

to the college a scholarship for Miss Jerram of £50 a year for two years with the condition that she will take geology as one subject. (Raisin 1915)

Then in 1916 she offers to contribute £50 a year to introduce a geography assistant lecturer to help with the geography curriculum during wartime.

There are many further examples throughout the *Bedford College London Magazine*, which started in December 1886, of her generosity to students, and her own personal correspondence demonstrates her acknowledged teaching ability.

Role model as a researcher. It was while an undergraduate that she developed an interest in microscopic petrology and gained a love of fieldwork. Both were to become her research interests. Catherine Raisin published over 22 journal articles, mostly on metamorphic facies (Raisin 1887, 1889, 1892, 1901*a*), igneous rocks (Raisin 1888, 1891, 1893*b*, 1897*a*; Bonney & Raisin 1894) and the petrology of serpentines (Raisin 1892, 1897*b*; Bonney & Raisin 1899, 1905). She was recognized as a leading expert on this last subject, often cooperating with her mentor Professor T. Bonney at University College (Burek 2004). Most of her important papers are listed in the references. Indeed her first paper, in 1887, on the metamorphic rocks of south Devon, was read to the Geological Society of London by Professor Bonney, not by herself, as women were not normally allowed into the meetings. It is interesting to note here that, in a role reversal, she tended to do most of the fieldwork and together they did much of the laboratory analysis work and writing, which he readily acknowledges:

I am bound to state that, though I have been once or twice on the ground since 1880 (when my paper was written) the whole of the laborious field-work for the present communication has been done by Miss Raisin, but that we are jointly responsible for the examination of specimens and the interpretation of results. (Bonney & Raisin 1899)

She thus travelled widely, which was unusual at the time for a woman, undertaking fieldwork abroad (see also Burek & Kölbl-Ebert 2007) as well as attending conferences overseas, such as the International Geological Congress in Toronto in 1913. In Figure 14 she is seen sitting with Rachael Workman, later to become Lady MacRobert. She was the first woman to be awarded money from the Lyell fund from the Geological Society of London (in 1893) for her work. However, like her first read paper, she could not receive it in person. Women did not attend or become fellows until 1919, 26 years later.

Fig. 14. Catherine Raisin at the International Geological Conference in Toronto in 1913 seated 3rd from left, second row.

She also worked on the microcrystalline formation of chert in Jurassic rocks and published an acclaimed paper in the *Proceedings of the Geologists' Association* (Raisin 1903). It was her careful field observation and meticulous petrological microscope laboratory work, which won her the admiration of her peers and students. It is significant that she was present at the centenary celebrations of the Geological Society of London, one of the few women geologists invited in her own right.

Except for a representative of a Belfast College, I was the only woman among the delegates and many would have regretted it if they had not included some representative of women geologists. (Raisin 1907 quoted in Crook 2001)

Catherine Raisin had the ability to ask questions that others could not (Burek 2003b).

What will happen to my collection when I retire? (Burek 2003a; Raisin 1916)

Who is going to continue the curatorship of the collection?

I might, at last, have looked forward to devoting time during the next four or five years, as I would wish, to some of my many specimens. But the Council have negatived [sic] this possibility. (Raisin 1915a)

Why is geology not taught in schools?

When do I have time to do research?

No other lecturer, holding as I do, that it is an essential for a college post, that research work should be carried on, would have such work interrupted first, by a struggle to maintain

Table 9. *Catherine Raisin's teaching legacy*

Name and Date	Further degrees		Achievements
F. J. Relf 1906	MSc (1922)		
Irene Lowe 1913	MSc (1920)		Demonstrator, Bedford College
Helen Muir-Wood 1919	MSc (1920)	DSc (1935)	1958 Lyell Medal Edited Brachiopoda section of the *Treatise on Invertebrate Palaeontology* 1960 OBE
Doris Reynolds 1920	MSc (1924)	DSc (1937)	1960 Lyell Medal Petrology

for many years two departments of the college, and then, secondly, when that extra burden was laid aside by the necessity of planning rooms and moving and arranging collections. Not I think would any other lecturer be likely to give to this last mentioned duty such an excessive amount of time as I have given... Much additional work has been laid upon me by the war. (Raisin 1915b)

Why do I get only one-year contracts?

I greatly regret that they have not allowed me the hope of retaining my appointment for the term of five years permitted by Statute. (Raisin 1915a)

Why are we so overworked and underpaid?

All these questions are as relevant today as they were over 100 years ago.

All Catherine Raisin's research students, Irene Lowe, Helen Muir Wood and Doris Reynolds, came towards the end of her career at Bedford College (Table 9) but she had established a good foundation for teaching and research at a difficult time for women in higher education.

Doris Reynolds

Doris Reynolds, the granite expert and later wife of Arthur Holmes, was one of Catherine Raisin's last students at Bedford College. Catherine Raisin is remembered by her most famous student, Doris Reynolds, as:

not only a stimulating and enthusiastic teacher, who worked ungrudgingly to promote their interests, but also as a generous, brave and sympathetic woman whom they loved. (Reynolds 1945)

She further recalls, in 1960, when receiving her Lyell medal from the Geological Society, the significant contribution that she felt Catherine Raisin gave:

It was Catherine Raisin who first interested me in geology, and from whom I inherited my love for petrology; I owe very much to her long memory and unbiased teaching. (Reynolds 1960)

Raisin was a true role model.

Emily Dix

Another strong role model originating from Bedford College was Emily Dix (1930–1947). After graduating from Swansea, she took up her position in 1930 against strong opposition. She worked closely with Leonard Hawkes. She was recognized as both an excellent teacher and researcher in the area of palaeobotany. Her story and contribution has been told elsewhere and will not be elaborated on here (Burek 2005; Burek & Cleal 2005).

Destinations of women geologists from Bedford and Newnham Colleges

An analysis of the destinations of graduates from the two institutions is difficult to gauge. Teaching is a top destination for many students, as would be expected. Other suitable professions open to women geologists included journalism, factory inspection and writing. Figures for Bedford College can only be determined for the whole group of graduates but those specifically identified as having studied geology are included in Figure 15.

A summary of destinations for the graduates of 1904 shows the importance of the new sanitary health course, and many girls went into this usual area or become factory inspectors (Table 10). Quotes from Bedford College Magazine include:

A very great number I am glad to say have married and gone out some of them to far countries.

Of the 8 factory inspectors 2 are our old students.

It was slightly easier for those from Newnham College, Cambridge. The records have allowed a detailed table of destinations after graduation to be built up (see Figure 16). As expected, the 1878–1939 data shows that the vast majority of students entered the field of education, sometimes teaching geology and always using the methods taught by their mentors, until they married. Even then, some carried on in the area of governance, both at a local and national level. These ladies then formed the role models for the next generation of teachers and geologists.

The importance of geology within geography as a school subject is highlighted in an article by Busk

WOMEN IN GEOLOGICAL HIGHER EDUCATION 33

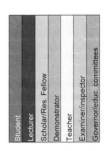

Fig. 15. Bedford College: Destinations geology students (1886–1932).

Table 10. *Bedford College: generic geology student destinations*

- Author (Mary Evans)
- Workers for public movements
- Artists
- Doctors
- Factory & sanitary inspectors
- Teachers & headmistresses
- Married

(1895). Here Miss Busk details the school curriculum for 6–18-year-olds and shows where geology is included. The curriculum for 6–9-year-olds was recommended by Archibald Geikie himself and he encouraged student-led questions. If we analyse the physical geography section, we find physical geology as an integral part for junior school (10–13-year-olds) and some implied in middle school (13–16-year-olds), with volcanic action listed. It is explicit for senior children (16–18-year-olds): 'physical geography ... scientific explanation of the more difficult physical phenomena including some elementary geology.' This need for geography teachers to have an understanding of geology in order to teach A-level equivalent geography helped to increase the number of students undertaking geology for a degree, as many of the school-leavers were destined to become teachers. Ethel Skeat is a good example, as shown in the paper by Burek & Malpas (2007).

Discussion

Most women at this time of evolving higher education for women went into teaching and served as role models. It was a socially acceptable opening for them. The difference between Newnham College, Cambridge, and Bedford College, London, is obvious and has been explored in this paper both from the point of view of location, clientele, lecturers and student destinations.

Bedford College, which, in 1849, was one of the first all-female colleges in the United Kingdom, explored and led the way for female higher education institutions, making mistakes and allowing others to learn from them. It established that:

- Entrance requirements were advisable.
- Female committees were unleadable because women at that time had no experience of such work, notwithstanding the fact that many dealt with extensive households later in life.
- Male council members became frustrated at the lack of female professionalism.
- Lack of an efficient infrastructure led to the college missing the first application for grants from the London Country Council.
- Having somewhere for day students to socialize was important and led to a better student experience.
- Up until 1894 all lecturers were part time and therefore perhaps lacked the loyalty needed for a successful college community.
- All lecturers were called professors, which is a difference from the terminology of today.

Conclusions

Bedford College had struggled to set the precedent in London for the higher education of women across all subjects. However, the success of geology at Bedford College owes no small debt to the perseverance and dedication of one strong-minded and determined lady: Dr Catherine Raisin.

Newnham College provided a vast number of long-serving teachers throughout the period studied and provided some valuable, albeit eccentric, role models. The importance of networks for these women geologists cannot be understated and the place that role models played is vital – and this remains the case today.

So what did it take to become a role model in late Victorian and early 20th century times?

If we look carefully at Catherine Raisin and Gertrude Elles, our role models are single, show strong determination and are prepared to stand up for their students. They would be called 'characters' as are most pioneers. They are intelligent women and have the respect of their peers in the subject of their choice, although here the subject is not important.

This study also points to other attributes. Perseverance in the face of adversity, patience and a sense of justice, while not solely the preserve of females, also helped. Lapworth recognized this, as did both Professor McKenny Hughes and Dr Marr.

The importance of good role models in education is adequately summarized by Baroness Nancy Seear of Paddington in 1982 in the House of Lords:

Writer after writer mentions a particularly understanding teacher or department head at a key stage in her career. Money also helps so does physical toughness. These women are the survivors, the fittest and the fortunate.

They are the role models for tomorrow's generations and carry a huge responsibility. So have all

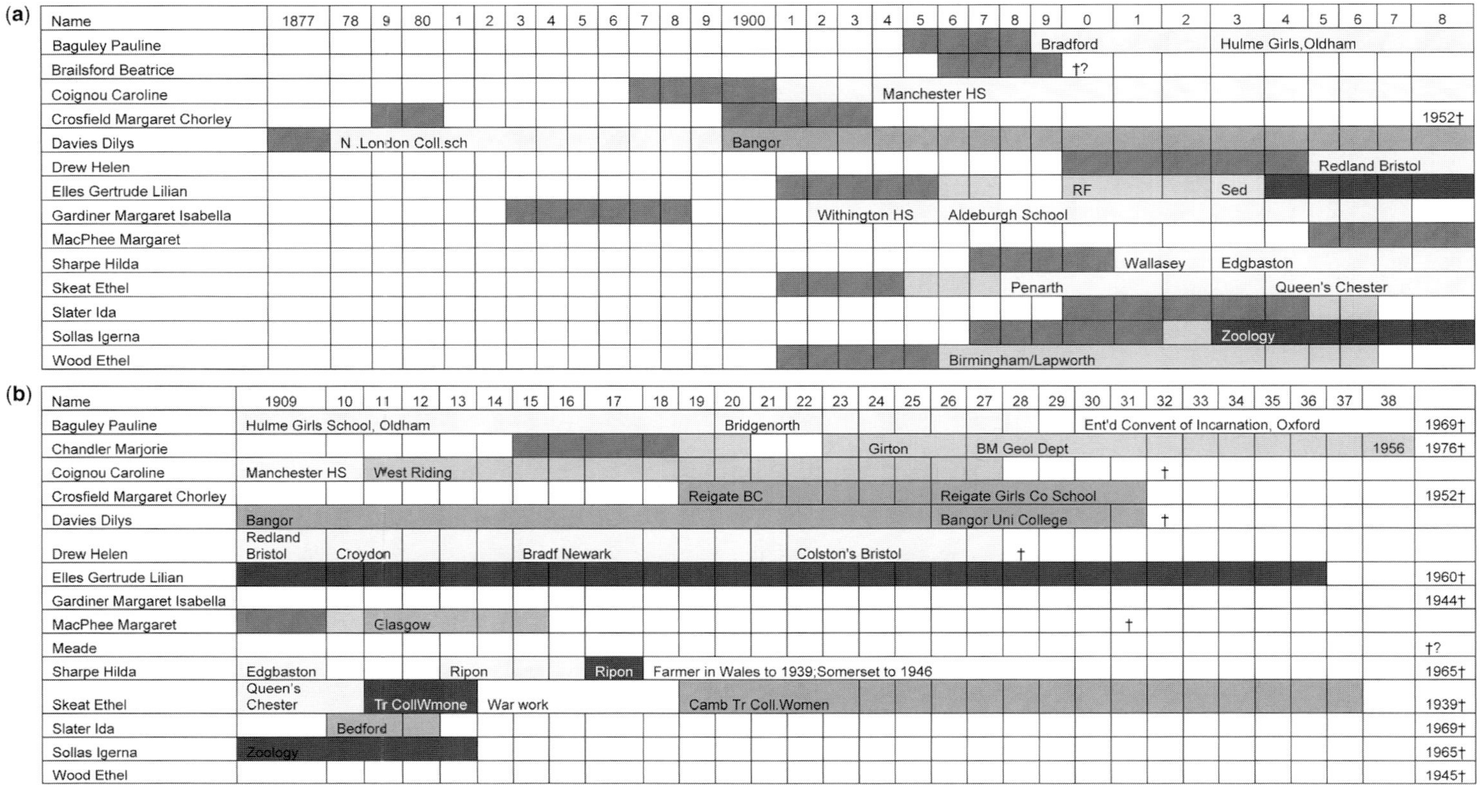

Fig. 16. (a) Newnham College geology student destinations (1877–1908); (b) Newnham College geology student destinations (1909–1938). Key as for Fig. 15. †, year of death.

teachers at all levels of education throughout history.

I would like to acknowledge the help of the archivists at the Sedgwick Museum, Royal Holloway College, Lady MacRobert Trust and Newnham College, and all the people that aided me on my travels, especially Kath Langley, Anne Esson, Jim Rose, Madeleine Montgomerie, Penny Durrell and my daughters, Frances and Veronica Cubitt, as well as my long-suffering son and husband. They have all helped me in my discovery of these forgotten women.

References

ADAMS, J. F. A. 1887. Is botany a suitable study for young men? *Science*, **9**, 117–118.

ANDREWS, E. A. 1881. Hurrah! We have won! *In*: PHILLIPS, A. (Ed.) *A Newnham Anthology*, The Principal and Fellows of Newnham College, Cambridge, 1988, 17–19.

ANON., 1888a. University of London. *Bedford College London Magazine*, **1**(May), 10.

ANON., 1888b. New laboratories for the college. *Bedford College London Magazine*, **1**(November), 11–12.

ANON. 1890. The Laboratories of Bedford College. *Bedford College London Magazine*, **1**(11), 15–16.

ANON. 1893. Presentation day. *Bedford College London Magazine*, **1**(June), 13.

ANON. (a former student) 1896. Extension of college premises. *Bedford College London Magazine*, **1**, 2–4.

ANON. 1898. Science. *Bedford College London Magazine*, **2**(Sept.), 16.

ANON. 1900. *Bedford College London Magazine*, **2**.

ANON. 1901. College notes. *Bedford College London Magazine*, **2**, 11.

BALL, M. D. 1988. Newnham scientists. *In*: PHILLIPS, A. (Ed.) *A Newnham Anthology*. The Principal and Fellows of Newnham College, Cambridge, 76–78.

BEDFORD COLLEGE. 1894. Calendar of Bedford College, London, 362–363.

BEDFORD COMMITTEE. 1915. '*Report of the Committee under by-law 29a, dated 27 March 1915.*' Bedford College Archives, Correspondence, 69.1.

BHW 1897. Science. *Bedford College London, Magazine*, (Dec.), 15–16.

BONNYCASTLE, J. 1796. *Introduction to Astronomy in a Series of Letters from a Preceptor to His Pupil*. Johnson & Co., London.

BONNEY, T. G. & RAISIN, C. A. 1894. On rocks and minerals collected by Mr W. M. Conway in the Karakoram Himalayas. *Proceedings of the Royal Society*, **4**, 350–468.

BONNEY, T. G. & RAISIN, C. A. 1899. On varieties of serpentine and associated rocks in Anglesey. *Quarterly Journal of the Geological Society, London*, **55**, 276–304.

BONNEY, T. G. & RAISIN, C. A. 1905. The microscopic structure of minerals forming serpentine, and their relation to its history. *Quarterly Journal of the Geological Society, London*, **61**, 690–711.

BOTT, A. 1932. *Our Mothers: A Cavalcade in Pictures, Quotations and Description of Late Victorian Women 1870–1900*. Victor Gollancz Ltd, London, 220 pp.

BUCKLEY, A. B. 1879. *The Fairy-land of Science*. H. M. Caldwell Co., New York.

BUREK, C. V. 2002. Women in Geoscience. *In*: LERNER, K. L. & LERNER, B. W. (eds) *World of Earth Science*. Gale Thomson, London.

BUREK, C. V. 2003a. Time to take responsibility for collections. *Earth Heritage*, **20**, 22–23.

BUREK, C. V. 2003b. Catherine Raisin: A Role model professional geologist. Women in the History of Geology 3. *Geology Today*, **19**(3), 107–111.

BUREK, C. V. 2004. Raisin, Catherine Alice (1855–1945). *Dictionary of Nineteenth Century British Scientists*, **4**, 1657–1658.

BUREK, C. V. 2005. Emily Dix, palaeobotanist – a promising career cut short. *Geology Today*, **21**, 144–145.

BUREK, C. V. & CLEAL, C. 2005. The life and work of Emily Dix (1904–1972). *In*: BOWDEN, A. J., BUREK, C. V. & WILDING, R. (eds) *History of Palaeobotany, Selected Essays*. Geological Society, London, Special Publications, **241**, 181–196.

BUREK, C. V. & KÖLBL-EBERT, M. 2007. The historical problems of travel for women undertaking geological fieldwork. *In*: BUREK, C. V. & HIGGS, B. (eds) *The Role of Women in the History of Geology*. Geological Society, London, Special Publications, **281**, 115–122.

BUREK, C. V. & MALPAS, J. A. 2007. Rediscovering and conserving the Lower Palaeozoic 'treasures' of Ethel Woods (neé Skeat) and Margaret Crosfield in northeast Wales. *In*: BUREK, C. V. & HIGGS, B. (eds) *The Role of Women in the History of Geology*. Geological Society, London, Special Publications, **281**, 203–226.

BUSK, H. 1894. Inaugural Lecture. *Bedford College London Magazine* **1**, 1–4.

BUSK, H. 1895. Geography as a school subject. *Bedford College Magazine*, **1**, 17–30.

CREESE, M. 1991. British women of the nineteenth and early twentieth centuries who contributed to research in the chemical sciences. *British Journal for the History of Science* **24**, 275–305.

CREESE, M. & CREESE, T. 2006. British women who contributed to research in the geological sciences in the 19th century. *Proceedings of the Geologists' Association*, **17**, 53–83.

CROOK, J. M. (ed.) 2001. *Bedford College University of London: Memories of 150 years*. Royal Holloway and Bedford New College, Surrey.

CROWTHER, C. 1896. Women on sufferance. *In*: PHILLIPS, A. (ed.) *A Newnham Anthology*. The Principal and Fellows of Newnham College, Cambridge, 1988, 37–39.

DALE, M. L. 1901. The Morton Sumner bequest. *Bedford College London Magazine*, **2**(45), 23–24.

EDGELL, B. 1945. Dr Catherine Raisin, Obituary. *Bedford College Old Students Association Journal*, 8–9. Royal Holloway University of London, Archives, BC AS 903/2.

ELLES, G. L. 1909. The relation of the Ordovician and Silurian rocks of Conway (North Wales). *Quarterly Journal of the Geological Society, London*, **65**, 169–194.

ELLES, G. L. 1922a. The graptolite faunas of the British Isles. Proceedings of the Geologists' Association, **33**, 168–200.

ELLES, G. L. 1922b. The Bala country: its structure and rock succession. *Quarterly Journal of the Geological Society, London*, **78**, 132–175.

ELLES, G. L. & SLATER, I. L. 1906. The highest Silurian rocks of the Ludlow district. *Quarterly Journal of the Geological Society, London*, **62**, 195–221.

ELLES, G. L. & WOOD, E. M. R. 1895. Supplementary notes on Drygill Shales. *Geological Magazine*, **2**, 216–249.

ELLES, G. L. & WOOD, E. M. R. 1901–1918. *Monograph of British Graptolites*. Parts 1–11. Palaeontological Society, London, Monographs, 1–539.

FERGUSON, J. 1748. *Young Gentlemen and Lady's Astronomy: Familiarly Explained in Ten Dialogues between Neander and Eudosia*. 6th Edn. James Williams, Dublin, 1778.

GRIMSHAW, M. E. 1988. Sunday nights in Kennedy. *In*: PHILLIPS, A. (ed.) *A Newnham Anthology*. The Principal and Fellows of Newnham College, Cambridge, 165–167.

HALLIDAY, D. L. 1988. Years of renaissance 1920. *In*: PHILLIPS, A. (ed.) *A Newnham Anthology*. The Principal and Fellows of Newnham College, Cambridge, 136–143.

HAWKES, L. 1946. Obituary Catherine Alice Raisin. Annual report of the Council. Proceedings of the Geologists' Association, **57**, 53–4.

HBS, 1897. *Bedford College London Magazine* **2**.

HIGGS, B. & WYSE JACKSON, P. N. 2007. The role of women in the history of geological studies in Ireland. *In*: BUREK, C. V. & HIGGS, B. (eds) *The Role of Women in the History of Geology*. Geological Society, London, Special Publications, **281**, 137–153.

HOLT, C. D. 1897. Letter from Newnham College 1889–1892. *In*: COCKBURN, E. D. (ed.) *Catherine Durning Holt*, 3rd Edn. The Principle and Fellows of Newnham College, Cambridge, UK, 50 pp.

HUGHES, T. & MCKENNY, 1892. *Letter to Mrs Sidgwick*. Sedgwick Club, Cambridge, Archives.

INGLEDEW, M. 1988. Mrs Palmer and Peile. *In*: PHILLIPS, A. (Ed.) *A Newnham Anthology*, The Principal and Fellows of Newnham College, Cambridge, 179–181.

JONES, A. F. 1894. Science side. *Bedford College London Magazine*, **1**(June), 8–9.

KINGSLEY, C. 1889. *Madam How and Lady Why, or First Lessons in Earth Lore for Children*. Facsimile 2005, Yesterday's Classics, Chapel Hill, North Carolina.

KÖLBL-EBERT, M. 2001. On the origin of women geologists by means of social selection: German and British comparison. *Episodes*, **24**(3), 182–193.

KÖLBL-EBERT, M. 2002, British geology in the early nineteenth century: a conglomerate with a female matrix. *Earth Sciences History*, **21**(1), 3–25.

LINDLEY, J. 1865. *'Ladies Botany; Or, A Familiar Introduction to the Study of the Natural System of Botany.'* HG Bohn, London.

L'OREAL, 2006. *Responsibility and Society*. World Wide Web Address: www.loreal.com/en_/_ww/dev_dur/part/programme.aspx

MARCET, J. 1809. *Conversations on Chymist*, 2nd American Edn. Hertford. Conn. James Humphrey Boston, Mass.

MCKENNY HUGHES, J. 1887. *'Lecture notes on the debate on women entry into University of Cambridge.'* Sedgwick Club, Cambridge, Archives.

PAGE, G. & SMITH, A. J. 2001. Geology. *In*: CROOK, J. M. (ed.) *Bedford College Memories of 150 Years*. Royal Holloway and Bedford New College, Surrey, 149–180.

PHILLIPS, A. (ed.) 1988. *A Newnham Anthology*. The Principal and Fellows of Newnham College, Cambridge, 278 pp.

RAISIN, C. A. 1887. Notes on the metamorphic rocks of South Devon. *Quarterly Journal of the Geological Society, London*, **43**, 715–733.

RAISIN, C. A. 1888. On some rock specimens from Somali Land. *Geological Magazine*, **5**, 414–418.

RAISIN, C. A. 1889. Devonian greenstones and chlorite schists of south Devon. *Geological Magazine*, **6**, 265–269.

RAISIN, C. A. 1891. Report on some rock specimens from the Kimberley diamond mines. *Geological Magazine*, **8**, 412–415.

RAISIN, C. A. 1892. So-called serpentines of the Lleyn. *Geological Magazine*, **9**, 408–413.

RAISIN, C. A. 1893. Award of the Lyell Geological Fund: Reply by Professor Bonney. *Proceedings of the Geological Society, London*, **49**, 42.

RAISIN, C. A. 1893b. On the so called spilites of Jersey. *Geological Magazine*, **10**, 59–64.

RAISIN, C. A. 1897a. On a hornblende-picrite from Zmutthal (Canton Valais). *Geological Magazine*, **iv**, 202–205.

RAISIN, C. A. 1897b. On the nature and origin of the Rauenthal serpentine, *Quarterly Journal of the Geological Society, London*, **53**, 246–268.

RAISIN, C. A. 1898. *'Letter to Dr Russell, 11 June 1898, Bedford College for Women.'* Royal Holloway, University of London, Archives, BC AR/150 D168.

RAISIN, C. A. 1901a. *'Letter to Council, 7 May 1901, Bedford College for Women.'* Archives, Royal Holloway, University of London, BC AR/150 D168.

RAISIN, C. A. 1901b. *'Letter for the Vice Principal, Bedford College for Women.'* Reprinted in Bedford College London Magazine, **2**(June), 10.

RAISIN, C. A. 1903. The formation of chert and its microstructures in some Jurassic Strata. *Proceedings of the Geologists Association*, **18**(2), 71–82.

RAISIN, C. A. 1906. *'Letter to Council, 25 May 1906, Bedford College for Women.'* Archives, Royal Holloway, University of London, BC AR/150 D168.

RAISIN, C. A. 1907. *'Natural Sciences versus Physical Sciences.'* Bedford College London Magazine, **X**(V), 22.

RAISIN, C. A. 1909. *'Letter to Council, 8 March 1909, Bedford College for Women.'* Archives, Royal Holloway, University of London, BC AR/150 D168.

RAISIN, C. A. 1910. *'Letter to Council, 21 February 1910, Bedford College for Women.'* Archives, Royal Holloway, University of London, BC AR/150 D168.

RAISIN, C. A. 1915a. *'Letter to Miss Tuke, 7 May 1915, Bedford College for Women.'* Archives, Royal Holloway, University of London, BC AR/150 D168.

RAISIN, C. A. 1915b. *'Letter to Miss Tuke, 7 July 1915, Bedford College for Women.'* Archives, Royal Holloway, University of London, BC AR/150 D168.

RAISIN, C. A. 1916. *'Letter to Miss Tuke, 25 January 1916, Bedford College for Women.'* Archives, Royal Holloway, University of London, BC AR/150 D168.

RAISIN, C. A. 1919. *'Personal correspondence to Major Darwin, 1 March 1919.'* Archives, Royal Holloway, University of London, BC AR/150 D168.

RAISIN, C. A. 1932. *'Personal correspondence to Miss Monkhouse, 12 August 1932.'* Archives, Royal Holloway, University of London, BC AR/150 D168.

REYNOLDS, D. 1945. Dr Catherine Alice Raisin, Obituaries. *Nature*, **156**(3959), 327–328.

REYNOLDS, D. 1960. Lyell medal reply. *Proceedings of the Geological Society of London*, 1580, 95–96.

RODERICK, G. W. & STEPHENS, M. D. 1972. *Scientific & Technical Education in 19th Century England*. David & Charles, Newton Abbot, 173 pp.

SALT, C. & BENNETT, E. 1986. *College Lives, Royal Holloway and Bedford New College*. University of London, 60 pp.

SEDGWICK CLUB. 1896. *'Minutes Book No VI. Jan 1893– Nov 1896.'* Sedgwick Club, Cambridge, Archives.

SHEFFIELD, S. LE-MAY. 2004. *Women and Science: Social impact and interaction*. ABC Clio Inc California, 409 pp.

SIDGWICK, E. M. 1892. *'Letter to Prof. Hughes.'* Sedgwick Club, Cambridge, Archives.

STORY-MASKELYNE, N. 1891. Opening of the Shaen wing by the Empress Frederick. *Bedford College London Magazine*, **1**(15), 1–3.

TUKE, M. J. 1939. *A History of Bedford College for Women 1849–1937*. Oxford University Press, Oxford, 364 pp.

WHITWORTH, M. 2000. *A History of Manchester University*. Manchester Museum, Manchester.

WILLMOT & EAST. 1946. *Personal correspondence on Catherine Alice Raisin bequest in her will*. Royal Holloway, University of London, Archives, BC AR/150 D168.

Fossil hunters, a cave explorer and a rock analyst: notes on some early women contributors to geology

M. R. S. CREESE

1650 Cambridge Road, Lawrence, KS 66044-2545, USA

Abstract: The work and achievements of nine women contributors to 19th- and early 20th-century developments in the geological sciences are sketched. Two of these women – Gordon Cumming and Gray – were Scottish, two – Owen and Maury – were from the United States, Cleve von Euler and Sahlbom were Swedish, and three – Pavlova, Solomko and Tsvetaeva – were Russian. Of these nine, seven worked in palaeontology (then and later the branch of the field most often taken up by women), Owen made her name primarily as a speleologist and Sahlbom was a rock and mineral analyst. The sketches are offered as additional material for the ongoing effort to uncover and assess the role played by women in early work in the sciences.

Contributions by women to 19th- and early 20th-century work in geology were not lacking, although very often absent from printed records and rarely noted in histories of geology until the emergence of the new field of history of women in science a mere 25 years ago. The picture is beginning to change, however; recent examples demonstrating the expanded coverage given to women can be found even in general, all-encompassing publications such as *The Oxford Dictionary of National Biography* and *The Dictionary of Nineteenth-Century British Scientists*, both of which appeared in 2004 and both of whose editors took pains to include many female subjects.

The fact that women were becoming involved in geological, particularly palaeontological, work from the early decades of the 19th century is hardly surprising (Fraser & Cleal 2007), given the great activity in what was then the new field of stratigraphic determination by fossil content; exciting discoveries were being made and, especially in the initial period, the possibility of taking part in some aspect of the endeavour without special training was open to those with the time and inclination.

In the earliest work the overall accomplishments of British women stand out in noteworthy contrast to the contributions made by women in other countries. Exploratory investigations in the United Kingdom in strata identification by fossil content carried out by mining engineers and surveyors in the 1820s and 1830s, along with regional stratigraphic projects of the gentlemen amateurs of the new London Geological Society, had aroused a great deal of interest in the general population; further, in response to popular demand, dependable introductory geology publications, both periodical articles and books, were becoming available (Issitt 2002). The feeling of the time was captured in a remark by Lady Gordon Cumming who is quoted by Lady Murchison as saying, 'I am so *envied* [original emphasis] for we live upon beds of fossil fish' (Collie & Diemer 1995, p. 15). And Rosina Zornlin (1839), an early English writer of popular science works, elaborated further in her well-researched book, *Recreations in Geology*:

The search for fossil remains has so much in it that is fascinating; the delight of meeting with rare, or even with well-known organic remains is so great; the possibility (open to all) of discovering some new species – some new link in nature's chain – so full of hope and expectation, that this science possesses a zest to which few others can lay claim.

Quoting from an article in the *Magazine of Popular Science*, Zornlin goes on to point out that even those whose observations are limited to the rocks of their home districts can make useful contributions in the form of important points of detail (Zornlin 1839, pp. iii–iv, lviii–lix).

Compared to the army of male amateurs involved in the work, the early 19th-century women fossil collectors in the United Kingdom were relatively few in number. Nevertheless, several made notable contributions, thanks in part to the long-established British tradition of decentralized amateur naturalist groups, often open to women, which encouraged independent local studies and provided a congenial social base from which to operate. One of the women amateurs was Lady Gordon Cumming of Altyre, on the shores of the Moray Firth, who collected fossil fishes from the Old Red Sandstone of the region in the 1840s; a second was Elizabeth Anderson Gray, remembered for her work in the rich Lower Palaeozoic strata along the Clyde coast from the 1880s until well after the turn of the century.

Somewhat in contrast to the United Kingdom, the United States and countries in continental Europe generally did not have a network of flourishing amateur groups. (Russia may be something of an exception. Local and regional groups there were active, although information about them appears to be hard to find; but see the note on Pavlova, below.) Throughout the 19th century, scientific work in France, Germany (see Kölbl-Ebert 2007) and Sweden for example, was largely in the hands of men who held posts in publicly supported institutions – universities and national museums. While the relatively substantial resources of these institutions facilitated the production of a great deal of high-quality research and the development of basic theory, the continental system had the side effect of largely excluding women. Their participation to any extent in scientific research had to wait until they entered the universities in the final decades of the century.

In the United States, as in continental Europe, women were slow to enter the field. A very few were studying geology in state universities by the 1880s but geology remained a marginal subject in the curricula of the early women's colleges until a vigorous programme was started at Bryn Mawr College, Pennsylvania, in the 1890s. Even then there was considerable opposition within the Bryn Mawr college administration; geology was considered to have 'no wide appeal to women' (Ogilvie 1945).

The British/Continental and British/North American contrasts are apparent even from an investigation based on numbers of geology papers by women published in scientific periodicals up to 1900 (Creese 1998, 2004a), that is, without taking into account the many women whose contributions came in the form of valuable fossil collections sold or given to well-known male geologists. British women, along with two or three Irish (see Higgs & Wyse Jackson 2007) dominated the field, out-publishing women from all other national groups. Their closest rivals were the women from Russia and Poland, who together produced one paper for every five from Britain/Ireland within the 1800–1901 time span; for the North Americans the ratio is about one to seven over the same period (Creese 2004a, pp. 209–215).

Since work on a number of the most prominent and most productive of the British women (such as Ogilvie Gordon, Elles, Raisin and Woods) is already fairly well advanced, with sketches, essays on particular aspects of their careers (see Burek 2007; Burek & Malpas 2007; Wachtler & Burek 2007) and, in a few cases, full-scale biographies appearing regularly, sketches of only two still somewhat neglected British women are offered here.

Two early Scottish fossil hunters

Lady Eliza Maria Gordon Cumming (c. 1798–1842). Lady Eliza Maria Gordon Cumming (c. 1798–1842) of Altyre House, Morayshire, was one of the many enthusiastic amateur collectors and illustrators who worked in all parts of Scotland from the 1820s (Andrews 1982). The north of the country is famous for its fossil fishes, the most important fauna of the Old Red Sandstone formations, and the quarries of the Moray Firth region provided a great deal of valuable material. Further, the discoveries there concerned no small backwater in the history of geology but played a major role in fixing the stratigraphic position of the Old Red Sandstone within the Devonian period.

Lady Gordon Cumming amassed a large collection of fossil fishes and, together with her eldest daughter, Lady Seymour, recorded her finds in many pen and watercolour sketches. She was a friend and correspondent of Buckland, Murchison and Agassiz, and provided them and others with important material. Murchison referred to her fossil collections many times in his correspondence with leading fellow male geologists (Collie & Diemer 1995, pp. 105, 157, 160, 161).

A grand-daughter of John Campbell, 5th Duke of Argyll, Eliza Maria married Sir William Gordon Cumming, 2nd Baronet of Altyre and Gordonstoun, in 1815. Vigorous, enterprising and endowed with considerable artistic talent, she had wide interests that included painting, gardening and horticulture, as well as current science. She designed the gardens at Altyre House, built up a large botanical collection and experimented with crossing to produce new varieties. The Gordon Cummings were prominent socially, spending the season in London and occasionally visiting Paris.

Lady Eliza Maria probably became interested in fossil fishes about 1839 after the discovery of particularly fine specimens in a lime quarry at Lethen Bar in neighbouring Nairnshire. She assembled a large collection from the material the quarrymen set aside for her; not lacking financial resources, after a time she may even have had the quarry worked specially to find more. In 1840, Agassiz, Buckland and probably Murchison visited Altyre House, examined the Gordon Cumming fossil collection and inspected fossil-rich sites in the vicinity. Agassiz named many of the Altyre House specimens and designated several as new species; a considerable number were reported in his monograph on the fossil fishes of the Old Red Sandstone (Agassiz 1844–45).

Much of the Altyre House collection is now housed in museums: the Royal Scottish Museum, Edinburgh, the British Museum (Natural History),

London, and the Institut de Géologie museum at the University of Neuchâtel. Several of Lady Eliza Maria's sketches, which she had hoped in vain to have published, are held in the Geological Society's archives.

Elizabeth Anderson Gray (1831–1924). Elizabeth Anderson Gray (1831–1924) is now recognized as one of the foremost Scottish fossil collectors of the late 19th and early 20th centuries (Cleevely *et al.* 1989; Creese 1998, pp. 290–291; Creese 2004*b*; Cleevely 2004). For more than 50 years she collected from the rich Lower Palaeozoic strata of the Girvan region along the Clyde coast and her materials were used by both stratigraphers and palaeontologists. Charles Lapworth, in his work on the 'Girvan Succession' referred extensively to her collections in his stratigraphical correlations.

Born in Alloway, Ayrshire, Elizabeth Anderson had little formal schooling but as a girl joined her father, Thomas Anderson, an innkeeper turned farmer, in his hobby of fossil collecting. When she married a Glasgow banker, Robert Gray, in 1856, she quickly expanded his already strong natural history interests to include fossil hunting. Within a few years he was exhibiting their jointly assembled collections at meetings of the Glasgow Natural History Society.

With time, Elizabeth Gray acquired a good general understanding of the significance of her fossils; a geology course for ladies at Glasgow University helped her considerably. Her methods were meticulous, her records precise, and she made sure that her materials were described by competent specialists (e.g., Longstaff 1924). After Robert Gray's death in 1887 the work continued with four daughters helping Elizabeth. The Gray collections, sold to institutions, were immense, a major one going to the British Museum in 1920 for £2250. Rich in type species, they are still considered important in studies of Ordovician fauna.

Over the years Elizabeth Gray's inspiration and her enthusiasm for extending her faunal lists, establishing occurrences and finding new species came from her ever-deepening awareness of the importance of her finds and their inter-relationships. Should she be classed as an 'amateur'? Yes, probably so. Gray was a skilled, competent and productive worker, but without ambitions to describe and publish her discoveries herself. Perhaps it is worth bearing in mind here an observation made by geology historian David Oldroyd. He pointed out that:

After the establishment of the [British Geological] Survey [in 1835], the surveyors regarded themselves as the 'professionals', while all other geologists, from university professors to local fossil collectors, were termed 'amateurs'. The amateurs did not reject the label. (Oldroyd 1996, p. 329, Note 28)

The radical changes in the connotations of these terms over the years make hazardous attempts at unqualified categorization of workers of that earlier time.

An American cave explorer and a petroleum geologist

In the United States, by the middle of the 19th century, women were attending public lectures on geology in the larger cities, a number of women, including Orra White Hitchcock and Sarah Hall (Creese 1998, p. 304, Note 1) were preparing technical illustrations for geology publications, and women authors had produced a few popular introductory texts. However, there is no evidence of significant numbers of early amateurs building up important fossil collections, as in the United Kingdom. Indeed, the first productive American woman geologist was university-trained Florence Bascom (1862–1945), a faculty member at Bryn Mawr College for women and the first woman to hold an assistant geologist appointment with the United States Geological Survey (Creese 1998, pp. 285–286; Clary & Wandersee 2007).

Luella Owen (1852–1932). Of Bascom's time but well separated from east-coast academic circles was the largely self-educated amateur Luella Owen (1852–1932) (Creese 1998, pp. 283–285). Born in what was then the rough, frontier outpost of St Joseph on the Missouri river, Luella was the daughter of a prominent Missouri lawyer. As a child she explored the caves in the bluffs along the river near her home and she never lost her interest in caves and caverns. By her early 20s, she had made contact with a number of geologists, including state geologists investigating the recent formations of the American Midwest. Her earliest publications are thought to have concerned glacial deposits and loess formations but, to please her father, she brought these out anonymously. Only after her father died in 1890 was she free to enter fully into cave exploration.

The 3500 caves in Missouri, plus several famous caverns in the region, offer ample scope for those with interests in speleology and Owen was one of several women associated with Missouri caves in the late 19th and early 20th centuries (Vineyard 1970). These women included amateur naturalist Ruth Hoppin who, in the late 1880s, made an important contribution to the Smithsonian Institution in Washington DC of specimens of both the blind Ozark cave fish, *Troglichthys* (now *Amblyopsis*) *rosae*, and the bristly cave crayfish, *Cambarus setosus* (Garman 1888–89). Also prominent were the Lynche sisters, Genevieve and Miriam, and the Mann sisters, Agnes, Margaret and Ada. Both the Lynches and the Manns, however, are remembered mainly for their work of cave management. Owners of the land on which the caves were found, they ran these as commercial ventures, maintaining them and keeping them open for public tours. The Lynches ran Marvel Cave from the late 1920s and the Manns Crystal Cave at about the same time. In each case the women inherited the business from fathers who had first opened the caves to visitors in the 1890s.

Luella Owen's work involved entering by rope and bucket and exploring by candlelight. Formations in the Ozark hills of southern Missouri and the Black Hills of South Dakota were her special interests. She was one of the early explorers of the system known as Grand Gulf Cave, a collapsed cavern system occupied by an underground river in southern Missouri. Here she used a small boat, but it is not known how far she penetrated along the nearly 13-km subterranean stretch to where the river resurfaces at Mammoth Springs, Arkansas. Her account from the 1890s remains of special interest since, in 1921, debris from a tornadic storm plugged a particularly narrow passageway, thereafter limiting exploration to a short distance. Later dye tracing conclusively demonstrated the underground connection to Mammoth Springs, but Owen's first-hand description would appear to be the only one readily available (Owen 1898a, 1970 reprint, pp. 99–100; Seeger 2002–03). Her journal articles, published in the late 1890s, discussed the region's more spectacular caverns. These appeared under her own name but in a Paris journal (Owen 1896, 1897, 1898b, 1899).

Owen's 1898 monograph *Cave Regions of the Ozarks and the Black Hills* (Owen 1898a), although intended for a general audience, remained a classic in Missouri speleology for many years. Written before there was general agreement that these caves were hollowed out by artesian flow, it presented Owen's own theory of speleogenesis. Using as an example the caves of the South Dakota Black Hills, she suggested that the region had gone through an earlier period of volcanic activity, similar to that in present-day Yellowstone, and that the caves had been formed in the Black Hills limestone by hot geyser action. The idea might not have appeared unreasonable were it not for her flawed basic assumption about the solubility characteristics of limestone.

In about 1900, considering herself too old for further cave exploration, she returned to her earlier studies of loess deposits, often making use of cuttings through the soft loess made during road and railway construction. Her reports on fossil contents and her ideas on loess deposition appeared in such journals as *American Geologist* and *Pan-American Geologist* between 1901 and 1926. It is satisfactory to note that she came down on the side of the now-accepted aeolian theory of deposition at a time when ideas of wind versus water origin were still being debated. Luella Owen's horizons were broad, however, and by no means limited to the geology of the American Midwest. She even made a round the world trip, during which, equipped with letters of introduction to geologists and geographers in several countries, she visited regions whose rock formations, especially loess deposits, were of particular interest to her. She also published work on the effects of rivers on land form, an interest stemming from her familiarity over many years with the whole length of the Missouri river basin. Her notable paper predicting the future importance of the Missouri river as a direct trade route between the American Midwest and Europe (initially presented at the Ninth Geographical Congress in Geneva, 1908) was first published in the influential geographical journal *Scottish Geographical Magazine* (Owen 1908).

Carlotta Maury (1874–1938). Carlotta Maury (1874–1938) was one of the earliest women to be employed as a professional scientist by an oil company, Royal Dutch Shell, Venezuela Division, which she joined in 1910 as consulting geologist and stratigrapher (Reeds 1938; Creese 1998, pp. 286–288; Anon. 2004). Born in Hastings-on-Hudson in 1874, Maury was descended on her mother's side from a member of the Portuguese nobility serving at the court of Emperor Dom Pedro I of Brazil, a connection which had an important influence on her career. She had a good education, with a year of post-graduate studies at the Sorbonne before completing PhD requirements at Cornell University in 1902. After a few years of teaching, she returned to field research, joining the team recruited by G. D. Harris, her Cornell adviser, for his on-going investigations of the oil-bearing strata of the Texas and Louisiana Gulf Coasts. The time was one of tremendous activity in oil exploration in the region, following the discovery of gusher wells about 50 km inland from the Texas coast. The team, led by Harris, then Louisiana's state geologist, provided the first substantial geological information about the Texas–Louisiana oil-producing areas. By microscopic examination of samples from wells in two important fields, Maury assembled data for a structure map of a large region. Based on her palaeontological findings, the final Harris-Maury substructure analysis has needed only minor correction in its details ever since (Maury 1910).

In 1910 Maury moved on to another field assignment, working as palaeontologist on a team employed by the General Asphalt Co., exploring promising areas of Trinidad and Venezuela; her discovery of Old Eocene beds in Trinidad with fossil faunas related to those of Alabama and Brazil was the first finding of Old Eocene in the entire Caribbean and northern South America region (Maury 1912).

With a short break for teaching at Huguenot College in Wellington, South Africa, Maury returned to the Caribbean in 1916 as leader of what is known as the Maury Expedition to the Dominican Republic. Despite the fact that the island was in the midst of internal political upheavals and was also suffering as a result of an invasion by United States Marines, Maury was undeterred. Her purpose was to determine exact stratigraphic sequences in the Miocene and Oligocene horizons in the exposed layers of fossil-rich sedimentary rocks of the island's Cibao Valley, and to establish for the region the line between these two horizons. The results – type sections and descriptions of fossils, including more than 400 new species – are now classic references (Maury 1917a, b, 1918). The first comprehensive report on the geology and fossils of the Cibao Valley, these papers form the foundation for the present international Dominican Republic Project, a large, multi-disciplinary research effort led by City College of New York's City University. Begun in about 1974, it involves field geologists, stratigraphers, geochronologists, biologists and palaeontologists. The aim is to understand patterns of evolution and biodiversity change in the Caribbean from the Miocene to the present using data from the relatively continuous record of ocean-bottom animals in the Cibao Valley rocks.

Maury, a recognized specialist in Antillean, Venezuelan and Brazilian fossil faunas, was an exceptionally able palaeontologist. She had a reputation for being extremely efficient and energetic; if she accepted an assignment she proceeded with speed and precision. As well as being consulting palaeontologist and stratigrapher to Royal Dutch Shell's Venezuela Division for more than 20 years, she was one of the official palaeontologists with the Geological and Mineralogical Service of Brazil, her work appearing in a steady succession of lengthy monographs and Mineralogical Service Bulletins between 1919 and 1937. Her studies on Venezuelan stratigraphy for Royal Dutch Shell took the form of confidential reports.

She delighted in the exotic, from the tropical forests of Venezuela to the changing sunset lights on the grey sandstones of South Africa's Drakensberg Mountains. An independent, freelance petroleum geologist, palaeontologist and stratigrapher with a cosmopolitan lifestyle, to a large extent she succeeded in ignoring the prevailing prejudice against professional women thanks to her widely acknowledged technical skill and ability, although her private financial resources undoubtedly increased her freedom of action.

Continental Europe: palaeontologists and a mineralogist

In the Old World, more geological research was reported by male workers in the French- and German-speaking countries than anywhere else, but published work by women in these regions was hardly noticeable; this was particularly the case in Germany and Austria. In Scandinavia, the output of publications by geologists was relatively slight overall but nevertheless, by the 1890s and in the opening decades of the 20th century, Sweden had a few notably impressive women working in the field; one of these was a palaeobotanist and another a mineralogist. Likewise, in late 19th-century Russia, a few ambitious women made creditable contributions to geological research (Creese & Creese 1994, 2006).

Sweden

Astrid Cleve (1874–1968). Astrid Cleve (1874–1968), palaeobotanist, was the daughter of Per Cleve, professor of chemistry at Uppsala University, and it was in her father's laboratory that Astrid received her early scientific training (Forin 1968; Creese 2004a, pp. 2–6). At the age of 16 years, she enrolled at the university, taking natural science courses; by the time she was 23 years old she had published several papers and successfully defended her doctoral dissertation, a study on Swedish plants. Her papers included reports of two summers of fieldwork in the mountains of northern Sweden where she investigated the then unknown diatom flora of the Arctic lakes.

Astrid Cleve then moved to an assistantship in the chemistry department at Stockholm Högskole (now Stockholm University), an institution of liberal outlook. She married chemist Hans von Euler in 1902 and, over the course of the next 10 years, not only produced five children but co-authored with von Euler about 16 papers in organic and biological chemistry, including preliminary studies on fermentation, research that evolved into the work that brought von Euler the Nobel prize in chemistry in 1929.

By 1912, however, the Cleve/von Euler partnership had ended in divorce; from then Astrid Cleve-von Euler supported her five children by high-school teaching and commercial laboratory work. She also returned to her earlier research interest, carrying out investigations for the Swedish

Hydrographical Biological Commission on the diatomaceous plankton in the waters around Stockholm.

Over the years Cleve-von Euler continued her research on both fossil and living diatoms, overcoming the constraints of limited financial resources and many demands on her time; coming from an academic background she was acutely conscious of the handicap of never having a university position as a base for her work. Nevertheless, in time she became the country's foremost expert on fossil and living forms of Baltic region diatoms. Her palaeobotanical studies continued for 40 years and, looking towards a broad comprehensive picture, she sought to integrate her observations into a general, theoretical framework of the Baltic region's recent geology. Her diatom studies of post-glacial sediments brought her to the problem of late- and post-Ice Age changes in the water level of the Baltic, changes that resulted from the periods of open and closed connection between the Baltic and the ocean. Her theories did not always agree with those of her Swedish and Finnish colleagues, but both in print and in her presentations to the Stockholm Geological Society she entered into debates and vigorously defended her interpretations of events.

Astrid Cleve-von Euler's greatest contribution was her tremendous output of taxonomic and systematic work. Her important diatom flora, 'Die Diatomeen von Schweden und Finnland', published between 1951 and 1955 in five parts, covered much of Sweden and all of Finland; containing descriptions of almost 1600 species, both fossil and living, it long remained a standard reference for Quaternary geologists working in the region (Cleve-von Euler 1951, 1952, 1953a,b, 1955).

Despite her lack of an academic position, Cleve-von Euler had many students and collaborators, people who came to her for consultation, discussion and advice, sometimes enjoying her warm hospitality over extended periods and receiving from her their first real instruction on diatoms. She also carried out a considerable amount of diatom analysis of sediments for geologists of the Swedish Geological Survey. In 1947/8, when she was in her early 70s, she gave a lecture course on diatomology at Uppsala University's Institute of Plant Ecology. In 1955, when she was 80 years old, Uppsala made her professor *honoris causa* in appreciation of her extensive research on diatoms.

As a footnote to Astrid Cleve-von Euler's story one might add that, just 2 years after her death, her second son, Ulf von Euler, shared the 1970 Nobel prize in medicine and physiology with Julius Axelrod and Sir Bernard Katz.

Naima Sahlbom (1871–1957). Mineralogist Naima Sahlbom (1871–1957) (Lundgren 2002; Andersson 2003; Creese 2004a, p. 8), daughter of civil engineer Gustav Sahlbom, was born in Stockholm and educated at the Wallinska girls' school, the first school in Sweden to offer girls instruction to university matriculation level. While there, her lively interest in science led her to her first laboratory experience as a volunteer in the Stockholm Water Department laboratory. Within a year or two, after study at the Stockholm Högskole, she had developed the interest considerably and become very proficient in mineral analysis. In 1894 the Stockholm Geological Society recognized her work in this area, accepting her as a member. Further periods of alternating study and employment followed, including work with the Geological Commission in Helsingfors and studies in Aachen, where she was able to pursue her growing interest in analysis of radioactive minerals and waters. The newly opened Neuchâtel University awarded her a doctorate in chemical physics in 1910.

After returning to Stockholm, Sahlbom was employed for a time as a chemical analyst, specializing in radioactivity estimations. In 1914, however, on the advice of Helge Backlund, a St Petersburg geologist and mineralogist of Swedish background, she opened her own laboratory, where she specialized in complete analyses of rocks, ores, and mineral and spring waters; the venture was pioneering for a woman at the time.

Sahlbom's skill and reliability were widely recognized and her considerable service to Swedish petrographers and mineralogists much valued. Bearing in mind the country's strong and impressive tradition in analytical chemistry, this should be recognized as no small achievement. Her own research publications were not many but she did bring out two reports on analyses of Swedish rocks and two on her extensive investigations correlating the radioactivity of spring waters and deep-bored wells in southern Sweden with the nature of the rocks through which they flowed (Sahlbom & Sjögren 1907; Sahlbom 1916).

As often happens, it was not chiefly for her scientific enterprise and achievements that Naima Sahlbom was most widely known and recognized, both in her lifetime and later. From her student days in Germany and Switzerland she had been much involved in the women's movement, contributing articles and letters to Swedish periodicals. During World War 1 and throughout the 1920s, in addition to running her laboratory without in any way relaxing her high professional standards, she worked hard in Swedish and international peace organizations. In particular, she vigorously supported efforts to ban chemical weapons, especially the military use of poison gas. Her award in 1946 of Sweden's Illis Quorum medal recognizing meritorious service was for this public work.

Russia

By Soviet times the number of women working in the Union's geological sciences, particularly palaeontology, was far greater than in any other country. For instance, as reported by D. V. Nalivkin in his 1979 book on Russia's early women geologists, in 1968 there were 1490 women palaeontologists in the USSR while in many other countries at that time there were fewer than 30 or 40. By then in the USSR women outnumbered men in biostratigraphy and palaeontology, the more difficult and complex the animal group the greater the likelihood of the research workers being women. However, very few women were employed in other areas, such as tectonics, and not many Soviet women went on to doctoral degrees in geology followed by teaching posts at universities (Nalivkin 1979).

The forerunners and pathfinders for these mid-20th-century Soviet women palaeontologists were a small group of late 19th-century women who, against the odds, succeeded in acquiring skill and competence in this field and carrying out valuable research. Barred from access to Russian universities, they either went to western Europe for training or educated themselves by informal means in their own country. Among the most outstanding were Maria Pavlova (née Gortynskaia), Evgenia Solomko and Maria Tsvetaeva, all of whom had begun their work by the early 1880s.

Maria Pavlova (née Gortynskaia) (1854–1938). The best remembered of Russia's early women geologists is undoubtedly Maria Pavlova (1854–1938) known especially for her work on tertiary vertebrates and for her substantial efforts to establish and develop the Museum of Palaeontology at Moscow University (Borisiak & Menner 1939; Nalivkin 1979).

The daughter of a district doctor, Maria Vasilievna Gortynskaia was born in the town of Kozel'sk, Ukraine, about 130 km north of Kiev. After studies at the Kiev Women's Institute she taught natural history for a few years before going to Paris in 1882. There she became one of a group of enthusiastic young students who were strongly influenced by Albert Gaudry, palaeontologist at the Muséum National d'Histoire Naturelle. An outstanding teacher and a brilliant lecturer, Gaudry rejected ideas of fixity of species, still defended by many, and vigorously expounded the new theory of the existence of evolutionary trends and intermediate forms through all of geological time. He was also among the leaders in establishing palaeontology as a distinct and separate science, rooted in both geology and biology. One of Pavlova's important contributions to 19th-century Russian palaeontology was to emphasize Gaudry's teachings, particularly his view of palaeontology as an independent science (Borisiak & Menner 1939, p. 78).

A year after returning to Moscow, Maria Gortynskaia married Aleksei Pavlov, geologist and palaeontologist at Moscow University, and it was only thanks to him and his colleague Vladimir Vernadski that she was able to gain access to the university's palaeontological and mineralogical collections. These she expanded considerably through her visits to museums in many parts of the country, from eastern Siberia to the Ukraine and the Caucasus; her careful examinations of the specialized local material in these regional collections not only produced useful data but inspired local workers to further investigations.

Pavlova's special interest was mammalian evolution, especially the development of the hoofed animals, particularly the horse family and the elephant group, but she also worked on other groups, including the rhinoceros and pig. In 1887 she brought out the first of her long series of articles on the palaeontological history of ungulates in America and Europe (Pavlova 1887, 1889, 1890, 1893, 1900). Written in French in a clear and very comprehendable style, these papers attracted considerable attention outside Russia. Her later important publications included *Fossil Elephants of Russia* (Pavlova 1910) and *Tertiary Mammals of New Russia* (Pavlova 1913, 1914); the short work, *Causes of Extinction of Animals in Past Geological Epochs* (Pavlova 1924), brought out her decidedly Darwinian outlook.

One of her most important contributions was in the unravelling of the complex history of the horse family, particularly her work in the early 1890s on the role of *Hipparion*, the small, three-toed horse that represents one of the secondary evolutionary lines of the family. Pavlova's emphasis on changes in tooth and skull structure in addition to foot development emphasized the need to investigate more than one morphological trait in order to establish a phylogenetic line (Pavlova 1892a, b).

As well as her research papers and monographs she published several works designed to introduce palaeontology to a wider audience. These included her series *Iskopaemie Slone [Fossil Elephants]* and translations into Russian of foreign works, such as Henry Hutchinson's *Extinct Monsters* (1893) and Melchior Neumayr's *die Stämme des Thierreichs* (1889).

Much of Pavlova's teaching was at the Lubianskii Courses for Women, Moscow, and Shaniavskii Moscow University. Nearly all Moscow-trained women geologists and palaeontologists from before the Revolution were her students; although few in number, several went on to make creditable contributions. With the coming of the Soviet era in 1917, Pavlova was appointed to a professorial post at Moscow University. A member of the Ukrainian Academy of Sciences from 1921, she became a corresponding member of the Academy of Sciences of the USSR in 1919 and an honorary member of the latter in 1930. Moscow University gave her a doctorate in zoology in 1916.

Somewhat less well known in the West than Pavlova are her contemporaries Evgenia Viktorovna Solomko and Maria Kuzminichna Tsvetaeva. Solomko is notable as the first Russian woman to be awarded a doctoral degree in geology. Tsvetaeva, a student of Moscow geologist Sergei Nikitin, worked with him for 20 years, to a large extent serving as his assistant.

Evgenia Solomko (1862–1898). Evgenia Solomko (1862–1898) was born in Yaroslavl', about 250 km northeast of Moscow (Anon. 1898–99; Nalivkin 1979, pp. 42–44). At the age of 16 years, after a period at the Catherine Institute for Women in Moscow, she studied for 5 years (1878–83) at the newly opened Bestuzhev Higher Courses for Women in St Petersburg.

A good student, Solomko was able to continue her training in the University of St Petersburg's Department of Geology and Palaeontology, where she first worked with the celebrated geologist and mineralogist Aleksandr Inostrantsev on problems in petrography. She took part in his field excursions in Olnetz province, Karelia, in the Caucasus and in the Ukraine, carrying out a study on volcanic rocks from Poltava province in east-central Ukraine. After this introductory period Solomko concentrated on palaeontology. Following the suggestion of Devonian specialist P. N. Veniukov she undertook the examination a difficult group of now extinct Devonian Hydrozoa, the stromatoporoids (reef-building animals). This was followed by an even more difficult group of Jurassic and Cretaceous corals of the Crimea. These groups were little known to Russian workers at the time, but with help from invertebrate palaeontologist and stratigrapher Konstantin Grewingk of the University of Dorpat, Estonia, and palaeontologist Karl von Zittel in Munich, she completed the study; her results appeared in two substantial monographs (Solomko-Sotiriadis 1887, 1888). While in Munich she attended von Zittel's palaeontology lectures (sitting in a neighbouring room suitably separated from the male students) and also worked extensively in the Munich Palaeontological Museum. As women were not yet eligible to receive academic degrees from Munich University, Solomko presented her dissertation research at the University of Zurich and there received a doctoral degree in 1887. At the end of that year she married the director of the Odessa Commercial College and appears to have given up geological research. She died 11 years later, at the age of 37 years.

Maria Tsvetaeva (b. 1854). After earlier studies at the First Moscow Gymnasium, Maria Tsvetaeva (b. 1854) enrolled in 1873 at Moscow's newly opened Lubianski Higher Courses for Women (Nalivkin 1979, pp. 37–40). Concentrating on the physical and mathematical sciences, she completed a full 4-year course at standard university level; her subjects included geology and mineralogy, taught by Moscow geologist Sergei Nikitin.

Throughout her working life, from her student days on, Tsvetaeva earned her living by teaching natural science at two Moscow gymnasiums. At the same time she stayed on as a student and co-worker of Nikitin. For about 15 years, during summers when she was free from teaching, she accompanied him on field trips, going on foot throughout almost the whole Volga basin, collecting palaeontological material and studying the geological structure of central Russia. Her structural work was incorporated into Nikitin's papers and her own publications were limited to palaeontological reports. These appeared as two monographs on her large collection of nautiloids and other cephalopod fauna of the Carboniferous limestones of central Russia. Tsvetaeva's monographs, which contained a vast amount of new material, were a substantial contribution to Russian palaeontology and remained important references for many decades (Tsvetaeva 1888, 1898). In addition, from 1884 to 1896, she assisted Nikitin in preparing the bibliographic guide, *Russian Geological References*, brought out under Nikitin's editorship by the recently established Russian Geological Committee.

A member of several Russian scientific societies, Tsvetaeva was also a participant in the International Geological Congress held in St Petersburg in 1897. Along with Pavlova, she had the notable honour of serving on the congress's organizational committee. She was an unassuming person but confident and painstakingly conscientious in her work; in her own time she was well known and respected in the Russian geological community, despite her lack of formal academic degrees. She has been described as Russia's first woman geologist, the more specific designation of palaeontologist being kept for Maria Pavlova (Nalivkin 1979, pp. 39–40).

Closing remarks

These sketches are presented as part of an on-going effort to uncover, and ultimately evaluate fairly, the role played by women in research in the sciences at a time when many still faced sizable handicaps to participating. Any significant overall analysis of this role depends on a comprehensive view which, in turn, depends on the building up of broad, country-wide and regional surveys to avoid the dangers inherent in generalizing from too limited samples.

Geology does not stand out among the sciences as a favourite field for women, falling noticeably behind the biological sciences and, except in the United Kingdom, behind astronomy and chemistry as well (Creese 1998, 2004a). It is perhaps worth noting that historian Oldroyd, in his 1996 overview, finds no women contributors to the development of the basic understanding of the earth before Danish seismologist Inge Lehmann (1888–1993). In 1936 Lehmann put forward the tentative proposal that there might be a solid core at the centre of the Earth, within the liquid core, the liquid core theory having held the field until then. Her conclusion came from painstaking, detailed analysis of a massive amount of seismic data from the 1929 Buller earthquake in New Zealand, collected from many laboratories. The only other woman idea-producer I noted in Oldroyd's book was the American Marie Tharp. Tharp had the job of collating data gathered on a 1952 Columbia University Lamont Observatory survey of the Pacific Ocean floor; she suggested that the mid-ocean ridges had 'rift valleys' running along their axes, an idea dismissed at the time as 'girl talk' by one of the expedition's leaders (Oldroyd 1996, pp. 237, 261–262, 344, Note 21).

However, accepting Oldroyd's picture of there being very few women among past leaders in developing the theoretical underpinnings of geology, should this absence of women be considered a matter of great concern in the history of women in the field? Even if there were no female counterparts of James Hutton or Baron George Cuvier, a few able and ambitious women, whose careers began during the 19th century, succeeded in producing contributions of permanent importance, contributions that have their part among the building blocks on which major theories are founded and also contributions that influenced and assisted large areas of research for many years.

References

AGASSIZ, J. L. R. 1844–45. *Monographie des Poissons Fossiles du Vieux Grès Rouge ou Systèm Dévonien des Îles Britannique et de Russie*. Agassiz, Neuchâtel; chez Jent & Gassmann, Soleure.
ANDERSSON, I. 2003. 'Women's Unarmed Uprising against War': a Swedish peace protest in 1935. *Journal of International Peace Research*, **40**, 395–412.
ANDREWS, S. M. 1982. *The Discovery of Fossil Fishes in Scotland up to 1845: With Checklist of Agassiz's Figured Specimens*. Royal Scottish Museum, Edinburgh.
ANON. 1898–99. [Biographical notice on Evgenia Solomko-Sotiriadis]. *Ezhegodnik po Geologii i Mineralogie Rossi*, **3**(1), 137.
ANON. 2004. The Dominican Republic Project at the City College, City University of New York. World Wide Web Address: http://condor.admin.ccny.cuny.edu/~rnehm/drp/drpoverview.htm
BORISIAK, A. A. & MENNER, V. V. 1939. Mariia Vasil'evna Pavlova. *Vestnik Akademii Nauk SSSR*, **6**, 78–80.
BUREK, C. V. 2007. The role of women in geological higher education – Bedford College, London (Catherine Raisin) and Newnham College, Cambridge, UK. *In*: BUREK, C. V. & HIGGS, B. (eds) *The Role of Women in the History of Geology*. Geological Society, London, Special Publications, **281**, 9–38.
BUREK, C. V. & MALPAS, J. A. 2007. Rediscovering and conserving the Lower Palaeozoic 'treasures' of Ethel Woods (neé Skeat) and Margaret Crosfield in northeast Wales. *In*: BUREK, C. V. & HIGGS, B. (eds) *The Role of Women in the History of Geology*. Geological Society, London, Special Publications, **281**, 203–226.
CLARY, R. M. & WANDERSEE, J. H. 2007. Great expectations: Florence Bascom (1862–1945) and the education of early US women geologists. *In*: BUREK, C. V. & HIGGS, B. (eds) *The Role of Women in the History of Geology*. Geological Society, London, Special Publications, **281**, 123–135.
CLEEVELY, R. J. 2004. Gray, [née Anderson], Elizabeth (1831–1924). *In*: MATTHEW, H. G. C. & HARRISON, B. H. (eds) *Oxford Dictionary of National Biography*, Oxford University Press, Oxford, **23**, 421.
CLEEVELY, R. J., TRIPP, J. P. & HOWELLS, Y. 1989. Mrs. Elizabeth Gray (1831–1924): a passion for fossils. *Bulletin of the British Museum of Natural History (Historical Series)*, **17**, 167–258.
CLEVE-VON EULER, A. 1951. Die Diatomeen von Schweden und Finnland I. *Kungliga Svenska Vetenskaps-Akademiens Handlingar*, Series 4, **2**, 162 pp.
CLEVE-VON EULER, A. 1952. Die Diatomeen von Schweden und Finnland V. *Kungliga Svenska Vetenskaps-Akademiens Handlingar*, Series 4, **3**, 152 pp.
CLEVE-VON EULER, A. 1953a. Die Diatomeen von Schweden und Finnland III. *Kungliga Svenska Vetenskaps-Akademiens Handlingar*, Series 4, **4**, 255 pp.
CLEVE-VON EULER, A. 1953b. Die Diatomeen von Schweden und Finnland II. *Kungliga Svenska Vetenskaps-Akademiens Handlingar*, Series 4, **4**, 158 pp.
CLEVE-VON EULER, A. 1955. Die Diatomeen von Schweden und Finnland IV. *Kungliga Svenska Vetenskaps-Akademiens Handlingar*, Series 4, **5**, 232 pp.
COLLIE, M. & DIEMER, J. 1995. Murchison in Moray. A Geologist on Home Ground. *Transactions of the American Philosophical Society*, New Series, **85**(3), 263 pp.
CREESE, M. R. S. 1998. *Ladies in the Laboratory? American and British Women in Science, 1800–1900*. Scarecrow Press, Lanham MD.

CREESE, M. R. S. 2004a. *Ladies in the Laboratory. II. West European Women in Science, 1800–1900*. Scarecrow Press, Lanham MD.

CREESE, M. R. S. 2004b. Gray, Elizabeth (née Anderson: 1831–1924). *In*: LIGHTMAN, B. (ed.) *The Dictionary of Nineteenth-Century British Scientists*. Thoemmes Continuum, Bristol, **2**, 826–827.

CREESE, M. R. S. & CREESE, T. M. 1994. British women who contributed to research in the geological sciences in the nineteenth century. *British Journal for the History of Science*, **27**, 23–54.

CREESE, M. R. S. & CREESE, T. M. 2006. British women who contributed to research in the geological sciences in the nineteenth century. *Proceedings of the Geologists Association*, **177**, 53–83.

FORIN, M.-B. 1968. Astrid Cleve von Euler. *Svensk Botanisk Tidskrift*, **62**(4), 549–564.

FRASER, H. E. & CLEAL, C. J. 2007. The contribution of British women to Carboniferous palaeobotany during the first half of the 20th century. *In*: BUREK, C. V. & HIGGS, B. (eds) *The Role of Women in the History of Geology*. Geological Society, London, Special Publications, **281**, 51–82.

GARMAN, S. 1888–89. Cave animals from south-western Missouri. *Bulletin of Comparative Zoology at Harvard College*, **17**, 225–240.

HIGGS, B. & WYSE JACKSON, P. N. 2007. The role of women in the history of geological studies in Ireland. *In*: BUREK, C. V. & HIGGS, B. (eds) *The Role of Women in the History of Geology*. Geological Society, London, Special Publications, **281**, 137–153.

HUTCHINSON, H. N. 1893. *Extinct Monsters. A Popular Account of some of the Larger Forms of Ancient Life*. D. Appleton & Co., New York.

ISSITT, J. R. 2002. Jeremiah Joyce: science educationist. *Endeavour*, **26**, 97–101.

KÖLBL-EBERT, M. 2007. The role of British and German women in early 19th century geology—a comparative assessment. *In*: BUREK, C. V. & HIGGS, B. (eds) *The Role of Women in the History of Geology*. Geological Society, London, Special Publications, **281**, 155–163.

LONGSTAFF, J. D. 1924. Descriptions of Gasteropoda, chiefly in Mrs. Robert Gray's collection from the Ordovician and Lower-Silurian of Girvan. *Quarterly Journal of the Geological Society*, **80**, 408–446.

LUNDGREN, A. 2002. Sahlbom, Naima. *In*: NILZÉN, G. (ed.) *Svenskt Biografiskt Lexikon*, Norstedts Tryckeri AB, Stockholm, **31**, 225–228.

MAURY, C. J. 1910. Stratigraphy of the Jennings oil field. *In*: HARRIS, G. D. (ed.) *Oil and Gas in Louisiana. US Geological Survey Bulletin*, **1910**, 56–60.

MAURY, C. J. 1912. A contribution to the paleontology of Trinidad. *Journal of the Academy of Natural Sciences of Philadelphia, Series 2*, **15**, 25–112.

MAURY, C. J. 1917a. Santo Domingo type sections and fossils. Part 1. *Bulletin of American Paleontology*, **29**, 244 pp.

MAURY, C. J. 1917b. Santo Domingo type sections and fossils. Part 2. Stratigraphy. *Bulletin of American Paleontology*, **30**, 43 pp.

MAURY, C. J. 1918. Santo Domingan paleontological explorations. *Journal of Geology (Chicago)*, **26**, 224–228.

NALIVKIN, D. V. 1979. *Nashi Pervye Zhenshchinye-geologi* [Our First Women Geologists]. Akademiia Nauk, Leningrad.

NEUMAYR, M. (1889). *Die Stämme des Thierreichs*. F. Tempsky, Wien.

NIKITIN, S. N. 1885–1897? *Russkaia Geologicheskaia Biblioteka. Bibliothèque Géologique de la Russie*. Eggers & Cie, St. Petersburg. [Supplement to *Bulletins du Comité Géologuqie*. Corp. author, Russia, Geologicheskii Komitet.]

OGILVIE, I. H. 1945. Florence Bascom, 1862–1945. *Science*, **102**, 320–321.

OLDROYD, D. R. 1996. *Thinking about the Earth: A History of Ideas in Geology*. Harvard University Press, Cambridge, Mass.

OWEN, L. A. 1896. Cavernes américaines. *Spelunca*, **2**, 8–13.

OWEN, L. A. 1897. Marble Cave (Missouri) et Wind Cave (Dakota). *Spelunca*, **3**, 22–31.

OWEN, L. A. 1898a. *Cave Regions of the Ozarks and Black Hills*. Editor Publishing Company, Cincinnati. Reprinted 1970, Johnson Reprint Corporation, New York.

OWEN, L. A. 1898b. La Caverne de Crystal. *Spelunca*, **4**, 77–81.

OWEN, L. A. 1899. Les cavernes de Ha Ha Tonka. *Spelunca*, **5**, 16–20.

OWEN, L. A. 1908. The Missouri river and its future importance to the nations of Europe. *Scottish Geographical Magazine*, **24**, 588–596.

PAVLOVA, M. V. 1910. *Les Éléphants Fossiles de la Russie. In Nouveau Mémoires de la Société des Naturalistes de Moscou*, 17 (1910–15), livre 2. J. N. Kouchnéerff & Cie. Moscou. Also published as M. V. Pavlova (1910), *Les Éléphants Fossiles de la Russie. In* M. V. Pavlova & A. P. Pavlov (1910–14). *Mammifères Fossiles de Russie*. J. N. Kouchnéreff & Cie, Moscou, 3 vols. [Additional popular form, *Iskopaemie Slone*].

PAVLOVA, M. V. 1913–14. *Mammiferes Tertiaires de la Nouvelle Russie. In Nouveaux Mémoires de la Société des Naturalistes de Moscou* **17** (1910–15) livre 4. J. N. Kouchnéreff & Cie, Moscou. Also published as M. V. Pavlova (1913–14). *Mammifères Tertiares de la Nouvelle Russie. In* M. V. Pavlova & A. P. Pavlov (1910–14). *Mamifères Fossiles de Russie*. J. N. Kouchnéreff & Cie, Moscou, 3 vols.

PAVLOVA, M. V. 1924? *Prichiny vymiraniia zhivotnykh v proshedshie geologischeskie epoki*. [*Causes of Extinction of Animals in Past Geological Epochs*.] Gos. izd-vo, Moskva. [First published in the series *Sovremennye problemy estestvoznaniia*. (1910–1919?). Gos. izd-vo, Moskva.]

PAVLOVA, M. V. 1887. Études sur l'histoire paléontologique des ongulées en Amérique et en Europe. *Bulletin de la Société Impériale des Naturalistes de Moscou, New Series*, **1**, 343–373.

PAVLOVA, M. V. 1889. Études sur l'histoire paléontologique des ongulées en Amérique et en Europe. *Bulletin de la Société Impériale des Naturalistes de Moscou, New Series*, **2**, 135–182.

PAVLOVA, M. V. 1890. Études sur l'histoire paléontologique des ongulées en Amérique et en Europe. *Bulletin de la Société Impériale des Naturalistes de Moscou, New Series*, **3**, 653–716.

PAVLOVA, M. V. 1892a. Notice sur l'*Hipparion crassum* du Roussilon. *Bulletin de la Société Impériale des Naturalistes de Moscou, New Series*, **5**, 161–164.

PAVLOVA, M. V. 1892b. Qu'est-ce que c'est que l'Hipparion? *Bulletin de la Société Impériale des Naturalistes de Moscou, New Series*, **5**, 410–414.

PAVLOVA, M. V. 1893. Études sur l'histoire paléontologique des ongulées en Amérique et en Europe. *Bulletin de la Société Impériale des Naturalistes de Moscou, New Series*, **6**, 137–221.

PAVLOVA, M. V. 1900. Études sur l'histoire paléontologique des ongulées en Amérique et en Europe. *Bulletin de la Société Impériale des Naturalistes de Moscou, New Series*, **13**, 268–328.

REEDS, C. A. 1939. Memorial to Carlotta Joaquina Maury. *Proceedings of the Geological Society of America for 1938*, 157–168.

SAHLBOM, N. 1916. Om radioaktiviteten hos svenska källvatten och des samband med geologiska förhällandena. *Arkiv för Kemi, Mineralogi och Geologi*, **6**(3), 52 pp.

SAHLBOM, N. & SJÖGREN, H. 1907. Undersökningar af radioaktiviteten hos svenska källvatten. *Arkiv för Kemi, Mineralogi och Geologi*, **3**(2), 28 pp.

SEEGER, C. 2002–03. Geologically Grand. The peculiar geology of Grand Gulf State Park. *Missouri Resources*, **19**(4), 20–22.

SOLOMKO-SOTIRIADIS, E. 1887. [Stromatopora des devonischen System Russlands.] *Verhandlungen der Russisch-Kaiserlichen Mineralogischen Gesellschaft zu St Petersburg*, **23**, 1–48.

SOLOMKO-SOTIRIADIS, E. 1888. Die Jura-und Kreidekorallen der Krim. *Verhandlungen der Russisch-Kaiserlichen Mineralogischen Gesellschaft zu St Petersburg*, **24**, 67–231.

TSVETAEVA, M. 1888. Cephalopodes de la section supérieure du calcaire carbonifère de la Russie centrale. *Trudi Geologicheskago Komiteta* [*Mémoirs du Comité Géologique*], St Petersburg, **5**(3), 58 pp. (Résumé 41–58).

TSVETAEVA, M. 1898. Nautiloidea et Ammonoidea de la section inférieure du calcaire carbonifère de la Russie centrale. *Trudi Geologischeskago Komiteta* [*Mémoirs du Comité Géologique*], St Petersburg, **8**(4), 46 pp. (Résumé. 35–46).

VINEYARD, J. D. 1970. Introduction. *In*: OWEN, L. A. 1898. *Cave Regions of the Ozarks and Black Hills*. Reprinted by Johnson Reprint Corporation, New York.

WACHTLER, M. & BUREK, C. V. 2007. Maria Matilda Ogilvie Gordon (1864–1939): Scottish researcher in the Alps. *In*: BUREK, C. V. & HIGGS, B. (eds) *The Role of Women in the History of Geology*. Geological Society, London, Special Publications, **281**, 305–317.

ZORNLIN, R. M. 1839. *Recreations in Geology*. J. W. Parker, London.

The contribution of British women to Carboniferous palaeobotany during the first half of the 20th century

H. E. FRASER & C. J. CLEAL

Department of Biodiversity & Systematic Biology, National Museum Wales, Cathays Park, Cardiff, CF10 3NP, UK (e-mail: chris.cleal@museumwales.ac.uk)

Abstract: During the first half of the 20th century, over a third of British palaeobotanists working on Carboniferous plants were women; neither before nor after this period have women played such a prominent role in this field. Few of these women were able to develop significant careers within the subject. They nevertheless produced some of the most innovative work in the field, pioneering work in plant phylogeny, cuticle studies, biostratigraphy, morphological variation, and anatomical thin sectioning. Two factors were critical for allowing this work to develop: the support of a small number of male colleagues, notably F. W. Oliver, W. H. Lang and D. H. Scott; and the existence of colleges that specifically supported women's education, including Newnham College (Cambridge), and Bedford, Royal Holloway, Westfield and University Colleges (London).

Although a number of British women made significant contributions to palaeontology during the 19th century (Creese & Creese 2006) Carboniferous palaeobotany in the United Kingdom was an exclusively male occupation. From 1900, however, several women in this country started to take an interest in Carboniferous plants. In some cases these female palaeobotanists have become well-known in the subject (e.g., Margaret Benson, Emily Dix, Marie Stopes), but others published just one or two papers, and then ceased to be active in Carboniferous palaeobotany. In all cases, it is arguable that they did not fulfil their scientific potential, at least in terms of research output, and it would be easy to dismiss their contribution to palaeobotany as being insignificant. However, if the papers themselves are read, they reveal some of the most innovative and thought-provoking work in the discipline for that time.

In this paper, we look briefly at the careers and scientific output of these women, in an effort to understand what they achieved. Although we have tried to identify all British women active in the field, we have been unable to devote the same space to each of them; most of the paper deals with eight women who made particularly significant contributions.

Margaret Benson (1859–1936)

Margaret Jane Benson (Fig. 1) was born on the 20th October 1859 in London, and moved to Hertford when she was about 12 years old (Creese 2004). Her father, who was an architect, was also an enthusiastic amateur field botanist who passed on this passion to his daughter. Between 1878 and 1879, she studied at Newnham College Cambridge, obtaining a certificate in classics (Cambridge did not award titular degrees to women until 1923, and full degrees not until 1948; Schmid 2001). She then went on to teach at Exeter High School. However, this seems to have been essentially a means to an end, as it allowed her to save money so that she could study science at University College London (UCL), which she entered in 1887. After obtaining her BSc in 1891, she started research on plant embryology, especially of hornbeams, with F. W. Oliver at UCL.

Oliver worked primarily on living plants, but he also had a significant interest in fossil plants. Although his palaeobotanical output was not prodigious, he made a number of key discoveries, most notably of the Carboniferous pteridosperms ('seed-ferns'), usually regarded as the first totally extinct group of plants to be identified from the fossil record (Oliver & Scott 1904; for a further discussion on this, see section dealing with Marie Stopes). Benson's research with Oliver was undoubtedly a major factor in drawing her towards palaeobotany.

After a short stint back at Newnham College, in October 1893, Benson was appointed head of the new Department of Botany at Royal Holloway College, the first woman in the United Kingdom to hold such a senior position in the field of botany (the first woman to achieve a comparable position in geology was Catherine Raisin, who became Head of Geology at Bedford College London, in 1890; Burek 2003, 2007). Benson retained this position until her retirement in 1922. Developing and then running the new department clearly took up much of her energy, and she travelled extensively (with her friend, the botanist Ethel Sargant) to obtain first-hand knowledge of

Fig. 1. Margaret Jane Benson. Portrait in the Archives of Royal Holloway, University of London (RHC PH/282/13). Reproduced with permission.

the best laboratory equipment to obtain, and to collect specimens for the new herbarium. She also spent much time helping other young women enter academic life. Nevertheless, she found time to develop her research interest in palaeobotany.

Given her original work on angiosperm embryology, it is not surprising that most of her palaeobotanical research centred on the anatomy of reproductive structures, especially of Carboniferous pteridosperms and lycophytes. This required the laborious cutting of thin sections of petrified plants, which she did using a gas-driven lapidary saw in a shed in the university grounds. Her first published contribution (Benson 1902a) was a short note on sporangial clusters that she suggested were attached to the plant that had stems, then known as *Lyginodendron oldhamium* Williamson and now known as *Lyginopteris oldhamia* (Binney) Potonié. This was a plant that had intrigued palaeobotanists for some time, as the stems appeared to show characters intermediate between ferns and cycad-like seed plants, but there was no unequivocal evidence of the reproductive structures to determine their systematic position. D. H. Scott (1900) had earlier suggested that similar sporangial clusters belonged to *Lyginopteris*, but this was based on compression fossils; Benson's anatomical evidence gave a much firmer basis to this model.

She later (Benson 1904) gave a far more exhaustive analysis of the sporangial clusters, which she named *Telangium*. By this time, *Lyginopteris* had been shown to be almost certainly an early seed plant (Oliver & Scott 1904) so the structure of these sporangial clusters (pollen organs) was now seen to be important for understanding the early evolution of seed plants. She noted many underlying similarities between the structure of *Telangium* and *Lagenostoma*, and used this to speculate on how ovules evolved from sporangial clusters by the outer sporangia becoming sterile and forming a protective sheath (nucellus) around the still-functional sporangium. The early evolution of the ovule remains a contentious issue (e.g., Rothwell & Scheckler 1988) but Benson's (1904) model remains one of the more likely explanations.

Benson (1922) later described other pollen organs from the Early Carboniferous (Viséan) Pettycur flora, which she interpreted as belonging to the lyginopterid plant that had the stems *Heterangium grievii* Williamson (for a summary of the Pettycur flora and its palaeobotanical significance see Cleal & Thomas 1995). She named them *Heterotheca grievii* Benson, but that generic name was pre-occupied and so it has been renamed in her honour as *Bensoniotheca grievii* (Benson) Mickle & Rothwell 1979. These Pettycur pollen organs look rather different to others associated with *Heterangium* stems (e.g., Jennings 1976) and it is possible that they were immature.

The only pteridosperm seed that she worked on in any detail was *Sphaerostoma ovale* (Williamson) Benson 1914, from the Pettycur flora. Although predating significantly the coal-ball seeds that Oliver & Scott (1904) had worked on, these Pettycur seeds show few of the primitive features that Gordon (1941) was able to demonstrate in the even older (Tournaisian) Oxroad Bay flora. Consequently, she was unable to develop her ideas about early seed evolution from these fossils.

Her earliest paper on Carboniferous lycophyte cones (Benson 1902b) was a brief note on *Lepidocarpon*. She stated that she intended to publish a more extensive account of these fossils, but this never materialized. Instead, her major study on lycophyte fructifications (Benson 1918) was on the cones of the *Sigillaria* plant. Her 1918 paper was the first published description of such cones preserved as petrifactions, for which she erected the morphogenus name *Mazocarpon* (Feng & Rothwell 1989, later named a species of *Mazocarpon* after her). Such cones had previously been reported as adpressions (for a summary of the then-available evidence; see D. H. Scott 1900) but these could not reveal the detailed structure of the sporangia. Most importantly, Benson found that the spores in her coal-ball specimens were embedded in large quantities of parenchymatous tissue. From her observations, she speculated on the relationship between the Palaeozoic arborescent lycophytes and the Recent *Isoetes*, with the Triassic *Pleuromeia* as a possible intermediate form. Benson (1920) later tried to argue that compressions of isolated sporophylls identified by Bassler (1919) as *Canthiophorus* were in fact *Mazocarpon*, but this is unlikely to be correct, at least in the case of the type species (B. A. Thomas & Brack-Hanes 1991).

Although her most important palaeobotanical work was on pteridosperms and lycophytes, she also looked at ferns and cordaites. Her most important fern paper (Benson 1911) revised the stem anatomy of the early Carboniferous fern-like plant *Botryopteris antiqua* Kidston from the Pettycur flora. Bringing her plant anatomist's eye to the problem, she was able to show that the plant was more complex than Kidston (1908) had originally envisaged, but she still had difficulties in understanding the overall structure of the plant. She was clearly being hindered by using ground thin sections. These were the only types of section through plant petrifactions available at that time, but they are not ideally suited to observing detailed structures and organ inter-relationships within relatively small plants. It was much later, following the development of the peel technique by Joy *et al.* (1956), that Holden (1962) was able to show that *B. antiqua* was a small scrambling fern in which the fronds often bore plantlets for vegetative propagation. Hence, *B. antiqua* had stems bearing fronds, which in turn bore stems, which not surprisingly puzzled Benson.

Benson (1912) briefly entered the world of cordaite leaf taxonomy, which to this day remains problematic. She described a new species, *Cordaites felicis*, and made some brief comments about the taxonomic heterogeneity of *Cordaites* as then understood (for comments on the continuing problems in this field see Trivett & Rothwell 1991).

After her retirement in 1922, she stopped scientific work for a time, devoting her energies to domestic and parish duties (Blackwell 1937). Some years later, however, she was encouraged by D. H. Scott to write up some of her earlier unpublished work on the root anatomy of the early Carboniferous pteridosperm *Heterangium* (Benson 1933). This seems to have rekindled her interest in the subject. Still concentrating on reproductive structures of the Carboniferous seed plants, she now started to investigate adpression fossils. It is tempting to think that this was because she no longer had access to research facilities to cut thin sections of petrifactions, but this is unlikely. She had retained contact with many of her old colleagues who were still working at institutions that would undoubtedly have provided her with the facilities. Rather, she seems to have realized that adpressions can provide significantly better information about the architecture of the plant structures, and that this would complement her earlier anatomical work (Benson 1935*a* & *c*). Also, she did not restrict her work to morphology, such as her work on sections through carbonaceous fossils loaned to her by T. G. Halle (Benson 1935*b*). She even continued with fieldwork when she was in her 70s. In the Natural History Museum, London, there is an unpublished manuscript in which she described a new fertile *Rhacopteris* that she collected from Teilia Quarry in North Wales in 1933.

Benson had the most successful career of any of the women palaeobotanists considered here. In 1904, she was among the first group of women to be elected as Fellows of the Linnean Society, and in 1912 she was appointed Professor of Botany at the University of London. This may not have been without some personal sacrifice, as she never married, although she remained close to her brothers and sisters, even into old age (Blackwell 1937). However, it is clear that she had great charm and was well liked by colleagues, many of whom were also personal friends (although, equally, she could be quite brusque with those she disagreed with scientifically – see, for instance, her comments about Bassler in her 1920 paper). She collaborated with several leading palaeobotanists of the time: in the early years with her mentor F. W. Oliver, as well as with D. H. Scott, one of the leading British anatomical palaeobotanists of the time. After retirement, she was given significant help by a number of colleagues to allow her to continue her work, such as T. G. Halle of Stockholm, W. Gothan of Berlin, and Emily Dix in London.

Benson's work was wide-ranging within Carboniferous palaeobotany, and is characterized by careful descriptions and illustrations (many of her papers were illustrated by drawings and wash-paintings [e.g., Fig. 2], which although not attributed, were almost certainly executed by her; painting was a hobby throughout her life, see Blackwell 1937). She was also not frightened to develop theoretical ideas based on her observations, such as her 1908 paper, in which she speculates on the phylogenetic significance of the sporangiophore in lycophytes, sphenophytes and ferns. Her ideas concerning the derivation of apparently different structures, such as the sporangial structures in lycophytes and the sori of ferns from homologous structures in common ancestors, still has resonances in modern-day debates on the cladistic analyses of these plants. However, she regularly seems to have narrowly missed major discoveries or made critical misinterpretations of the fossils. This clearly was not due to a lack of ability or understanding of the science. To understand what

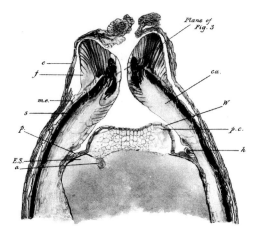

Fig. 2. Typical wash illustration by Margaret Benson illustrating one of her palaeobotanical papers, in this case the apical part of the Early Carboniferous seed *Sphaerostoma ovale* taken from her 1914 paper.

(like Williamson and Scott) used, most of the major discoveries had already been made; one of the last major discoveries was Oliver & Scott's (1904) of the pteridosperms. Coal-ball studies did not really start to flourish again until the 1960s, with the development of the peel-method by Joy *et al.* (1956; although in fact a modification of the method developed earlier by Walton 1928*a*) that allowed much finer serial sectioning, and thus the revealing of much finer detail of the plant tissue. It was notable that, when she did return to palaeobotany towards the end of her life, it was to the study of adpressions and not petrifactions, the study of the former having been relatively neglected in the United Kingdom and for which new techniques had recently been developed (e.g., Lang 1926; Halle 1933). Sadly, these new developments occurred too late in her life for her to take full advantage of them; Benson died aged 77 years at Highgate, Middlesex on 20th June 1936 (Creese 2004).

happened, it is important to see her research in the context of the science of the time. She was investigating Carboniferous plant petrifactions at the end of a period of major progress in that field, notably by W. C. Williamson and D. H. Scott (Watson 2005; Wilding 2005). Within the constraints of the thin-section method that Benson

Agnes Arber (1879–1960)

Agnes Arber is perhaps best known for her work on living plants, and among palaeobotanists as being the wife of E. A. Newell Arber (Fig. 3). However, she made her own contribution to palaeobotany

Fig. 3. Agnes Arber and her husband E. A. Newell Arber, photographed in about 1916. Taken from Schmid (2001) and reproduced with permission of Oxford University Press.

during the early 20th century, which merits consideration in this paper. There have been numerous biographical accounts of Agnes Arber, one of the most recent being by Schmid (2001). A full list of her publications is given by M. A. Arber (1968).

She was born Agnes Robertson on 23rd February 1879, in Primrose Hill, London, into a middle-class intellectual family: her father was Henry Robert Robertson, an author and member of the group of social-realist painters known as 'The Idyllists'. She first attended UCL, where she was awarded a first-class honours BSc in 1899. She then studied at Newnham College, Cambridge, gaining a first-class honours in 1902 (although, being a woman, she was not awarded a degree). She then returned to London, working with Ethel Sargant in her laboratory for a year, and then teaching at UCL until 1909, where she came into contact with F. W. Oliver. The University of London awarded her a degree of DSc in 1905 for her research on *Torrey*.

During her years at Cambridge, Agnes had met Newell Arber, and in 1906 they became engaged, eventually marrying in 1909. As her husband had a position at the University of Cambridge, she had to resign her job in London and continue her research from home at Cambridge. In 1911, she was appointed to a position at the Balfour Botanical Laboratory at Newnham College, where she stayed until its closure in 1927. A. C. Seward, who was then Professor of Botany at Cambridge, tried to find facilities for her to continue to work in the university but she ultimately decided to set up her own private laboratory at her home. Here, she continued to work until about a year before her death on 22 March 1960.

Agnes Arber was a passionate and prolific researcher, publishing over 200 books and papers (M. A. Arber 1968). However, her palaeobotanical output forms only a very modest proportion of this, all dating from between 1910 and 1914, the years when she was married to Newell Arber (Fig. 3). It is tempting to relate her stopping palaeobotanical work with the death of her husband Newell Arber, in 1918, but there is no documentary evidence that this was the case. In fact, her first papers were based on work that she undertook at UCL under F. W. Oliver. Also, although Agnes did not publish any palaeobotanical papers under her own name after his death, she posthumously edited for publication four papers and a book by her husband (E. A. N. Arber 1919, 1920, 1921*a* & *b*, 1922). Her failure to pursue palaeobotany was probably just that her research interests changed at that time towards the morphology of living plants.

Agnes Arber's first palaeobotanical papers (A. Arber 1910*a*, *b*) were a study of petrified ovules known as *Mitrospermum compressum* (Williamson) A. Arber. It was based on a wide range of specimens that she had obtained from many of the leading British palaeobotanists of the day, including Margaret Benson. Her extremely detailed description of the anatomy showed that these ovules had a fundamentally different vasculature to typical *Cardiocarpon* (to which they had originally been assigned), and so she established the new generic name *Mitrospermum*, which is still used today. She expressed some reservation as to whether they were cordaitalean, as generally assumed, noting that comparable platyspermic ovules had been reported with the pteridosperm foliage *Dicksonites*. In fact, the consensus today is that they are cordaitalean, probably belonging to cones known as *Gothania* Hirmer (e.g., see Trivett & Rothwell 1991). She also discussed the 'problem' that there seemed to be many more 'species' of ovules than of leaves in the cordaitales. She suggested that this may have been because the 'species' of leaves were morphologically defined, whereas those of the ovules were anatomically defined, a point which has only recently been confirmed by cuticles studies on this foliage (e.g., Šimůnek 2000).

Agnes Arber's next major palaeobotanical contribution were papers on the Carboniferous lycopsid cones then known as *Lepidostrobus* (A. Arber 1913, 1914*a*). This work was carried out in the Sedgwick Museum at Cambridge, although, like her *Mitrospermum* work, it was based on specimens from a wide range of sources. She described two new species, *Lepidostrobus binneyanus* and *L. gracilis*, and two new forms of *L. oldhamius* Williamson, f. *minor* and f. *pilosus* (these are all true *Lepidostrobus* species in the sense of Brack-Hanes & Thomas 1983). *L. gracilis* A. Arber has since been renamed *Lepidostrobus arberi* Jongmans 1930, as the name was pre-occupied.

Other than establishing these new taxa, Arber addressed two main problems. Firstly, she demonstrated significant variation in the sporangiophores, both within the cones of a single species and in different positions in the same cone, and that it was therefore necessary to base identifications on more than one thin section. Secondly, she discussed the functional significance of different parts of the cone. This sort of approach was relatively new in botanical studies on living plants, and was virtually unheard of in palaeobotany. She made comparison with work on the cones of living lycophytes and showed how the Carboniferous cones had had to develop features to account for their dramatically larger size. The work clearly shows the influence of F. O. Bower, who was developing analogous ideas at Glasgow, although there is no evidence that they ever met.

Agnes Arber's (1914b) last paper was a short note describing some fine anatomical details of the medullosalean ovule *Trigonocarpus*. Perhaps most important was that she confirmed that the nucellus and integument were completely free above the chalaza, a character now regarded as diagnostic of this group of plants. At the start of the paper, she noted that she observed these features while cutting sections through coal-balls, suggesting that she was actively involved in the research of these fossils. However, nothing more appears in the literature.

It clearly cannot be said that Agnes Arber did not fulfil her potential in research; it is merely that it was not in palaeobotany. Her palaeobotanical papers are characterized by extremely detailed description of large numbers of specimens, and of careful analyses of what she saw. But it has to be said that none of the papers represent major advances in the subject. Like Margaret Benson, she may have been experiencing the limitations of the ground thin sections with which she had to work, and which restricted the sort of details that could be observed. If we compare them with the results of her work on modern plant anatomy, it is not difficult to understand why she found the latter a more exciting and fulfilling subject.

Marie Stopes (1880–1958)

Of all the women dealt with in this paper, Marie Stopes (Fig. 4) is by far the best-known outside of palaeobotany, mainly through her work as a pioneer of women's rights and family planning. There are several extensive biographies on her, the most recent by Rose (1992), but these mainly emphasise her non-geological activities. However, she started her career as a palaeobotanist and achieved considerable status in the field during the first quarter of the 20th century. Much of this work was on Mesozoic floras, especially on bennettitaleans and early angiosperms (Chaloner 2005), but she also undertook a number of investigations into Carboniferous floras.

Marie Charlotte Carmichael Stopes was born on 15 October 1880 in Edinburgh. Her father Henry, an enthusiastic archaeologist, encouraged Marie to partake in his digs and to help catalogue his collection. She was initially educated at home by her mother, but at the age of 12 years (by which time the family had moved to London) she entered the North London Collegiate School. At the age of 18 years she was awarded a scholarship to enter UCL. She originally intended to major in chemistry, but that department refused to accept her so she took botany as her main subject, studying under the noted palaeobotanist F. W. Oliver. Clearly, she was formidably intelligent, graduating in only 2 years with honours in botany and geology (she was only able to do this by also registering for night school classes at Birkbeck College at the same time as taking the daytime classes at UCL).

It was during these early years at UCL with Oliver that Stopes made her earliest and arguably most important palaeobotanical discovery, although the one for which she received the least acknowledgement (Pearson 2005). Oliver had discovered that *Lagenostoma*, one of the anatomically preserved ovules found in the coal-balls of northern England, was covered by distinctive glands. Stopes was then studying microscope slides in the W. C. Williamson Collection, and found essentially identical glands on fronds now known as *Lyginopteris*. This was the key evidence that Oliver & Scott (1904) used to argue that the seeds were attached to fronds of a superficially fern-like appearance, and that they were in effect a 'missing link' between ferns and seed plants. Although this 'missing link' hypothesis is now regarded as wrong, the discovery of this group of totally extinct seed plants in the Carboniferous fossil record was ground-breaking for its time. Stopes evidently played a critical part in this discovery – not only finding the glands on the foliage, but also providing some of the vital drawings of the ovules in the Oliver & Scott (1904) paper (see Fig. 5).

Oliver and Scott acknowledge in their paper the help that she provided, but she was not included as a co-author. It is unlikely that Oliver and Scott were by nature unfair in this way, or that they regarded assistance by a young woman student as not meriting full recognition; as we show in the present paper, Oliver in particular was highly supportive of his palaeobotanical students, many of whom were women. Also, even when young, Stopes was probably not the type of person who would remain quiet if she perceived that she had been the victim of unfair treatment. It was probably just that, although Stopes had given some technical support during the investigation, it had been Oliver and Scott who recognized the importance of the glands; the existence and discovery of the pteridosperms was almost exclusively their discovery, and Stopes's role was relatively peripheral. Nevertheless, her involvement in this groundbreaking discovery was almost certainly a factor that drew her into palaeobotany as a career, and which clearly coloured some of her thinking in her subsequent papers.

After graduating from UCL, Stopes went to Munich as a research student and, again, after only 2 years she presented and defended her PhD thesis (in German) on cycad seed structure and function. Although cycads are today relatively

Fig. 4. Marie Stopes: (**a**) as a young girl, pictured with her younger sister Winnie; (**b**) and (**c**) as a young woman; (**d**) in middle age; (**e**) receiving an honorary doctorate in 1948. Reproduced with permission from Marie Stopes International UK organization.

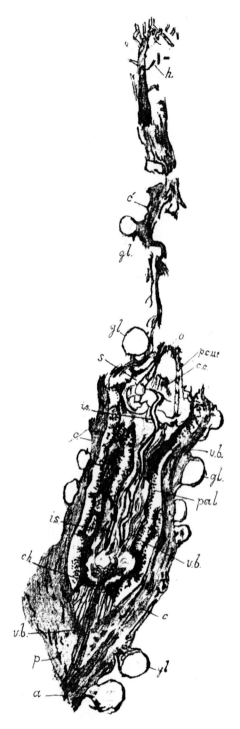

Fig. 5. One of the drawings by Marie Stopes of a section through a *Lagenostoma* ovule showing the tell-tale glands. The drawing was figured in the paper by Oliver & Scott (1904, pl. 10, fig. 34). Reproduced with permission of the Royal Society of London.

small group of plants, their seeds are similar in many ways to those of the Carboniferous pteridosperms. A more detailed investigation of cycad seeds was thus an obvious first step in developing a better understanding of the Carboniferous pteridosperms. While at Munich she also published her first two palaeobotanical papers: both were studies on Carboniferous plant anatomy, based on work that she had done under Oliver at University College: one on cordaite foliage, the other on calamite roots (Stopes 1903*a*, *b*). Perhaps more significantly, however, she met the Japanese botanist Kenjiro Fuji, a noted authority on *Ginkgo*. Her affair with Fuji and its consequences for the development of her personal life have been well documented in the various biographies of Marie Stopes, but it also had a major impact on the development of her research, as it resulted in a swing away from Palaeozoic studies towards Mesozoic palaeobotany, and in particular the early evolution of the angiosperms.

On returning to the United Kingdom in 1904, she was appointed Demonstrator in Botany at Manchester University under F. E. Weiss, where she was the first female member of staff (Watson 2005). It was during this time at Manchester that she met Robert Falcon Scott, just before his ill-fated expedition to Antarctica. She tried to persuade him to take her with him to Antarctica. He refused, but promised to bring back some plant fossils and, when his body was eventually found, specimens of *Glossopteris* were with his baggage (Seward 1914).

Among the students at Manchester at this time was D. M. S. Watson. Watson was later to become a noted authority on vertebrate palaeozoology, becoming Professor of Zoology at UCL, but during his Manchester years his main interests were in palaeobotany, in particular coal-ball studies. Like Stopes, before becoming involved in geology he had studied chemistry. It is not totally surprising, therefore, that the two collaborated in the first major study on coal-ball taphonomy, involving geochemical, petrographical and stratigraphical analyses (Stopes & Watson 1908). Coal-ball palaeobotany had been extensively studied at Manchester during the late 19th century, notably by W. C. Williamson, but the understanding of how these plant petrifactions were formed was very limited. It had been noted that, at least in northern England, coal-balls occur in just one coal seam, and that this seam was overlain by shale containing abundant remains of marine shells. For some, this was evidence that the coal was formed from plant remains that had drifted into a marine environment, whereas others regarded the plant material as being *in situ* peat that had been flooded by a marine transgression. Stopes & Watson (1908) showed that the peat was indeed *in situ*, and that the petrifaction of the plant tissue was due to infiltration by minerals from the marine flood water; this model of coal-ball formation remains widely accepted today (A. C. Scott *et al.* 1996). In addition, they showed that the vegetation preserved in coal-balls was of notably different composition to that preserved as compressions in the roof shales of the coals. This had never been properly appreciated before, and represented a major advance in thinking; the Carboniferous coal forests were not a homogeneous mixture of the plants represented in the forest record, but a dynamic complex of habitats.

During her 3 years at Manchester, she published two papers on Carboniferous plant anatomy. One was a short note on wound reaction in a *Calamites* stem (Stopes 1907); the other a more extensive study on a specimen of the fern *Tubicaulis*, which had been found in a nodule ('bullion') from the shales above the coal-ball-bearing seam in Lancashire (Stopes 1906*a*). She noted many of the unusual features of this Carboniferous fern and suggested that it may indicate a link between the ferns and pteridosperms. This is perhaps not surprising as the pteridosperms were the palaeobotanical 'hot topic' of the day, and she must still have remembered vividly her part in their discovery when she was at UCL. The problem was that no reproductive organs were then known for *Tubicaulis* and her interpretation was based exclusively on the anatomy of the stem and frond. When reproductive organs were found for *Tubicaulis* (Eggert & Delevoryas 1967), they were found to have nothing to do with pteridosperms, but were similar to those of the extant fern family the Gleicheniaceae. Today, these plants are recognized as showing a curious mixture of primitive and advanced features and are normally assigned to their own family, the Sermeyaceae.

Kenjiro Fuji had followed Stopes to Manchester and stayed with her for a few months. After returning to Japan, he sent her samples of what proved to be petrified Cretaceous angiosperms. The importance of this material was obvious and resulted in a major change in her research interests away from Carboniferous palaeobotany. At the end of her contract with Manchester in 1907, she moved to Japan to join Fuji. This was obviously partly for personal reasons (she was still having an affair with Fuji, who by that time had divorced his wife), but there was also a good scientific basis for her visit, which was funded by the Royal Society.

Stopes returned to the United Kingdom in 1909, her affair with Fuji having come to an end, and was appointed lecturer at Manchester, a post arranged for her by Weiss (Watson 2005). This coincided with the appointment of W. H. Lang to a chair in the same department and resulted in Manchester becoming arguably the most active British palaeobotanical institution of its day. However, there is no evidence that Stopes and Lang collaborated in any way, Stopes continuing to concentrate on Mesozoic studies. It was during this time that she published her 'popular' palaeobotanical book *Ancient Plants* (Stopes 1910). Some years earlier she had published a similar work on living plants (Stopes 1906*b*), which met with considerable success, and she clearly intended to do the same in palaeobotany. There had been no comparable palaeobotanical publications up to that date and her style of writing was perfectly suited to achieve her goals; *Ancient Plants* continued to be an influential book for many years, introducing many young people to the study of fossil plants.

Her complicated personal life at this time resulted in her spending much time in London and she was granted research facilities at UCL. According to Watson (2005), in July 1910 her doctor advised her no longer to live in Manchester for health reasons and her post there was terminated in November of that year. She continued to publish on palaeobotany, mainly Mesozoic, for another 10 years, but her only major contribution to Palaeozoic palaeobotany during these years was her study on the Canadian 'Fern Ledges' (Stopes 1914; see also Falcon-Lang & Miller 2007). Although Stopes had little experience working on Carboniferous adpression floras, her monograph provided an excellent overview of this flora and was well-received by the scientific community of the time. Significantly, she recognized that the flora included some upland taxa, which allowed her to develop ideas concerning the palaeoecology of the coal forests, in particular the relationship between the vegetation of the fully wetland habitats and that of the drier habitats surrounding the basin.

The 'Fern Ledges' monograph was the last of Stopes's published contributions to Carboniferous palaeobotany, but she clearly retained an interest in the subject for some years to come. This is most clearly seen in her collaboration with the Welsh palaeobotanist David Davies (B. A. Thomas 1986). She probably made contact with Davies through Robert Kidston, with whom she had corresponded during her Canadian studies. Davies had been collecting plant fossils from the South Wales Coalfield for some years, and had received substantial help from Kidston with the identification of the fossils. Stopes encouraged Davies to expand his collecting activities, the result being the massive collection now stored in the National Museum Wales in Cardiff. The goal was partly to improve the correlation of the coal seams, but Stopes was also clearly aiming to follow up some of her ideas on the palaeoecology of the coal forests that she had developed in her work on coal-balls and on the 'Fern Ledges'. According to Thomas (1986), Davies was captivated by Stopes and gave the impression that he was doing the collecting for her. However, from about 1920, Stopes seems to have lost interest, presumably becoming increasingly distracted by her non-geological activities. Nevertheless, Davies continued with the project and some of the palaeoecological results that Stopes helped bring about were eventually published in Davies's classic 1929 paper (the unpublished part of Davies's work was summarised posthumously by North 1935).

Stopes eventually left UCL in 1920, to pursue her interests in family planning and women's issues. In her later years there, she became involved in coal studies, and in particular in helping develop a terminology for coal petrography which is still used today (Stopes & Wheeler 1918). Even after leaving UCL she retained some interest in coal studies, but her formal contributions were limited.

Although Marie Stopes's research output in Carboniferous palaeobotany was not enormous, her published papers almost all contain novel and important insights, in particular into coal-forest ecology. She was arguably the first to view the coal forests as a living ecosystem rather than just as a series of dead fossils – an approach that was not really emulated again for nearly half a century, except in Davies (1929) study, which Stopes did much to instigate. It would seem, however, that Carboniferous studies did not appeal to her much as Mesozoic palaeobotany, especially the origin of angiosperms and other seed plants. Why this was so is not clear; she had been closely involved with the discovery of the pteridosperms, one of the most important palaeobotanical discoveries of the 20th century and one would have thought that this would have drawn her to develop her Carboniferous interests further. She also held professional positions at two of the leading institutions for Carboniferous palaeobotanical studies: UCL and Manchester. But maybe this was the problem: Stopes was a strong personality with an independent spirit, and the very fact that she was working in institutions with established Carboniferous research programmes may have driven her to follow a different path. The origin of the angiosperms has been a 'magnet' to many palaeobotanists and the potential importance of the early petrified angiosperms given to her by Fuji must have been an irresistible attraction.

Isabel Browne (1881–1947)

Lady Isabel Mary Peyronnet Browne (Fig. 6a) was clearly a remarkable woman born to a remarkable family. Alone among the women dealt with here, she received no formal scientific education and never became a professional scientist. Nevertheless, she made several major contributions in botany and palaeobotany.

She was born in London on 6 November 1881 and died on 8 June 1947. Her father was the third Marquess of Sligo (Fig. 6c) and her mother was the daughter of Baron Jules de Peyronnet (Fig. 6d). She was one of twin girls (Fig. 6b), her sister being christened (with a certain lack of imagination) Mary Isabel; and were their only children. As her father was a semi-invalid during her childhood, her early years were greatly influenced by her mother, who was described as having 'a vivid and charming personality'. Isabel and her sister were encouraged, from the age of 9 years, to study botany, zoology, geology and astronomy. She was an avid reader in literature of an 'unusually solid kind' and was also keenly interested in French history and politics of the post-1815 period. Browne was a very humanitarian person, which can be seen in her devotion to tracing wounded or missing soldiers in Flanders in World War 1; this work earned her the OBE. During the World War 2, she spent her time caring for a group of evacuees in her home. Lady Isabel was a very social person and would be seen at many of the dances of the time. She played tennis and hockey most days but always ensured that her mornings were kept free for her true passion–botany.

Browne was arguably more of a botanist than a palaeobotanist. She was inspired to start studying plants seriously from the age of 15 years, when her tutor gave her D. H. Scott's book *Structural Botany (Flowering Plants)* as a birthday present. A few years later, she heard a lecture by Scott at UCL on the lycophytes, and from thereon her interests became focused mainly on the so-called 'lower plants' (i.e., lycophytes, sphenophytes and ferns). She spent several years reading on the subject in the library of the Botany Department of the British Museum (Natural History), from which arose her first scientific publication – a synthesis of the phylogenetic relationships of the 'lower plants' (Browne 1908, 1909). Given that she had had no formal education in botany and had no experience of botanical laboratory work, this is a quite startling work. It presents no new data on the subject, but provides a most lucid and well-argued account of the evolutionary relationships between the main groups of these plants. Lady Isabel's twin sister, Mary (see A. Arber 1947, p. 154) claimed that her

unusual capacity for weighing evidence may have been, in part, an inheritance from the galaxy of lawyers among her forbears, for nearly all of the men of her mother's family were of this profession

Fig. 6. Lady Isabel Browne: (a) portrait at about 19 years old, by Flora Russell; (b) with her twin sister Mary, photographed when young; (c) her father (3rd Marquess of Sligo) and (d) mother. Reproduced with permission of Robert F. Quick, Chief Constable of Surrey Police Headquarters, Guildford, where these portraits are now kept.

Whether or not this was the reason, her 1908 and 1909 papers, which integrated evidence from both living and fossil plants, provided a remarkably robust starting point for her future botanical investigations (Browne 1908, 1909).

It became clear to her that, if she was to take the subject any further, she needed more practical experience in botany. To achieve this, in 1907 she joined the Department of Botany of UCL, under F. W. Oliver. Here, she was taught microtoming techniques by Agnes Arber, which allowed her to start research on the detailed anatomy of living *Equisetum*, a subject that she continued to investigate for the rest of her life (see A. Arber 1947 for an incomplete list of Browne's publications). However, it was not just the scientific wonders of the stained thin sections that inspired her; she compared them in beauty to the paintings of the Flemish masters, which she saw at an exhibition in 1927.

Remarkably detailed though these descriptions of *Equisetum* cones were, she was never interested in the subject in a merely abstract way. Throughout, her concern was to improve the understanding of the evolutionary history of these plants. It is not surprising, therefore, to find that, in the 1920s, she turned to palaeobotany, investigating the detailed anatomy of Carboniferous calamite cones. The work was not initially undertaken in Britain but in France, in particular on ground thin sections in the collections of the Muséum d'Histoire Naturelle, Paris, and the Société d'Histoire Naturelle, Autun. Comparable British material would have been available to her in London, but it appears that she was in these days a regular visitor to Paris for non-scientific reasons, and it may have been as convenient for her to work on the French collections.

Her initial papers were highly detailed descriptions of various calamite cones preserved in these collections (Browne 1923, 1925a, b). From the start of her palaeobotanical investigations, she was concerned with the problem of the origin of the bracts in the Carboniferous cones, the feature which most obviously separates them from the living *Equisetum* (she had also discussed these ideas in her 1908 paper). Her thoughts became crystallized in a later paper (Browne 1926), where she suggested that the bracts were neither lobes of the sporangiophores nor sporangiophores that had become sterile – the view most widely accepted in those days. Rather, she suggested that the bracts were homologous to sterile leaves that became increasingly intercalated between the sporangiophores until there were alternating whorls of sporangiophores and bracts. As a consequence, the modern *Equisetum* seems to bear a closer resemblance to the early Carboniferous *Archaeocalamites*, and the Late Carboniferous calamites seem to be a more derived evolutionary side-branch that subsequently became extinct (see also more recent comments by Good 1971). She also queried the relationship between the calamites and sphenophylls, which most authorities of the time regarded as closely related. While agreeing that they had their origins in a common ancestor, she argued based on a detailed comparison of their cone anatomies that the two groups (clades, in modern parlance) diverged very early in their phylogenetic history (cf. comments by Stein *et al.* 1984).

Browne (1933) next turned to the problem of the enigmatic group of plants known as the Noeggerathiales. This was not an attempt to revise the group; she relied for her knowledge of it mainly on the work of Němejc (1928, 1931). Her main concern was to examine the new evidence that had recently come out on these plants and its effect on the classification of the group of spore-bearing plants as a whole. Despite a very careful analysis of the data, Browne was unable to come to any clear conclusion but this is not surprising.

The Noeggerathiales remain one of the most enigmatic Carboniferous plant groups, often ignored in phylogenetic analyses (e.g., Kenrick & Crane 1997); even the most recent re-evaluation of the group remained equivocal as to its systematic position (Šimůnek & Bek 2003).

Browne's (1935) final paper with a palaeobotanical slant differs from the others in that it dealt with the phylogeny of foliage within the spore-bearing plants. Again essentially a review paper, it nevertheless distils a vast array of information from both British and Continental sources, and attempts to come to a conclusion as to the patterns of evolution within these plants. The evidence was, and still is, equivocal on the details of these patterns, and so again Browne does not come to any clear-cut conclusions. However, one of the most insightful comments is right at the end of the paper, where she notes that the key to the problem probably lies in the 'anomalous ill-preserved or incompletely understood forms' and that details of their structure must be incorporated into phylogenetic analyses – a point which many present-day palaeobotanists often ignore.

Although Lady Browne published no further scientific papers, she maintained an interest in botany and palaeobotany for the rest of her life. There is a letter in the archives of the Linnean Society, dated 22 March 1946, asking if they had received a copy of the new French palaeobotanical book by L. Emberger, *Les Plantes fossils dans leur rapports avec les végétaux vivants*. Until World War II, she lived in the centre of London at Cadogan Square, within easy walk of both the British Museum (Natural History) and the Linnean Society. By 1946, however, she had moved out of London to the family's country home at Losely Park, near Guildford (known as Mount Browne), presumably to get away from the bombing, and here she remained for the rest of her life. She died in 1947, apparently the last of her family. Her father and mother had died in 1896 and 1927 respectively, and there is no record of her sister Mary still being alive. Mount Browne was sold to the Surrey Police Force to form part of their headquarters and it is in a small museum there that the portraits shown in Figure 6 are now kept.

While Browne was more than capable of laboratory work in both palaeobotany and botany, her real strength lay in synthesizing information. She was amazingly focused in her work. From the start, she decided to work on horsetails, and that is what she stuck at for most of her scientific life. This of course limited her publication output: she produced only 23 papers over a span of 32 years (it should be noted that A. Arber's 1947 list of her publications omits four review papers by Browne, published in 1910, 1911, 1913 and 1914). Even for those days, that would be regarded as a low level of output for an active professional scientific researcher. However, she was of course not a professional; she was a lady of ample independent means who could do what she wanted at the pace she wanted. For her, science was balanced against the social life that someone in her position was expected to lead. During both World Wars, her humanitarian activities restricted the time she could spend on science, and in 1916 resulted in her having to decline the offer of co-authoring a book with D. H. Scott. But the very fact that Scott invited her to join him in writing a book shows that Browne was no dilettante; she was a respected scientist, who made a major contribution in her field specialization.

Lucy Wills (1888–1964)

Cuticle studies are often the only way of getting anatomical evidence from plant fossils. They have proved an important part of Mesozoic and Tertiary palaeobotanical studies, but Carboniferous palaeobotanists were much slower to appreciate their potential importance. There were some early records that cuticles could be prepared from plant fossils of this age (e.g., Zeiller 1890; Huth 1912), but these records were only really intended to show that cuticles were present, rather than to study them systematically. The earliest attempt to investigate Carboniferous cuticles properly was in a paper published in the *Geological Magazine* of 1914, by Lucy Wills (Fig. 7). Wills has been something of an enigma in the palaeobotanical world; her paper is very widely quoted, but what happened to her subsequent to its publication has until now been unknown.

Lucy Wills was born in Sutton Coldfield on 10 May 1888, the third child of William Leonard Wills and Gertrude Annie Johnston. William was an engineer who owned a firm in Birmingham, manufacturing edge tools. However, there was also clearly a spirit of adventure in the family: Lucy's great uncle, Sir Alfred Wills, was one of the first great Victorian mountaineers and a founder of the Alpine Club; and his grandson, Edward Felix Norton, led the 1924 Everest expedition. Lucy's elder brother, Leonard Johnston Wills, went on to become Professor of Geology at the University of Birmingham.

Lucy Wills's education began in a small private school and continued at Cheltenham College for Young Ladies. She then proceeded to Newnham College, Cambridge, where she studied botany and geology, gaining a double first-class honours degree in botany and geology in 1911. While at Cambridge, she inevitably came into contact with A. C. Seward and, most importantly, H. H. Thomas, two of the leading British palaeobotanists of their time. The contact with Thomas proved particularly important, as he had been instrumental in developing the study of cuticles from fossil plants. Although he had published some work on Carboniferous palaeobotany (H. H. Thomas 1909a & b, 1910), he only seems to have studied cuticles from Mesozoic plant fossils (H. H. Thomas & Bancroft 1913; H. H. Thomas 1928). However, he seems to have successfully encouraged Wills to try to prepare cuticles from Carboniferous fossils, which had been provided by her brother Leonard. This was probably the result of an undergraduate project, as there is no evidence that she studied for a higher degree while at Cambridge.

Having independent means, Wills travelled to South Africa after graduating, to join a Cambridge friend, Margaret Hume, who was lecturing there. It seems to have been while she was in South Africa, that she wrote up and submitted for publication the results of her student work on Carboniferous cuticles (Wills 1914). The resulting paper represented a major advance in Carboniferous palaeobotanical studies. She described cuticles from medullosalean pteridosperms and cordaites, noting the presence of stomata, trichomes and hydathodes. She discussed their potential for phylogenetic studies, noting the similarity between the stomata of the medullosaleans and those of the Mesozoic cycads, thus providing evidence for what is now the accepted view that these two groups were related. She also discussed the ecological implications of her results, especially the presence of hydathodes that suggest very high

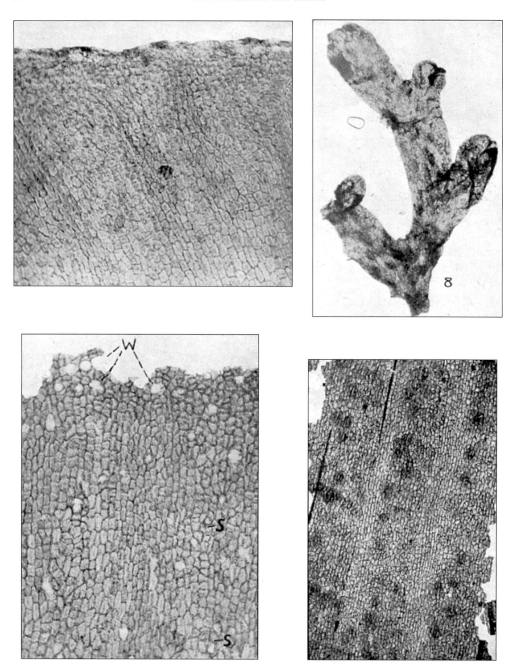

Fig. 7. Part of plate 31 in Lucy Wills (1914) pioneering paper on Carboniferous cuticles. Reproduced with permission from Cambridge University Press.

humidity in the coal forests. Of especial interest was a specimen of a thalloid structure, which she interpreted as a possible bryophyte. This was later re-investigated by Walton (1925b, 1928b) in his pioneering study on Carboniferous bryophytes. In his view, Wills's specimens were the first documented examples of Carboniferous liverworts, which he named *Hepaticites willsii* in her honour. Wills also described a pecopterid fern. Sadly, this latter fossil was not illustrated, but it is worth noting that most specialists have regarded pecopterids as being impossible for cuticle studies, and only

Fig. 8. Lucy Wills: (**a**) as a Labour Party councillor, photographed outside the Chelsea Council Offices. Reproduced from *The History of Folic Acid* by Hoffbrand & Weir, in the *British Journal of Haematology*, **113**, 579–589 (permission granted by Blackwell Science Ltd); (**b**) relaxing in the garden (photo reproduced from the James Lind Library).

very recently has some progress been made in their study (Pšenička & Bek 2003).

Despite the importance of this paper, it proved to be her only published palaeobotanical work. There was certainly considerable potential for further work in the field. She had demonstrated the potential of cuticles for helping improve the understanding of Carboniferous plants, a fact that would have been no surprise to anyone studying the plant life of any other geological time. She also clearly stated that she had made preparations of other material, including cordaite seed cuticles, and spores from lycophyte cones, and that she intended to investigate these more fully in the future. However, events prevented this from happening. After Wills's pioneering work, interest in Carboniferous cuticle studies became more or less dormant for nearly half a century, until the work of the German palaeobotanist Manfred Barthel (1961, 1962), (the only notable exceptions being three papers by Florin 1925*a* & *b*, 1926).

Wills's cuticle slides were eventually deposited at the Lapworth Museum of the University of Birmingham, where her brother Leonard had been appointed and later became Professor of Geology. In the archives of the Lapworth Museum, there are letters from John Walton, then of Manchester University, who had had her slides on loan for several years during the 1920s as part of his studies on Carboniferous bryophytes (it was Walton who showed Edith Bolton how to prepare cuticles; see later). He noted that, by the time he had started to study them, the Wills's glycerine jelly preparations were starting to dry out and so he remounted them in Canada balsam.

Wills was still in Cape Town when World War 1 started, and she remained there for about a year, working as a volunteer nurse. When she returned to England, she enrolled in the London School of Medicine for Women (later to become the Royal Free Hospital School of Medicine). She had initially intended to specialize in psychiatry but, after graduating in 1920, she found her interests lay more towards the biological aspects of medicine. With this in mind, she joined the Department of Chemical Pathology at the Royal Free Hospital and travelled to India in the late 1920s and early 1930s. There she worked on nutritional anaemias that were widespread among women working in the textile industry, especially those who were pregnant. She discovered that yeast extracts (e.g., Marmite) provided a cheap source of vitamin B12 and was a highly effective remedy, saving many lives. She also argued that there was another factor responsible for this type of anaemia, later shown to be a lack of folic acid, which became referred to as the 'Wills Factor'.

Wills continued to work for the Royal Free Hospital, initially in a voluntary capacity, but in 1928 she was offered the position of part-time biochemist. She remained committed to researching in India, producing a long list of scientific publications (for a list of some these publications, see Roe 1978, and for an assessment of her scientific contribution to medicine, see Hoffbrand & Weir 2001). When World War 2 broke out, she returned to England and took up a position as full-time pathologist in the Emergency Medical Service. She was then in charge of pathology at the Royal Free Hospital and, by the end of the war, had established the first Haematology Department there. She retired in 1947 but continued research into the effects of nutrition on health in Fiji and Africa.

Lucy Wills was clearly a remarkable person. She was very hard working and, although compassionate to people's failings, was impatient with laziness and ideas that she considered had not been well thought through. This was combined with considerable physical energy. She was a keen nature rambler and mountain climber, and made early use of skis to cross snow-laden countryside, which she did with great enjoyment. She always chose to ride a bicycle rather than drive a car, and would be seen 'arriving at the Royal Free Hospital on her bicycle with gloves fixed on the handlebars, when the other physicians came in large cars' (Roe 1978). During her early career, she observed much poverty in India, which no doubt was partly responsible for her very strong convictions on social issues. In the last decade of her life, she became a Labour Party councillor for the Borough of Chelsea (Fig. 8a), enabling her to continue dedicating the remainder of her life to the social issues of the day. Clearly, palaeobotany's loss was the wider world's gain.

Edith Bolton (1893–1974)

Edith and Alice Bolton were twin daughters born in Manchester in 1893 to Herbert and Jennie Bolton (Roger Clark pers. comm.; 1901 census return). In 1898, Herbert Bolton (1863–1936) was appointed Curator and Secretary of the Bristol Museum & Library, becoming Curator of the Bristol Natural History Museum in 1904, and then Director of Bristol City Museum & Art Gallery from 1911, until he retired in 1930 due to ill-health (the titles of these posts reflect the changes in the organization of the Museum over this period). He was most notable, scientifically, as the leading British expert of his time on Carboniferous insects (H. Bolton 1921), and there can be little doubt that his interest in palaeontology rubbed off on his daughter Edith.

Edith graduated from Bristol University in 1916 and soon joined the staff of the City Museum & Art Gallery. Despite her essentially geological background, she was appointed Assistant Curator in Botany. Little is known about her activities in the Museum during this time, her name rarely appearing in its Annual Reports. However, she is reported in the *Proceedings of the Bristol Naturalists' Society* as giving a number of talks to that group, all on geological themes. Also during this time, she published her first paper: a review of the fossil flora and fauna of the Lower Coal Measures (E. Bolton 1919).

In 1919 or 1920, she was awarded an MSc by the University of Bristol, based on work that she had done on the Carboniferous Limestone under S. H. Reynolds (*Report on the Geological Department [including Geography] for the Session 1919–20* – Loeffler pers. comm.). Her thesis was entitled '1) Geology of the Wickwar, Chipping Sodbury Area. 2) Some *Syringopora* of the Bristol District'. The work was done partly in collaboration with another woman geologist, Margaret C. Tuck, with whom she jointly presented a paper to the Geological Society of London (E. Bolton & Tuck 1921), although the full text was never published. Tuck later continued to work in this area and published a complete description of the Lower Carboniferous geology of the Wickwar–Chipping Sodbury area (Tuck 1925). Tuck fully acknowledged Bolton's contributions, but by this time Bolton seems no longer to have been interested in working on the Carboniferous Limestone.

In 1919, presumably after submitting her MSc thesis, Edith Bolton resigned from the Museum to take up a 2-year post at Armstrong College, Newcastle. While there, she became interested in the palaeobotany of the Coal Measures of the Newcastle and Durham Coalfield, and later published a review of the subject (E. Bolton 1926). There was little tradition of palaeobotanical work at Armstrong College, but her interest was probably sparked by finding that part of the James Hutton Collection was still stored there (this collection was the basis of one of the most famous British palaeobotanical monographs of the 19th century; Lindley & Hutton 1831–37). As pointed out by Bolton in her review paper, there had been remarkably little work on the palaeobotany of this coalfield, other than by Hutton, and reviews of Hutton's collection by Howse (1890) and Kidston (1891). She reported that she had discussed the age of the various coal seams with Robert Kidston, although curiously she omits mention of Kidston's (1922) review of the palaeobotany of the coalfield. Equally curious is that Kidston (1922) makes no mention of Bolton's collecting in the coalfield, despite the fact that he must have been aware of her activities there. As will be shown later, this type of failure in communication seems to have been a regular feature in Bolton's research.

Edith Bolton's position at the museum in Bristol had been kept open during her time at Newcastle, and she returned to her position as Assistant Curator of Botany in 1921. In contrast to her earlier years at the museum, she now seems to have become more active in museum life. The annual reports of the museum record how she started to revamp the botanical and palaeobotanical displays, and she regularly gave public talks on various botanical themes, often to children's groups. It was during this period of enhanced activity that she wrote her most important palaeobotanical papers.

While living in Newcastle, she had become particularly interested in the Phoenix Brickworks at Crawcrook, County Durham, where she found a well-preserved siderite nodule biota. Although the preservation of the fossils here (both plants and animals) is exceptionally fine, analogous to that found at Coseley (Kidston 1914) and Mazon Creek in Illinois (Darrah 1969), to this day the Phoenix Brickworks fossil biota has not been fully described. Bolton found that remains of medullosalean pteridosperm foliage of the 'Neuropteris'-type was particularly abundant here and this formed the basis of her most important palaeobotanical study (E. Bolton 1925).

The main issue that she addressed in her 1925 paper was the variation in pinnule morphology in different positions of the frond and between different fronds. These fronds were originally large and complex, but the preserved fossils are almost always relatively small fragments of pinna or even isolated pinnules. Trying to build up a biologically meaningful species concept from such fragments remains one of the critical problems in medullosalean taxonomy, and to this day Bolton's paper is one of the few serious attempts to address the issue (perhaps the most notable other study was by Scheihing & Pfefferkorn 1980).

Her approach was based on the recognition of a series of pinnule morphotypes, which she defined by the pinnule morphology seen in the type specimens of some of the well-known neuropteroid species. This is rather a 'geological' approach to the problem and compares with the morphotype approach used by A. E. Trueman to resolve another palaeontological problem concerning morphological variation – that of the Late Carboniferous non-marine bivalves (Trueman & Weir 1946–56) – and which in turn had evolved out of the classic work on Jurassic ammonite faunas. Bolton's analysis was clearly literature-based and there can be no questioning the breadth of her direct knowledge of that literature. The problem was that it is difficult to interpret the types purely from illustrations, especially in the early 19th-century monographs; the venation and even the pinnule morphology was to varying extents stylized in these works (Cleal et al. 2005). She also failed to realize the importance of frond architecture in classifying these fronds, although this is more understandable: it was not for another 15 years that the taxonomic importance of this feature became properly understood (Gothan 1941). The result of these confusions was that Bolton over-synonymized not only species of the same genus but also species that are now recognized to belong to different genera as now classified (for summary of the present-day taxonomy of the neuropteroids see Cleal & Shute 1995).

The nomenclatural confusion that resulted from Bolton's analysis has meant that her work has been largely dismissed by subsequent authors (e.g., comments by Crookall 1955–76 and Laveine 1967). This is unfortunate as it ignores the undoubted insights that Bolton had into the problem of morphological variation in these fossils – insights that were in fact far in advance of those of the other British palaeobotanists studying such fossils at this time (e.g., Robert Crookall and Emily Dix), and arguably of many palaeobotanists working today. It is tempting to suggest that the underlying problem was her lack of contact with the more experienced palaeobotanists working at the time, and it is noticeable that she gives no acknowledgements in her 1925 paper. When she returned to Bristol in 1921, she would have found Robert Crookall studying at the Botany Department of the University; the University awarded him a PhD in 1925 based on his studies on the palaeobotany of the Bristol–Somerset Coalfield. However, there is no evidence that Bolton had any contact with Crookall. Interestingly, in her 1929 paper, Bolton acknowledges Arthur Trueman for some of the material that she studied. It was not until 1933 that Trueman transferred from Swansea to Bristol (George 1974), but he may already have been developing contacts with Bristol in the mid-1920s, which may explain Bolton's contact with him; his influence certainly seems to be present in the morphotype approach that she used in her paper.

It was perhaps unfortunate that Bolton did not develop contacts with the palaeobotanists then working in continental Europe, who were starting to address similar issues, and whose work has ultimately proved more successful in developing a practical taxonomy (contrast the situation with Emily Dix, discussed later). Paul Bertrand at Lille, in particular, was aware of these problems and his now classic monograph on the neuropteroids of the Saar–Lorraine Coalfield demonstrate an alternative (and ultimately more successful) approach to dealing with these fossils (Bertrand 1930). If Bolton had attended the 1930 Cambridge International Botanical Congress, as did Dix, and met her continental colleagues, maybe her palaeobotanical career would have been more successful. As will be shown later, however, Bolton had other, more personal things on her mind in 1930.

A year later, she published a review paper on the Carboniferous macrofloras of the Northumberland and Durham Coalfield (E. Bolton 1926). Unfortunately, there are no illustrations and so it is impossible to check the identifications. The neuropterid identifications were clearly based on her 1925 paper, and so over-represent the true diversity of this group. Other identifications are also suspect, such as of *Dicksonites plueckenetii* (Sternberg) Sterzel from the middle Duckmantian Bensham Seam (this species is not known to range below the upper Asturian Substage). Nevertheless, the paper does give a useful insight into the diversity of the Carboniferous macrofloras from this coalfield and is still one of only two papers attempting to provide a synopsis (the other being by Kidston 1922).

Bolton's final palaeobotanical paper was a short note in the *Annals of Botany* dealing with neuropterid cuticles (E. Bolton 1929). She noted that Wills (1914) had described such cuticles from neuropterids, and that Florin (1926) had subsequently obtained them from cyclopterids. She makes a number of comments about the similarities in the features that she could see in her cuticles and those prepared by Wills and Florin, but no detailed descriptions or illustrations are provided. Most frustratingly, she does not even state from where her specimens originated. Perhaps the most interesting thing in this paper is that she states that John Walton of Glasgow had shown her how to prepare the cuticles – the only other palaeobotanist other than Kidston with whom there is evidence that she had personal contact.

Edith Bolton resigned her post of Assistant Curator in Botany on the 14 June 1930 to get married at the relative late age (for the time) of 37 years to Charles Helsby; the *Annual Reports of Bristol Museum and Art Gallery* for 1930 record that she 'received the best wishes of the [museum] committee on the occasion of her marriage'. We know little of her post-museum life beyond a report of her death in the local newspaper for Colwyn Bay in North Wales (information kindly provided by the librarian of that town):

On July 30, peacefully in a nursing home, of 10 Church Drive, Rhos on Sea, Edith Helsby, the beloved wife of the late Charles Helsby and the dear companion of Renee Kelly. Service and committal took place at Colwyn Bay Crematorium on Friday, August 2.

From this we conclude that there were no children.

Was it coincidental that she left the museum in the same year as her father retired? We cannot say. All that we can say is that Edith Bolton clearly had considerable insight into the problem of morphological variation, but that she did not fulfil her potential, which was at least partly due to her lack of contact with more experienced colleagues.

Emily Dix (1904–72)

Together with Margaret Benson, Emily Dix was one of two women to develop a significant career as a Carboniferous palaeobotanist during the first half of the 20th century. However, Dix's route to and through the subject was quite different. Burek & Cleal (2005) provide a detailed account of Dix's life and career, and the following is merely an abstract.

Dix was born and educated in South Wales, graduating in geology from University College Swansea in 1925. She then continued at Swansea to research the geology of the western part of the South Wales Coalfield under A. E. Trueman, being awarded an MSc in 1926. Her initial studies were as a field geologist, rather than as a laboratory palaeontologist, which meant that she had to work in the very male-orientated environment of the mining industry. However, her gender seems to have been no significant obstacle to her work, and she was held in high regard by her professional colleagues in both industry and academia.

Dix initially studied all aspects of the Late Carboniferous biotas in South Wales (e.g., Dix & Trueman 1924, 1928; Dix 1928, 1934; Dix & Pringle 1929; see Burek & Cleal 2005 for a full list of her publications), but her interests soon started to swing more towards the floras. This was probably partly because the non-marine bivalves, which were then the main tool used for stratigraphical correlations of British Late Carboniferous sequences, were being studied by a number of palaeontologists, but the floras were being relatively ignored. Palaeobotanical biostratigraphy thus offered her more opportunities for developing innovative research. At that time, the leading stratigraphical palaeobotanist, Robert Crookall (1931, 1932, 1938), was still using the traditional approach to the subject developed by Robert Kidston (1905), and this was not providing the sort of stratigraphical resolution being achieved by the non-marine bivalves. Dix's innovation was to apply a more modern biostratigraphical approach to the macrofloras, clearly inspired by the developments made by her mentor, Trueman (George 1974). Her seminal 1934 paper on the South Wales Coalfield macrofloras introduced the concept of defining biozones purely on the stratigraphical ranges of the fossils, and quite independent of the lithology of the sequence in which they occurred (Fig. 9). Such an approach was not new with fossil animals (e.g., Lapworth 1880) but had never before been used in palaeobotany. It represented a major advance in plant biostratigraphy in the United Kingdom, and it was more than 40 years before similar work was carried out in this country (e.g., Cleal 1978).

The turning point in Dix's career came in 1930. In that year, she was appointed Lecturer in Palaeontology at Bedford College in London, a position that she held for the rest of her working life. It was also the year that she attended the International Botanical Congress in Cambridge, where she met for the first time many of the leading European experts in Late Carboniferous plant biostratigraphy, including W. Gothan, P. Bertrand, W. J. Jongmans and A. Renier. She soon established close friendships with these older colleagues, who helped her significantly with her work during the 1930s. They met again at the 1935 International Carboniferous Congress at Heerlen (in the Netherlands) and the photographs taken of them together there is clear evidence of this friendship (Fig. 10). Crookall never seems to have established a comparable rapport with these continental experts, and this is reflected in the way that Dix's research developed, whereas Crookall's, frankly, did not.

In the second half of the 1930s, Dix's research started to swing away from South Wales as she became interested in the definition of the Westphalian–Stephanian boundary (Dix 1935, 1937). This was a problem that was impossible to resolve with the non-marine bivalves, and her palaeobotanical interests made Dix well placed to make a major contribution. The problem would not have been solvable with data obtainable from the British Coalfields; it was ultimately the successions in northern Spain that held the key to the problem (e.g., Wagner & Winkler Prins 1979, 1985). However, the major work on these Spanish successions was initiated after World War 2 by her friend W. J. Jongmans, and it is not difficult to imagine that, if things had happened differently, Dix could well have collaborated with him on the project.

Inevitably, Dix's life was disrupted by the war. Along with the rest of Bedford College, she was evacuated to Cambridge. Away from her collection of fossils and library (the latter was badly damaged by bombing) and unable to do fieldwork, it was almost impossible for her to do any new research. Any contact with her continental colleagues was also severed, which clearly caused her considerable distress (Morey & Lyons 1995). After the war, she should have been able to pick up the threads of her work, but tragedy struck. In 1944, she started to show signs of mental illness and, by mid-1945, she was being treated in a mental hospital. Initially, there was hope that she would be able to return to work, but by mid-1946 it was clear that this would be impossible, and her contract with Bedford College was terminated in 1947. She was only 43 years old, but would never work again. She spent some time at a Quaker hospital in York, before eventually dying in 1972 in Swansea.

Dix's long list of scientific publications is a testament to her energy and ability. She revolutionized palaeobotanical biostratigraphy in the United Kingdom and was far in advance of her time in this. However, there can be no doubt that, if illness had not struck, she would have contributed much more, most notably to the problem of the Westphalian–Stephanian boundary.

Mary Calder (?1906–1992)

Mary Gordon Calder (Fig. 11) is the most recent of the palaeobotanists dealt with in this paper, but about whom the least is known. Much of the following biographical details are based on a brief note in the *University of Glasgow Newsletter* (No. 181 for June 1996), information from a temporary exhibition on her work provided by J. Liston (Hunterian Museum), a paper by Watson (2005), and details gleaned from Calder's own papers. She was born in about 1906 at Uddingston, in Lanarkshire, Scotland. Her father, William Calder, was a buyer, later (1931–2) becoming general manager of a warehouse. Unfortunately, Mary contracted polio myelitis during her childhood, leaving her dependent on leg callipers for the rest of her life.

Despite this disability, in 1924, at the age of 18 years, she entered Glasgow University to study pure science. Her interests were mainly in the biological and chemical sciences and, in 1929, she graduated with honours in botany (she may have been influenced in this by her mother, who was reputedly a passionate amateur botanist). After graduation, she went on to do research at Glasgow, initially under the head of department James Drummond, and was awarded her PhD on 30 June 1933. Her research was in two parts. One was on the 'Morphology of adnation in the tomato plant', which was clearly Drummond's main interest. However, in 1930, Calder seems to have come under the influence of John Walton, who had just

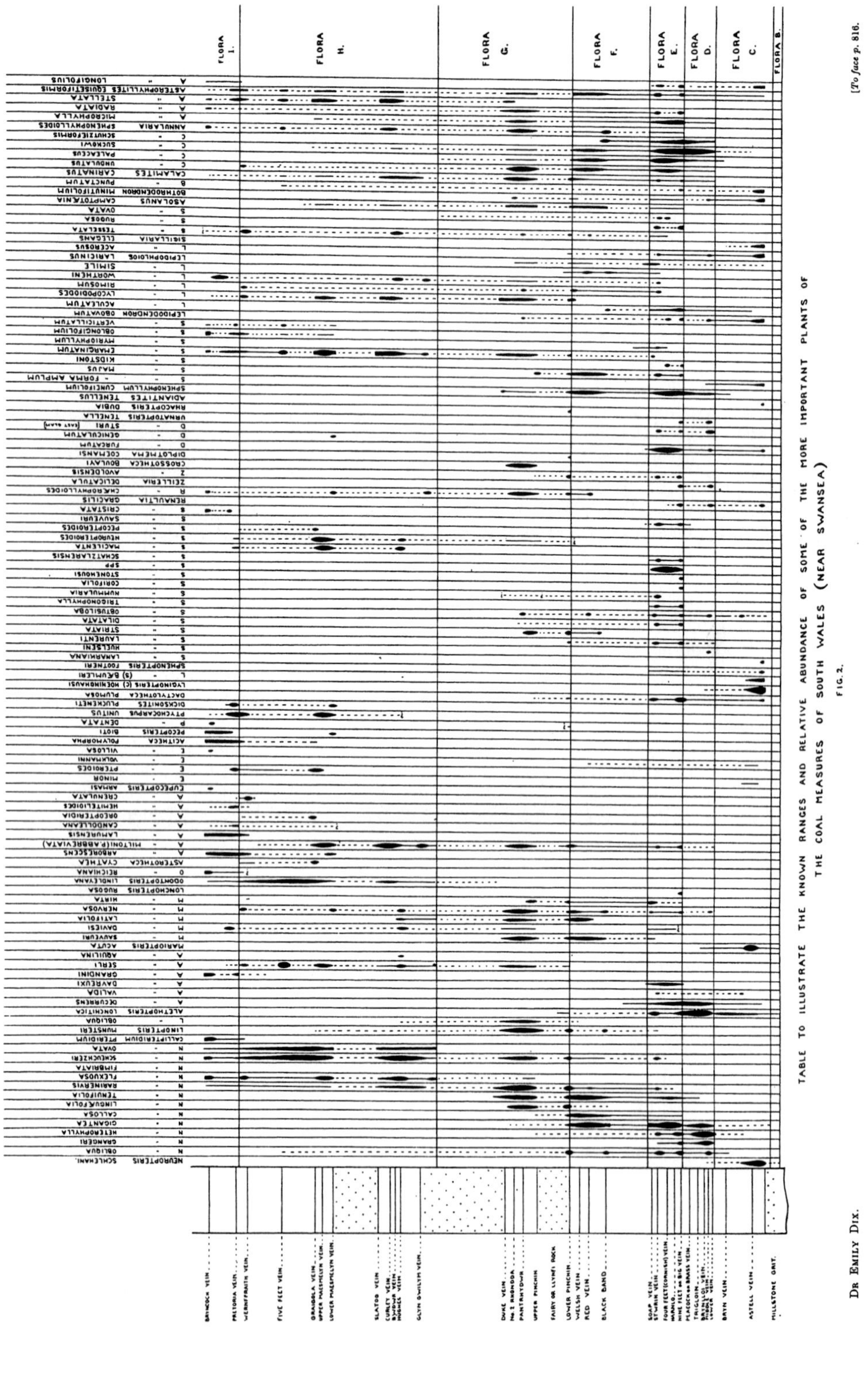

Fig. 9. Range chart from Dix's (1934) paper. Dix was the first to use stratigraphical range charts in palaeobotany, and revolutionized the way plant fossils were used to correlate strata.

Fig. 10. Emily Dix: photographs taken during the 2nd International Carboniferous Congress (1935). Originally reproduced in Morey (1995); reproduced here with permission of Elsie Morey and the United States Geological Survey.

Fig. 11. Mary Calder: (**a**) Portrait as a young woman, probably when she graduated in 1929 (image provided by Jeff Lister of the Hunterian Museum); (**b**) portrait in later life, during her years at Manchester (image provided by Joan Watson of Manchester University).

then moved to Glasgow from Manchester. Walton seems to have encouraged her to study plant fossils and the second part of her PhD thesis consisted of a detailed catalogue of Robert Kidston's substantial collection of coal-ball slides. She did not pursue her work on tomatoes, which was clearly Drummond's main interest, any further and, as far as we are aware, it was never published. Calder was clearly more interested in the palaeobotanical work, and it is probably no coincidence that two of her three examiners were palaeobotanists: John Walton, from her own department, and W. H. Lang from Manchester.

Calder continued to work at Glasgow, initially as a research fellow and then later (? from 1936) as a lecturer, and this provided her with the opportunity to pursue her palaeobotanical interests, in collaboration with John Walton (although the two never co-authored). Kidston's substantial collection of coal-ball slides, which he himself did little work on, provided an excellent resource for Calder to start an intensive programme of research on Carboniferous plant anatomy. Her initial work was on the stem anatomy of arborescent lycophytes, and she published the results in four papers (Calder 1933a, b, 1934a, b). Most dealt with Late Carboniferous species and were mainly concerned with refining the identifications and interpretations made by Kidston. However, the third paper (1934b) dealt with Early Carboniferous lycophytes and made important comparisons of anatomy with the Late Carboniferous species. None of the papers included major new discoveries, but they are characterized by very careful and detailed descriptions. It is important to remember that she was following after the detailed studies by W. C. Williamson and D. H. Scott, and had to deal with ground thin sections, which limited the possibilities of making substantial new discoveries in these fossils (see also comments earlier in this paper on Margaret Benson's coal-ball work).

Her next paper saw a change in direction in both the plant group described (pteridosperms) and the technique used (Calder 1935). Although her sections were partly based on Kidston's thin sections, she also prepared new sections herself, using the cellulose peel method developed by Walton (1928a). This was one of the first published papers to use this revolutionary new technique, which opened up the study of plant petrifactions to obtaining significantly more refined structural detail than was available from ground thin sections. It allowed her to observe details of the structure of the vascular trace along the frond rachises, a feature that has proved to be a considerable taxonomic significance.

Unfortunately, Calder did not continue with her peel studies, and her last published paper from the 1930s returned to re-examining the ground thin sections in the Kidston Collection (Calder 1938). Part of the paper dealt with a detailed examination of pteridosperm stem and petiole anatomy, but the most significant contribution was the description of two very early ovules, which she called *Calymmatotheca kidstonii* Calder and *Samaropsis scotica* Calder (the latter species also being based on material from the D. H. Scott Collection). They are Tournaisian in age, and remain among the oldest known anatomically preserved seed plant ovules known anywhere in the world. Calder seems not to have realized the potential significance of the material, probably partly because the available thin sections managed to miss the pollen chamber in the distal part of the ovules. Based on what could be seen in her ground thin sections, they did not seem to differ substantially from ovules that were known from the Late Carboniferous coal-balls. It was not until Long (1959, 1960) obtained fresh examples and applied the Joy *et al.* (1956) modification of Walton's peel technique which Calder helped pioneer, that the true significance of this material was realized. Long showed that these primitive ovules, which he renamed *Genomosperma kidstonii* (Calder) Long and *Lyrasperma scotica* (Calder) Long, were substantially different from their Late Carboniferous counterparts: they lacked a micropyle, pollen capture instead being facilitated by an apical elongation of the nucellus, known as a lagenostome or salpinx. This distinctive feature of Tournaisian ovules has proved critical in the subsequent development of ideas concerning the early evolution of seed plant reproduction, now often referred to as hydrasperman reproduction (Rothwell 1986).

In 1940, Calder moved south to London, where she joined Westfield College for women as a lecturer. Her position here was presumably mainly as a teacher; although she was at Westfield for some 10 years, she published no scientific work while there. We hear nothing more of her until 1950, when she was appointed Senior Lecturer in Palaeobotany at Manchester University. Her appointment to such a senior position, on the face of it, seems remarkable. She had not published any scientific work for well over a decade, either in palaeobotany or any other field. Even prior to that, her publication record had not been that substantial. Undoubtedly, the fact that her old mentor John Walton had close contacts with Manchester (he had been a lecturer there before moving to Glasgow) may have played a part in the decision to employ her. Also, W. H. Lang knew of her work (he had been an examiner for her PhD) and would undoubtedly have had an influence on the appointment; the post was, in effect, to act as the belated successor to Lang, who had retired in 1940.

This position provided Calder with ample opportunity to develop her interests in Carboniferous plant petrifactions, but she did not take advantage of it. Her only major scientific publication from the post-war years was a monograph on the Mesozoic petrified conifers of Patagonia (Calder 1953). This study was largely based on material donated to the British Museum (Natural History) in 1936 by Franz Mansfeld. Calder had no previous experience of working on Mesozoic plants. However, she was the most experienced British palaeobotanist of the time in the study of ground thin sections of plant petrifactions, and so was probably an obvious person for the British Museum (Natural History) to invite to deal with the collection.

But why did she publish nothing more on Carboniferous plants? She had experience of Walton's peel method, which had the potential for revolutionizing our understanding of these fossils, but she seems to have made no effort to pursue this field. If she had, she would have found herself at the forefront of Carboniferous palaeobotanical studies, at a time when the subject started to expand as North American palaeobotanists adopted the peel method (for a summary of this early American work on coal-balls see Phillips 1980). Perhaps Calder found it difficult to acclimatize herself to Manchester life. In Glasgow, she was remembered for her

... serene, philosophical approach and her delightful sense of humour [with a] wide range of knowledge and understanding in fields different from her own.

In Manchester, in contrast, Watson (2005) recounts that she left behind

... a reputation in both departments of being elusive and with the claim that she had a hiding place; Botany thinking she must be in Geology whilst Geology were thinking she must be in Botany.

It is also noticeable that, although she officially retired in 1966, she had ceased to be at the university from 1964.

She retired to live at Milngavie near Glasgow, and died in 1992. She left a substantial bequest to Glasgow University, which was used to fund an extensive refurbishment of laboratories, seminar facilities and a computing cluster in the IBLS (Institute of Biomedical and Life Science) Division of Infection & Immunity. This resulted in a plaque being unveiled to her in the Joseph Black Building at the University, and will probably be a greater memorial than her palaeobotanical output.

Others

We have summarized the life and work of the most significant women Carboniferous palaeobotanists of this time, but they were not the only ones involved in the subject. Ten other women published at least one paper in Carboniferous palaeobotany (Table 1) during this time, nearly all on coal-ball petrifactions.

Of these, four were students of F. W. Oliver at UCL. All four went on to successful careers in Recent botany, either as lecturers in one or other of the London women's colleges or, in the case of Winifred Brenchley, as Head of Botany at

Table 1. *Other British women palaeobotanists who studied Carboniferous plants during the first half of the 20th century*

Name and dates	University education	Subjects and dates of palaeobotanical papers	Main research interests	Career
Ainsworth, F. Hilda (1891–1971)	Nottingham University	Lycophyte cones (1936)	?	?
Bancroft, Nellie	Nottingham University	Fern stems (1911) and Pteridosperm stems (1914)	Mesozoic cuticles	Researcher at Newnham College, Cambridge (with H. H. Thomas)
Berridge, Emily M.	?	Lycophyte cones (1905)	Recent plants, including *Ephedra*	?
Brenchley, Winifred Elsie (1883–1953)	UCL (F. W. Oliver)	Lyginopterid stems (1913)	Weeds and weed killers	Head of Botany at Rothamsted Experimental Station
De Fraine, Ethel	UCL (F. W. Oliver)	Medollosalean stems (1912, 1914)	Saltmarsh plants	Lecturer, Westfield College, London
Halket, Ann C.	UCL (F. W. Oliver)	Roots of cordaites (1930) and lyginopterids (1931)	Saltmarsh plants	Lecturer, Bedford College, London
Lindsey, Marjorie	Manchester (W. H. Lang)	Lycophyte stems (1915)	?	?
Prankerd, Theodora Lisle (1878–1939)	UCL (F. W. Oliver)	Lyginopterid ovules (1912)	Recent ferns and aquatic plants	?
Scott, Henderina ('Rina') Victoria	Jodrell Laboratory, Kew (D. H. Scott)	Lycophyte megaspores (1906), ferns (1908) and fungi (1911)	Recent plants	?
Wigglesworth, Grace	Manchester (W. H. Lang)	Calamite roots (1905)	?	Assistant Keeper of Botany, Manchester

Rothamsted Experimental Station and a renowned expert on weeds and weed killers. The fact that they briefly studied palaeobotany is probably of relatively minor significance; we get the impression that Oliver encouraged most of his students to undertake a project in palaeobotany and, if the results were of interest, for them to be published.

Another two of these women came from W. H. Lang's department in Manchester, Lang being particularly supportive of women's education (Watson 2005). The most notable was Grace Wigglesworth, who became Assistant Keeper of Botany at Manchester Museum from 1906 to 1944, and was an active lecturer to the public on palaeobotany during that time. Watson (2005) reported that she did not publish on palaeobotany while at the museum (although there was at least two botanical paper; Wigglesworth 1905, 1937). Nothing is known of the subsequent history of the other Manchester woman palaeobotanist, Marjorie Lindsey.

Another two women Carboniferous palaeobotanists of this time were originally students of Nottingham University. There was a recognized Carboniferous palaeobotanist there at this time. H. S. Holden. Curiously, however, neither of the Carboniferous papers published by these women make any reference or acknowledgement to Holden, which would have been expected if they were students of his. Of these women, one (Rina Scott) was the one-time student and later wife of the eminent Carboniferous palaeobotanist D. H. Scott. The other, Emily Berridge, is an enigma as none of her papers (either palaeobotanical or botanical) that we have seen give any hint of where she was based or who she worked with.

Discussion

Women clearly played a major role in the development of Carboniferous palaeobotany during the first half of the 20th century. We have been able to identify 18 British women who published at least one scientific paper on this subject during this time. This compares with 33 British men: E. A. N. Arber, R. Crookall, D. Davies, H. Duerden, W. T. Gordon, W. Hemmingway, G. Hickling, H. S. Holden, A. R. Horwood, T. Johnson, R. Kidston, W. S. Lacey, W. H. Lang, D. G. Lillie, J. Lomax, A. G. Long, R. C. Mclean, A. J. Maslen, L. R. Moore, F. W. Oliver, T. G. B. Osborne, E. J. Salisbury, D. H. Scott, A. C. Seward, A. G. Tansley, D. Thoday, H. H. Thomas, R. D. Vernon, James Walton, John Walton, D. M. S. Watson, F. E. Weiss and G. Wild (drawn from relevant volumes of *Annals of Botany*, *New Phytologist* and *Botanical Journal of the Linnean Society*, and the bibliography at the end of Crookall, 1955–76).

In other words, well over one third of all British palaeobotanists studying the Carboniferous at this time were women – a truly remarkable figure given the prevailing attitudes towards 'career women'. Over half of the women went on to have an established career in botany or palaeobotany in either universities or museums, which is not dramatically less than the three-quarters of the above men who did the same. The real difference lies in the formal status that these women achieved within the scientific establishment. Only two women, Benson and Brenchley, achieved a senior position of professor or equivalent (i.e., 11% of the total women) in contrast to 15 male professors (Gordon, Holden, Lang, Mclean, Moore, Oliver, Osborne, Salisbury, Scott, Seward, Tansley, Thoday, Walton and Weiss; i.e., 42% of the total); in addition, D. M. S. Watson also became a professor, but of zoology.

The importance of women to Carboniferous palaeobotany does not just lie in the number of them involved at this time. Most of the truly innovative ideas and techniques that were developed at this time came from or were developed by women, including cuticle studies (Wills), early evolution of seeds and pollen organs (Benson), biostratigraphy (Dix), plant palaeoecology (Stopes), understanding morphological variation (Bolton), sphenophyte phylogeny (Browne), and acetate peels for studying petrifactions (Calder). If we compare this to the scientific output of the contemporaneous men, the women can be seen to be far ahead of their time.

Are there any common themes to the lives and careers of these women that help us understand their achievements? The most obvious is that very few ever married. Of the women for whom we have any biographical details, only one seems to have married (Bolton) and marriage marked the end of her career. However, we should not be surprised at this, given the social attitudes of the time. It would have been a problem for any woman trying to develop a career; the then expected duties of a wife, especially if children were involved, made any professional activity difficult, although not impossible given the right circumstances (cf. the career of geologist Maria Ogilvie; Burek 2006, as summarized by Creese & Creese 2006).

Two other important points are revealed by this study. The first is the importance of the university colleges that had been established during the late 19th century specifically for women, most notably Newnham College, in Cambridge, and Royal Holloway, Bedford and Westfield Colleges in London (see Burek 2007). Also in this class is

UCL, which although not an exclusively women's institution, was the first university in England to welcome women on equal terms with men; they had awarded full degrees to women since 1878, whereas Cambridge University, for instance, did not award full degrees to women until 1948. Nearly all of the women came through one or other of these institutions and, without them, it is difficult to see how they could have achieved what they did (see Burek 2007; Creese & Creese 2006, p. 67 on the importance of these institutions for the movement of women into geology as a whole).

Secondly, a number of male palaeobotanists of this time were particularly supportive of women's education and helped them to develop careers in academia. Without the help of these male colleagues, it is likely that many of the women palaeobotanists would have struggled to develop their careers. The name of F. W. Oliver of UCL, has regularly cropped up in the stories that we have told. W. H. Lang was also regarded as supportive of women in science (Watson 2005), although, for reasons that are unclear, he ultimately proved less successful than Oliver in helping the women who passed through his department to develop long-term careers. D. H. Scott was also supportive of women's education (Andrews 1980), but for most of the time covered in this paper he did not work within any formal institution and so was less able to provide practical support for women. Nevertheless, Scott clearly encouraged many women working in palaeobotany during this time, including his wife. Today, such a positive attitude towards helping women would not be seen as being too unusual, but things were very different prior to World War 2.

During the second half of the 20th century, Carboniferous palaeobotany again became a male-dominated subject. Carboniferous palaeobotany generally underwent something of a decline in the United Kingdom during this time; the attitude seemed to be that 'it had all been done' in the subject and there were more exciting plant fossils to study in other parts of the geological column. Nevertheless, a number of British men continued to be active in the field, most notably K. C. Allen, P. D. W. Barnard, R. M. Bateman, M. C. Boulter, W. G. Chaloner, C. J. Cleal, R. Crookall, T. De Witt West, H. J. Falcon-Lang, J. Franks, I. Glasspool, A. R. Hemsley, F. A. Hibbert, H. S. Holden, T. P. Jones, J. A. Knight, W. S. Lacey, A. G. Long, N. P. Rowe, A. C. Scott, C. H. Shute, B. A. Thomas and John Walton. While several women undergraduate and postgraduate students have undertaken Carboniferous palaeobotanical studies and published some papers (e.g., Denise Crampton, Alison Dytko, Kate Habgood, Alison Hill, Karen Logan, Helen Purdy, Gill Rex, Rachel Wood), none have subsequently become established in the field (Margaret Collinson published two papers on Carboniferous palaeobotany [Chaloner & Collinson 1975a & b] but her subsequently successful career was based mainly on work on Tertiary palaeobotany).

So, why did no British women develop careers in Carboniferous palaeobotany after World War 2? One reason may have been the demise of women-only university colleges; although Newnham College remains women only, Royal Holloway, Bedford (now part of Royal Holloway) and Westfield (now part of Queen Mary College) have become mixed gender. Because these institutions were organized in such a way as to facilitate the education of young women, they would undoubtedly also have provided an atmosphere that would have been more conducive to allowing women academics to develop their careers. The demise of such institutions after World War 2 meant that women now had mostly to compete with male colleagues for facilities, grants and status. The gender differences in approach to such matters would have placed women at a disadvantage in the male-dominated universities. It has been possible for some British women to succeed in palaeobotany, as is seen in the careers of Dianne Edwards, Joan Watson and Margaret Collinson, but they worked mainly in the Devonian/Silurian, Cretaceous and Tertiary, respectively. There were, however, no senior British women palaeobotanists still interested in Carboniferous plants by that time to help them: Benson had died, Dix had retired due to ill health, and Calder became inactive. This consequential lack of senior female supporters and role models in a male-dominated field can only have been discouraging to any young women looking at Carboniferous palaeobotany as a field to enter.

And finally, does the evidence presented in this paper support the suggestion given in the introduction that women did not fulfil their scientific potential within Carboniferous palaeobotany? There can be little doubt that these women could have achieved more in this field. However, we have found little evidence to suggest that women had been seriously hindered from developing as palaeobotanists by either personal or institutional prejudice. In many cases, especially those who were students of F. W. Oliver, palaeobotany was part of their education as botanists: to be done once but not further pursued. Palaeobotany just does not seem to have been regarded as a sufficiently attractive or important subject to draw them in. This is easy to understand in the case of Lucy Wills and Marie Stopes: their contributions to medical science and family planning, respectively, were far more important to the wider community than any work on Carboniferous plants could have been. The same can probably be said for the work by Winifred Brenchley on weed killers. However, although many of the other women continued to work in academic botany, they evidently failed to see the contribution that palaeobotany could make to understanding plants. Only the remarkable Isabel Browne, Emily Dix and Margaret Benson produced scientific papers that can be regarded as having had a significant long-term impact on the field, and any further expansion in their work was limited, in Browne's case, by her commitment to humanitarian work and an active social life (she was, after all, a wealthy 'amateur'; A. Arber 1947), and in Dix's case by serious ill health. The failure of Carboniferous palaeobotany to 'sell' itself as a worthwhile pursuit for men as well as women really reflects failures by the senior (and all male) academics then active in the field; to an extent, the subject still suffers from this today.

We are extremely grateful to a number of colleagues who have provided us with background information used in this paper: B. Chaloner and Margaret Collinson (Royal Holloway & Bedford New College), R. Clark (Bristol City Museum & Art Gallery), J. Clatworthy (University of Birmingham), Gina Douglas (Linnaean Society), Julie Douglas (Marie Stopes International), Dianne Edwards (University of Cardiff), H. Falcon-Lang (University of Bristol), J. Lister (Hunterian Museum), E. Jarzembowski (Maidstone Museum), Liz Loeffler (University of Bristol), P. Lyons, Elsie Morey, C. Shute (Natural History Museum, London), Anne Thomas and staff (Colwyn Bay Library), B. Thomas (University of Aberystwyth) and Joan Watson (Manchester University).

References

AINSWORTH, F. H. 1936. The anatomy of a cone-bearing axis of *Lepidodendron wortheni* Lesquereux. *Annals of Botany*, **50**, 511–517.
ANDREWS, H. N. 1980. *The Fossil Hunters. In Search of Ancient Plants*. Cornell University Press, Ithaca and London, 421 pp.
ARBER, A. 1910*a*. A note on *Cardiocarpon compressum*, Will. *Proceedings of the Cambridge Philosophical Society*, **15**, 393–394.
ARBER, A. 1910*b*. On the structure of the Palaeozoic seed *Mitrospermum compressum* (Will.). *Annals of Botany*, **24**, 491–509.
ARBER, A. 1913. On the structure of *Lepidostrobus laminatus* sp. nov. *Report of the Meeting of the British Association for the Advancement of Science (Dundee, 1912)*, 674.
ARBER, A. 1914*a*. An anatomical study of the Palæozoic cone-genus *Lepidostrobus*. *Transactions of the Linnean Society of London, Botany Series*, **8**, 205–238.
ARBER, A. 1914*b*. A note on Trigonocarpus. *Annals of Botany*, **28**, 195–196.
ARBER, A. 1947. Isabel Browne (obituary). *Proceedings of the Linnean Society*, **159**, 154–158.

ARBER, E. A. N. 1919. Remarks on the organization of the cones of *Williamsonia gigas* L. & H. *Annals of Botany*, **33**, 173–179.
ARBER, E. A. N. 1920. On the fossil flora of the southern portion of the Yorkshire Coalfield. Part II. Derbyshire. *Proceedings of the Yorkshire Geological Society*, **19**, 350–358.
ARBER, E. A. N. 1921*a*. *Devonian Floras: A Study of the Origin of Cormophyta*. University Press, Cambridge.
ARBER, E. A. N. 1921*b*. A sketch of the history of palaeobotany, with special reference to the fossil flora of the British Coal Measures. *In*: SINGER, C. (ed.) *Studies in the History and Method of Science*. Vol. 2. Clarendon Press, Oxford, 472–489.
ARBER, E. A. N. 1922. Critical studies of coal-measure plant-impressions. *Botanical Journal of the Linnean Society*, **46**, 171–217.
ARBER, M. A. 1968. List of published works of Agnes Arber, E. A. N. Arber and Ethel Sargant. *Journal of the Society for the Bibliography of Natural History*, **4**, 370–384.
BANCROFT, N. 1914. Pteridosperm anatomy and its relation to that of the cycads. *New Phytologist*, **13**, 41–68.
BANCROFT, N. 1915. A contribution to our knowledge of *Rachiopteris cylindrica*, Will. *Annals of Botany*, **29**, 531–565.
BARTHEL, M. 1961. Der Epidermisbau einiger oberkarbonischer Pteridospermen. *Geologie*, **10**, 828–849.
BARTHEL, M. 1962. Epidermisuntersuchungen an einige inkohlten Pteridospermenblättern des Oberkarbons und Perms. *Geologie*, **11**, 1–140.
BASSLER, H. 1919. A sporangiophoric lepidophyte from the Carboniferous. *Botanical Gazette*, **68**, 73–108.
BENSON, M. J. 1902*a*. The fructification of *Lyginodendron oldhamium*. *Annals of Botany*, **16**, 575–576.
BENSON, M. J. 1902*b*. A new lycopodiaceous seed-like organ. *New Phytologist*, **1**, 58–59.
BENSON, M. J. 1904. *Telangium scotti*, a new species of *Telangium* (*Calymmatotheca*) showing structure. *Annals of Botany*, **18**, 161–176.
BENSON, M. J. 1908. The sporangiophore – a unit of structure in the pteridophyta. *New Phytologist*, **7**, 143–149.
BENSON, M. J. 1911. New observations on *Botryopteris antiqua* Kidston. *Annals of Botany*, **25**, 1045–1057.
BENSON, M. J. 1912. *Cordaites felicis*, sp. nov. a cordaitean leaf from the lower coal measures of England. *Annals of Botany*, **26**, 201–207.
BENSON, M. J. 1914. *Sphærostoma ovale* (*Conostoma ovale et intermedium*, Williamson), a Lower Carboniferous ovule from Pettycur, Fifeshire, Scotland. *Transactions of the Royal Society of Edinburgh*, **50**, 1–15.
BENSON, M. J. 1918. *Mazocarpon* or the structural *Sigillariostrobus*. *Annals of Botany*, **32**, 569–589.
BENSON, M. J. 1920. Cantheliophorus, Bassler: new records of *Sigillariostrobus* (*Mazocarpon*). *Annals of Botany*, **34**, 135–137.
BENSON, M. J. 1922. *Heterotheca grievii*, the microsporange of *Heterotheca grievii*. *Botanical Gazette*, **74**, 121–142.
BENSON, M. J. 1933. The roots and habit of *Heterangium Grievii*. *Annals of Botany*, **47**, 313–315.

BENSON, M. J. 1935a. The fructification, *Calathiops Bernhardti*, n. sp. *Annals of Botany*, **49**, 155–160.

BENSON, M. J. 1935b. The new evidence of isospory in Palaeozoic seed plants. *New Phytologist*, **34**, 92–96.

BENSON, M. J. 1935c. The ovular apparatus of *Sphenopteridium affine* and *bifidum* and of *Diplopteridium (Sphenopteridium) teilianum* (Walton). *New Phytologist*, **34**, 232–244.

BERRIDGE, E. M. 1905. On two new specimens of *Spencerites insignis*. *Annals of Botany*, **19**, 273–280.

BERTRAND, P. 1930. Bassin houiller de la Sarre et de la Lorraine. I. Flore fossile. 1. Fascicule Neuroptéridées. Études des Gîtes Minéraux de la France, 58 pp.

BLACKWELL, E. M. 1937. Dr. Margaret Benson (Obituary). *Proceedings of the Linnean Society of London*, **1936–7** (4), 186–189.

BOLTON, E. 1919. Lower coal measures in relation to fossil plants and animals. *Proceedings of the Bristol Naturalists' Society, Fourth Series*, **5**, 30–38.

BOLTON, E. 1925. A critical study of certain species of the genus *Neuropteris* Brongn. *Journal of the Linnean Society, Botany*, **47**, 295.

BOLTON, E. 1926. Fossil flora of the Northumberland and Durham Coalfield. *Transactions of the Natural History Society of Northumberland, Durham and Newcastle-upon-Tyne, New Series*, **6**, 167–181.

BOLTON, E. 1929. On the cuticle of certain species of *Neuropteris* Brongn. *Annals of Botany*, **43**, 414–415.

BOLTON, E. & TUCK, M. C. 1921. The Carboniferous Limestone of the Wickwar–Chipping Sodbury area (Gloucestershire). *Abstracts of the Proceedings of the Geological Society of London*, **1063**, 30–31.

BOLTON, H. 1921. *A Monograph of the Fossil Insects of the British Coal Measures*. Paleontographical Society Monographs, **73**, 156 pp.

BRACK-HANES, S. D. & THOMAS, B. A. 1983. A re-examination of *Lepidostrobus* Brongniart. *Botanical Journal of the Linnean Society*, **86**, 125–133.

BRENCHLEY, W. E. 1913. On branching specimens of *Lyginodendron oldhamium*, Will. *Botanical Journal of the Linnean Society*, **41**, 349–356.

BROWNE, I. M. P. 1908. The phylogeny and interrelationships of the Pteridophyta. *New Phytologist*, **7**, 93–113, 150–166, 181–197, 230–253.

BROWNE, I. M. P. 1909. The phylogeny and interrelationships of the Pteridophyta. *New Phytologist*, **8**, 13–31, 51–72.

BROWNE, I. M. P. 1910. On a little known type of vascular structure re-described by P. Bertrand. *New Phytologist*, **9**, 188–191.

BROWNE, I. M. P. 1911. Graf Solms Laubach on the petioles attributed to *Cladoxylon* and *Medullosa (Steloxylon) ludwigi*. *New Phytologist*, **10**, 70.

BROWNE, I. M. P. 1913. A colonial contribution to our knowledge of the genus *Lycopodium*. *New Phytologist*, **12**, 222–225.

BROWNE, I. M. P. 1914. A new fern-like stem, described by Count Solms-Laubach. *New Phytologist*, **13**, 93–96.

BROWNE, I. M. P. 1923. Note sur les bractées de *Palaeostachya gracilis* Ren. *Bulletin du Muséum d'Histoire Naturelle*, **29**, 541–542.

BROWNE, I. M. P. 1925a. A note on *Calamostachys tuberculata* Stbg. *New Phytologist*, **24**, 305–308.

BROWNE, I. M. P. 1925b. Notes on the cones of the *Calamostachys* type in the Renault and Roche Collections. *Annals of Botany*, **39**, 313–358.

BROWNE, I. M. P. 1926. A new theory of the morphology of the Calamarian cone. *Annals of Botany*, **41**, 301–320.

BROWNE, I. M. P. 1933. The Noeggerathiae and Tingiae. The effects of their recognition upon the classification of the Pteridophyta: an essay and review. *New Phytologist*, **32**, 344–358.

BROWNE, I. M. P. 1935. Some views on the morphology and phylogeny of the leafy vascular sporophyte. *Botanical Review*, **1**, 383–404, 427–447.

BUREK, C. V. 2003. Catherine Raisin, a role-model professional geologist. *Geology Today*, **19**, 107–111.

BUREK, C. V. 2006. Dame Maria Matilda Ogilvie Gordon. *Teaching Earth Sciences*, **30**, 42–44.

BUREK, C. V. & CLEAL, C. J. 2005. The life and work of Emily Dix (1904–1972). *In*: BOWDEN, A., BUREK, C. V. & WILDING, R. (eds) *History of Palaeobotany: Selected Essays*. Geological Society, London, Special Publications, **241**, 181–196.

CALDER, M. G. 1933a. Note on the Kidston Collection of fossil plant slides. No. I. The anatomy of the axis of *Lepidodendron Brownii* Unger sp., with special reference to the relationship between this stem and *Lepidostrobus Brownii* Unger sp. *Transactions of the Royal Society of Edinburgh*, **57**, 547–555.

CALDER, M. G. 1933b. Note on the Kidston Collection of fossil plant slides. No. II. The anatomy of the axis of *Bothrodednron mundum* Williamson sp. *Transactions of the Royal Society of Edinburgh*, **57**, 665–673.

CALDER, M. G. 1934a. Note on the Kidston Collection of fossil plant slides. No. III. Some points on the anatomy of *Sigillaria elegans* Brongniart. No. IV. On the nature of the corona and its relationship to the leaf-traces in the Lepidodendreæ and Sigillariaæ, with special reference to certain 'diploxyoid' specimens in the Kidston Collection. *Transactions of the Royal Society of Edinburgh*, **58**, 49–62.

CALDER, M. G. 1934b. Note on the Kidston Collection of fossil plant slides. No. V. On the structure of two Lower Carboniferous lepidodendroid stems, one of the *Lepidophloios wünschianus* type and the other of the *Lepidodendron fuliginosum* type. No. VI. On the structure of two lepidodendroid stems from the Carboniferous flora of Berwickshire. *Transactions of the Royal Society of Edinburgh*, **58**, 113–124.

CALDER, M. G. 1935. Further observations on the genus *Lyginorachis* Kidston. *Transactions of the Royal Society of Edinburgh*, **58**, 549–559.

CALDER, M. G. 1938. On some undescribed species from the Lower Carboniferous flora of Berwickshire; together with a note on the genus *Stenomyelon* Kidston. *Transactions of the Royal Society of Edinburgh*, **59**, 309–331.

CALDER, M. G. 1953. The coniferous petrified forest of Patagonia. *Bulletin of the British Museum (Natural History), Geology Series*, **2**, 97–138.

CHALONER, W. G. 2005. The palaeobotanical work of Marie Stope. *In*: BOWDEN, A. J., BUREK, C. V. & WILDING, R. (eds) *History of Palaeobotany: Selected*

Essays. Geological Society, London, Special Publications, **241**, 127–135.
CHALONER, W. G. & COLLINSON, M. E. 1975a. An illustrated key to the commoner British Upper Carboniferous plant compression fossils. *Proceedings of the Geologists' Association*, **86**, 1–44.
CHALONER, W. G. & COLLINSON, M. E. 1975b. Application of SEM to a sigillarian impression fossil. *Review of Palaeobotany and Palynology*, **20**, 85–101.
CLEAL, C. J. 1978. Floral biostratigraphy of the upper Silesian Pennant Measures of South Wales. *Geological Journal*, **13**, 165–194.
CLEAL, C. J. & SHUTE, C. H. 1995. A synopsis of neuropteroid foliage from the Carboniferous and Lower Permian of Europe. *Bulletin of the British Museum (Natural History), Geology Series*, **51**, 1–52.
CLEAL, C. J. & THOMAS, B. A. 1995. *Palaeozoic Palaeobotany of Great Britain*. Geological Conservation Review Series, Chapman & Hall, London, **9**, 295 pp.
CLEAL, C. J., LAZARUS, M. & TOWNSEND, A. 2005. Illustrations and illustrators during the 'Golden Age' of palaeobotany: 1800–1840. *In*: BOWDEN, A. J., BUREK, C. V. & WILDING, R. (eds) *History of Palaeobotany: Selected Essays*. Geological Society, London, Special Publications, **241**, 41–61.
CREESE, M. R. S. 2004. Benson, Margaret Jane (1859–1936). *In*: HARRISON, B. (ed.) *Oxford Dictionary of National Biography*. Oxford University Press, Oxford.
CREESE, M. R. S. & CREESE, T. M. 2006. British women who contributed to research in the geological sciences in the nineteenth century. *Proceedings of the Geologists' Association*, **117**, 53–83.
CROOKALL, R. 1931. A critical revision of Kidston's Coal Measure floras. *Proceedings of the Royal Physical Society*, **22**, 1–34.
CROOKALL, R. 1932. The relative value of fossil plants in the stratigraphy of the Coal Measures. *Memoirs and Proceedings of the Manchester Literary and Philosophical Society*, **76**, 91–122.
CROOKALL, R. 1938. On floral subdivisions in the Yorkian Stage in South Wales. *Geological Magazine*, **75**, 346–352.
CROOKALL, R. 1955–76. *Fossil Plants of the Carboniferous Rocks of Great Britain* [Section 2]. Memoirs of the Geological Survey of Great Britain, Palaeontology, **4**, 1004 pp.
DARRAH, W. C. 1969. *A Critical Review of the Upper Pennsylvanian Floras of Eastern United States with Notes on the Mazon Creek Flora of Illinois*. W. C. Darrah, Gettysburg, 220 pp.
DAVIES, D. 1929. Correlation and palæontology of the Coal Measures in east Glamorganshire. *Philosophical Transactions of the Royal Society of London, Series B*, **217**, 91–153.
DE FRAINE, E. 1912. On the structures and affinities of *Sutcliffia*, in the light of a newly discovered specimen. *Annals of Botany*, **26**, 1031–1066.
DE FRAINE, E. 1914. On *Medullosa centrofelis*, a new species of *Medullosa* from the Lower Coal Measures. *Annals of Botany*, **28**, 251–264.
DIX, E. 1928. The Coal Measures of the Gwendraeth Valley and adjoining areas. *Proceedings of the South Wales Institute of Engineers*, **44**, 423–510.
DIX, E. 1934. The sequence of floras in the Upper Carboniferous, with special reference to South Wales. *Transactions of the Royal Society of Edinburgh*, **57**, 789–838.
DIX, E. 1935. Note on the flora of the highest 'Coal Measures' of Warwickshire. *Geological Magazine*, **72**, 555–557.
DIX, E. 1937. The succession of fossil plants in the South Wales Coalfield with special reference to the existence of the Stephanian. *2e Congrès Stratigraphie du Carbonifère, Heerlen, 1935 Compte rendus*, **1**, 159–184.
DIX, E. & PRINGLE, J. 1929. On the fossil Xiphosura from the South Wales Coalfield with a note on the myriapod *Euphoberia*. *Summary of Progress of the Geological Survey*, **1928** (2), 90–114.
DIX, E. & TRUEMAN, A. E. 1924. The correlation of the Coal Measures in the western portion of the South Wales Coalfield. Part II. The Coal Measures of north Gower. *Proceedings of the South Wales Institute of Engineers*, **40**, 353–383, 390–395.
DIX, E. & TRUEMAN, A. E. 1928. Marine horizons in the Coal Measures of South Wales. *Geological Magazine*, **65**, 356–363.
EGGERT, D. A. & DELEVORYAS, T. 1967. Studies of Paleozoic ferns: *Sermeya*, gen. nov. and its bearing on filicalean evolution in the Paleozoic. *Palaeontographica, Abteilung B*, **120**, 169–180.
FALCON-LANG, H. J. & MILLER, R. F. 2007. Marie Stopes and the Fern Ledges of Saint John, New Brunswick. *In*: BUREK, C. V. & HIGGS, B. (eds) *The Role of Women in the History of Geology*. Geological Society, London Special Publications, **281**, 227–245.
FENG, B.-C. & ROTHWELL, G. W. 1989. Microsporangiate cones of *Mazocarpon bensonii* (Lycopsida) from the Upper Pennsylvanian of the Appalachian Basin. *Review of Palaeobotany and Palynology*, **57**, 289–297.
FLORIN, R. 1925a. Zur Kenntnis der paläozoischen Pflanzengattung *Dolerophyllum* Saporta. *Svensk Botanisk Tidskrift*, **19**, 171–191.
FLORIN, R. 1925b. Zur Kenntnis einiger Cyclopteriden des Oberkarbons. *Geologiska Föreningens i stockholm Förhandlingar*, **47**, 223–244.
FLORIN, R. 1926. Über einige Blätterabdrücke vom *Cyclopteris*-Typus aus dem Karbon und Perm. *Arkiv für Botanik*, **20-A**, 1–19.
GEORGE, T. N. 1974. Fossil molluscs and molluscs in biostratigraphy: the geological work of A. E. Trueman. *In*: OWEN, T. R. (ed.) *The Upper Palaeozoic and Post-Palaeozoic Rocks of Wales*. University of Wales Press, Cardiff, 1–30.
GOOD, C. W. 1971. The ontogeny of Carboniferous articulates: *Calamostachys binneyana*. *Botanical Gazette*, **132**, 337–346.
GORDON, W. T. 1941. On *Salpingostoma dasu*: a new Carboniferous seed from East Lothian. *Transactions of the Royal Society of Edinburgh*, **60**, 427–464.
GOTHAN, W. 1941. Paläobotanische Mitteilungen. 5. Die Unterteilung der karbonischen Neuropteriden. *Palaeontologische Zeitschrift*, **22**, 421–438.
HALKET, A. C. 1930. The rootlets of *Amyelon radicans*, Will.: their anatomy, their apices and their endophytic fungus. *Annals of Botany*, **44**, 865–905.

HALKET, A. C. 1931. A note on the origin of lateral roots and the structure of the root-apex of *Lyginopteris oldhamia*. *New Phytologist*, **31**, 279–283.

HALLE, T. G. 1933. The structure of certain fossil spore-bearing organs believed to belong to pteridosperms. Kungliga Svenska Vetenskapsakademiens Handlingar, Series 3, **12** (6), 103 pp.

HOFFBRAND, A. V. & WEIR, D. G. 2001. The history of folic acid. *British Journal of Haematology*, **113**, 579–589.

HOLDEN, H. S. 1962. The morphology of *Botryopteris antiqua*. *Bulletin of the British Museum (Natural History), Geology Series*, **5**, 359–380.

HOWSE, R. 1890. Contributions towards a catalogue of the flora of the Carboniferous System of Northumberland and Durham. Part I. Fossil plants from the Hutton Collection. *Natural History Transactions of Northumberland, Durham, and Newcastle-on-Tyne*, **10**, 19–151.

HUTH, W. 1912. Die fossile Gattung *Mariopteris* in geologischer und botanischer Beziehung. *In*: POTONIÉ, H. (ed.) *Abbildungen und Beschreibungen fossiler Pflanzenreste (Lieferung VIII)*. Königlich Preussischen Geologischen Landesanstalt, Berlin, 1–88.

JENNINGS, J. R. 1976. The morphology and relationship of *Rhodea*, *Telangium*, and *Heterangium*. *American Journal of Botany*, **63**, 1119–1133.

JONGMANS, W. J. 1930. *Fossilium catalogues. II: Plantae. Pars 16: Lycopodiales III*. W. Junk, Berlin, 327–650.

JOY, K. W., WILLIS, A. J. & LACEY, W. S. 1956. A rapid cellulose peel technique in palaeobotany. *Annals of Botany, New Series*, **20**, 635–637.

KENRICK, P. & CRANE, P. R. 1997. *The Origin and Early Diversification of Land Plants. A Cladistic Study*. Smithsonian Institution, Washington DC, 441 pp.

KIDSTON, R. 1891. Notes on the Palæozoic species mentioned in Lindley and Hutton's 'Fossil Flora'. *Proceedings of the Royal Physical Society of Edinburgh*, **10**, 345–391.

KIDSTON, R. 1905. On the divisions and correlations of the upper portion of the Coal-Measures, with special reference to their development in the Midland counties of England. *Quarterly Journal of the Geological Society*, **61**, 308–321.

KIDSTON, R. 1908. On a new species of *Dineuron* and of *Botryopteris* from Pettycur, Fife. *Transactions of the Royal Society of Edinburgh*, **46**, 361–364.

KIDSTON, R. 1914. On the fossil flora of the Staffordshire Coal Fields. Part III. The fossil flora of the Westphalian Series of the South Staffordshire Coal Field. *Transactions of the Royal Society of Edinburgh*, **50**, 73–190.

KIDSTON, R. 1922. List of fossil plants from the Upper Carboniferous rocks of the Northumberland and Durham Coalfield and their bearing on the age of the coalfield. *Summary of Progress of the Geological Survey of Great Britain and the Museum of Practical Geology*, **1921**, 129–145.

LANG, W. H. 1926. A cellulose-film transfer method in the study of fossil plants. *Annals of Botany*, **40**, 710–711.

LAPWORTH, C. 1880. On the geological distribution of the Rhabdophora. Part III. Results. *Annals and Magazine of Natural History, Fifth Series*, **6**, 185–207.

LAVEINE, J.-P. 1967. Contribution a l'étude de la flore du terrain houiller. Les Neuroptéridées du Nord de la France. Études Géologiques pour l'Atlas Topographie Souterraine, **1** (5), 344 pp.

LINDLEY, J. & HUTTON, W. 1831–37. *The Fossil Flora of Great Britain*. W. J. Ridgeway and Sons, London, Vols 1–3, 640 pp.

LINDSEY, M. 1915. The branching and branch shedding of *Bothrodendron*. *Annals of Botany*, **29**, 223–230.

LONG, A. G. 1959. On the structure of *Calymmatotheca kidstoni* Calder (emended) and *Genomosperma latens* gen. et sp. nov. from the Calciferous Sandstone Series of Berwickshire. *Transactions of the Royal Society of Edinburgh*, **64**, 29–44.

LONG, A. G. 1960. On the structure of *Samaropsis scotica* Calder (emended) and *Eurystoma angulare* gen. et sp. nov., petrified seeds from the Calciferous Sandstone Series of Berwickshire. *Transactions of the Royal Society of Edinburgh*, **64**, 261–280.

MICKLE, J. E. & ROTHWELL, G. W. 1979. *Bensoniotheca*, a new name for *Heterotheca* Benson. *Taxon*, **28**, 591.

MOREY, E. D. 1995. William C. Darrah's European experience in 1935: palaeobotanical connections and stratigraphic consequences. *In*: LYONS, P. C., MOREY, E. D. & WAGNER, R. H. (eds) *Historical Perspectives of Early Twentieth Century Carboniferous Paleobotany in North America*. Geological Society of America, Memoir, **185**, 23–34.

MOREY, E. D. & LYONS, P. C. 1995. William Culp Darrah (1909–1989): a portrait. *In*: LYONS, P. C., MOREY, E. D. & WAGNER, R. H. (eds) *Historical Perspectives of Early Twentieth Century Carboniferous Paleobotany in North America*. Geological Society of America, Memoirs, **185**, 1–22.

NĚMEJC, F. 1928. A revision of the Carboniferous and Permian flora of the coal districts in Central Bohemia. *Palaeontographica Bohemica*, **12**, 1–82.

NĚMEJC, F. 1931. The morphology and systematic relations of the Carboniferous Noeggerathiae with regard to the 'genera' *Tingia* and *Plagiozamites* of eastern Asia. *Preslia*, **10**, 111–114.

NORTH, F. J. 1935. The Coal Forests – and David Davies, Gilfach Goch. *Proceedings of the South Wales Institute of Engineers*, **51**, 271–300.

OLIVER, F. W. & SCOTT, D. H. 1904. On the structure of the Palaeozoic seed *Lagenostoma Lomaxi*, with a statement of the evidence upon which it is referred to *Lyginodendron*. *Philosophical Transactions of the Royal Society of London, Series B*, **197**, 193–247.

PEARSON, H. L. 2005. Marie Stopes and a century of seed ferns. *The Linnean*, **21**, 25–28.

PHILLIPS, T. L. 1980. Stratigraphic and geographic occurrences of permineralized coal-swamp plants – Upper Carboniferous of North America and Europe. *In*: DILCHER, D. L. & TAYLOR, T. N. (eds) *Biostratigraphy of Fossil Plants*. Dowden, Hutchinson & Ross, Stroudsburg, 25–92.

PRANKERD, T. L. 1912. On the structure of the Palæozoic seed *Lagenostoma ovoides*, Will. *Botanical Journal of the Linnean Society*, **40**, 461–490.

PŠENIČKA, J. & BEK, J. 2003. Cuticles and spores of *Senftenbergia plumosa* (Artis) Bek and Pšenička from the Carboniferous of the Pilsen Basin, Bohemian Massif. *Review of Palaeobotany and Palynology*, **125**, 299–312.

ROE, D. A. 1978. Lucy Wills (1888–1964). A biographical sketch. *Journal of Nutrition*, **108**, 1379–1383.

ROSE, J. 1992. *Marie Stopes and the Sexual Revolution.* Faber & Faber, London.

ROTHWELL, G. W. 1986. Classifying the earliest gymnosperms. *In*: SPICER, R. A. & THOMAS, B. A. (eds) *Systematic and Taxonomic Approaches in Palaeobotany.* Systematics Association, London, Special Volume, **31**, 137–162.

ROTHWELL, G. W. & SCHECKLER, S. E. 1988. Biology of ancestral gymnosperms. *In*: BECK, C. B. (ed.) *Origin and Evolution of Gymnosperms.* Columbia University Press, New York, 85–134.

SCHEIHING, M. H. & PFEFFERKORN, H. W. 1980. Morphologic variation in *Alethopteris* (pteridosperms, Carboniferous) from St. Clair, Pennsylvania, USA. *Palaeontographica, Abteilung B*, **172**, 1–9.

SCHMID, R. 2001. Agnes Arber née Robertson (1879–1960): fragments of her life, including her place in biology and in women's studies. *Annals of Botany, New Series*, **88**, 1105–1128.

SCOTT, A. C., MATTEY, D. P. & HOWARD, R. 1996. New data on the formation of Carboniferous coal balls. *Review of Palaeobotany and Palynology*, **93**, 317–331.

SCOTT, D. H. *Structural Botany. Flowering Plants.* A&C Black, London.

SCOTT, D. H. 1900. *Studies in Fossil Botany*, 1st edn. A&C Black, London, 533 pp.

SCOTT, R. 1906. On the megaspore of *Lepidostrobus foliaceus*. *New Phytologist*, **5**, 116–119.

SCOTT, R. 1908. On *Bensonites fusiformis*, sp. nov., a fossil associated with *Stauropteris burntislandica* P. Bertrand, and on the sporangia of the latter. *Annals of Botany*, **22**, 683–687.

SCOTT, R. 1911. On *Traquairia*. *Annals of Botany*, **25**, 459–467.

SEWARD, A. C. 1914. Antarctic fossil plants. British Antarctic ('Terra Nova') expedition, 1910. *British Museum (Natural History), Report, Geology*, **1**, 1–49.

ŠIMŮNEK, Z. 2000. Cuticles of *Cordaites* from the Westphalian, Stephanian and Autunian of the Bohemian Massif (Czech Republic). *Acta Palaeobotanica*, **40**, 25–34.

SIMUNEK, Z. & BEK, J. 2003. Noeggerathiaceae from the Carboniferous basins of the Bohemian Massif. *Review of Palaeobotany and Palynology*, **125**, 249–285.

STEIN, W. E., WIGHT, D. C. & BECK, C. B. 1984. Possible alternatives for the origin of the Sphenopsida. *Systematic Botany*, **9**, 102–118.

STOPES, M. C. 1903a. On the leaf structure of *Cordaites*. *New Phytologist*, **2**, 92–98.

STOPES, M. C. 1903b. The 'epidermoidal' layer of calamite roots. *Annals of Botany*, **20**, 792–794.

STOPES, M. C. 1906a. A new fern from the Coal Measures: *Tubicaulis sutcliffii* sp. nov. *Memoirs and Proceedings of the Manchester Literary and Philosophical Society*, **50**, 1–34.

STOPES, M. C. 1906b. *The Study of Plant Life for Young People.* Blackie, London.

STOPES, M. C. 1907. Note on a wounded calamite. *Annals of Botany*, **21**, 277–280.

STOPES, M. C. 1910. *Ancient Plants.* Blackie, London.

STOPES, M. C. 1914. *The 'Fern Ledges' Carboniferous flora of St John, New Brunswick.* Geological Survey of Canada, Memoirs, **41**, 142 pp.

STOPES, M. C. & WATSON, D. M. S. 1908. On the present distribution and origin of the calcareous concretions in coal seams, known as 'Coal Balls'. *Philosophical Transactions of the Royal Society, London, Series B*, **201**, 1–90.

STOPES, M. C. & WHEELER, R. V. 1918. *The Constitution of Coal.* HMSO, London (on behalf of the Department of Scientific and Industrial Research).

THOMAS, B. A. 1986. *In Search of Fossil Plants: The Life and Work of David Davies (Gilfach Goch).* National Museum of Wales, Cardiff, Geological Series, **8**, 54 pp.

THOMAS, B. A. & BRACK-HANES, S. D. 1991. A re-examination of *Cantheliophorus* Bassler. *Neues Jahrbuch für Geologie und Paläontologie, Abhandlungen*, **183**, 203–215.

THOMAS, H. H. 1909a. On a cone of *Calamostachys binneyana* (Carruthers) attached to a leafy shoot. *New Phytologist*, **8**, 249–260.

THOMAS, H. H. 1909b. On a specimen of the cone *Calamostachys binneyana* (Carr.). *Proceedings of the Cambridge Philosophical Society*, **15**, 236–238.

THOMAS, H. H. 1910. On the assimilating tissue of some Coal Measure plants. *Proceedings of the Cambridge Philosophical Society*, **15**, 413–415.

THOMAS, H. H. 1928. Further observations on the cuticle structure of Mesozoic cycadean fronds. *Botanical Journal of the Linnean Society*, **48**, 389–415.

THOMAS, H. H. & BANCROFT, N. 1913. On the cuticles of some Recent and fossil cycadean fronds. *Transactions of the Linnean Society of London, 2nd Series, Botany*, **8**, 155–204.

TRIVETT, M. L. & ROTHWELL, G. W. 1991. Diversity among Paleozoic Cordaitales. *Neues Jahrbuch für Geologie und Paläontologie, Abhandlungen*, **183**, 289–305.

TRUEMAN, A. E. & WEIR, J. 1946–56. *A Monograph of British Carboniferous Non-marine Lamellibranchia.* Palaeontographical Society Monographs, 271 pp.

TUCK, M. C. 1926. The Avonian succession between Wickwar and Chipping Sodbury. *Proceedings of the Bristol Naturalists Society*, **6**, 237–249.

WAGNER, R. H. & WINKLER PRINS, C. F. 1979. The Lower Stephanian of western Europe. *8e Congrès International de Stratigraphie et de Géologie du Carbonifère, Moscow, 1975, Compte rendus*, **3**, 111–140.

WAGNER, R. H. & WINKLER PRINS, C. F. 1985. Stratotypes of the two lower Stephanian stages, Cantabrian and Barruelian. *10e Congrès International de Stratigraphie et de Géologie du Carbonifère, Madrid, 1983, Compte rendus*, **4**, 473–483.

WALTON, J. 1925. Carboniferous Bryophyta I. *Annals of Botany, New Series*, **39**, 563–572.

WALTON, J. 1928a. A method of preparing fossil plants. *Nature*, **122**, 571.

WALTON, J. 1928b. Carboniferous Bryophyta. II. Hepaticae and Musci. *Annals of Botany, New Series*, **42**, 707–716.

WATSON, J. 2005. One hundred and fifty years of palaeobotany at Manchester University. *In*: BOWDEN, A. J., BUREK, C. V. & WILDING, R. (eds) *History of Palaeobotany: Selected Essays*. Geological Society, London, Special Publications, **241**, 229–257.

WIGGLESWORTH, G. 1905. The papillae in the epidermoidal layer of the calamitean root. *Annals of Botany*, **18**, 645–648.

WIGGLESWORTH, G. 1937. South African species of *Riella*, including an account of the developmental stages of three of the species. *Botanical Journal of the Linnean Society*, **51**, 309–332.

WILDING, R. 2005. D. H. Scott and A. C. Seward: modern pioneers in the structure abd architecture of fossil plants. *In*: BOWDEN, A. J., BUREK, C. V. & WILDING, R. (eds) *History of Palaeobotany: Selected Essays*. Geological Society, London, Special Publications, **241**, 153–160.

WILLS, L. 1914. Plant cuticles from the coal-measures of Britain. *Geological Magazine*, **51**, 385–390.

ZEILLER, R. 1890. *Bassin houiller et permien d'Autun et d'Epinac: flore fossile*. Études deos Gîtes Minéraux de la France, **2** (1), 304 pp.

The role of women in British Quaternary science

J. K. HART

School of Geography, University of Southampton, Southampton, SO17 1BJ, UK
(e-mail: jhart@soton.ac.uk)

Abstract: This study represents the first attempt to record the lives of women who have contributed to the field of Quaternary science. A list of women who lived or studied in the United Kingdom and Ireland was drawn up from suggestions provided by members of the British Quaternary Research Association (QRA). These were divided into three categories: pre-20th century, 20th century and retired. The lives of some of these women are described. It was particularly interesting that in the pre-20th-century and 20th-century categories, the number of geomorphologists was equal to the number of archaeologists/palaeoecologists. In addition, a study was undertaken of women physical geographers in British geography departments, which was compared with the results from a similar survey in 1996. In this way, recent changes (the last 10 years) in women in Quaternary science could also be evaluated. It was shown that although there was an increase in the number of women physical geography academics since 1996, the actual percentage of women had gone down, and although the percentage of women professors had risen, men's chance of reaching this position was still much higher. However, it is argued that the key element of women in Quaternary science, both today and in the past, is their enthusiasm for fieldwork, and it is this passion that enabled them to continue their studies whatever the current social conventions.

The realization that huge ice sheets once covered the landscape of northern Europe and North America ('the Ice Age'), initially proposed by Louis Agassiz in 1837, and its associated implication concerning climate change, was one of the 'big discoveries' in geology over the last 200 years. Women have been very active in the study of this time period, the Quaternary, since those early days, including Elizabeth C. Cary (1822–1907) (Fig. 1), who married Louis Agassiz in 1850 and assisted him on his expeditions (including the Thayer Expedition to Brazil in 1865–66 and the Hassler Expedition through the Magellan Straits in 1871–72), as well opening a girls, school to fund his career, and later opening the 'Harvard Annex' for women (which became Radcliffe College) (Tharp 1959). However, the role of many women in Quaternary science has been underestimated because many were unable to pursue a professional academic career because of social barriers, but they still made a vital contribution to the subject.

Quaternary science covers all aspects of the environment and climate change over the last 2 million years. Subjects range from archaeology (in particular human evolution), palaeoecology and palaeoentology to geomorphology and sedimentology. In this paper, I start by investigating the role of women within the different fields of Quaternary science, describe the lives of some key women, and investigate the position of women in Quaternary science today. The aim is not to set up a historiography of 'famous' women in line with 'famous' men (as warned against by Monk 2004), but to celebrate women's successes and make the academic community aware of their contribution to the rich tapestry of Quaternary science.

Women in Quaternary science: the last 200 years

Although Quaternary science covers a large range of disciplines (geology, geography, environmental science, botany, zoology and archaeology), women have tended to be concentrated in a few of these. Women were not allowed to join the 'professional' organizations of the Geological Society of London until 1919 and the Royal Geographical Society until 1913. However, they were able to join 'amateur' societies, such as the Botanical Society of the British Isles from its beginning in 1836, and the Geologists' Association from its initiation in 1855.

There has been no specific study of women in Quaternary science today or in the past, because the subject spans so many different disciplines. Probably the closest recent survey was of women physical geographers in 1996 (Dumayne-Peaty & Wellens 1998). They showed the overall participation of women within British geography departments, and the particular fields of study in which these women were engaged. A similar study was

From: BUREK, C. V. & HIGGS, B. (eds) *The Role of Women in the History of Geology*. Geological Society, London, Special Publications, **281**, 83–95. DOI: 10.1144/SP281.5
0305-8719/07/$15.00 © The Geological Society of London 2007.

Fig. 1. Elizabeth C. (Cary) Agassiz (1822–1907), co-founder and first president of Radcliff College (now Radcliffe Institute for Advanced Study at Harvard) and wife and assistant to Louis Agassiz, early proponent of the 'Ice Age'. (From Tharp 1959).

carried out in the United States in 1995/6 by Luzzadder-Beach & Macfarlane (2004), who came to similar conclusions. Physical geography is similar to Quaternary science in that it covers a very broad range of environmental subjects, and so is a good proxy to use. The Dumayne-Peaty and Wellens survey also showed the numbers of postgraduate students in 1996 and their subject areas. It was decided to use this study as the basis for a series of comparative studies of women in the past, and in 2006 (10 years on).

In order to reconstruct the nature of women's participation in Quaternary science, a questionnaire was sent around to all members of the British Quaternary Research Association (QRA) (with over 1000 members) to ask them to contribute the names of female Quaternary scientists who had influenced them, either personally or through their research. A total of 58 names were provided, but it was decided to restrict this survey to the 49 women who were either born or studied in the United Kingdom or Ireland. This was because the QRA is based in the United Kingdom and so its members preferentially selected local scientists. Two examples of Dutch-speaking women Quaternary scientists that were suggested included Jacoba Hol (1886–1964), who was very influential in the development of physical geography and geomorphology in the Netherlands, and the first woman to be appointed to a chair at Utrecht University in 1946 (Hautvast 2002); and Antje Schreuder (1887–1952), a palaeontologist who worked as a museum curator at the Amsterdam Natural History Museum and Teylers Museum Haarlem and specialized in the study of Quaternary small vertebrates, particularly beavers (van Veen 2004).

The women included in this study were divided into three distinct groups (Table 1):

1. Pre-20th century – these were women that generally had no formal university education nor salaried position.
2. 20th century – these women ranged from the non-salaried assistant to salaried professional as the century progressed.
3. Retired – these women, who were slightly younger or contemporaneous with the second group, are still alive, but experienced similar life histories to the second group.

This list of names cannot possibly include everyone, and by the nature of the data collection, preferentially includes the most 'famous' and most recent. However, since it is the first time such a survey has been undertaken, it probably reflects a representative sample, and is a good beginning on which to build.

These women were then grouped into similar study areas to those suggested by Dumayne-Peaty & Wellens (1998). The botanical subjects of archaeology, palaeoecology and palaeontology are grouped together as archaeology/palaeoecology; sedimentology and geomorphology are grouped together as geomorphology; environmental management and geochemistry are grouped together under environmental management; and geochronology is classified under 'other'.

It is often assumed that women have been concentrated within the biological side of the subject; this was not found in this study. The women Quaternary scientists were involved in palaeoecology/archeology or geomorphology in equal numbers both during the pre-20th century and 20th century, although palaeoecologists predominate in the 'retired' category. This will be discussed in more detail below.

Pre-20th century

Prior to the 20th century, women's access to science was limited, and science was usually a 'hobby' for intelligent wealthy women. However, as Creese & Creese (2006) have described, British women were able to make a remarkable contribution to the subject of geology as a whole (given the

Table 1. *Women Quaternary scientists born or who studied in the United Kingdom and Ireland (suggested by Quaternary Research Association members)*

Pre-20th century	20th century	Retired
Dorothea Bate 1878–1951	Robin Andrews 1909–99	Hillary Birks
Charlotte Eyton 1839–1917	Muriel Arber 1913–2004	Anne Bonney
Elizabeth Hodgson 1813–77	Kathleen B. Blackburn 1892–1968	Anne Connolly
Mary Caroline Hughes 1862–1916	Monica Cole 1922–94	Camilla (Lambert) Dickson
Nina Layard	Shirley Coryndon 1926–76	Jill Eddison
Dorothy Liddell	Miller Dixon	Elizabeth Haworth
Eleanor Reid 1860–1953	Suzanne Lawless Duigan 1924–93	Sarah (Sally) Holland
Mrs S. Scott	Gillian E. Groom	Cuchlaine A. M. King
Gertrude Skeat 1865–1939	Jean Mary (Clark) Grove 1927–2001	J. P. Lishman
Mary Somerville 1780–1872	Alice Garnett	Joyce Brown Macpherson
Gertrude Caton-Thompson 1888–1985	Mary Leakey 1913–96	Margaret Marker
Mary Sydney Thompson ?–1923	Nora McMillan	Theya Molleson
	Hilda Parks ?–1994	Syvia Peglar
	Dorothy Peake 1915–2002	Winifred Pennington
	Catherine Raisin 1855–1945	Maura Scannell
	Dora Smee	Jane Soons
	Marjorie Sweeting 1920–1994	Dorothy Trotman
	Mabel Tomlinson	Judy Turner
		Sybil Watson

social barriers of the time), including Quaternary science. To illustrate this, the lives of three contrasting women are discussed below: Mary (Fairfax) Somerville, mathematician and physical geographer; Mary Caroline (Weston) Hughes, Quaternary molluscan expert; and Dorothea Bate, Pleistocene vertebrate palaeontologist (also see Shlinder 2007).

Mary Somerville (1780–1872). Although Mary (Fairfax) Somerville (1780–1872) (Fig. 2) had virtually no formal education (1 year at boarding school), she became interested in mathematics as a teenager (with the help and encouragement of family and friends). In 1804, Mary married Captain Samuel Greig and within 3 years had two children, but had little time or support to develop any scientific ideas. However, in 1807, she was widowed and, in 1812, married William Somerville, who encouraged her scientific endeavours. She then had four more children, translated Pierre-Simon Laplace's *Mécanique Céleste* under the title of *The Mechanism of the Heavens* (Somerville 1831) and went on to write three further books of her own. In 1848, she wrote *Physical Geography* which is the first British book with that title (Somerville 1848). This book was a critical success (with six editions) and was used in schools and universities until the beginning of the 20th century. Maddrell (2004) argues that this was an important academic advance in geography at a time when the Royal Geographical Society was preoccupied with exploration and colonial expansion. Mary Somerville was awarded a Victoria medal by the Royal Geographical Society in 1869, at a time when women could not be members. Neeley (2001) suggests she was able to attain high

Fig. 2. Mary (Fairfax) Somerville (1780–1872), author of the classic book *Physical Geography* (1848) (photo from http://www-history.mcs.st-andrews.ac.uk/history/PictDisplay/Somerville.html).

status within the 19th-century society 'through careful management of her gender identity' and by producing 'rich, readable, and authoritative accounts of science'.

Mary Hughes (1862–1916). Mary Caroline (Weston) Hughes (1862–1916), led a life more typical of late Victorian women interested in science. She was married to Thomas McKenny Hughes, who was the Woodwardian Professor of Geology (1873–1917) and who supervised the building of the Sedgwick Museum in Cambridge. She accompanied her husband on his geological excursions, and published with him (Hughes & Hughes 1890) and on her own (McKenny-Hughes 1888).

Dorothea Bate (1878–1951). Dorothea Bate (1878–1951) was one of earliest women to work for the Natural History Museum, London, and was a pioneer of the study of the Pleistocene fauna of the Mediterranean islands and the Middle East and one of the founders of modern archaeozoology. However, she also had little formal education. In particular, she researched the dwarf elephant (*Elephas cypriotes*) and dwarf hippopotamus (*H. minutus*), and an extinct goat-like antelope (*Myotragus*) (Bate 1901, 1902, Bate 1905; Garrod & Bate 1937). She also worked alongside many renown archaeologists, such as Louis and Mary Leakey.

Overview. These three women reflect the different ways in which women could participate in science within the pre- and early 20th century, and it is interesting that both Mary Somerville and Dorothea Bate have been the subject of recent biographies (Neeley 2001; Shindler 2005). The 'career' of Mary Caroline Hughes is probably the most typical of Victorian women geologists, although she was unusual in that she was able to write under her own name and with her husband. She was a positive influence over the growing number of women students studying geology at Cambridge (even if they were not allowed to graduate until 1920), by attending field trips and often acting as a chaperone to her husband's students (see Burek 2007). This resulted in a very active, Cambridge-based group of university-trained women geologists who were to have a very important influence in geology in the 1890s (Creese & Creese 2006). Mary Somerville was unusual in being able to pursue her own career as well as motherhood, while Dorothea Bate represents an early example of a professional woman at a time when the combination of motherhood and career was virtually impossible.

Gould (2000) shows how contemporary and later reports of both Mary Somerville and Mary Caroline Hughes stressed their 'femininity' as well as their scientific achievements. Delamont (1989) argues that this 'double conformity' approach was used by women to combat attacks from society that scientific activities were 'bad for women's health' or would 'masculinize' them, and used this to their own advantage. Mary Somerville was noted for remaining 'modest and diffident of her own powers' (Walford 1898), and Fawcett (1889) argued that a bust produced after her death (now in Girton College library) reflected that 'the fine and delicate lines of her beautiful face offer to the students of the College a worthy ideal of complete womanhood, in which intellect and emotion balance one another and make a perfect whole'. Mary Caroline Hughes was described by 'journalist' Eliza Linton in 1890 as follows:

> It is doubly delightful to meet with one who is not only an accomplished scientist, but also a charming woman – who can tell all about a fossil with a ponderous name, yet knows how to dress with perfection, and does not disdain feminine adornments. Learned women need such a beautiful example as our 'Professoress'. (Linton 1890)

Dorothea Bate was born later than the other two, and so had more opportunities open to her, but was still nevertheless a pioneer. She was not part of an established scientific or geological family and was probably only able to pursue her career due to strength of character as well being single. It is very interesting that the Natural History Museum in 2004–06 had a 'Dorothea Bate Gallery character', that is, an actress dressed as Ms Bates who 'wanders through the galleries sharing stories of her fossil finds with all she meets', which is advertised as family entertainment. This is in contrast to the previous century when she actually worked at the Museum, when she would probably have been regarded as unusual and not a suitable role model!

The 20th century

Women geologists in the 20th century began to have more opportunities. Women were awarded degrees and single women were beginning to have employment as academics, research assistants, technicians and museum curators (all roles covered in this volume). Here some examples are highlighted, including geomorphologists Marjorie Sweeting, Margaret Marker, Gillian Groom and Jean Grove; sedimentologist Mabel Tomlinson; archaeologist Mary Leakey; and palaeoecologists Kathleen Blackburn and Suzanne Duigan.

Marjorie Sweeting (1920–94). Marjorie Sweeting (1920–94) (Fig. 3) was probably the most well known of British women geomorphologists in the

Fig. 3. Marjorie Sweeting (1920–94) (photo from http://www.karst.edu.cn/exchange/images/fig-sweeting.htm).

20th century. Her most influential book *Karst Landforms*, (Sweeting 1972), was read by geomorphologists across the world, and during her lifetime she wrote over 70 publications and supervised over 30 postgraduates. On her retirement in 1987, the leading British geomorphological journal *Earth Surface Processes and Landforms* printed a list of her publications, and a special issue of *Zeitschrift für Geomorphologie* on 'Tropical and subtropical karst: essays dedicated to the memory of Dr Marjorie Sweeting' was published in 1997 (Williams 1997). She won a series of medals from the Royal Geographical Society, including the Gill Memorial medal in 1955 and the Busk Medal in 1980 (Viles 1996, 1997) and, after her death, the British Geomorphological Research Group Dissertation Prize was named in her honour. She spent most of her professional life as an academic in the Oxford Geography Department, and remained single.

Margaret Marker. One student who was influenced by Marjorie Sweeting was Margaret Marker, who went to South Africa in the 1960s, where she studied karst geomorphology and climate change (e.g., Marker & Sweeting 1975, 1983), and published over 22 publications on the subject. In 1996, she was awarded the Gold Medal of the Society of South African Geographers in recognition for her outstanding research.

Gillian Groom. Another academic influenced by Marjorie Sweeting was Gillian E. Groom, who worked in the University of Swansea Geography Department (Fig. 4). Her research included Spitsbergen glaciers (Sweeting & Groom 1956; Groom 1959), the raised shorelines of Gower (Groom 1971) and karst limestone (Sweeting *et al.* 1965).

Jean Grove (1927–2001). In contrast, Jean Mary (Clark) Grove (1927–2001) (Fig. 5), a contemporary of Marjorie Sweeting at Cambridge, had a career more typical of women during this period. Jean was able to combine motherhood with an active research and teaching career. In 1960, Jean was elected to a Fellowship at Girton College Cambridge (a women's college until 1979), where she became Director of Studies in Geography until her retirement in 1994. In this role, she was able to influence the careers of many students with her passion for geography and particularly fieldwork (Allen 2003). Her initial research was in glacial erosional processes in Jotunheimen,

Fig. 4. Gillian E. Groom (photo provided by Dr Colin Rouse).

Fig. 5. Jean Mary (Clark) Grove (1927–2001) (photo provided by Dick Grove).

Norway (Grove 1960). However, she is most widely known for her book *The Little Ice Age* (Grove 1988). This was the first detailed reconstruction of glacial fluctuations of the last thousand years from all over the world, and has become one of the most widely cited texts on recent climate change.

Mabel Tomlinson. Mabel Tomlinson was a renowned Quaternary sedimentologist and stratigrapher who worked in the Midlands but was an amateur geologist (Fig. 6). She received a BA Combined Arts in 1916, a BSc in Geological Science in 1922 and an MSc in 1923 (all three degrees from the University of Birmingham). In 1929 she was awarded her PhD and in 1936 she was awarded a DSc (both from the University of Birmingham). She was a schoolteacher who taught for over 40 years at the Yardley Grammar School, Birmingham (1917–1959). In 1961, she was awarded the Henry Stopes Memorial Fund and Medal of the Geologists' Association and the R. G. H. Worth prize of the Geological Society of London. Her key papers on the terraces of the Warwickshire Avon were based on research she carried out in her 'spare' time by bicycle (Tomlinson 1925, 1935).

Mary Leakey (1913–96). Mary (Nicol) Leakey (1913–1996) is probably the most reknowned of women Quaternary scientists during the 20th century. Although she attended lectures in geology and archaeology at the University of London, she did not attain a formal degree, and at the age of 17 years began writing to archaeologists to offer her services as an assistant. She was initially given a job by Dorothy Liddell, where she excavated and sketched the finds at Windmill Hill for publication. She was then invited to illustrate a book by Dr Gertrude Caton-Thompson *The Desert Fayoum* (Caton-Thompson & Gardner 1934), and then one by her future husband, Louis Leakey, *Adam's Ancestors* (Leakey 1953).

In 1934, Mary joined Dorothy Liddell's excavations of Hembury Fort, Devon, and later that year she began her own excavation near Clacton, Essex (Oakley & Leakey 1937). After this, Mary and her husband began excavations at the Olduvai Gorge in Tanzania, where they discovered some of the earliest hominid bones and tools, including a 1.75 Ma *Australopithecus boisei* skull, a *Homo habilis* skull and hand bones, and a 1 Ma *Homo erectus* cranium. At Laetoli in Tanzania she found 3.75 Ma hominid fossils, and between 1978

Fig. 6. Mabel Tomlinson (photo taken with permission from *Yardleian* (1959), magazine of the Yardley Grammar School, Birmingham).

Suzanne Duigan (1924–93). Suzanne (Lawless) Duigan (1924–93) (Fig. 7) was born in Australia and originally studied at the University of Melbourne before coming to the University of Cambridge to study for her PhD with Harry Godwin in the Subdepartment of Quaternary Research. On returning to Melbourne University, she took up a lectureship and carried out research on pollen morphology, Quaternary palynology of Lake Mountain and Macquarie Island, and the Tertiary palynology of the Victorian brown coals on which she worked closely with Isabel Cookson (Duigan 1951, 1966; Watson 2005). After her death, a special issue of the *Australian Journal of Botany* was dedicated to her memory (Kershaw 1997).

Overview. Women Quaternary scientists of the 20th century had more opportunities than their Victorian predecessors, but they were still very few in number. Mary Leakey had no formal university education nor academic position and worked alongside her husband in a similar way to Mary Caroline Hughes, but was able to develop her own reputation as a scientist. It is interesting that, in her early years, she was helped by women archaeologists from a previous (Victorian) generation. Mabel Tomlinson, in contrast, was able to have a career as a schoolteacher, but continued her research on Quaternary sedimentology. However, the most significant change was the fact that women were able to

and 1981 she found a 3.6 Ma bipedal hominid footprint trail left in ashes (*Australopithecus afarensis*). Mary Leakey was the author of numerous publications concerning her finds, and was awarded numerous honorary degrees and awards. In 1955, Mary and Louis Leakey were awarded the Stopes Medal from the Geologists' Association and in 1969 the Prestwich Medal from the Geological Society of London (Morell 1995).

Kathleen Blackburn (1892–1968). Kathleen Blackburn (1892–1968) worked in the Botany Department of Armstrong College (now the University of Newcastle) and introduced the technique of pollen analysis into the United Kingdom after visits from Swedish botanist G. Erdman in the 1920s (Lunn 2004). She developed microscopy for pollen analysis, as well as being the first person to discover sex chromosomes in plants (Blackburn & Harrison 1924). She reconstructed the late-glacial and post-glacial pollen record for Northumbria (Raistrick & Blackburn 1931a, b; Blackburn 1953) as well as working on the glaciation of the North Pennines with Arthur Raistrick (Raistrick & Blackburn 1938).

Fig. 7. Suzanne (Lawless) Duigan (1924–93) (photo from http://www.ctie.monash.edu.au/hargrave/duigan_suzanne_bio.html).

develop careers within academia, and a few even attained chairs.

The Subdepartment of Quaternary Research (formed in 1948) at Cambridge provided career opportunities for many women during the late 20th century, many of whom are now retired (see Table 1). Academics and PhD students included Anne Stevens, Judy Turner, Camilla (Lambert) Dickson and Hilary Birks. In addition, women had other roles within the organization, including research assistants/technicians who counted the fossil samples and curated the collections. These included Robin Andrews, Marie Ransom, Mary Pettit and Sylvia Peglar.

Retired 20th-century Women Quaternary Scientists

Winifred Tutin. Another group of women that need to be mentioned are the 'retired' group, many of whom are of the same generation as the women mentioned above. One woman who has played a vital role in Quaternary science during the 20th century is Winifred (Pennington) Tutin, who is the only female Quaternary Fellow of the Royal Society. She is renowned for her research on the ecological history of Lake Windermere, demonstrating a late-glacial oscillation in northern England for the first time (Pennington 1970). When she worked as a lecturer at the University of Leicester she taught Quaternary science as an interdisciplinary subject in the 1960s and 1970s. On her retirement a book was published in her honour on lake sediments (Haworth & Lund 1984).

Cuchlaine King. Another key retired Quaternary scientist is Cuchlaine King (1922–), renowned for her classic geomorphology text books (King 1972; Embleton & King 1975), as well as her research in glacial geomorphology (e.g., King & Ives 1954; King 1959). Sack (2004) records how she was initially taught at Cambridge by Frank Debenham, who was a member of Scott's Antarctica expedition in 1910–12, as well as glaciologist Vaughan Lewis. Once King had an academic position at the University of Nottingham, she organized her own expeditions to Iceland (beginning in 1953), Norway and other Arctic environments, and become Professor of Physical Geography in 1969.

Overview. It seems, because of their relatively high numbers, women archaeologists and palaeoecologists were able to provide a support network for other women. In particular, the women palaeoecologists at the Subdepartment of Quaternary Research provide such a model. A further example of this is the 'famous four' of palaeoecology: Winifred Tutin, Robin Andrews, Judith Turner and Anne Bonney, who worked together on Lake Windermere. However, such examples in geomorphology and sedimentology are rare. Except for the support network provided by Marjorie Sweeting for her students, most women geomorphologists tended to operate on their own and, apart from a very few exceptions, needed to be single and very strong willed to be able to be 'successful' within the academic sphere.

Women in Quaternary science today

In order to record the changes in women physical geographers in British geography departments since 1996 (Dumayne-Peaty & Wellens 1998), a study was made in 2006 from the geography department websites provided by the Royal Geographical Society. Although there have been a number of changes in geography departments over the last few years, with some amalgamation of geography and geology departments and geography and environmental science departments, these changes were quite small, and so a comparison over the last 10 years can be made. Figure 8 shows the fields in which women studied in 1996. Women tended to represent approximately 20% of physical geography academic staff in most fields. It can be seen that only in palaeoecology did women make up over 50% of the total postgraduates, and so it is in this area that we would expect improvements in 2006.

Between 1996 and 2006, the number of women physical geography academics in British university departments has risen from 107 in 1996 to 136 in 2006. However, as a percentage of the total staff, this is a decrease: women represented 21% of all physical geography academic staff in 1996, but only 18% in 2006. In addition, these figures hide a huge variation, with some departments having no women physical geography academic staff (e.g., Aberdeen) and one department having 50% (Birkbeck College). Dumayne-Peaty & Wellens (1998) also suggested that women were clustered in the post-1992 universities. However, in 2006, there was no overall relationship between the number of women in a department and the 2001 Research Assessment exercise rating, although the least women are found in the highest grade (grade 5*) (Fig. 9).

Figure 10 shows the changing status of women over the last 10 years. Whereas women were preferentially clustered at the lowest levels 10 years ago, the number of women professors has now risen. However, in 2006, 23% of all male physical geography academics, but only 15% of women physical geography academics, had chairs.

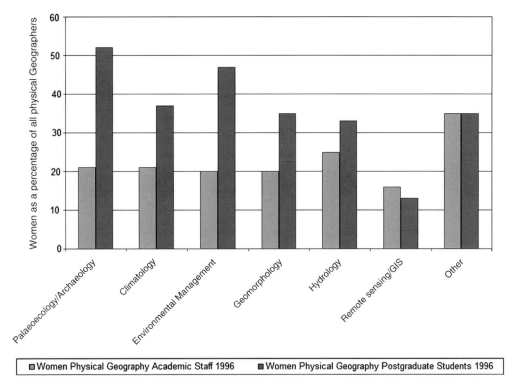

Fig. 8. Fields of study of women physical geography academic staff and postgraduate students in British university geography departments in 1996 (from Dumayne-Peaty & Wellens 1998) (women as a percentage of all physical geographers).

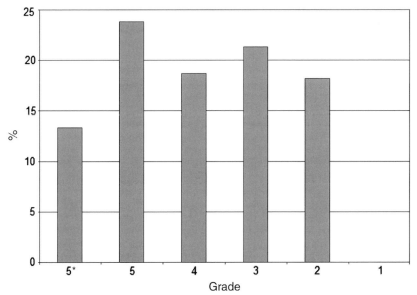

Fig. 9. Percentage of women physical geography academic staff against 2001 Research Assessment Exercise grade (5*, highest grade; 1, lowest grade).

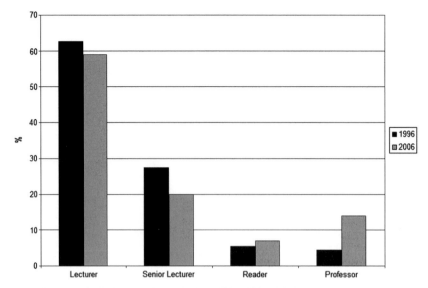

Fig. 10. Status of women physical geography academic staff in 1996 and 2006, as a percentage of the total women physical geography academics.

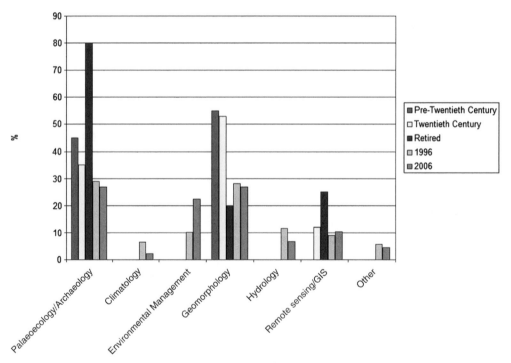

Fig. 11. Changes in women's research areas as a percentage of the total women in the survey for each time period. ('Pre-20th century', '20th century' and 'Retired' based on figures from QRA survey; '1996' figures from Dumayne-Peaty & Wellens (1996); '2006' figures from university geography department websites).

Figure 11 shows the areas of study over the last 200 years (as a percentage of the total at the five different time zones). From this a number of conclusions can be made:

1. Palaeoecology/Archeology has not always been the dominant field for women in Quaternary science, and during the pre-20th and 20th centuries it was equal with geomorphology. Palaeoecologists predominate in the 'retired' category, this is either because palaeoecologists live longer or (more likely) these figures reflect an increase in women in this field through the later part of the 20th century.
2. Over time, as the subject areas have broadened, so the percentage of women studying palaeoecology/archaeology and geomorphology have fallen in number, although the subjects still remain popular. In particular, in recent years there has been a large rise in the number of women researching environmental management and geochemistry.
3. Unfortunately, the high numbers of postgraduate students in palaeoecology from 1996 did not become academics. This is part of the wider problem in science, where women are 'leaking' from the system at all levels but particularly at the postgraduate/post-doctoral to lecturer transition (Rosser 2004).

Discussion

An investigation of the role of women within Quaternary science has shown how women were a vital, but often underestimated component. Although it was assumed the women would initially preferentially cluster on the botanical side of the subject, as this was considered a suitable hobby for Victorian women, this study has shown that women were active in both the botanical and the geomorphological side of the subject, and only in recent years have women been able to branch out into newer fields, such as remote sensing and environmental management.

For geography, it has been argued (mostly by human geographers) that fieldwork in physical geography is unattractive to women and 'remains a masculinist endeavour' (Nairn 1996; Madge & Bee 1999), and although this has been countered by Bracken & Mawdsley (2004), few women physical geographers have addressed this issue. The results from the study of the women in Quaternary science can contribute to this debate. This study suggests that it is the attraction of fieldwork that has enabled the women in this study to undertake their research. This research was often done as part of a family enterprise and/or under conditions which were thought unsuitable for women at the time. It is the very fact that they wanted to study the environment that made them geologists/Quaternary scientists/physical geographers; otherwise they would have chosen a different field of study. Sack (2004) also refers to this in her study of women physical geographers from the 1980s, where she quotes Cuchlaine King as saying 'a real appreciation of the value and beauty of the landscape' was essential for a successful career in geomorphology.

Although, women's situation has improved since Victorian times, there are still barriers to women's careers, as illustrated above. The problem of 'double conformity' is still an issue (Madge & Bee 1999). Women are still judged by their appearance, and fieldwork in Quaternary subjects is still regarded as an unusual interest for women (by society), while the women themselves regard the fieldwork as one of its major attractions.

Conclusion

This study is an important first step in the rediscovery of women who have made a contribution to Quaternary science. It has shown that they were active in both the botanical and geomorphological side of the subject, although more recently there seems to have been a stronger support network on the botanical side. The lives of a number of these women have been outlined, with unpaid family assistants dominating in the early years and professional women in the later. Using women physical geography academics in British university geography departments as a proxy for women Quaternary scientists, the changes over the last 10 years could be documented. Although, the number of women physical geographers has increased and there are improvements in women's academic status, these changes are very slow and women are still underrepresented in the subject.

I would like to thank Cynthia Burek for inviting me to be involved in this project and her continued support at all stages of manuscript preparation; to the members of the QRA who sent me information; and Stewart Campbell for helpful suggestions to the text.

References

BATE, D. M. A. 2003. Jean Grove. World Wide Web Address: http://www.quaternary.group.cam.ac.uk/history/others/Grove.html.
BATE, D. M. A. 1901. A short account of a bone cave in the Carboniferous limestone of the Wye valley. *Geological Magazine*, **8**, 101–6.
BATE, D. M. A. 1902. Preliminary note on the discovery of a pigmy elephant in the Pleistocene of Cyprus.

Proceedings of the Royal Society of London, **71**, 498–501.

BATE, D. M. A. 1905. Further note on the remains of *Elephas cypriotes* from a cave-deposit in Cyprus. *Philosophical Transactions of the Royal Society*, **197B**, 347–60.

BLACKBURN, K. B. 1953. A long pollen diagram from Northumberland. *Transactions of the Northern Naturalists Union*, **2**, 40.

BLACKBURN, K. B. & HARRISON, J. W. H. 1924. A preliminary account of the chromosomes and chromosome behaviour in the Salicaceae. *Annals of Botany*, **38**, 361–378.

BRACKEN, L. & MAWDSLEY, E. 2004. 'Muddy glee': rounding out the picture of women and physical geography fieldwork. *Area*, **36**, 280–286.

BUREK, C. V. 2007. The role of women in geological higher education – Bedford College, London (Catherine Raisin) and Newnham College, Cambridge, UK. *In*: BUREK, C. V. & HIGGS, B. (eds) *The Role of Women in the History of Geology*. Geological Society, London, Special Publications, **281**, 9–38.

CATON-THOMPSON, G. & GARDNER, E. W. 1934. *The Desert Fayum*. Royal Anthropological Institute, London, 456 pp.

CREESE, M. R. S. & CREESE, T. M. 2006. British women who contributed to research in the geological sciences in the nineteeth century. Proceedings of the Geologists' Association, **117**, 53–83.

DELAMONT, S. 1989. *Knowledgeable Women: Structuralism and the Reproduction of Elites*. Routledge, London, 398 pp.

DUIGAN, S. L. 1951. Catalogue of the Australian Tertiary Flora. Proceedings of Royal Society of Victoria, **69**, 5–13.

DUIGAN, S. L. 1966. The nature and relationships of the Tertiary brown coal flora of the Yallourn area in Victoria, Australia. *The Palaeobotanist*, **14**, 191–201.

DUMAYNE-PEATY, L. & WELLENS, J. 1998. Gender and physical geography in the United Kingdom. *Area*, **30**, 197–205.

EMBLETON, C. & KING, C. A. M. 1975. *Glacial and Periglacial Geomorphology*. Vol. 2. Edward Arnold, London, 203 pp.

FAWCETT, M. G. 1889. *Some Eminent Women of Our Time: Short Biographical Sketches*. MacMillan, London.

GARROD, D. A. E. & BATE, D. M. A. 1937. The Stone Age of Mount Carmel. *Quarterly Journal of the Geological Society of London*, **96**, 567–74.

GOULD, P. 2000. The Use of Role Models to Promote Women's Work in Science. *British Society for the History of Science, Science Communication, Education, and the History of Science*, conference at the Royal Society, London: 12–13 July. www.bshs.org.uk/conf/2000 science comm/papers/gould.doc

GROOM, G. E. 1959. Niche glaciers in Bünsow Land, Vestspitsbergen. *Journal of Glaciology*, **3**, 369–375.

GROOM, G. E. 1971. Geomorphology. *In*: BALCHIN, W. G. V. (Ed.) *Swansea and Its Region*. British Association for the Advancement of Science, Swansea, 29–40.

GROVE, J. M. 1960. A study of Veslgjuv-breen. *In*: LEWIS, W. V. (ed.) *Norwegian Cirque Glaciers*. R.G.S. Research Series, London, **4**, 69–82.

GROVE, J. M. 1988. *The Little Ice Age*. Cambridge University Press, Cambridge.

GROVE, J. M. & SWITSUR, R. 1994. Glacial geological evidence for the medieval warm period. *Climatic Change*, **26**, 143–169.

HAUTVAST, S. 2002. Wegwijzer in de wetenschap: prof. dr. Jacoba Hol (1886–1964). Nijmegen, Verloren, 78 pp.

HAWORTH, E. Y. & LUND, J. W. G. 1984. *Lake Sediments and Environmental History*. Leicester University Press, Leicester, 411 pp.

HUGHES, T. M. & HUGHES, M. C. 1909. *Cambridgeshire*. Cambridge County Geographies, Cambridge University Press, Cambridge, 256 pp.

KERSHAW, A. P. 1997. (ed.) Australasian Palaeoclimates: Refinement of Estimates from Palaeobotanical data (Volume dedicated to the memory of S. L. Duigan). *Australian Journal of Botany*, **45** (3), 359–617.

KING, C. A. M. 1959. Geomorphology in Austerdalen, Norway. *The Geographical Journal*, **125**, 357–369.

KING, C. A. M. 1972. *Beaches and Coasts*. Edward Arnold, London, 570 pp.

KING, C. A. M. & IVES, J. D. 1954. Glaciological observations on some of the outlet glaciers of south-west Vatnajökull, Iceland, 1954. Part I: Glacier regime. *Journal of Glaciology*, **2** (18), 563–569.

LEAKEY, L. 1953. *Adam's Ancestors: The Evolution of Man and His Culture*. Harper and Row, New York.

LINTON, E. L. 1890. A Pleasure Party. *Queen*, 2 August. [Cutting held in Sedgwick Club Archive, Earth Sciences Library, University of Cambridge.]

LUNN, A. 2004. *Northumberland*. New Naturalist Series, Collins, London, 328 pp.

LUZZADDER-BEACH, S. & MACFARLANE, A. 2004. The environment of gender and science: status and perspectives of women and men in physical geography. *Professional Geographer*, **52** (3), 407–424.

MADDRELL, A. 2004. Complex locations: historiography, feminism and difference, *Geography and Gender Reconsidered*, Women and Geography Study Group CD, 84–94.

MADGE, C. & BEE, A. 1999. Women, science and identity: interviews with female physical geographers. *Area*, **31**, 335–38.

MARKER, M. E. & SWEETING, M. M. 1975. Large scale subsurface weathering features (pinnacles) in the northern Cape Province. *Actes Comm. Phenomen. Karstiques Collogue*, 53–69.

MARKER, M. E. & SWEETING, M. M. 1983. Karst development on the Alexandria Coastal Limestone, Eastern Cape Province, S.A. *Zeitschrift fur Geomorphologie*, **27**, 21–38.

MCKENNY-HUGHES, Mrs T. 1888. On the Mollusca of the Pleistocene gravels in the neighbourhood of Cambridge. *Geological Magazine*, **25**, 193–199.

MONK, J. 2004. Women, gender, and the histories of American geography. *Annals of the Association of American Geographers*, **94**, 1–22.

MORELL, V. 1995. *Ancestral Passions: The Leakey Family and the Quust for Humankind's Beginnings*. Simon and Schuster, New York.

NAIRN, K. 1996. Parties on geography field trips: embodied fieldtrip. *New Zealand Women's Studies Journal*, **12**, 88–97.

NEELEY, K. A. 2001. *Mary Somerville: Science, Illumination and the Female Mind*. Cambridge University Press, Cambridge, 280 pp.

OAKLEY, K. P. & LEAKEY, M. 1937. Report on excavations at Jaywick Sands, Essex (1934) with some observations on the Clactonian industry and on the fauna and geological significance of the Clacton Channel. *Proceedings of the Prehistoric Society*, **3**, 217–260.

PENNINGTON, W. 1970. Vegetation history in the northwest of England: a regional synthesis. *In*: WALKER, D. & WEST, R. (eds) *Studies in the Vegetation History of the British Isles*. Cambridge University Press, Cambridge, 41–79.

RAISTRICK, A. & BLACKBURN, K. B. 1931a. The late-glacial and post-glacial period in the north Pennines. III. The post-glacial peats. *Transactions of the Northern Naturalists Union*, **1**, 79.

RAISTRICK, A. & BLACKBURN, K. B. 1931b. Pollen analysis of the peat on Heathery Burn Moor, Northumberland. *Proceedings of the University of Durham Philosophical Society*, **8**, 351.

RAISTRICK, A. & BLACKBURN, K. B. 1938. Linton Mires, Wharfedale. Glacial and post-glacial history. *Proceedings of the Durham Philosophical Society*, **10**, 24.

ROSSER, S. V. 2004. *The Science Glass Ceiling*. Routledge, New York, 192 pp.

SACK, D. 2004. Experiences and viewpoints of selected women geomorphologists from the mid-20th century. *Physical Geography*, **25**, 438–452.

SHINDLER, K. 2005. *Discovering Dorothea: The Life of the Pioneering Fossil-Hunter Dorothea Bate*. HarperCollins, London, 309 pp.

SHINDLER, K. 2007. A knowledge unique: the life of the pioneering explorer and palaeontologist, Dorothea Bate (1879–1951). *In*: BUREK, C. V. & HIGGS, B. (eds) *The Role of Women in the History of Geology*. Geological Society, London, Special Publications, **281**, 295–303.

SOMERVILLE, M. 1831. *Mechanism of the Heavens*, London, 816 pp.

SOMERVILLE, M. 1848. *Physical Geography*, London, **1**, 432 pp.; **2**, 476 pp.

SWEETING, M. 1972. *Karst Landforms*. London, MacMillan, 362 pp.

SWEETING, M. & GROOM, G. 1956. Notes on glacier fluctuations in Bünsow Land, central Vestspitsbergen. *Journal of Glaciology*, **19**, 640–41.

SWEETING, M. M., GROOM, G. E., WILLIAMS, V. H., PIGOTT, C. D., INGLE-SMITH, D. & WARWICK, G. T. (eds) 1965. Denudation in Limestone Regions: A Symposium. *Geographical Journal*, **131**, 34–56.

THARP, L. H. 1959. *Adventurous Alliance: The Story of the Agassiz Family of Boston*, Little, Brown & Co., Boston.

TOMLINSON, M. E. 1925. River terraces of the lower valley of the Warwickshire Avon. *Quarterly Journal of the Geological Society of London*, **81**, 136–63.

TOMLINSON, M. E. 1935. The superficial deposits of the country north of Stratford on Avon. *Quarterly Journal of the Geological Society of London*, **91**, 425.

WALFORD, L. B. 1898. *Four Biographies*. Blackwood, London.

WATSON, J. 2005. One hundred and fifty years of palaeobotany at Manchester University. *In*: BOWDEN, A. J., BUREK, C. V. & WILDING, R. (eds) *History of Palaeobotany*. Geological Society, London, Special Publications, **241**, 229–258.

WILLIAMS, P. W. (ed.) 1997. Tropical and Subtropical Karst: Essays Dedicated to the Memory of Dr. Marjorie Sweeting. *Zeitschrift für Geomorphologie, Supplementbände*, **108**.

VAN VEEN, J. C. 2004. The history of the palaeontological – mineralogical cabinet of the Teylers Museum, Haarlem, The Netherlands. *In*: VELDMEIJER, A. J., VAN ROODE, S. M. & HENSE, A. M. (eds) *The PalArch Foundations Newsletter*, **1**, 7–20.

VILES, H. A. 1996. Obituary: Marjorie Sweeting 1920–1994. *Transactions of the Institute of British Geographers*, **21**, 429–432.

VILES, H. A. 1997. A lifetime of landforms: Marjorie Sweeting's work on tropical and subtropical karst. *In*: WILLIAMS, P. W. (ed.) *Tropical and Subtropical Karst: Essays Dedicated to the Memory of Dr. Marjorie Sweeting*. *Zeitschrift für Geomorphologie, Supplementbände*, **108**, 1–4.

The quiet workforce: the various roles of women in geological and natural history museums during the early to mid-1900s

PATRICK N. WYSE JACKSON[1] & MARY E. SPENCER JONES[2]

[1]*Department of Geology, Trinity College, Dublin 2, Ireland (e-mail: wysjcknp@tcd.ie)*
[2]*Department of Zoology, Natural History Museum, Cromwell Road, London, SW7 5BD, UK (e-mail: msj@nhm.ac.uk)*

Abstract: When examining the work of women in geology during the 18th and 19th centuries, one can broadly, and perhaps crudely, divide those women with geological interests into two broad groups: firstly, the geological wife, sister, or daughter, and the museum assistant; and secondly the museum user, the academic and the museum research scientist. It was not until the close of the 19th century that women began to have a role, albeit minor, in museum education. Women typically were employed in the major national or university museums as preparators, illustrators or assistants, and this trend continued until the 1930s. These women received little academic credit for their research as it was frequently incorporated into the publications of the men for whom they worked. Adelaide Quisenberry worked for Ray Bassler at the Smithsonian Institution in Washington DC, while Edith Goodyear was employed as an assistant to Edmund Garwood at University College, London. Edith later published an important paper in 1924, co-authored with Garwood. From 1964, Veronica Burns, a women of a very modest character, spent 17 years at Trinity College, Dublin, where as curator she diligently amassed many important specimens (some of which were used by colleagues in publications) and published two papers in her own right. By the 1930s, women were being appointed to serious research positions in some museums – among the earliest such appointment was that of Helen Muir-Wood, a specialist in Upper Palaeozoic brachiopods to the staff of the British Museum (Natural History). Anna Birchall Hastings was also appointed by the same institution to work on Recent bryozoans, but upon her marriage was required to relinquish her post, even though she had married a museum palaeontologist; he remained in post while she became a volunteer.

Introduction

In the 1850s and 1890s there was a rather romantic view of the subjects to which women could apply themselves. Natural history was one such subject that attracted many women, and they joined natural history and field clubs in their thousands, and this period was characterized by the craze of collecting ferns (Allen 1969). Other topics investigated by women included mapping the distribution of glacial erratics (Higgs & Wyse Jackson 2007), algae and Lepidoptera, and a number of female authors, such as Margaret Gatty (1809–73) (Rauch 2004) and Arabella Buckley, produced books on natural history topics that sold in their thousands and appealed in particular to the female reader (Barber 1980; Burek 2007). Gatty's *Parables from Nature* (1855–71), which ran to five volumes and many editions, was in this genre, while her book on British algae was a serious account of that subject (Gatty 1863).

However, the scientific efforts and contributions made by women at this time were often ignored or scorned by the established male-dominated scientific organizations. Visually the labours/research of these women was presented in one of two ways. Firstly, they could be portrayed in a dream-like diaphanous manner, as exemplified in the painting *Nature Study* by William Stephen Coleman (1829–1904), which was subsequently engraved by the celebrated Chartist William James Linton (1812–97) (see Barber 1980, p. 135). Coleman wrote on natural history and illustrated various books written by other people, including a number of the Reverend. John's books, but he is now best known in most circles as a talented ceramic artist who established Minton's Art Pottery Studio in Kensington Gore, close to the British Museum (Natural History) (Baker 2004). A second type of representation of the scientific role of women was to portray them in a light-hearted (although today we would call this a rather sexist) way, which probably reflected their thoughts that their activities were of little scientific value, but kept the men amused. The cartoonist John Leech (1817–64) who started life as a medical student but left the medical profession to follow an artistic career, submitted over 3000 drawings to the magazine *Punch*. One of his better-known drawings appeared in 1857 and was entitled *Common Objects at the*

Seaside. Here Leech depicts a beach being combed by two men and fourteen women, the latter in wide skirts. One of the former is holding a collecting net. Perhaps Leech was commenting on the intellect of the women: he may have considered that women would find it difficult to look for interesting specimens and use the collecting equipment at the same time. Perhaps this conclusion is fanciful; there is no empirical evidence to suggest that women made better or worse collectors of natural history specimens than did their male counterparts.

By the final decades of the 19th century women were beginning to gain a foothold in natural history museum circles. While they had for some considerable time before this donated collections to these museums (see Cleevely 1983), they were finally allowed to partake in museum work or research on material that was housed, or subsequently deposited, in museums. For convenience in this paper we have divided these women into five categories:

1. the geological wife/sister/daughter;
2. the museum assistant aide (whether paid or not, or volunteer;
3. women who used museums for research but who had no particular institutional affiliation;
4. the academic;
5. the museum research scientist.

We can place these categories into two larger groups. The first groups comprises categories one and two. Here women were most active between 1900 and the 1930s, although some continued in these 'subservient' roles into the last decades of the 20th century. The second group comprises categories three to five, in which women working between the 1920s and 1950s broke free and developed their own research or academic careers. In general these categories illustrate: firstly, a stepwise evolution from museum or research assistant to independent research scientist; secondly, a slow acceptance (by their male colleagues) of the scientific work of women in museums; and finally, scientific independence of women from male museum workers.

These groupings, developed below, cannot be said to be based on any formal museum structures and, as can be seen, the boundaries between them were often open and fuzzy, and some women moved from group to group at various times in their lives. In order to illustrate the conditions of employment, we have chosen to highlight the work of a number of individuals in each grouping. It would be a mammoth and encyclopaedic task to document the work of all museum women in the period under review. [See Orr 2007, on the situation in France at the beginning of the 19th century.]

The quiet workforce

The geological wife/sister/daughter

Pre-1900. Many female relatives of male scientists have contributed to the scientific output of their menfolk, be it in a supportive (hidden) role, providing meals and taking responsibility for rearing the children, or in a more visible way. In the geological world it was Charlotte Murchison (1788–1869) who pushed her husband Roderick to take the first tentative steps on the road that led to him becoming Director of the Geological Survey of Great Britain in 1855, and she accompanied her husband and Charles Lyell on geological trips, when she would have collected material and drawn specimens (Kölbl-Ebert 2007). For 33 years, until her death in 1862, John Phillips relied on his sister Anne for domestic support as well as assistance in the field (Morgan 2006). Mary Buckland (née Morland) (1797–1857) had to cope with the eccentricities of both her husband William, Professor of Geology at Oxford, and her son Frank, the noted zoologist, while all the time providing a comfortable and suitable environment for their work, as well as acting as a field assistant and conservator, repairing and cleaning her husband's specimens (Kölbl-Ebert 2002). Charlotte Owen, we are told, had to deal with the stench of decaying carcasses of animals brought into her home by her husband Richard, the palaeontologist and anatomist (Barber 1980).

By and large, women were good draftsmen and colourists and we see evidence of this throughout the geological literature. Many daughters acted as illustrators or assistants for their fathers; for example, Anna Children drew illustrations for John George Children's (1777–1851) volume *Lamarck's Genera of Shells*, translated from the French, published in 1823. At least one of George Busk's daughters was a painter and helped him in his studies. She also painted several portraits, including that of her father and that of George James Allman, both of which now hang in the Linnean Society. The artwork of Blanche Cole (née Vernon), wife of the English petrologist Grenville Arthur James Cole, can be found in his papers and books, where she skillfully illustrated the petrographic textures of some igneous rocks (see Cole 1902, frontispiece). The contributions of a number of British women is discussed and illustrated more fully in Kölbl-Ebert (2002).

Post-1900. In this paper we focus on the contributions of the two daughters of the geologist William Johnson Sollas (1849–1936), who was a major figure in British and Irish geological circles in the 20th century. He was successively Professor of Geology at Bristol, Trinity College, Dublin, and Oxford, and applied himself to a diverse array of geological subjects: glacial deposits, plesiosaurs, anthropology, among others (Wyse Jackson 2004). Married twice, he had two daughters by his first wife, and they became closely involved with a number of his geological projects.

Hertha Beatrice Coryn Sollas (b. 1875), his elder daughter, was educated at Newnham College, Cambridge, and received a PhD from the University of Heidelberg for a thesis on Oliver Goldsmith's influence on 18th-century Germany (H. B. C Sollas 1903). She was clearly a fluent German-speaker, and took on the responsibility for translating into English the mammoth work *Das Antlitz der Erde* written by the Austrian geologist Eduard Suess. This was not an easy task, given the size of the original work and the geological complexity of its content. The work was directed by her father, who ensured that the accuracy in geological terms of the original was retained, and it appeared in English as *The Face of the Earth*, in five volumes (Suess 1904–24). She also provided an English translation of Hans Przibram's (1874–1944) work on embryology published in Cambridge in 1908 (Przibram 1908). He was a professor of zoology in Vienna and a leading researcher in experimental biology. He emigrated to Holland in 1939, where he was subsequently arrested by the Nazis, and he died at Theresienstadt concentration camp at Terezín (now in the Czech Republic) in the spring of 1944.

Like her sister, Igerna Brünhilda Johnson Sollas (1877–1965) was educated at Alexandra School and College, Dublin, and then at Newnham College, Cambridge, where she was a Fellow and Lecturer in Zoology between 1903 and 1904 and 1906 and 1913 (Creese & Creese 2006; Burek, 2007). She also served on the Evolution Committee of the Royal Society in about 1902, and she produced a short history of Newnham College in 1912 (I. B. J. Sollas 1912) but seems to have disappeared from academic life after 1914. What happened to her is not clear; she applied for a university job as a demonstrator in zoology in 1911, but was rejected as she was deemed over-qualified (Packer 1997, p. 94).

The younger Sollas sister is now best remembered for a paper written with her father in which they carried out a ground-breaking study on the Devonian fish *Palaeospondylus gunni* Traquair (W. J. Sollas & I. B. J. Sollas 1904). He had established a methodology for the production and study of serial sections taken from fossils which allowed researchers to determine the internal structure of fossils (W. J. Sollas 1904; Forey 2005). Based on the sections of the fossil fish, Sollas & Sollas produced wax models which illustrated in detail its 3-D structure for the first time (Fig. 1). In their paper of 1904 they illustrated all the minor bones of the fish, to which they gave Greek-derived names. Nearly 100 years later their material was re-examined, a new digital 3-D reconstruction technique has provided new information, and *Palaeospondylus gunni* has been reinterpreted as being the larval stage of the lungfish, possibly *Dipterus valenciennesi* Sedgwick & Murchison, 1828 (Thomson et al. 2003; Thomson 2004).

Also with her father she published papers on *Lapworthura*, a Silurian brittle-star (I. B. J. Sollas & W. J. Sollas 1912), and on the skull of a *Dicynodon* (I. B. J. Sollas & W. J. Sollas 1914), whose structure they also revealed using serial sections. In her own right she published on the radula of molluscs (I. B. J. Sollas 1907a) and contributed a number of papers to the Royal Society, all of which were communicated by her father: on chitin (I. B. J. Sollas 1907b), and on a Carboniferous brittle-star (I. B. J. Sollas 1914). Like her father, she published on sponges, contributing to the *Cambridge Natural History* series (I. B. J. Sollas 1906), and she was honoured when the demosponge genus *Igernella* was named for her in 1905 by the French spongiologist Émile Topsent (1862–1951) in the Parisian journal *Archives de Zoologie Expérimentale et Générale*.

The museum assistant, aide or volunteer

Pre-1900. Natural history museums as we now know them grew out of the development of private collections assembled by the nobility and royalty of Europe during the 1700s and early 1800s. By the mid-1800s many of these collections were incorporated into the holdings of newly established public museums that sprang up throughout the United Kingdom, Europe and elsewhere. These new centres of learning and entertainment became very popular with the general public and attracted both men and women in nearly equal measure. However, it was the men who were employed as curators, and it was they who published on the material that they or their female colleagues collected.

Post-1900. It was not until the close of the 19th century that women began to have a role in museums, albeit minor. Women typically were employed in the major national or university museums as mounters of herbarium specimens, library cataloguers, preparators, illustrators, or assistants to the male academics and curators, and this trend continued until the 1930s and beyond (see Snell & Tucker 2003, for a number of photographs of such women engaged by the British Museum (Natural History)). These women received

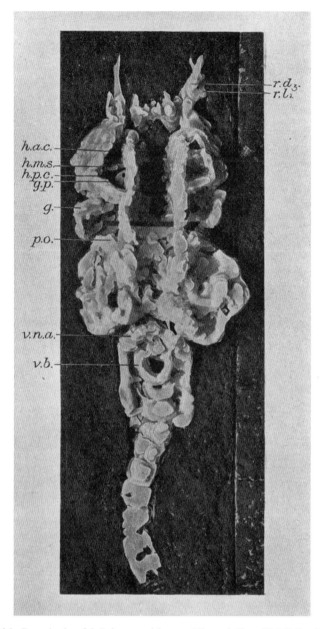

Fig. 1. Wax model of the Devonian lungfish *Palaeospondylus gunni* Traquair (from W. J. Sollas & I. B. J. Sollas 1904, Pl. 16, Fig. 3).

little, if any, academic credit for their research as it was frequently incorporated into the publications of the men for whom they worked.

Next we briefly examine the work of some women firstly in the university sector and secondly, at the Natural History Museum (London) and the Smithsonian Institution, Washington DC.

Women in the university sector

Pre-1900. From the mid-1800s, many universities maintained natural history collections, either together within a university museum or separated out into specialist subject museums. Until the post-World War II expansion of the universities, many scientific subjects were taught by a handful of academics, presided over by a chair-holder. Throughout the United Kingdom and Ireland, geology was firmly held in the grasp of male professors who were often 'one-man departments': this was true at Oxford and Cambridge and in all the universities in Ireland. Women were often employed to look after the geological collections, and probably paid a small salary. This situation continued up until relatively recently in Trinity College, Dublin, where Veronica Burns was employed as an Attendant (see below) but was *de facto* curator of the Geological Museum.

In 1841, University College, London, became one of the first universities in England to appoint a Professor of Geology, and the first to admit women to its degree courses in 1878 (Hinsley 1999; Burek 2007). The first Professor of Geology was Thomas Webster but he died 3 years after being appointed.

Post-1900. One of Webster's successors was Edmund Johnston Garwood (1869–1949; Professor from 1901 to 1931), who had developed a significant reputation for his work on the Carboniferous successions of Lancashire and Yorkshire. He also was a mountaineer of some repute, having climbed in the Himalayas (Hudson 1950). Soon after joining University College, London, he appointed Edith Goodyear as his assistant, and she looked after the Geological Museum in the college for many years (Kirk 1994). She held the degree of BSc, and was later elected a Fellow of the Geological Society (Muir-Wood 1959). Goodyear was responsible for cataloguing (and probably providing initial identifications) Garwood's growing collections, and it is likely that she accompanied him into the field. While invariably most of Goodyear's research found its way into Garwood's publications, she did publish one joint paper with him. This was an important and long paper on the Lower Carboniferous around Settle, which was published by the Geological Society on 25 July 1924 (Garwood & Goodyear 1924) (Fig. 2). Today, Edith Goodyear is commemorated by a room named after her in University College (Wendy L. Kirk pers. comm. 2006).

Garwood encouraged female students to attend his geological courses and taught Marie Stopes (Fraser & Cleal 2007) and Helen Muir-Wood (see below) among others. Stopes went on to make a name for herself in the diverse areas of palaeobotany and birth control (Chaloner 2005; Burek 2007). Muir-Wood carved out a career for herself as a palaeobrachiopodologist (see below). One of Garwood's students did not have an illustrious geological career, but she did make a now-significant contribution to the field of bryozoology. She was Madeline Munro, a woman about whom we unfortunately know very little. She was a student of Garwood who worked on Lower Carboniferous bryozoans, and we believe identified much of his palaeontological material. In 1912 she published her own work as an appendix to one of his papers (Munro 1912) – perhaps he or the Geological Society of London did not allowed it to be published as an independent paper. In this paper she erected the trepostome bryozoan genus *Stenophragma*, named for material collected from Carboniferous strata around Ravenstonedale in Yorkshire; later her generic name was discovered to be preoccupied by an Austral-Oceanic fly, and today is a junior synonym of *Stenophragmidium* Bassler, 1952. Munro corresponded with Ray Smith Bassler (1878–1961) of the Smithsonian Institution and exchanged material with him. In the 1940s she donated her specimens and notebooks to the Natural History Museum in London and these have proved invaluable during the course of a revision of her genus (Cleary & Wyse Jackson 2007). Munro was probably typical of the many female students under the tutelage of Garwood and male academics in other institutions in that they acted as research assistants and did some independent research, but this was often in palaeontological (see Fraser & Cleal 2007) and soft rock topics, which was rarely published. Nevertheless, in the years immediately preceding World War I, it is clear that women were beginning to establish themselves in the research realm (Burek, 2007).

In 1964 in Trinity College, Dublin, the Professor of Geology, Robert George Spencer Hudson, appointed Veronica Burns as a photographer on a low grade from which she was never adequately advanced (Higgs & Wyse Jackson 2007). Soon afterwards she also took responsibility for the Geological Museum, even though the Professor was officially the curator of its collections (Wyse Jackson 1992). Her interests included graptolites, minerals and astronomy. She was a quiet unassuming woman about whom little was known when she died. At birth she was fostered but only traced by her natural family after 21 years (Wyse Jackson 1998a & b). She curated the mineral collections to a very high standard and was also a tireless fossil collector. In her 17 year tenure she did little research (as she was not expected to carry any out), but she did co-author two papers that described the graptolites she had collected from poorly fossiliferous successions in north County Dublin and adjacent districts in County Meath (Rickards *et al.* 1973; Burns & Rickards 1993). Veronica dedicated her weekends to collecting when others would not have bothered. Veronica also collected many thousands of fossils

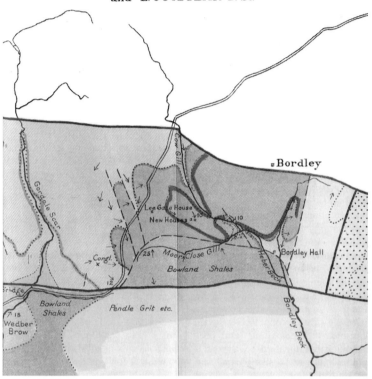

Fig. 2. Portion of the geological map of the Carboniferous succession near Settle, Yorkshire, published in 1924 by Edmund Garwood and Ethel Goodyear.

from the Lower Carboniferous of Dublin, particularly from the Waulsortian build-up at Feltrim Hill; her labours can be reflected in the faunal lists published in Hudson *et al.* (1966). These colleagues named the brachiopod *Plicochonetes burnsi* for her, but unfortunately the name was never validated through formal designation, description and publication. On her retirement in 1980, she continued to entertain school parties in the museum until 5 years before her death in 1998. Veronica was typical of those women associated with university museums five decades earlier. She was a wonderful warm lady, and the last of those women who could be considered as being a 'paid volunteer'.

Women in the 'national' museums

Museum of Science and Art, Dublin. In Dublin in the 1890s the major collection was the one assembled by the Royal Dublin Society, which, in 1877, was re-branded as the Museum of Science and Art. By and large, until the late 1970s, when the ban on married women working in the public sector was lifted, all women that were employed by this institution were spinsters (see Higgs & Wyse Jackson 2007). Most were employed in fairly menial positions: as labellers, cataloguers, model builders or cleaners, although there were two notable exceptions. Matilda Cullen Knowles (1864–1933) was employed as a Temporary

Assistant in 1902, promoted to Assistant in 1907 and remained in this post until her death 26 years later. She carried out a considerable amount of research on lichens and mapped out the distribution of 800 species in Ireland (Knowles 1929; Scannell 1997). Jane Stephens (b. 1879) was, like Igerna Sollas, interested in sponges, about which she published a number of important papers, including those on marine and freshwater sponges as part of the celebrated Clare Island Survey (Stephens 1912*a* & *b*). She was employed in 1905 as a Technical Assistant and later promoted to Assistant Naturalist, but had to relinquish her job in 1920 when she married Robert Francis Scharff, the Acting Director – a post he held until he retired the following year (Praeger 1949).

National Museum of Natural History (Smithsonian Institution), Washington DC. During the first three decades of the 20th century, a host of women were employed as secretaries, artists and assistants (or 'aides' – a politically correct sounding title for the latter) in the Department of Geology of the National Museum of Natural History (Smithsonian Institution) in Washington DC. Ray Smith Bassler spent his childhood in Cincinnati where he developed, unsurprisingly, a love of Ordovician fossils. From an early age he came under the influence of one of the foremost palaeontologists in the United States at the time: Edward Oscar Ulrich. Later, this pair supplemented their incomes by dealing in fossils, particularly in thin sections, which they made themselves. Bassler went on to become an expert on fossil and Recent Bryozoa, authoring the first volume of the *Treatise on Invertebrate Paleontology*, and was appointed Assistant Curator to the museum in 1904. He eventually reached the position of Head Curator in the Department of Paleontology in 1929 and held this position for 19 years until his retirement (Caster 1965).

During his tenure, Bassler was assisted by various women, who acted as curators, proofreaders and artists, and other women were employed in other capacities within the Department of Geology. Francisca Wieser was an artist and photographic assistant who spent at least a decade in the Department of Geology, and her skilled work adorns many of Bassler's monographs (Bassler 1911). Adelaide Corinna Quisenberry (b. 1881) was a native of Lexington, Fayette, Kentucky, and the eldest of four children. She was employed as Bassler's aide (Sanner 2002) and was largely responsible for assembling his major work on Early Tertiary bryozoans, co-authored with the Frenchman Ferdinand Canu (Canu & Bassler 1920); (Fig. 3). She quit the Museum in 1918 when she married John Bernard Reeside Jr, a US Geological Survey palaeontologist who was later head of Survey palaeontology at the museum. The following year, the new bride was replaced as Bassler's secretary and aide by Jessie G. Beach, who remained at Bassler's side until 1954. Given her long employment by the Museum it is reasonable to conclude that she never married. Florence George Wall was a Secretary in the Department of Geology between 1910 and 1914 but left when she married a fellow museum employee. Margaret W. Moodey worked in the Department between 1910 and the 1930s and was also given curatorial responsibility for the gem collection.

Volunteers were very important to the workings and well-being of various natural history museums and a number of female volunteers made significant contributions through their role as collectors of specimens. In the United States, Mary Agnes Chase (1863–1969) collected plants in South America in the 1920s. She was a botanist with the US Department of Agriculture, and her later museum role came as Honorary Curator of Grasses at the National Herbarium in the Smithsonian Institution.

British Museum (Natural History), London. In England, the British Museum (Natural History), now known as the Natural History Museum, in London also benefited from the work of unpaid female collectors. Two notable women in this regard spring to mind. Mary Henrietta Kingsley (1862–1900) reached a crisis in her life when she was 31 years old: her father died suddenly, followed shortly afterwards by his invalid wife, whom Mary had looked after. To escape her grief Mary took up travelling, and planned to visit Africa to complete a book that her anthropologically minded father had begun. She also decided to study zoology and, following the advice of Albert Gunther of the British Museum (Natural History), assembled a fine collection of fishes. She returned to Africa for a longer trip that lasted 2 years and returned with many species of fish and rare reptiles. Mary Kingsley died in 1900 in South Africa, where she was nursing soldiers wounded during the Boer War (McGirr 2000), but her legacy can be found in the new Spirit Building at the Museum, where her fishes 'swim', pickled in glass jars.

Lucy Evelyn Cheeseman (1881–1969) had from childhood wished to become a veterinary surgeon but this career path was denied to her. Instead she turned her hand to entomology and found a position at Regent's Park Zoo. Late in life she undertook several expeditions to Papua New Guinea and the New Hebrides (1929–34) and the insects that she collected during these trips can now be found in drawers in the Natural History Museum at South

Fig. 3. Adelaide Quisenberry and Ray Smith Bassler, *c.* 1919. She is holding the manuscript of Canu and Bassler's monograph on the Early Tertiary bryozoans of North America. Note the French and American flags on the table, which indicate the nationality of the collaborators (from Sanner 2002; courtesy of the National Museum of Natural History, Washington, DC).

Kensington. For her contribution to entomology Lucy Evelyn Cheeseman received the OBE.

During the period under discussion, 1890–1930, numerous women also worked on unofficial basis in the British Museum (Natural History); Stearn (1998) outlines the work and contributions of a number of such women. John Edward Gray (1800–75) is generally acknowledged as having founded the zoological collections at the Museum, where he was Keeper of Zoology (Cleeveley 1983, p. 134). His wife, Emma, had a keen interest in algae and may well have been responsible for starting that collection at the Museum. Gulielma Lister (1860–1949) and her father Arthur Lister (1830–1908) were both unofficial museum workers. He published a book on Myxomycetes in 1894, and she was responsible for editing two later editions of the work. Ethel Sarel Barton (1864–1922) worked in the Botany Department, where she researched alongside the cryptogram botanist Antony Gepp (1862–1955). Not only did they publish a number of joint papers, but they also married in about 1900. He remained working in the Museum until the age of 93 years, when he died following a fall down a flight of stairs at the South Kensington building. Annie Lorrain Smith (1854–1937), who worked on lichens, was an unofficial worker or 'Acting Assistant' in Botany between 1889 and 1933. Miriam Rothschild worked on her father's collection of fleas, which was later deposited in the Museum in 1913. Her uncle donated Tring Zoological Museum to the Natural History Museum. Eleanor Mary Reid (1860–1953), wife of Clement Reid of the Geological Survey of Great Britain, and Marjorie E. J. Chandler are remembered for their joint work *The London Clay Flora* that appeared in 1933 (Reid & Chandler 1933); both were associates of the Palaeontology Department.

Another associate was Sidnie Manton (1902–79), who along with her sister Irene (1904–88) were two of the foremost female scientists active in the United Kingdom during the 20th century. Both were amongst the first women to be elected Fellows of the Royal Society: Sidnie in 1948 and Irene in 1961 (Mason 1995). Sidnie was educated in Cambridge and found employment at Queen Mary College, London; she later married John Phillip Harding, Keeper of Zoology at the British Museum (Natural History). Her sister worked at the University of Leeds and her major research interest was in the British flagellates.

Although the majority of these women were employed in natural history museums in clerical positions, or were otherwise associated with, but not employed by these institutions, they succeeded in making major contributions to geology and natural history in the early to middle-decades of the 1900s. Their importance cannot, and should not be understated; their physical legacy can be found in the collections, labels and accession catalogues found in museums on either side of the Atlantic Ocean, and their intellectual legacy lies in the many scientific reports published more often, but not exclusively, by their male employees.

1920s–1950s: Emergence of professional female museum scientists

From the 1920s there emerged a new breed of female museum worker, who could for the first time be considered to be professional research scientists. Why was this? Certainly in the United Kingdom, this can be directly attributed to two factors. The first was the influence of the women's colleges. In the 1840s a number of these had opened in England: Queen's College, London (in 1847) and Bedford College, London (in 1849) (Burek 2007), and these were followed by further examples: Girton College, Oxford, and Newnham College, Cambridge. In the United States, similar women's colleges were providing an education that included geology on the curriculum (Rossiter 1981; Clary & Wandersee 2007). As we have seen, University College, London, was the first university in the United Kingdom to admit women to its degree courses (in 1878) and, by the end of that century and the early years of the next, a number of universities had begun to follow their lead, including Liverpool, Leeds and Manchester (Hinsley 1999), although some would not confer degrees. Oxford and Cambridge resisted conferring degrees on women until 1919 and 1948, and it was for this reason that, between 1904 and 1907, many Oxbridge women travelled to Dublin to take University of Dublin degrees (Parkes 2004; Higgs & Wyse Jackson 2007).

The second reason that women began to gain a foothold in museums at this time was due to the wholesale slaughter of young men during World War 1. Today, many institutions still carry plaques detailing the service and sacrifice made by its employees during this conflict. In Dublin, Colin Selbie was an Assistant Naturalist employed by the Museum of Science and Art, but his career as a crustacean biologist was cut short when he was mown down on the Somme in 1916 at the tender age of 26 years (Praeger 1949). Arnold Lockhart Fletcher was employed by the geophysicist John Joly at Trinity College, Dublin, but he lost his life on the fields of France in 1917 (Wyse Jackson 2006). The loss to science of many thousands of young men was immeasurable, and women soon began to fill the vacancies created by this loss.

In the United Kingdom some women over 30 years of age had gained the right to vote in 1918, and the following year the Geological Society elected its first female Fellows. On 19 April 1919 the following eight women were proposed for election to Fellowship: Margaret Chorley Crosfield, Gertrude Lilian Elles, Maria Matilda Ogilvie Gordon, Mary Sophia Johnston, Jane Longstaff, Lady Rachel Workman McRobert, Mildred Blanche Robinson and Ethel Gertrude Woods. Fifty years later, nearly 220 women had been elected as Fellows (Wallace 1969, p. 212).

Museum users with no particular institutional affiliation

Natural history museums reached the height of their popularity after the mid-1800s, by which time many provincial towns in the United Kingdom and the United States had opened museums, often in association with scientific, philosophical or literary societies (Knell 2000). Women certainly made up a significant proportion of visitors, and they also donated material to the collections: Etheldred Benett (1776–1845) (Torrens 1985; Torrens *et al.* 2000; Burek 2001) assembled a fine collection dominated by Cretaceous fossils, and her material found its way into museums in Bristol, Folkstone, Leeds, London, Warwick and York (Cleevely 1983), and even as far away as Philadelphia (Spamer *et al.* 1989). In Dublin, the university museum benefited from the generosity of women who donated, among other collections, examples of fossil wood from Antigua (Ball 1846). This pattern of female donation was repeated elsewhere (see Cleevely 1983 for a listing of donations of fossil material made to museums, and Stace *et al.* 1987 for a similar listing of geological donations made to Scottish institutions).

As we have seen, women were not generally employed in a scientific role in these museums but contributed much in less prestigious positions. However, by the 1880s, some women were forging research niches for themselves, even though they did not hold official museum positions. The research that they carried out was based either on material that they had collected themselves or was available in natural history museums.

As an example of these women who had no affiliation to any particular institution or museum, but who were research users of their collections, we highlight the work of Mary Jane Longstaff (née Donald) (1855–1935). Jane Donald was born in Carlisle on 27 August 1855 and educated in London and the Carlisle School of Art. She was encouraged to study gastropods by the geologist John George Goodchild (1844–1906) of the Geological Survey (McConnell 2004), and she dedicated herself to this group, becoming an expert on Palaeozoic gastropods (Cox 1936; McConnell 2004; Valier 2004) and publishing 20 papers on them between 1885 and 1933 many of which appeared in the *Quarterly Journal of the Geological Society of London*. In general, her papers were taxonomic in thrust and, while they might appear to be a bit old-fashioned for latter-day palaeobiologists, she did document in great detail this hitherto neglected group. Apart from a monograph by Batten (1966) on the gastropod fauna of the Hotwells Limestone in Somerset, and some more recent papers on the large flanged gastropod *Phanerotinus* (Morris & Cleevely 1981; Wyse Jackson *et al.* 2000), the literature on Carboniferous gastropods of the United Kingdom and Ireland is sparse. Consequently Longstaff's work remains highly significant. Longstaff was a seriously committed gastropodologist who even went as far as to breed giant land snails in her home. Given her later work on the gastropods of the family Murchisoniidae (Longstaff 1926), it was appropriate that she should have received the Murchison Fund from the Geological Society in 1898. She was in the first group of ladies admitted as Fellows of the Geological Society of London in 1919 (Wallace 1969), a Fellow of the Linnean Society in 1906, and a member of the Geologists' Association from 1883 (Cox 1936). In 1906, when she was 51 years old, she married the entomologist George Blundell Longstaff (1848–1921), after which she travelled extensively collecting Recent and fossil shells. Following her death in Bath in 1935 these were presented to the British Museum (Natural History) by her nephew (Cox 1936). Additional collections are in the Tullie House Museum and Art Gallery, Carlisle, the Hunterian Museum in Glasgow, the Royal Scottish Museum, Edinburgh, and in the collections of the British Geological Survey (Cleevely 1983, 188).

The academic

In the United States between the years 1894 and 1904 four women's colleges appointed females to academic posts (Rossiter 1981). Elizabeth Florette Fisher (1873–1941) taught at Wellesley College from 1894; Florence Bascom (1862–1945) (Clary & Wandersee 2007) at Bryn Mawr from 1895; Ida Ogilvie (1874–1963) at Barnard College from 1903, and Mignon Talbot (1869–1950) at Mount Holyoak College from 1904. Although this appeared to be a major advance, in fact the colleges produced few geological graduates, and Margaret Rossiter (1981) has attributed this to the generally marginalized status of geology in the American educational system at the time. Nevertheless, these pioneering geological academics carried out some major research and were certainly museum users. Talbot was largely responsible for building up a major geological collection at Mount Holyoak, which was unfortunately destroyed in a fire in 1916 (Elder 1982). Undaunted, she built up a new collection. Fisher obtained a considerable reputation for her work on the Texas oilfields, while Blascom was an expert in mineralogy and petrology. She was the first woman to be elected to the Council of the Geological Society of America in 1924 (Schneiderman 1997).

In the United Kingdom, Catherine Alice Raisin (1855–1945) was appointed Head of the Geology; Department at Bedford College, London in 1890 (for more details of her life and petrological work see Burek 2003; Creese & Creese 2006, Burek 2007). In Cambridge the foremost female palaeontologist at the end of the 19th century was Gertrude Lilian Elles (1872–1960). Elles was educated at Cambridge and remained there for the rest of her career. She made her reputation researching graptolites, and with the Cambridge-educated Ethel Wood (later Dame Ethel Shakespear); (1871–1945) authored an important monograph on these colonial organisms (Elles & Wood 1901–18). Elles provided the taxonomic descriptions and Wood the illustrations. Elles had been inspired to geological study by Thomas McKenny Hughes, Professor of Geology at Cambridge and, following the completion of her studies, was appointed to a position at Newnham, and ended up vice-Principal (Creese & Creese 2006). A ScD was conferred upon her by Trinity College, Dublin, on 6 July 1905 – she was one of the 'Steamboat ladies' (Parkes 2004; Higgs & Wyse Jackson 2007) – and eventually, in 1947, she was given another ScD by her own university. It was only in 1926 that she was appointed to a lectureship (Creese & Creese 2006). She became Reader in 1936. She had a long association with the Sedgwick Museum where she did most of her work on graptolites. One of her successors in Cambridge, Barrie Rickards, recalled that:

'She studied her specimens at one end of a long bench. As each job was completed and her notes prepared, that batch of specimens was pushed to her right hand side, towards the other end of the table. Those specimens already at the far end of the table fell on the floor and accumulated in a heap.' (Rickards 1999)

Although this sounds horrific, she carried out excellent work and was rewarded for it with the Lyell Fund (in 1900), and then the Murchison Medal (in 1919) was awarded to her by the Geological Society of London.

Ethel Wood was appointed assistant to Charles Lapworth at Birmingham in 1896 and remained until 1906. She was awarded the Wollaston Fund in 1904, and the Murchison Medal in 1920, and held the degree of DSc. By 1920 she had moved from academic circles and had turned to humanitarian work, in particular relating to the welfare of soldiers wounded in World War 1. For this she was made a Dame (Creese & Creese 2006).

These academic women, and those that followed, constituted a new emerging breed of professional woman who worked in geological museums, not in menial or subservient roles, but as acquirers of the collections, and as research scientists. They gained a foothold in museums a decade before women began to find employment in the major scientific museums in the United Kingdom and the United States (see below).

The museum research scientist

The beginnings of women being employed as research scientists in their own right can be traced back to an indomitable woman collector/researcher associated with the British Museum (Natural History). She was Dorothea Minola Alice Bate (1878–1951) who, in 1898, walked into the hallowed halls and demanded a job (Shindler 2005, 2007). Fifty years later she had left a legacy of subfossil material from various Mediterranean islands, as well as copious notebooks and scientific findings. She was ostensibly employed from 1914 as a preparator and was paid on the basis of the number of specimens that she completed. However, she also reported on her own finds and later reported on material collected by others. She was an invaluable member of the Museum staff, albeit employed on a *pro rata* basis without the benefits of security. This perhaps suited her, as it allowed her to spend a great deal of time collecting and exploring. In 1924 she was appointed to the non-official position of Curator of Aves and Pleistocene mammals and paid on the same basis as before (Shindler 2005). She had a most impressive publication record that spanned the years 1901 to 1955. Shindler (2005) lists 59 scientific papers and articles, all but one of which were solely authored, and she noted that Dorothea also wrote many unpublished reports on collections now in the Museum in London. It was a record that her fully tenured male colleagues would have been proud to claim as their own.

Here, to illustrate the work of the female research scientist, we focus primarily on the British Museum (Natural History), which acts as a proxy for the changes in employment practices seen elsewhere from the 1920s. Much of the credit in getting women employed on the scientific staff has been given to William Thomas Calman (1871–1952), who became Keeper of Zoology 1927. Charles Tate Regan (1878–1943) was the Director at this time.

Helen Marguerite Muir-Wood (1895–1968) was educated at Bedford College, London, where she did a first degree before gaining an MSc for a thesis on iron ores (Williams 1969). She then transferred to University College, London, where, under the guidance of Edmund Garwood, she researched Carboniferous brachiopods. She joined the British Museum (Natural History) in a part-time capacity in 1919, and the following year was elevated to the rank of part-time assistant. She was given charge of the brachiopod collections in 1923 and appointed to the permanent staff in 1936. She rose to be Deputy Keeper in the Department of Palaeontology in 1955 and continued in this position until her retirement 5 years later. Like many colleagues, she continued to commute to South Kensington in retirement and there she continued her researches. She died in 1968.

Muir-Wood was honoured by the Geological Society, which awarded her its Lyell Medal in 1958. This was wholly merited as she had become the premier researcher on brachiopods of her generation in the United Kingdom. (Fig. 4). In the 1930s she authored a major monograph on the brachiopods of the Fuller's Earth (Muir-Wood 1936). In 1951 she visited Washington DC, where she worked with the premier American brachiopodologist Gustav Arthur ('Coop') Cooper (1902–2000) (see Brunton 2001), and this led to their co-authored volume on the productoids (Muir-Wood & Cooper 1960). She also contributed to the revision of Part H (on brachiopods) of the *Treatise on Invertebrate Paleontology* (Muir-Wood & Williams 1965).

Other female permanent members of staff included Ethelwynn Trewavas (1900–93), who was born in Penzance, Cornwall, and worked from 1925 in an unofficial capacity in the Department of Zoology, where she specialized in the study of African freshwater fish. There she assisted Regan. In 1935 she was officially employed as a fully fledged staff member of the Museum. Isabella Gordon (1901–88) was a permanent member of staff who worked on crustaceans, retiring from the Museum in 1966. Jessie M. Sweet (1901–79) served on the staff as curator of the mineral collection in the Department of Mineralogy for many

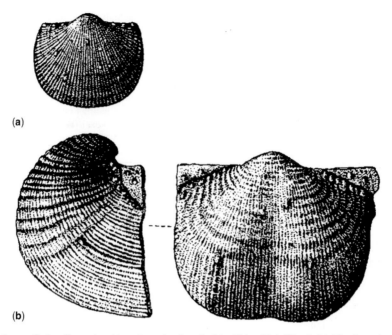

Fig. 4. Two Lower Carboniferous brachiopod species described by Helen Muir-Wood. (**a**) '*Productus*' *craigmarkensis*; (**b**) *Antiquatonia hindi* (from Anon. 1975; courtesy of the Natural History Museum, London).

years from the 1930s. She is perhaps best known for her suite of papers on 18th-century mineralogists and geologists, published in *Annals of Science*, the most notable of which concerned Robert Jameson (Sweet 1963, 1967; Sweet & Waterston 1967). Jessie Sweet was awarded the MBE in 1961 and is commemorated in the mineral name 'Sweetite', a zinc hydroxide from Derbyshire (Clark *et al.* 1984). Susan Finnegan (b. 1903) was appointed in 1927 and spent most of her time researching acarids (mites) until she left in 1936. This was precipitated by her marriage to W. Campbell Smith, of the Department of Mineralogy in the Museum. He remained in post. Another permanent member of staff who had to resign on her marriage (again to a member of staff at the Museum) was the bryozoologist Anna Birchall Hastings (1902–77). She had been appointed assistant to the Director, Sydney Frederic Harmer (1862–1950). Later, following his retirement, she took full responsibility for the Recent bryozoans. Among her publications was an account of the bryozoans from the area around the Panama Canal (Hastings 1930) and a major monograph on the bryozoans from the Antarctic, collected during the *Discovery* expedition (Hastings 1943). For the latter she had to examine over 2300 slides. Although she had to resign when she married Henry Dighton Thomas in 1942, she did not change her work practices and continued work in an unpaid research capacity, an arrangement that continued for many years.

Another woman employed by a natural history museum in the 1930s was Nora Fisher McMillan (1908–2003). She was born in Belfast and, at an early age, came to the attention of the photographer and malacologist Robert John Welch (1859–1933). He encouraged her early research with the Belfast Naturalists' Field Club (see Evans & Turner 1977, p. 6). She moved to Liverpool and joined the staff of the Liverpool Museum in 1933, and this association was to continue until her death, albeit interrupted between 1937, on her marriage, and 1954 when she was allowed to return (Edmondson 2006). 'Mrs Mac' as she was invariably known, was a larger than life character who for many years kept goats (Brack 1976). She was a prolific author with over 400 publications to her name. She was interested primarily in post-glacial freshwater Mollusca, but also in the history of natural history, and was the author of the successful popular books *British Shells* (McMillan 1968) and *The Observer's Book of Seashells of the British Isles* (McMillan 1977). She was President of the Conchological Association of Great Britain and Ireland, elected a Member of the Royal Irish Academy in 1970 (Brack 1976), was conferred an Honorary MSc by the university of her adopted city in 1991, and received an MBE in 1992 (Edmondson 2006).

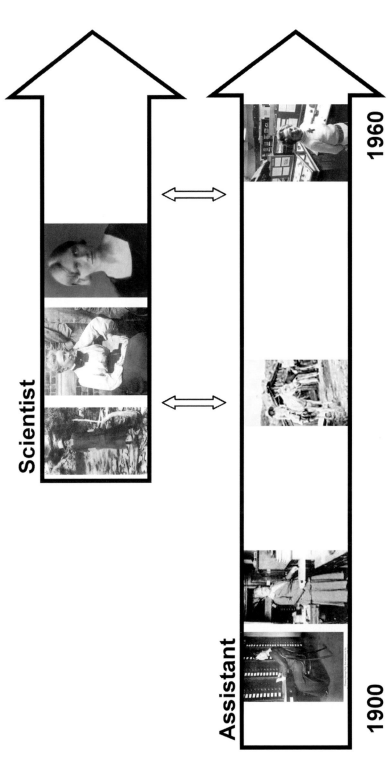

Fig. 5. Diagram illustrating women's status in geological museums, 1900–60. The upper arrow represents the emergence of women as fully employed museum research scientists or academics. The lower arrow represents a situation where women generally acted as lowly paid assistants or volunteers. The small interconnecting arrows represent cross-over between the strands. The women illustrated are: (from top left), Winifred Goldring, Gertrude Elles and Anna Hastings; (from bottom left) Florence Wall, Edith Goodyear, Mary Agnes Chase, and Veronica Burns.

In the United States, the 'professionalization' and acceptance of female scientists was paralleling the changes in attitudes to women in the United Kingdom. By the mid-1930s they were gaining a foothold as professional geologists and museum research staff. Their story is epitomized by Winifred Goldring (1888–1971), the daughter of an orchid specialist formerly employed at the Royal Botanic Gardens in Kew. She eventually became the State Paleontologist of New York in 1939, the first female President of the Paleontological Society in 1949, and the Vice President of the Geological Society of America the following year. One wonders, however, why she was never elected President of the latter organization.

Her father emigrated to the United States in 1879 when he was offered the position of curator of an enthusiast's orchid collection. Later he established his own business as a floral plantsman and supplier (Aldrich & Leviton 2005). Winifred entered Wellesley College (a women's college) where she was taught by Elizabeth Florette Fisher (1873–1941). She graduated with a BSc in 1909 and an MSc in 1912, and for the next 2 years taught at a teacher's college in Boston. In 1914 she was invited to design and set up some displays for the New York State Museum and remained there for the rest of her career. By 1916 she had commenced the palaeontological research on Devonian crinoids for which she is now primarily remembered. Her first major work was a 670-page monograph in Devonian crinoids that was illustrated with 60 plates (Goldring 1923). These were executed not by her, but by the artist George Barkentin; their collaboration is an interesting example of gender reversal, given the usual role of the women as illustrators and men as the scientists. Between 1926 and 1954 Winifred published a further 18 papers on crinoids, and she was responsible for erecting 24 genera and over 100 species (Aldrich & Leviton 2005).

Conclusion

In the last 160 years women have contributed to the work of natural history and geological museums in the United Kingdom and Ireland and in the United States in many different ways. During the Edwardian period, at the beginning of the 20th century, numerous women were employed as assistants and aides to male research and curatorial staff, and although they carried out much research it was rarely published under their names. In the late 1920s and early 1930s a shift in the attitudes towards female employees in museums occurred, and women were employed as research scientists for the first time. This came about as a result of World War I, which saw the decimation of a generation of young men, and because many women had a scientific background gained at one of the women's colleges that provided a third-level of education. Women were needed to staff the museums and it was clear that their intellect was no less than their male colleagues. This acceptance of women as equals with their male counterparts was no slower in coming within museum circles. This shift in attitude paralleled the general trend in society as a whole, a trend which had its greatest impact when women in the United Kingdom gained the vote in 1919.

Women with museological interests can be divided into two main groups: the first can be regarded as containing non-professional women made up of the wife, sister, or daughter who assisted the work of male relatives, as well as the lowly paid museum assistant. The second consists of women who carried out research and published it under their own names. It comprises the non-affiliated museum user, and the professional women: the academics, and the museum research scientists.

Whatever labels one may wish to pin to the bosom of these museum ladies, it is clear that they made major contributions to natural history and geological research and to science as a whole through their work in museums, and their contribution should not be understated.

This paper was presented at the History of Geology Group (HOGG) meeting, *The Role of Women in the History of Geology*, on 28 November 2005 at the Geological Society, London. We thank Cynthia Burek (University of Chester) and Bettie Higgs (University College, Cork) for their kind invitation to address this meeting and to contribute to this volume. We are grateful to the following persons who supplied pictures and information, or gave us access to archives in their care: Jo Ann Sanner (Smithsonian Institution, Washington, DC), Wendy Kirk (University College, London), Michele Aldridge (California Academy of Sciences), A. Leviton (California Academy of Sciences), N. T. Monaghan (National Museum of Ireland), J. Edmondson (Liverpool Museums), I. Wallace (Liverpool Museums) and P. Taylor (Natural History Museum, London).

References

ALDRICH, M. L. & LEVITON, A. E. 2005. Winifred Goldring (1888–1971): New York paleontologist. *Northeastern Geology and Environmental Sciences*, **27**, 229–238.

ALLEN, D. E. 1969. *The Victorian Fern Craze: A History of Pteridomania*. Hutchinson, London.

ANON, 1975. *British Palaeozoic Fossils*. 4th edn. British Museum (Natural History), London.

BAKER, A. P. 2004. Coleman, William Stephen (1829–1904). *Oxford Dictionary of National Biography*, Oxford University Press, Oxford, **12**, 550.

BALL, R. 1846. *First Report on The Progress of the Dublin University Museum; January 1846*. M. H. Gill, Dublin.

BARBER, L. 1980. *The Heyday of Natural History 1820–1870*. Doubleday and Co. Inc., New York.

BASSLER, R. S. 1911. *The Early Paleozoic Bryozoa of the Baltic Provinces*. United States National Museum Bulletin, **77**, 382 pp.

BATTEN, R. L. 1966. The Lower Carboniferous gastropod fauna from the Hotwells Limestone of Compton Martin, Somerset. Parts 1 & 2. Palaeontographical Society, London, Monographs, **119**, 1–52; **120**, 53–109.

BRACK, A. 1976. Many a Mollusc. *Cheshire Life*, August, 56–57.

BRUNTON, C. H. C. 2001. Gustav Arthur Cooper (1902–2000). *The Geological Curator*, **7**, 181–182.

BUREK, C. V. 2001. The first lady geologist, or collector par excellence? *Geology Today*, **17**, 192–194.

BUREK, C. V. 2003. Catherine Raisin – a role model professional geologist. *Geology Today*, **19**, 107–111.

BUREK, C. V. 2007. The role of women in geological higher education – Bedford College, London (Catherine Raisin) and Newnham College, Cambridge, UK. *In*: BUREK, C. V. & HIGGS, B. (eds) *The Role of Women in the History of Geology*. Geological Society, London, Special Publications, **281**, 9–38.

BURNS, V. & RICKARDS, R. B. 1993. Silurian graptolite faunas of the Balbriggan Inlier, counties Dublin and Meath, and their evolutionary, stratigraphical and structural significance. *Proceedings of the Yorkshire Geological Society*, **49**, 283–291.

CANU, F. & BASSLER, R. S. 1920. *North American Early Tertiary Bryozoa*. United States National Museum Bulletin, **106**, 879 pp.

CASTER, K. E. 1965. Memorial to Ray S. Bassler. *Geological Society of America Bulletin*, **76**, 167–174.

CHALONER, W. G. 2005. The palaeobotanical work of Marie Stopes. *In*: BOWDEN, A. J., BUREK, C. V. & WILDING, R. (eds) *History of Palaeobotany: Selected Essays*. Geological Society, London, Special Publications, **241**, 127–135.

CLARK, A. M., FEJER, E. E., COUPER, A. G. & JONES, G. C. 1984. Sweetite, a new mineral from Derbyshire. *Mineralogical Magazine*, **48**, 267–269.

CLARY, R. M. & WANDERSEE, J. H. 2007. Great expectations: Florence Bascom (1842–1945) and the education of early US women geologists. *In*: BUREK, C. V. & HIGGS, B. (eds) *The Role of Women in the History of Geology*. Geological Society, London, Special Publications, **281**, 123–135.

CLEARY, D. M. & WYSE JACKSON, P. N. 2007. Stenophragmidium Bassler, 1952. (Trepostomida: Bryozoa) from the Carboniferous of Ireland and Britain. *Irish Journal of Earth Sciences*, **25**, 1–25.

CLEEVELY, R. 1983. *World Palaeontological Collections*. British Museum (Natural History) and Mansell, London.

COLE, G. A. J. 1902. *Aids in Practical Geology*. 4th edn, revised. Charles Griffin & Co., London.

C[OX], L. R. 1936. Mrs. Mary Jane Longstaff (née Donald). *Proceedings of the Geologists' Association*, **47**, 97.

CREESE, M. R. S. & CREESE, T. M. 2006. British women who contributed to research in the geological sciences in the nineteenth century. *Proceedings of the Geologists' Association*, **117**, 53–83.

ELDER, E. S. 1982. Women in early geology. *Journal of Geological Education*, **30**, 287–293.

ELLES, G. L. & WOOD, E. M. R. 1901–18. *A Monograph of British Graptolites. Parts 1–11*. Palaeontographical Society, London, Monographs, 539 pp.

EDMONDSON, J. 2006. Nora McMillan (1908–2003). World Wide Web Address: http://www.shnh.org. [Synopsis of obituary forthcoming in *The Linnean*.]

EVANS, E. E. & TURNER, B. S. 1977. *Ireland's Eye: The Photographs of Robert John Welch*. Blackstaff Press, Belfast.

FOREY, P. L. 2005. Slicing through history. *Set in Stone, Natural History Museum Palaeontology Departmental Newsletter*, **3**(1), 5–7.

FRASER, H. E. & CLEAL, C. J. 2007. The contribution of British women to Carboniferous palaeobotany during the first half of the 20th century. *In*: BUREK, C. V. & HIGGS, B. (eds) *The Role of Women in the History of Geology*. Geological Society, London, Special Publications, **281**, 51–82.

GARWOOD, E. J. & GOODYEAR, E. 1924. The Lower Carboniferous succession in the Settle district and along the line of the Craven faults. *Quarterly Journal of the Geological Society of London*, **80**, 184–273.

GATTY, M. 1863. *British Seaweeds: Drawn from Professor Harvey's 'Phycologia Britannica'. With Descriptions, An Amateur's Synopsis, Rules for Laying Out Sea-weeds, an Order for Arranging Them in the Herbarium, and An Appendix of New Species*. Bell & Daldy, London.

GOLDRING, W. 1923. *The Devonian Crinoids of the State of New York*. New York State Museum, Memoirs, **16**, 670 pp.

HASTINGS, A. B. 1930. Cheilostomatous Polyzoa from the vicinity of the Panama Canal collected by Dr. C. Crossland on the cruise of the S. Y. 'St. George'. *Proceedings of the Zoological Society of London*, **47**, 697–740.

HASTINGS, A. B. 1943. Polyzoa (Bryozoa). *Discovery Reports*, **22**, 301–510. Cambridge University Press.

HIGGS, B. & WYSE JACKSON, P. N. 2007. The role of women in the history of geological studies in Ireland. *In*: BUREK, C. V. & HIGGS, B. (eds) *The Role of Women in the History of Geology*. Geological Society, London, Special Publications, **281**, 137–153.

HINSLEY, A. 1999. In a class of their own: the first women university students. *Family History Monthly*, **40**, 46–56.

H[UDSON], R. G. S. 1950. In memoriam. Edmund Johnston Garwood 1864–1949. *Proceedings of the Yorkshire Geological Society*, **38**, 1–4.

HUDSON, R. G. S., CLARKE, M. J. & SEVASTOPULO, G. D. 1966. A detailed account of the fauna and age of a Waulsortian knoll reef limestone and associated shales, Feltrim, Co. Dublin, Ireland. *Scientific Transactions of the Royal Dublin Society*, **2**(A), 251–272.

KIRK, W. L. 1994. *A History of the Geology Collections at U.C.L.* University College London.

KNELL, S. J. 2000. *The Culture of English Geology, 1815–1851: A Science Revealed Through its Collecting*. Ashgate, Aldershot.

KNOWLES, M. C. 1929. The Lichens of Ireland. *Proceedings of the Royal Irish Academy*, **38B**, 179–434.

KÖLBL-EBERT, M. 2002. British geology in the early nineteenth century: a conglomerate with a female matrix. *Earth Sciences History*, **21**(1), 3–25.

KÖLBL-EBERT, M. 2007. The geological travels of Charles Lyell, Charlotte Murchison and Roderick Impey Murchison in France and Northern Italy (1828). *In*: WYSE JACKSON, P. N. (ed.) *Geological Travellers – On Foot, Bicycle, Sledge or Camel: The Search for Geological Knowledge.* Geological Society, London, Special Publications, **287**, 109–117.

LONGSTAFF, J. 1926. A revision of the British Carboniferous Murchisoniidae. *Quarterly Journal of the Geological Society of London*, **82**, 526–555.

MCCONNELL, A. 2004. Longstaff [née Donald], Mary Jane (1859–1952). *Oxford Dictionary of National Biography*, Oxford University Press, Oxford, **34**, 412.

MCGIRR, N. 2000. *Nature's Connections: An Exploration of Natural History.* The Natural History Museum, London.

MCMILLAN, N. F. 1968. *British Shells.* F. Warne, London.

MCMILLAN, N. F. 1977. *The Observer's Book of Seashells of the British Isles.* F. Warne, London.

MASON, J. 1995. The Women Fellows' Jubilee. *Notes and Records of the Royal Society of London*, **49**, 125–140.

MORGAN, N. 2006. Anne Phillips and the mystery of the Malverns. *Geoscientist*, **16**(3), 6–7, 12–15.

MORRIS, N. J. & CLEEVELY, R. 1981. *Phanerotinus cristatus* (Phillips) and the nature of euomphalacean gastropods. *Bulletin of the British Museum (Natural History), Geology*, **35**, 195–212.

MUIR-WOOD, H. M. 1936. *A Monograph on the Brachiopoda of the British Great Oolite Series., Part I The Brachiopoda of the Fuller's Earth.* Palaeontographical Society, London, Monographs, **89**, 144 pp.

M[UIR]-W[OOD], H. M. 1959. Edith Goodyear. *Proceedings of the Geological Society of London*, **1572**, 147–148.

MUIR-WOOD, H. M. & COOPER, G. A. 1960. Morphology, Classification, and Life Habits of the Productoidea (Brachiopoda). Geological Society of America, Memoirs, **81**, 447 pp.

MUIR-WOOD, H. M. & WILLIAMS, A. 1965. Strophomenida. *In*: MOORE, R. C. (ed.) *Brachiopods. Part H.* Geological Society of America, Inc, Boulder and University of Kansas Press, Lawrence, H361–H521.

MUNRO, M. 1912. Description of some new forms of Trepostomatous Bryozoa from the Lower Carboniferous rocks of the north-western province. *Quarterly Journal of the Geological Society of London*, **68**, 574–579.

PACKER, K. 1997. A Laboratory of one's own: the life and works of Agnes Arber, F. R. S. (1879–1960). *Notes and Records of the Royal Society of London*, **51**, 87–104.

ORR, M. 2007. Keeping it in the family: the extraordinary case of Cuvier's daughters. *In*: BUREK, C. V. & HIGGS, B. (eds) *The Role of Women in the History of Geology.* Geological Society, London, Special Publications, **281**, 277–285.

PARKES, S. M. (ed.) 2004. *A Danger to the Men? A History of Women in Trinity College Dublin 1904–2004.* Lilliput Press, Dublin.

PRAEGER, R. Ll. 1949. *Some Irish Naturalists.* Dundalgan Press, Dundalk.

PRZIBRAM, H. 1908. *Embryogeny: An Account of the Laws Governing the Development of the Animal Egg as Ascertained through Experiment.* Cambridge University Press, Cambridge. [Trans. Sollas, H. B. C.]

RAUCH, A. 2004. Gatty, Margaret (née Scott: 1809–73). *In*: LIGHTMAN, B. (ed.) *Dictionary of Nineteenth-Century British Scientists.* Thoemmes Continuum, Bristol, 761–764.

REID, E. M. & CHANDLER, M. E. J. 1933. *The London Clay Flora.* British Museum (Natural History), London.

RICKARDS, R. B. 1999. A century of graptolite research in Cambridge. *The Geological Curator*, **7**, 71–76.

RICKARDS, R. B., BURNS, V. & ARCHER, J. B. 1973. The Silurian sequence at Balbriggan, Co. Dublin. *Proceedings of the Royal Irish Academy*, **73B**, 303–316.

ROSSITER, M. W. 1981. Geology in nineteenth-century women's education in the United States. *Journal of Geological Education*, **29**, 228–232.

SANNER, J. 2002. CANU AND BASSLER. *In*: WYSE JACKSON, P. N. & SPENCER JONES, M. E. (eds) *Annals of Bryozoology.* International Bryozoology Association, Dublin, 243–250.

SCANNELL, M. 1997. Inspired by lichens. *In*: MULVIHILL, M. (ed.) *Stars, Shells and Bluebells: Women Scientists and Pioneers.* Women in Technology and Science, Dublin, 84–97.

SCHNEIDERMAN, J. S. 1997. Rock stars: a life of firsts: Florence Bascom. *GSA Today*, **7**, 8–9.

SHINDLER, K. 2005. *Discovering Dorothea: The life of the Pioneering Fossil-hunter Dorothea Bate.* HarperCollins, London.

SHINDLER, K. 2007. A knowledge unique: the life of the pioneering explorer and palaeontologist, Dorothea Bate (1878–1951). *In*: BUREK, C. V. & HIGGS, B. (eds) *The Role of Women in the History of Geology.* Geological Society, London, Special Publications, **281**, 295–303.

SNELL, S. & TUCKER, P. 2003. *Life Through a Lens: Photographs from the Natural History Museum 1880 to 1950.* Natural History Museum, London.

SOLLAS, H. B. C. 1903. *Goldsmiths Einfluss in Deutschland im 18. Jahrhundert: Inaugural-Dissertation zur Erlangung der Doktorwürde einer Hohen Philosophischen Fakultät der Ruprecht-Karls-Universität zu Heidelberg vorgelegt.* K. Rössler, Heidelberg.

SOLLAS, I. B. J. 1906. Porifera. *Cambridge Natural History.* Vol. 1. Macmillan, London.

SOLLAS, I. B. J. 1907a. The molluscan radula. *Quarterly Journal of Microscopical Science*, **51**, 115–136.

SOLLAS, I. B. J. 1907b. On the identification of chitin by its physical constants. *Philosophical Transactions of the Royal Society of London, Series B*, **79**, 474–481.

SOLLAS, I. B. J. 1912. *The Story of Newnham College.* W. Heffer and Sons Ltd, Cambridge.

SOLLAS, I. B. J. 1914. On *Onychaster*, a Carboniferous brittle-star. *Philosophical Transactions of the Royal Society of London, Series B*, **204**, 51–62.

SOLLAS, I. B. J. & SOLLAS, W. J. 1912. *Lapworthura*: a typical brittlestar of the Silurian age; with suggestions for a new classification of the Ophiuroidea. *Philosophical Transactions of the Royal Society of London, Series B*, **202**, 213–232.

SOLLAS, I. B. J. & SOLLAS, W. J. 1914. A study of the skull of a *Dicynodon* by means of serial sections.

Philosophical Transactions of the Royal Society of London, Series B, **204**, 201–225.

SOLLAS, W. J. 1904. A method for the investigation of fossils by serial sections. *Philosophical Transactions of the Royal Society of London, Series B*, **196**, 259–265.

SOLLAS, W. J. & SOLLAS, I. B. J. 1904. An account of the Devonian fish *Palaeospondylus gunni* Traquair. *Philosophical Transactions of the Royal Society of London, Series B*, **196**, 267–294.

SPAMER, E. E., BOGAN, A. E. & TORRENS, H. S. 1989. Recovery of the Etheldred Benett collection of fossils mostly from Jurassic–Cretaceous strata of Wiltshire England: analysis of the taxonomic nomenclature of Benett (1831), and notes and figures of type specimens contained in the collection. *Proceedings of the Academy of Natural Sciences of Philadelphia*, **141**, 115–180.

STACE, H. E., PETTITT, C. W. A. & WATERSTON, C. D. 1987. *Natural Science Collections in Scotland*. National Museums of Scotland, Edinburgh.

STEARN, W. T. 1998. *The Natural History Museum at South Kensington: A History of the Museum, 1753–1980*. Natural History Museum, London.

STEPHENS, J. 1912a. Clare Island Survey. Marine Porifera. *Proceedings of the Royal Irish Academy*, **31B**(59), 1–42.

STEPHENS, J. 1912b. Clare Island Survey. Fresh-water Porifera. Proceedings of the Royal Irish Academy, **31B**(60), 1–18.

SUESS, E. 1904–24. *The Face of the Earth*. Vols 1–5. Clarendon Press, Oxford. [Trans. Sollas, H. B. C.]

SWEET, J. M. 1963. Robert Jameson in London, 1793. Excerpts from Robert Jameson's 'Journal of a Voyage from Leith to London 1793'. *Annals of Science*, **19**(2), 81–116.

SWEET, J. M. 1967. Robert Jameson's Irish journal, 1797. Excerpts from Robert Jameson's 'Journal of my Tour in 1797'. *Annals of Science*, **23**, 97–126.

SWEET, J. M. & WATERSTON, C. D. 1967. Robert Jameson's approach to the Wernerian theory of the earth, 1796. *Annals of Science*, **23**, 81–95.

THOMSON, K. S. 2004. A palaeontological puzzle solved? *American Scientist*, **92**(3), 209.

THOMSON, K. S., SUTTON, M. & THOMAS, B. 2003. A larval Devonian lungfish. *Nature*, **426**, 833–834.

TORRENS, H. S. 1985. Women in Geology. 2 – Ethedred Benett. *Open Earth*, **21**, 12–13.

TORRENS, H. S., BENAMY, E., DAESCHLER, E. B., SPAMER, E. E. & BOGAN, A. E. 2000. Etheldred Benett of Wiltshire, England, the first lady geologist: Her fossil collection in the Academy of Natural Sciences of Philadelphia, and the rediscovery of 'lost' specimens of Jurassic Trigoniidae (Mollusca: Bivalvia) with their soft anatomy preserved. *Proceedings of the Academy of Natural Sciences of Philadelphia*, **150**, 59–123.

VALIER, H. K. 2004. Longstaff, Mary Jane (née Donald) (1856–1935). *In*: LIGHTMAN, B. (ed.) *Dictionary of Nineteenth-Century British Scientists*. Thoemmes Continuum, Bristol, 1260–1261.

WALLACE, P. 1969. Fifty years of feminine fellowship. *Proceedings of the Geological Society of London*, **1658**, 209–214.

W[ILLIAMS], A. 1969. Muir-Wood, Helen Marguerite (1895–1968). *Proceedings of the Geological Society of London*, **1655**, 123–125.

WYSE JACKSON, P. N. 1992. The geological collections of Trinity College, Dublin. *The Geological Curator*, **5**, 263–274.

WYSE JACKSON, P. N. 1998a. Veronica Burns 1914–1998, Curator Geological Museum, Trinity College, Dublin 1964–1980. *The Geological Curator*, **6**, 339–340.

WYSE JACKSON, P. N. 1998b. Veronica Conroy Burns (1914–1998). *Irish Naturalists' Journal*, **26**, 1–4.

WYSE JACKSON, P. N. 2004. Thomas Hawkins, Lord Cole, William Sollas and all: casts of Lower Jurassic marine reptiles in the Geological Museum, Trinity College, Dublin, Ireland. *The Geological Curator*, **8**, 11–18.

WYSE JACKSON, P. N. 2006. *The Chronologers' Quest: Episodes in the Search for the Age of the Earth*. Cambridge University Press, Cambridge.

WYSE JACKSON, P. N., NUDDS, J. R. & HIGGS, B. 2000. *Phanerotinus cristatus* (Phillips, 1836): a giant flanged gastropod from the Lower Carboniferous of Ireland. *Irish Journal of Earth Sciences*, **18**, 113–122.

The historical problems of travel for women undertaking geological fieldwork

C. V. BUREK[1] & M. KÖLBL-EBERT[2]

[1]*Centre for Science Communication, Department of Biological Sciences, University of Chester, Parkgate Road Chester, CH1 4BJ, UK (e-mail: c.burek@chester.ac.uk)*

[2]*Jura-Museum Eichstätt, Willibaldsburg, 85072 Eichstätt, Germany (e-mail: Koelbl-Ebert@jura-museum.de)*

Abstract: From unsuitable clothes to lack of chaperones, from sexual harassment to lack of proper funding, throughout history women geologists have encountered difficulties travelling to their field locations or working in the field, whether these locations were close by or abroad. From Etheldred Benett to the present day, problems were often sociological and political as well as logistical. Most early women geologists were able to avoid many difficulties because they were protected through working locally, where their high social standing was known and respected, or because they worked in a team with husband, father or brother. However, the problem developed virulence in the second half of the 19th century, when women started to appear as students and professionally trained geologists. The single travelling woman geologist had to face discriminating attitudes, ranging from pity to disregard and even to sexual harassment. Benevolent society also had its problems with these women, when, for example, professors needed their wives as chaperones to take women students on field trips.

While women geologists out in the field certainly had and have to face problems because of their gender, those problems have also been used as an argument against employment of female geologists out of paternalistic concern, i.e. a discriminating strategy which is possibly the most difficult to reject.

This paper explores the difficulties of women geologists undertaking fieldwork from the late 18th to the mid-20th century by looking at specific examples and deducing the social and logistic reasons behind these problems.

It is a well-known saying within geology that the best geologist is the one who has seen the most rocks. Rocks need to be interpreted *in situ* and, in quantity, cannot be moved any distance, so it requires the student and researcher to travel to them. Therefore, unlike some other sciences, geology usually requires outdoor activities. Originally this was perceived as a masculine occupation requiring stamina and fitness, the ability to travel and stay in remote, inhospitable and often hazardous environments. The same is also true for the study of physical geography (Bracken & Mawdsley 2004). This was even more important in earlier times, before the extensive use of computers and of petrological or geochemical laboratory techniques. Thus, fieldwork is the foundation of nearly everything that is later done in the laboratory or the office. For the woman geologist, this simple truth sometimes causes problems. From unsuitable clothes to lack of chaperones, from sexual harassment to lack of proper funding, throughout history women geologists have encountered difficulties travelling to their field locations or working in the field, whether these locations were close by or abroad.

Historically, travelling women are by no means an exclusively recent event. Women used to travel vast distances as pilgrims; however they usually travelled in groups, as is shown, for example, in Chaucer's 14th century *Canterbury Tales*, and were often accompanied and 'protected' by male relatives. Mary Wollstonecraft's (author of the *Vindication of the Rights of Women*, 1792) well-publicized voyage to Scandinavia with just a maid and her baby in 1795 (Wollstonecraft 1796), however, was still an unusual event (Todd 2000). While ostensibly on business for her husband, as an observant traveller she made one or two geological observations by remarking on the rocks of the lakeshore. However, these were merely reflections on passing and cannot be taken as anything more.

Solitary girls and women collecting fruit or wood in the forests or working as shepherdesses also used to be a common sight in previous centuries. However, what was normal for working-class women, such as professional fossil collector and dealer Mary Anning (1799–1848) of Lyme Regis, caused difficulties for upper-class women. Most of the early women geologists of the late 18th and early 19th centuries came from this latter social background. Nevertheless, British women, in particular those on 'The Grand Tour', travelling along the Rhine Valley to Italy, accustomed contemporary men and women in Germany to the sight of travelling females (Pelz 1991).

It was speculated by German feminists (Pelz 1991) that female mobility was enhanced by the introduction of public transportation via a regular coach network, starting in the 18th century. This shifted travel from horseback to carriages, which was a great advantage to women travellers. By travelling in a 'closed box', women could move around in mobile, private homes, thus travelling and, so to speak, staying at home at the same time. Later, the introduction of the railway solved part of the conflict between social respectability and travel.

In the United Kingdom, this was particularly true of the early Geologists' Association's field trips. The first field trip, to Folkestone on 9 April 1860, was made by train, leaving at 08.35 am and arriving at noon. The second was to Maidstone on 19 June 1860, to inspect the Hythe beds, which had yielded the type specimen of *Iguanodon mantelli* in 1834. The annual report of that year (Geologists' Association, 1860, p. 57) states 'these excursions gave great satisfaction to members', and this, of course, included females who had been encouraged to join the Association from its inception in 1858 (Sweeting 1958; Green 1989). One of the first photos to survive to record this type of travel on the newly opened railway line to Hertford was not taken until 54 years later in 1914 (Fig. 1). Later, cycling excursions were introduced by the society but these proved much less popular. An early attempt to encourage a cycling branch by Professor J. F. Blake in 1898 succeeded at first with at least eight cycling expeditions between 1899 and 1905 (Green 1989). However, these were not a success and were abandoned in 1907 (Sweeting 1958). Field trips by bus were popular, as is shown in Figure 2, a photo from 6 June 1914 on a trip to Newland's Corner, near Guildford, Surrey.

Early women geologists

Early female geologists, mostly British (see Kölbl-Ebert 2001*a*, 2002), were able to avoid some difficulties by following one of two strategies. They either used to work locally, where their high social standing was known and respected or they worked in a team with their husband or brother. Stratigrapher and palaeontologist Etheldred Benett (1775–1845) (Fig. 2), the unmarried, financially independent daughter of a Wiltshire country squire, is an example of the former category. She spent much of her time collecting fossils and on stratigraphical work. Her layer-by-layer description of the strata and fossils of Upper Chicksgrove quarry at Tisbury on the estate of her brother, is one of the earliest detailed quarry sections published in England (Burek 2001). Many of her specimens are depicted in Sowerby's *Mineral Conchology of Great Britain* (1812–20). She also contributed stratigraphical information to Greenough's *Geological Map of England and Wales*, and published a stratigraphically arranged catalogue of the specimens in her collection, including a number of new species, which were first illustrated by her (Torrens 1985; Torrens *et al.* 2000; Burek 2003*a*). While Miss Benett could venture out for fieldwork on her 'homeground', where she was known and respected as the squire's daughter, she inevitably had to face problems as soon as she left familiar paths.

A lady going into the quarries is a signal for the men begging money for beer, and the few times I have been there [i.e., in Portland] I never got a specimen worth bringing home. All my

Fig. 1. Geological excursion by train in 1914 by the Geologists' Association to Hertford.

Fig. 2. Bus excursion by the Geologists' Association in 1914 to Newland's Corner, Guildford, Surrey.

Portland fossils have been purchased in Weymouth!. (Letter to Gideon Mantell. 2 November 1835, quoted after Cleevely 1998)

The second category, the use of their own wife or sister as a geological assistant (Fig. 3) was quite a common feature in the first half of the 19th century:

After the last Geological Society meeting of the spring season, the leading researchers gathered up their hammers and their wives and set off on extensive stratigraphical tours. (Secord 1990)

We should remember that the French Revolution had brought in a new fashion in clothes, which for a few decades allowed fashionable and wealthy women to move around in comparatively comfortable dresses. Thus the fashion of the first three decades of the 19th century contributed in no small way to their 'mobility'. On the other hand, the renewal of corseting and the crinoline, both of which became fashionable during the 1850s were compared by Roderick Murchison, whose wife accompanied him on fieldtrips as long as her health permitted, with 'cages' and 'horrid iron girdles round their [i.e. women's] legs' (Geological Society of London: M/J7, p. 71]. However, during 1851, the introduction in the cycling community, of the divided skirt or 'bloomer' named after Amelia Bloomer, caused a sensation. However, it soon became acceptable and thus the fashion in society again allowed greater freedom of movement.

Female students and professionals

While the early female geologists could manage quite well, as long as they adhered to the rules of their society, the problem of whether geological fieldwork was seemly for women developed virulence in the second half of the 19th century, when women started to appear as students and professionally trained geologists. In 1894, Professor H. B. Woodward of the Geological Association feared 'that women, when mapping in quarries and railway-cuttings, might encounter men, and not always polite men and even cock-fighting in Glamorganshire' (Appleby 1979).

This was a problem, which 80 years earlier did not seem to concern Lady Mary Cole, who mapped for William Buckland in Glamorganshire, South Wales (Gordon 1894).

Fig. 3. Mrs McKenny Hughes (front row, 4th from left) helping her husband as an assistant on a field trip to the Malverns in 1913.

An example from *Queen* magazine highlights the attitudes (Linton 1890). The piece, entitled 'A pleasure party' concerns college girls visiting the Lake District as part of a mixed group of students. The landlady describes them as 'abnormal specimens of humanity'. She was very dubious about having a group of college geology students, which included women and was seriously concerned about their behaviour. However,

The rivals on the mountainside become once more squire and dame within the four walls of the house.

She finds it

... strange at first to hear the long hard Latin names flow as glibly over those young lips as if they had learned them at their mother's knee.

She is also relieved that:

Rivals in science and each eagerly desirous of scientific finds on the fells, where all fend for themselves on independent lines ... at table, in the drawing room, in all the acts and observances of social life, the old order reigneth.

In order to avoid more damaging gossip, professors had to take their wives as chaperones on fieldtrips. The landlady quoted above became enchanted by the professor's wife.

She is young, pretty, distinguished in name and race ... She is as modest as she is learned and as learned as she is womanly .... She does not despise the natural leadership of men nor seek to turn society and the world upside down that women may usurp function for which they are naturally unfit.

The problem of decency in the late 19th century was no small matter. Archibald Geikie had to deal with it, when he was appointed Professor of Geology at the University of Edinburgh in 1871. He was one of the first professors to offer geology courses for women outside the university in rooms rented by the Ladies' Educational Association. He took his female students on geological field trips, the first of which took place on 14 December 1872. This was regarded as a novelty in the Scottish educational system. In later years, Geikie regularly took female students on field trips, which lasted several days, and on which his wife and another older woman accompanied him as chaperones (Fenton & Fenton 1952).

When in the summer of 1891, Professor Ferdinand von Richthofen took the young British palaeontologist and alpine geologist Maria Matilda Ogilvie Gordon (1864–1939) (Fig. 4) for a month-long excursion to south Tyrol, his wife accompanied them to prevent etiquette breaches (Creese 1996; Burek 2005; Watchler & Burek 2007).

Fig. 4. Maria Ogilvie Gordon.

Taking a wife into the field, however, was only possible if the wife was sympathetic, and thus it was often the professors that were married to wives with progressive thoughts about women's education who tended to be rather liberal themselves as to the problem of accepting female students. In 1892, Professor McKenny Hughes took an excursion to the Malverns accompanied by his wife Mary, who was interested and later researched in geology. Out of the 23 students, 9 were female (39%) including Ethel Wood, Ethel Skeat, Margaret Crosfield, Elizabeth Dale and Clara Dale, Gertrude Elles, Margaret Skeat, Dorothy Alford and Margaret Alford. The Zurich geology professor Albert Heim, for example, who liberally admitted women to his courses, was married to physician Dr Marie Heim-Vögtlin (1845–1916), the first woman to enter the Faculty of Medicine at the University of Zurich (Katharina von Salis, Zurich, pers. comm.; Ogilvie & Harvey 2000).

In the United Kingdom, with the cloister-like colleges at the old, prestigious universities, there had been fewer problems, because in all-female colleges, such as Bedford College, London, there were often female lecturers and staff (Burek 2007). They frequently took their students on field trips and even ventured overseas (Fig. 5). For example, by 1936, Emily Dix was regularly taking her Bedford College students abroad but in the company of Head of Department, Leonard Hawkes (Burek & Cleal 2005). This was carrying on the fieldwork tradition started by Catherine Raisin 40 years before.

The female staff sometimes adopted another strategy to overcome society's problems with female students in the field. Dr Catherine Raisin for example, first female Head of Department of Bedford College, London, would take 'her girls' out into the field until male students arrived, at which point she would march her girls out of the quarry or away from the cliff section (Burek 2003a). Professor Lapworth often took mixed classes out into the field (Fig. 6). However, this can still be a problem today, and most geology departments include a female member of staff on geological excursions if there are female students present (personal experience, C. Burek).

An unusual case of male–female cooperation in fieldwork emerged in 1887 when Catherine Raisin published her paper on pioneering work mapping metamorphic facies (Burek 2003a). Her paper followed up on work by Professor Bonney in south Devon and Salcombe, Devon. She made two visits to the area to collect and study for this first paper and a pattern started to emerge where Catherine did the fieldwork and Professor Bonney did the microscopy. Catherine travelled widely to collect their specimens as far afield as Vosges Mountains, the Eifel region, Brenner Pass, and other areas including Finland and Sweden. (Burek 2003b). He acknowledged this in 1899 (Bonney & Raisin 1899). This arrangement was extremely unusual at the time and is testimony to the esteem in which both professor and colleagues held Catherine Raisin.

On the other hand, many females suffered from lack of cooperation out of society's concern with decency. The above-mentioned Maria Ogilvie Gordon had to travel by herself or with her family (Burek 2005; Wachtler & Burek 2007) and recalled in her response to receiving the Lyell medal from the Geological Society of London 'it was a lonely furrow that I ploughed in my fieldwork abroad' (Ogilvie Gordon 1932). In 1891, when female students were not yet permitted to attend German or Austrian universities, she became a private student of Professor von Zittel (Munich) and was obliged to listen to his lectures from a side-room as she was not allowed to sit in the lecture hall together with the male students. The social corset of her time excluded her from the network that promoted her male colleagues: 'I was not under the eye of any Professor. There was no one to include me in his official round of visits among the young geologists in the field' (Ogilvie Gordon 1932).

The problem of social decency continued well into the 20th century, although matters improved slowly

Fig. 5. Bedford College field trip to Antrim, Easter 1937.

but steadily. In the United States, in the 1930s, Helen Tappan and her future husband Alfred Löblich met as students at the University of Oklahoma. The two students did most of their fieldwork together, with Helen's father acting as chauffeur and chaperone. This was the start of a fruitful collaboration, which, contrary to that of many other couples, survived marriage and children. From the 1950s onwards, husband and wife collaborated on many joint projects, such as the Foraminifera volumes of the *Treatise on Invertebrate Paleontology*. They managed to travel and work together, since their four children, who accompanied them, were attended by a grandmother (Löblich 1993).

Carlotta Joaquina Maury (1874–1938) was a veritable female geological globetrotter. She received her geological education at Radcliffe College (United States). She was palaeontologist to A. C. Veatch's geological expedition to Venezuela (1910–11), organized and conducted the Maury expedition to the Dominican Republic (1916), worked as consulting palaeontologist and stratigrapher for an oil company in Venezuela, and was official palaeontologist to Brazil (1914). Additionally, she was employed as lecturer and later as professor at various colleges and universities in North America and South Africa (Ogilvie 1986).

Fig. 6. Professor Lapworth (top hat and beard) with a mixed Geologists' Association group in the field.

Fieldwork and the job situation

In the 1950s, there were:

... plenty of women geologists, and good ones, in the Texas–Oklahoma–Kansas area, of course. However, most stay in town or back at the home offices of the oil companies. All except Mrs. Bitgood. You can usually find her, accompanied by her big German shepherd, Susie, and perhaps her 11-year-old son, Charles Decker, out where the wells are drilling. She loves it. (Frizzel 1993)

Ellen Posey Bitgood (died 1982), daughter of a tool driller, grew up close to the oil business, which enabled her to feel comfortable around the oilrigs, radiating self-confidence and natural authority without being offensive to the workers on the drill site. It was probably this which made this graduate of the University of Oklahoma special, although she must have also impressed the men by being better than everybody else, be it geologically or physically (Frizzel 1993).

As with much oral history, this little story about the exceptional Mrs Bitgood reminds us that the examples, which are remembered, are of those of 'winners' and not of the 'losers'. The problems women geologists allegedly have to face in the field still affects their job situation today. This can be seen where these problems have actively been used as argument against employment of women geologists out of good-intentioned paternalistic concern. This becomes a discriminating strategy which is possibly the most difficult to reject, because many would-be employers or colleagues of women geologists are unaware of their discriminating attitude:

'No, we cannot give the job to this woman. She is too young and tender, the strain of the field work would be too much for her'.

She wouldn't like the rough treatment by the workers on the building-site.

(Quotes courtesy of members of the Commission for Equal Opportunity at a German university, collected during the 1990s; see also Kölbl-Ebert 2001a). Thus, historically, women, if they were employed, tend to be found more often in office- or laboratory-based positions, where they worked with museum collections, for example, Tilly Edinger (Hofer 1969), or as human 'computers', such as Inge Lehmann (Kölbl-Ebert 2001b and references therein) or Marie Tharp (Barton 2002).

A previous edition of this paper was first read as 'The problems of women as geological travellers prior to 1939', at the INHIGEO International Commission of the History of Geological Sciences 28th Symposium on the Theme of Geological Travellers, Dublin, July 2003. We would like to acknowledge Royal Holloway College for permission to use the photographs from Bedford College archive and Margaret Carrick of the Geologists' Association for access to the Geologists' Association's fieldwork photographs. The photo of Maria Ogilvie is courtesy of M. Wachtler. We would like to thank Patrick Wyse Jackson for his encouragement to include this paper in the book rather than elsewhere and to the referees who made some valid suggestions. Finally we would like to thank all those pioneering women who have trod carefully along the footpaths of time!

References

APPLEBY, V. 1979. Ladies with hammers. *New Scientist*, 29 November, 714–715.

BARTON, C. 2002. Marie Tharp, oceanographic cartographer, and her contributions to the revolution in the Earth sciences, *In*: OLDROYD, D. R. (ed.) *The Earth Inside and Out: Some Major Contributions to Geology in the Twentieth Century* Geological Society, London, Special Publications, **192**, 215–228.

BONNEY, T. G. & RAISIN, C. A. 1899. On varieties of serpentine and associated rocks in Anglesey. *Quarterly Journal of the Geological Society of London*, **55**, 276–304.

BRACKEN, L. & MAWDSLEY, E. 2004. 'Muddy Glee' rounding out the picture of women and physical geography fieldwork. *Area*, **36**, 280–286.

BUREK, C. V. 2001. The first lady geologist or collector par excellence? Women in the History of Geology II. *Geology Today*, **17**, 192–194.

BUREK, C. V. 2003a, Catherine Raisin, a role model professional geologist. *Geology Today*, **19**(3), 107–111.

BUREK, C. V. 2003b. Time to take responsibility for collections. *Earth Heritage*, **20**, 22–23.

BUREK, C. V. 2005. Who were they? The lives of geologists. 5. Dame Maria Matilda Ogilvie

Gordon – a Britisher – and a woman at that (1864–1939). *Teaching Earth Sciences*, **30**, 42–44.

BUREK, C. V. & CLEAL, C. J. 2005. The life and work of Emily Dix (1904–1972). *In*: BOWDEN, A. J., BUREK, C. V. & WILDING, R. (eds) *History of Palaeobotany, Selected Essays*. Geological Society, London, Special Publications, **241**, 181–196.

BUREK, C. V. 2007. The role of women in geological higher education – Bedford College, London (Catherine Raisin) and Newnham College, Cambridge, UK. *In*: BUREK, C. V. & HIGGS, B. (eds) *The Role of Women in the History of Geology*. Geological Society, London, Special Publications, **281**, 9–38.

CREESE, M. R. S. 1996. Maria Ogilvie Gordon (1864–1939). *Earth Sciences History*, **15**, 68–75.

CLEEVELY, R. 1998. Picture quiz [featuring Miss Etheldred Benett (1776–1845)]. *The Linnean*, **14**, 4–9.

FENTON, C. L. & FENTON, M. A. 1952. *Giants of Geology*. Doubleday, New York.

FRIZZEL, M. V. 1993. One of the smartest women in Texas. *Earth Scientist*, Fall Issue, 22–23.

GEOLOGICAL SOCIETY, London, Library. Murchison Papers: M/J7.

GEOLOGISTS ASSOCIATION 1860. Annual report. *Proceedings of the Geologists Association*, **1**, 57.

GORDON [E. O.] 1894. *The Life and Correspondence of William Buckland, D.D., F.R.S., Sometime Dean of Westminster, Twice President of the Geological Society, and First President of the British Association*. John Murray, London.

GREEN, C. P. 1989. Excursions in the past: a review of the field meeting reports in the first one hundred volumes of the *Proceedings of the Geologists' Association*. *Proceedings of the Geologists' Association*, **100**, 17–29.

HOFER, H. 1969. In memoriam Tilly Edinger. *Morphologisches Jahrbuch*, **113**, 303–317.

KÖLBL-EBERT, M. 2001a. On the origin of women geologists by means of social selection: German and British comparison. *Episodes*, **24**, 182–193.

KÖLBL-EBERT, M. 2001b. Inge Lehmann's paper: 'P' (1936). *Episodes*, **24**, 262–267.

LINTON, E. L. 1890. A pleasure party. *Queen*, August.

LÖBLICH, H. T. 1993. Reminiscence of scientific careers: Helen Tappan Löblich and Alfred R. Löblich, Jr. *Earth Scientist*, Fall Issue, 10–11.

MCKENNY HUGHES, T. 1892. Excursion album of Prof. Hughes, Malvern, 1892. Sedgwick Museum, Cambridge, Archive of the Sedgwick Club.

OGILVIE GORDON, M. 1932. President's medal presentation: Mrs Ogilvie Gordon's reply. *Proceedings of the Geological Society of London*, **88**, 49–50.

OGILVIE, M. B. 1986. *Women in Science. Antiquity through the Nineteenth Century*. MIT Press, Cambridge, Massachusetts.

OGILVIE, M. B. & HARVEY, J. (eds) 2000. *The Biographical Dictionary of Women in Science. Pioneering Lives from Ancient Times to the Mid-20th Century*. Vols 1 & 2. Routledge, New York.

PELZ, A. 1991. Reisen Frauen anders Von Entdeckerinnen und reisenden Frauenzimmern. *In*: BAUSINGER, H., BEYRER, K. & KORFF, G. (eds) *Reisekultur. Von der Pilgerfahrt zum Modernen Tourismus* Verlag C. H. Beck, München, 174–178.

SECORD, J. 1990. *Controversy in Victorian Geology: The Cambrian–Silurian Dispute*. Princeton University Press, Princeton.

SWEETING, G. S. 1958. *The Geologists' Association 1858–1958*. Benham & Co., Colchester.

TODD, J. 2000. *Mary Wollstonecraft: A Revolutionary Life*. Phoenix Press, London.

TORRENS, H. S. 1985. Women in geology. 2. Etheldred Benett. *Open Earth*, **21**, 12–13.

TORRENS, H. S., BENAMY, E., DAESCHLER, E. B., SPAMER, E. E. & BOGAN, A. E. 2000. Etheldred Benett of Wiltshire, England, the first lady geologist. Her fossil collection in the Academy of Natural Sciences of Philadelphia, and the rediscovery of 'lost' specimens of Jurassic Trigoniidae (Mollusca: Bivalvia) with their soft anatomy preserved. *Proceedings of the Academy of Natural Sciences of Philadelphia*, **150**, 59–123.

WACHTLER, M. & BUREK, C. V. 2007. Maria Matilda Ogilvie Gordon (1864–1939): a Scottish researcher in the Alps. *In*: BUREK, C. V. & HIGGS, B. (eds) *The Role of Women in the History of Geology*. Geological Society, London, Special Publications, **281**, 305–317.

WOLLSTONECRAFT, M. 1792. *A Vindication of the Rights of Women*. *In*: RHYS, E. (ed.) Science, Everyman's Library, J. M. Dent & Sons Ltd. Reprinted *c.* 1929, London.

WOLLSTONECRAFT, M. 1796. Letters from Norway, Sweden and Denmark. *In*: MORRIS, M. (ed.) *The Virago Book of Women Travellers*. Virago Press, London, 1993, 8–13.

Great expectations: Florence Bascom (1842–1945) and the education of early US women geologists

R. M. CLARY[1] & J. H. WANDERSEE[2]

[1]*Department of Geosciences, Mississippi State University, P.O. Box 5448, Mississippi State, MS 39762-5448, USA*

[2]*Department of Educational Theory, Policy, and Practice, Louisiana State University, Peabody Hall, Room 223 F, Baton Rouge, LA 70803, USA*

Abstract: Florence Bascom, the first woman to be awarded a PhD from Johns Hopkins University and the first professional US woman geologist, gained access through male connections into what was perceived to be a man's discipline. Her entrance into geology was not straightforward, but instead proceeded fairly erratically through an interwoven network of male acquaintances. Although Bascom's participation in professional circles garnered her acceptance by her male associates, it was her long academic career at Bryn Mawr College that made an impact on education and secured a supply of women in the discipline. Basing her instruction upon high expectations for herself and her students, innovative teaching techniques, and the pure joy she derived from geological investigation and geological thought, Bascom served as mentor to many in the succeeding generation of US women geologists. It was largely through Bascom's efforts that more women were able to pursue higher education in geology and excel at it in the first half of the 20th century.

During the late 1800s, the cultural environment of the United States was conducive to scientific endeavours, but only if the participants were of the male gender. Female scientists were an oddity; if women were employed within the sciences, they occupied secondary roles with little hope of advancement. Women also were barred from admittance to most universities until the latter part of the 19th century. When the momentous occasion came, and women were allowed into higher education programmes, the women faced yet another obstacle to their general education: some disciplines were considered to be in the realm of women, while others, particularly most physical sciences, were considered exclusive to men. Although some women achieved visibility in the biological sciences and astronomy, the geological realm was essentially devoid of women. In the United States, the woman who challenged this scenario was Florence Bascom (1862–1945), the 'Pioneer of Women Geologists' (Fig. 1).

Florence Bascom became the first woman to whom Johns Hopkins University awarded the PhD degree. She was also the first woman to be hired by the United States Geological Survey (USGS) as an assistant geologist. Although some modern accounts depict a pivotal moment in which Bascom decided to make geology her life profession, it appears instead that her entrance into the geological discipline was erratic and partially based on the presence of male acquaintances. Bascom entered into new geological territories with a male contact already firmly in place. In order to be successful as a geologist, Bascom had to participate in the professional organizations and conduct fieldwork like a 'regular fellow'; an academic career at a women's college was unlikely to garner the respect of her male associates. However, it was through her academic career that Bascom had the most lasting impact on the discipline of geology. She trained and mentored many in her successive generation of US women geologists while at Bryn Mawr College, having single-handedly founded the geology department there. In Bascom's teaching pedagogy, three themes emerge: Bascom had high expectations of herself and others, her teaching techniques were innovative, and a spirit of pure joy pervaded her geological investigation and geologic thought.

Role of male acquaintances in career decisions

Florence Bascom, the youngest of five children, was fortunate to have been born into a family which encouraged women's entrance into mainstream US society (Fig. 2). John Bascom (1827–1911), a professor of philosophy, and her mother Emma Curtiss Bascom, a women's rights activist in the suffrage movement, were certainly more tolerant and liberal-minded than the average US parents.

From: BUREK, C. V. & HIGGS, B. (eds) *The Role of Women in the History of Geology.* Geological Society, London, Special Publications, **281**, 123–135. DOI: 10.1144/SP281.8
0305-8719/07/$15.00 © The Geological Society of London 2007.

Fig. 1. Florence Bascom in her field outfit, holding a Brunton compass (Sophia Smith Collection, Smith College).

The greater parental influence on Florence Bascom's decisions appears to be from her father. Smith (1981, p. 15) reported that Bascom 'sought the company of her high-minded father. They understood each other'. Bascom also later remarked that her father was 'all that I needed, association with him gratified all I wanted or needed so far as men were concerned' (Smith 1981, p. 24).

Although Bascom emerges as a pioneer for US women in geology, her entrance into the field was not made without extensive networking. She relied on previous male contacts to gain entrance into a man's discipline. Her movement into geology was punctuated, and major educational and career moves were predicated by having a male contact in place within an institution. Her father emerges as the first who influenced her education and career decisions. He was followed by his employees at the University of Wisconsin: Roland Irving and Charles Van Hise; a colleague of Irving's, George Huntington Williams; a friend of the Bascom family, Edward Orton; and a crystallographer with whom Williams had some connection, Victor Goldschmidt (Fig. 3).

Male mentor 1: John Bascom (1827–1911)

When Florence was only 12 years old, John Bascom accepted a position as the President of the University of Wisconsin and moved his family from their native Massachusetts. The move was important for Florence: it was at the University of Wisconsin that Bascom began her post-secondary education in 1877. Although the University of Wisconsin began admitting women in 1872, co-education did not mean equality. Men and women were not allowed in the library on the same days, and women had to defer to men if a classroom was full. The gymnasium was almost the exclusive domain of men (Bascom 1925). Bascom remarked in an undergraduate essay that the boundaries for the sphere of women 'seem too narrow for womankind. They are constantly being blamed for "going out of their sphere"' (Arnold 1999, p. 165). It is ironic perhaps, that although Florence was encouraged to move into the sphere of men, her older sister Jennie became the family's social 'sacrificial lamb' and never moved beyond traditional social and domestic roles, in spite of college experience.

Geology did not claim Bascom's interest immediately. The first degrees she received were two bachelor's degrees in arts and letters in 1882. She then spent a year after graduation in Madison, Wisconsin, engaged in 'social' activities (Arnold 1999) (Fig. 4). It was her father who encouraged Bascom to study geology. John Bascom suggested that Florence return to the University of Wisconsin and 'take a thorough course in geology under Professor Irving' (Arnold 1999, p. 167). However, Bascom applied to Hampton Institute in Virginia, one of the first colleges for blacks and a pioneering institution for Native American education. When she learned that she had not been accepted, Bascom returned to Wisconsin and continued her education, earning a bachelor's degree in science (1884), which would eventually be followed by a master's degree in geology (1887).

Even so, a geological career trajectory was not evident after Bascom's first science degree. Although it has been reported that her interest in geology was heightened by a trip to Mammoth Cave with her father (Reynolds 1999), this appears speculative. Smith (1981) noted only that Bascom and her father had travelled some distance 'to perhaps visit a cave' after her bachelor's degree was awarded. If this was a pivotal moment in Bascom's life choice, it is not evidenced by her next action. She re-applied to Hampton Institute and was accepted. Bascom taught for a year, but did not seem to relish the

Fig. 2. The Bascom family, *c.* 1880. From left are John Bascom, Florence, Jennie, George, and Emma Curtiss Bascom (Sophia Smith Collection, Smith College).

experience. After returning home in June 1885, she then accompanied a group toward the Great Sioux Reservation, as far as the Missouri River. Although later she claimed that her father provided the only male association she ever needed, Bascom's infatuation with a member of the party was documented in her diaries (Arnold 1999).

Smith (1981, pp. 20–21) reported that Bascom's

... interest in this branch of natural sciences had first been aroused years earlier when, on a drive with her father, they were accompanied by one of his friends, Dr Edward Orton, geologist at Ohio State University ... It was he, while driving with John Bascom and his daughter, who pointed out to her an aspect of the landscape which she had not understood. From that day she had been stimulated to learn more about the making of a landscape, to understand the constitution of the earth, and geologic processes.

Rossiter (1981) also believed that Orton (1829–99) should be credited for Bascom's interest in geology; he eventually re-emerged as an important male

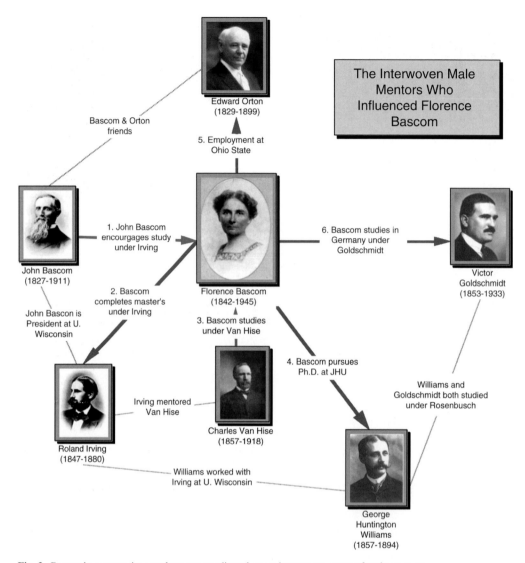

Fig. 3. Bascom's entrance into geology was predicated on an interwoven group of male mentors.

mentor to Bascom. Once the commitment began, Bascom felt that geology was the:

... most important of the sciences, to some extent including the others, and that it was really the culmination of all science. (Ogilvie 1945, p. 320)

Male mentors 2 and 3: Roland Duer Irving (1847–80) and Charles Robert Van Hise (1857–1919)

Bascom returned to the University of Wisconsin in 1885 to study microscopy, again at the encouragement of her father. Her geological interests turned to petrology and structural geology, and her thesis work in geology focused on petrography, specifically the sheet gabbros of Lake Superior. Her thesis advisor, Roland Duer Irving, was a pioneering petrographic microscopist and had previously served as a mentor to Charles Robert Van Hise. Bascom (1925) reported that Irving 'by training and temperament' was 'disinclined to coeducation'. It was Van Hise, as assistant professor, who probably worked more closely with Bascom and therefore had more influence on Bascom as far as her eventual teaching methods.

Most modern geologists rally around the slogan that 'geology is best taught in the field'. However,

Fig. 4. After receiving two bachelor's degrees, Bascom spent a year engaged in social activities before returning to the University of Wisconsin. In this undated photograph, Bascom (left) is shown painting (Sophia Smith Collection, Smith College).

Bascom, mostly out of social considerations, did not accompany male geologists in fieldwork. It seems necessary that Bascom's geological training at the University of Wisconsin centered on petrographic analyses that could be conducted in the laboratory with hand samples and thin sections. Bascom (1925, p. 305) would later remark that the earlier days at the university were characterized by 'no manifestation ... of the spirit of research' methods (Knopf 1946). However, this was destined to change, as she was introduced during these years to George Huntington Williams, a professor on the faculty of Johns Hopkins (Arnold 1999).

Following her master's degree in microscopic petrography, Bascom did not immediately seek her doctoral degree. Her father left the presidency at the University of Wisconsin, and she followed her family back to Massachusetts. Following a brief stint as a teacher in a local school, she then taught science and mathematics for 2 years at Rockford College in Illinois, labouring under a heavy teaching load. Smith (1981) noted that the ebb in family finances facilitated this move, and was further aided by Bascom's restless, ambitious, independent and strong-minded spirit. Even though Bascom managed to set up a good teaching laboratory at Rockford, it was not long before she decided to return to school. This time, she chose an institution that did not admit women.

Male mentor 4: George Huntington Williams (1856–94)

Daniel C. Gilman, then President of Johns Hopkins, opposed the co-education of women at his university. However, Florence Bascom petitioned for admission by noting the limited opportunities for advanced petrographical study, as well as the prestige of Johns Hopkins (Arnold 2000). Her professors at the University of Wisconsin bolstered her application with letters of support and recommendation, while Edward Griffin, a Johns Hopkins professor and personal friend of John Bascom, championed her cause. Florence's application was approved, and she was allowed to attend classes in 1891. However, women were not allowed to enroll as regular students and were absent from the roster of students, which resulted in their being denied scholarships and fellowships. Furthermore, since women were considered to be a potential distraction to male students, they were required to sit behind a screen in the corner (Arnold 1984).

Bascom perhaps migrated to Johns Hopkins because of George Huntington Williams, her future doctoral advisor. Williams was not an unfamiliar face, having worked with Irving at the University of Wisconsin when Bascom was finishing her master's degree. Williams had trained in

Germany under the pioneering crystallographer, Rosenbusch, and this connection indirectly propelled Bascom to later study crystallography in Heidelberg. It was at Johns Hopkins that she finally had an opportunity to conduct research in the field, although she was moving against the established norms of acceptable lady-like behavior. (Williams' wife accompanied them on their first field experience.) Bascom's research again centred on microscopic techniques, and she was able to verify that the Precambrian rocks of Maryland's South Mountain were altered volcanic rocks instead of having the presupposed sedimentary origin. Although she was not guaranteed a degree when she entered Hopkins, Bascom's admission to candidacy was supported by George Huntington Williams, who used some of her fieldwork in his own reports (Rossiter 1982). She was unanimously passed in oral examination and received her PhD in 1893, the first woman to be awarded that degree by the institution. (Bascom was not, however, the first woman to be awarded a geology PhD, nor the first Hopkins woman to *earn* the doctoral degree: Christine Ladd-Franklin was denied a PhD after completing the doctoral requirements in 1882. Although her dissertation was published by the university in 1883, she was not awarded a PhD until 1926.) Bascom's PhD did not facilitate the admission of women at Johns Hopkins: a hostile response arose from the rest of the faculty, with physicist Henry Rowland publicly denouncing women candidates. Another 14 years were to pass after Bascom received her PhD before Johns Hopkins officially admitted women to the university (Rossiter 1982).

Male mentor 5: Edward Francis Baxter Orton (1829–99)

Bascom traveled to Ohio State University, degree in hand, to work with Edward Orton. If Orton had provided the initial inspiration for Bascom to study geology, it is fitting that she returned to her first mentor after earning her final degree. Although Bascom's appointment at a major university was unprecedented at the time, it received little attention (Arnold 2000). While at Ohio State University, she was involved in laboratory work, fieldwork and teaching; Bascom even managed to put inorganic geology on a firm basis (Arnold 2000). In spite of her duties and accomplishments, Bascom left Ohio State University in 1895 for Bryn Mawr College, the women's counterpart of Johns Hopkins. Arnold (1984) reported that Bascom was lured away from Ohio State University by the opportunity to conduct original research and teach higher level geology classes.

Male mentor 6: Victor Mordechai Goldschmidt (1853–1933)

Bryn Mawr would be Bascom's academic home for the remainder of her career. It was mostly through her efforts that the geology department at Bryn Mawr developed into a programme that achieved international respect. Although the President of Bryn Mawr, M. Carey Thomas, threatened to reduce geology to the status of an elective instead of a major in 1899, Bascom held fast to her principles and succeeded in keeping geology as an active programme. It would not be her only confrontation with Thomas, however. By 1903, Bascom had acquired an associate who relieved her of her palaeontology and stratigraphy requirements (Ogilvie 1945). In 1906, she was appointed to a full professorship. When Thomas later threatened to eliminate the geology department because of insufficient numbers, Bascom's students came to the rescue, raising $750 000 (Rossiter 1981).

During the 1906/7 academic year, Bascom took a sabbatical leave and travelled to Heidelberg, Germany, to study crystallography under Victor Goldschmidt (Fig. 5). It is probably not coincidental that Bascom chose to study in Heidelberg, just as her previous Johns Hopkins mentor, George

Fig. 5. Florence Bascom poses at the door of the laboratory in Heidelberg (Sophia Smith Collection, Smith College).

Huntington Williams, had done. There was a further connection between Williams and Goldschmidt: both men had studied crystallography methods under the pioneering crystallographer Rosenbusch. Bascom's year of study under Goldschmidt proved fruitful for both professional research analyses and teaching techniques. Bascom left Germany at the end of the sabbatical with firm friendships in place. She exchanged letters with Victor Goldschmidt and his wife for years, and their correspondence provides insights into Bascom's academic life at Bryn Mawr College and her professional work with the USGS.

The need for professional geological recognition and USGS field experience

A second theme to emerge in Florence Bascom's geological career was the need for a dual role in both professional and academic circles. Bascom was not content to focus solely on academia, and her involvement in geology quickly expanded to encompass membership in professional organizations and employment as a field geologist. Smith (1981, p. 5) commented on Bascom's twofold character, and remarked that Bascom provided

> ... a vivid example of a scientist committed not only to her career as a teacher, but as one very much involved in applying her skill to extending our knowledge of the world.

Bascom sought to expand her career through involvement in the Geological Society of America (GSA) and the USGS, and she won accolades from both.

Bascom's milestones with the Geological Society of America

Fortunately for Bascom, the GSA was more liberal than its British counterpart. The first woman fellow was elected when the GSA was only 1 year old, solely on the basis of receiving her doctorate. It is often incorrectly reported that Bascom was the first elected fellow of the GSA; that milestone was accomplished in 1889 by Mary Emilee Holmes, who received her PhD in paleontology from the University of Michigan. In 1894, Florence Bascom became the second elected fellow of the GSA, and was likewise elected to the position solely on the basis of her PhD. However, unlike Holmes, Bascom became the first woman fellow of the GSA to practise within the field of geology, and was often acclaimed as 'the dean of women geologists' (Eckel 1982, p. 37). Indeed, Emmanuel de Margerie would recall that Bascom was the only woman in attendance at the 1897 International Geological Congress in St Petersburg who had earned the right to be there (Smith 1981). Although de Margerie perhaps erred by not recognizing the attendance and contributions of two Russian women geologists, his statement serves as testimony to Bascom's importance within the international geological community. Bascom recorded further milestones with the GSA: she was the first woman to present a paper before the Geological Society of Washington (in 1901), to serve on the GSA Council (1925–27), and to serve as a GSA officer (Bascom served as the second Vice President in 1930).

The United States Geological Survey

Membership of the GSA was not sufficient to fulfil Bascom's professional ambitions. In 1896, she became the first woman to be hired as an assistant geologist by the USGS. In 1909, she was promoted to full geologist. A pattern quickly emerged for Bascom: she conducted fieldwork in the summer months, focusing on the mid-Atlantic Piedmont area of the Appalachian Mountains. In winter she would analyse the hand samples and thin sections of the crystalline rocks with the petrographic microscope. Winter was also the time when Bascom would write her reports. In all, Bascom wrote approximately 40 articles, with the majority of her USGS work represented by seven major folios, which she co-authored with her USGS colleagues. Bascom's work in the Appalachians has survived controversy and is still of value today.

Bascom's USGS reports, issued from 1909 to 1938, are characterized by succinctness and clarity of approach. Knopf (1946) reported that Bascom's incisive writing was similar to her conversational style, which was characterized as forceful, clear-cut and even caustic at times. This concise approach emerged early in her geological writings; in 1893, her report on the volcanic rocks of South Mountain was clear, crisp and organized with bulleted lists (Bascom 1893). On several occasions, Bascom sent copies of her work to her former teacher, Victor Goldschmidt. He praised Bascom's efforts, and remarked how professional and painstaking her USGS studies were (Arnold 1993). Goldschmidt noted that Bascom's publication in *The American Mineralogist* was 'so short, so simple, so clear and understandable' (Arnold 1993, p. 217).

Bascom was territorial in her fieldwork and disliked others trespassing within a research area that she considered to be her own (Arnold 1983). Therefore, an interesting scenario played out when Bascom's interpretation of the age of the Wissahickon mica schist was challenged by two of her former students: Anna Jonas Stose (1881–1974) and Eleanora Bliss (later Knopf) (1883–1974), who proposed an age of Precambrian for the mica schist and the presence of the Martic Overthrust (Arnold 1983). Even when publicly breaking with her students, Bascom remained succinct; she offered only 'an alternative hypothesis tentatively held by the writer' (Bascom & Stose 1932). Bascom maintained her clear organization and proceeded to list four observations in support of her hypothesis, concluding that

... any one of these observations taken singly is capable of another interpretation, but the combined observations seem to the writer to point to similar relations between the mica schist and Cambrian and Ordovician formations and to the Ordovician age of the schist. (Bascom & Stose 1932, p. 5)

Eventually, Benjamin Miller (1935, p. 755) challenged the analysis of Stose and Bliss, and concluded

Bascom's views concerning the age of these schists should not have been set aside.

Bascom's original interpretation prevailed, with eventual modifications.

Victor Goldschmidt was aware of the huge time allotments the USGS required of Bascom through her correspondence. Although he advised her to give up her work with the survey and focus upon her academic career at Bryn Mawr, this was probably never a serious consideration for Bascom. It was in the USGS that Bascom was at the same level with the male geologists, partaking of the same research opportunities and field excursions. She remarked in a letter to Goldschmidt that giving up her work with the USGS 'is like giving up my profession' (Arnold 1993, p. 199).

It should be acknowledged that, although Bascom could participate in similar field and research opportunities as male geologists, there was no equality between the genders. A discrepancy among salaries was obvious; even male field assistants earned 33% more than their female counterparts (Arnold 1983). When Bascom hired both male and female field assistants, Edgar Wherry earned $4 a day, while Eleanora Bliss earned $3 a day. Partly because of the inequalities, some women scientists tried to ignore their gender in order to avoid attention; Hild (1999) noted that this left them without alternative support in a difficult work environment. Bascom tried to avoid drawing attention to her gender, as evidenced by the author's name on her professional papers: she chose to use 'F. Bascom' (Bascom 1902). Her attempts at gender anonymity were not always successful, however; in the 1902 *Maryland Geological Survey*, one colleague, Edward Bennett Mathews, referred to the 'recalculations of *Miss Bascom*' [italics added].

Bascom's influence on early US women geologists

Florence Bascom served as role model for the next generation of women geologists for both the completion of her doctoral degree, and the application of her scientific knowledge in the field. However, Bascom's greatest influence would be as the educator of the succeeding generation of US women geologists at Bryn Mawr. It was at Bryn Mawr that Bascom established a geology programme that earned international respect. She stated, in a letter to Professor Hermann Fairchild:

I have always claimed that there was no merit in being the only one of a kind ... I have considerable pride in the fact that some of the best work done in geology today by women, ranking with that done by men, has been done by my students ... these are all notable young women who will be a credit to the science of geology. (Bascom 1932, p. 110)

Bascom spent her academic career at Bryn Mawr in relentless pursuit of producing future women geologists. This was a monumental undertaking. Often, if a woman was serious about a career in science, she was forced to choose between marriage and her work. Florence Bascom chose her career in geology and never married. She once commented that a fine palaeontologist was 'lost at present to the science by marriage', but (faintly) commended another woman who married a fellow geologist and 'evidently does not intend that her activities shall be swallowed up in marriage' (Bascom 1932, p. 111).

Bascom founded the geology department at Bryn Mawr by herself, starting with rooms on the fourth floor of a new science building that had been initially set aside for future use as a museum and physiological psychology lab (Arnold 1984). She immediately set to work converting the rooms, and within 2 years managed to procure over 1000 geological specimens from over 100 donors (Rossiter 1981). Without delay, Bascom developed a curriculum worthy of an institution of higher geological study. By 1899, she had secured a microscope and a collection of rocks and thin sections, and was teaching petrography (Ogilvie 1945).

Her women students were trained in the field, in the laboratory and by lecture, although Bascom wished that more time could be given to field excursions (Arnold 1984). Bascom mourned the lack of geological journals and vertebrate palaeontological specimens, but partly addressed the latter by bringing in the famous vertebrate palaeontologist and evolutionist Edward Drinker Cope (1840–97) as a guest lecturer.

In spite of Bascom's efforts, the geology department of Bryn Mawr never became wildly successful in terms of its numbers. Bascom was forced to battle with the President of Bryn Mawr over the possible dissolution of her geology department; she also was thwarted at times in her efforts to conduct fieldwork, being forced to remain on the campus during commencement week (Rossiter 1982). She saw her geology enrollments drop in 1904–06, while the college as a whole enjoyed an expansion of 50%; the introductory survey courses in particular dropped to half (Rossiter 1981). This may not be unusual, considering that those students who did not exhibit a true love for geology may not have wanted to experience Bascom's strong opinions and uncompromising scientific attitude.

However, when viewed in the light of the large percentage of women geologists that the programme trained, its importance extends beyond the relatively few females who graduated. Geology largely remained the domain of men. In 1946, the report on the state of women in geology began with the observation that geology 'is generally considered a field for men' (Heminway 1946, p. 1281). Even today, the Bryn Mawr geology department consists of only four professorships, all of which are held by men. Rossiter's (1981) examination of the early 20th-century volumes of *American Men of Science* (1906, 1910 and 1921) revealed that Bryn Mawr awarded 4 of the 18 bachelor's degrees of the women geologists listed; in second place was Ohio State University, which awarded three. It is probable that Mignon Talbot, one of the graduates of Ohio State University, was acquainted with Florence Bascom, having conducted fieldwork with Edward Orton (Rossiter 1981).

Sixteen of the women listed in *American Men of Science* had earned doctorates. Bryn Mawr and Columbia University topped this list, each having awarded three of the doctoral degrees. Just as there was an overlap between Bryn Mawr and Ohio State, there was also an overlap between Bryn Mawr and Columbia: Ida Ogilvie, who had received her bachelor's degree under Bascom, was one of Columbia's doctoral successes. The impact of the Bryn Mawr geology department on the education of women geologists in the first half of the 20th century cannot be denied.

The educational programme that Bascom developed at Bryn Mawr multiplied her influence in the discipline. She became a mentor to many of the next generation of women geologists and worked as the superior to at least one man. Edgar Theodore Wherry (1895–1982) served under Florence Bascom as a junior geologist of the USGS during summer mapping projects. Although Arnold (1983) noted that this must be the first example of a US male geologist with a female mentor, Wherry was already employed as a mineralogy instructor at Lehigh University (Pennsylvania) following his doctoral degree, so the degree of mentorship must be questioned. (In 1931, Wherry and Bascom, together with Stose and Jonas, coauthored *US Geological Survey Bulletin* 828 as a result of their fieldwork). Without question, Bascom mentored numerous students at Bryn Mawr. Some of the women who blossomed geologically under Bascom's tutelage include: glacial geomorphologist Ida Ogilvie (1874–1963), who founded the geology department at Barnard College; Julia Anna Gardner (1882–1960), a successful palaeontologist; petrologists Eleanora Bliss (Knopf) (1883–1974) and Anna Jonas (Stose) (1881–1974), both employed by the USGS; and Mary Porter (1886–1980), who became a successful and respected crystallographer (Fig. 6). Throughout Bascom's teaching career, three themes emerged in her teaching pedagogy that contributed to her success: the high expectations and goals that she set for herself and others; her innovative instructional approaches to teaching geology; and the pure joy that she found in learning geology and conducting fieldwork.

High expectations and goals for herself and others

Like many educators, Bascom may have reflected what she was taught. Her first geology influences, at the University of Wisconsin, were Roland Irving and Charles Van Hise. It was Van Hise, as assistant professor, who probably worked more closely with Bascom and therefore, had more influence on her

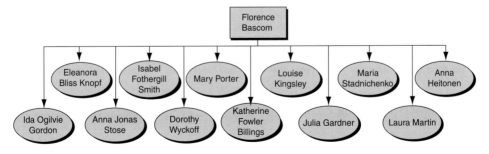

Fig. 6. Florence Bascom's greatest and lasting influence was in the education of many of the first US women geologists.

as far as teaching methods were concerned (Knopf 1946). Van Hise exhibited the ability to relegate and categorize data in order that a pattern could be recognized; he was also demanding of himself and others in the production of results (Arnold 1984). It may not be coincidental that Bascom herself was later described by students as 'rigorous, incisive, and consistent' (Schneiderman 1997).

Bascom set high goals for herself and her students. (Knopf 1946, p. 171) recalled that she once suggested starting fieldwork a little later in the morning; her proposition met with 'vigorous negation', and Knopf had 'many times been grateful for the vigor of my early training that took long hours as a matter of course'. Although Bascom may have been a 'tough taskmaster', students were nonetheless fortunate that they had come from Bryn Mawr; this fact carried weight among outside geology departments (Dott 2003). Ogilvie (1945, p. 321) likewise noted that Bascom put

... the maximum material into her courses, and she required the maximum of effort on the part of her students... She had so much respect for her subject and for real scholarship that she put into her teaching that earnestness of purpose that was always characteristic of her. The result was that her elementary class was always small, never numbering more than about thirty, but her advanced classes grew. Soon graduate students were coming to her from all parts of this country and from foreign countries as well, and her students went out to all parts of the world.

Other students concurred that Bascom

... was thorough, uncompromising in her standards of scholarship. (Smith 1981, p. 35)

Perhaps former student Dorothy Wyckoff summarized students' feelings best:

She expected of her students clear and honest thinking, not by precept so much as by example. (Smith 1981, p. 35)

The challenge of Bascom's interpretation of the Wissahickon mica schist by her former students should appear, therefore, as a triumph of her teaching: Bascom obviously trained her students to think independently, and to communicate their findings.

Innovative instructional approaches

When she arrived at Bryn Mawr, Bascom designed and developed the geology department by herself, complete with a collection of specimens, journals and maps. Through George Huntington Williams, her doctoral advisor at Johns Hopkins, Bascom's graduate geology programme developed into a thorough integration of classwork and field study. These practices were taken up by Bascom in her own teaching at Bryn Mawr (Fig. 7).

Bascom's reliance on fieldwork for women students was unusual, since most courses prepared women for indoor occupations, such as cartography, editorial work, laboratory positions and teaching (Coates 1986). Although geology was still a man's realm, many of the early women geologists specialized in palaeontology, whose research could be conducted without a significant field component. However, Bascom modelled her programme on the parallel programme for male geologists, regardless of the social norms for women students. She made arrangements with local railroads to provide her students with free transportation to their field excursions in New Jersey (Rossiter 1981). Bascom's students comprised most of the few successful women field geologists prior to World War II.

In addition to employing highly integrated field methods with her female students, Bascom also incorporated petrographical innovations within her teaching. When she studied under Victor Goldschmidt in Heidelberg, she learned to use the two-circle goniometer. Bascom then brought the instrument back to Bryn Mawr for use with her own students (Arnold 1993). Bascom's exchanged correspondence with Goldschmidt reveal that he was quite impressed with the achievements that she had attained with her students in the study of crystallography at Bryn Mawr (Arnold 1993). Ogilvie (1945, p. 321) noted

It is characteristic of her that when the universal stage became an adjunct of the petrographic microscope she mastered its use and restudied many of her slides.

Florence Bascom never stopped learning, and she never failed to bring her newly acquired knowledge back into the classroom.

The pure joy of geology

Although Bascom's entrance into the geological discipline was not a straightforward path, and although she relied heavily on positioned male contacts when making career decisions, Bascom eventually came to claim geology as her own. In her farewell dinner at Bryn Mawr in 1928, Bascom reiterated the draw of geology:

This is the life, to plunge into the welcome isolation of the field, to return to the stimulating associations of Bryn Mawr, to observe and in part to clear up geologic phenomena, to return to the exposition and interpretation of geologic phenomena. (Smith 1981, p. 32; Arnold 2000)

Furthermore, it was the research involved in the discipline that engaged Bascom, not the isolated facts that emerged:

The fascination of any search after truth lies not in the attainment, which at best is found to be very relative, but in the pursuit, where all the powers of the mind and character are brought into play and are absorbed in the task. One feels oneself in contact with something that is infinite and one finds a joy that is beyond expression in 'sounding the abyss of science' and the secrets of the infinite mind. (Smith 1981, p. 33)

Fig. 7. Florence Bascom with other women in the field on a Grand Canyon expedition, 1906 (Sophia Smith Collection, Smith College).

It was through geology that Bascom measured her success; she remarked that

... the selection of work in which one delights, and a diligent adherence to it, are main ingredients of success. (Reynolds 1999, p. 17)

Fortunately for her students, Bascom did not seek to contain the joy of geology within herself, but instead sought to inspire others. Goldschmidt poignantly summarized Bascom's teaching, in the midst of one of her battles with the Bryn Mawr administration. He said

I am convinced that with your energy and love for the subject you will meet their wishes by yourself: to introduce your students to the field and even make some into followers. (Arnold 1993, p. 201)

Indeed, many of the women geologists in the first part of the 20th century were followers of Florence Bascom.

Conclusions

Florence Bascom was a geologist of many firsts. Her accolades include being the first woman to whom Johns Hopkins awarded a PhD, the first professionally practising fellow of the GSA, the first woman officer of the GSA, and the first woman geologist employed by the USGS. Although the stories of Bascom's immediate draw to geology after an inspirational event are fictionalized at best, Bascom did eventually claim the discipline as the measure of her life's success.

Bascom's entry into the field of geology was accomplished partly through her integrated network of male acquaintances. It is perhaps fitting that Bascom relied on male influences in her career decision since she remarked that she preferred the company of men to women and found their conversation stimulating. Women, on the other hand, 'indulge in small talk' (Smith 1981, p. 24). Bascom was also displeased with the all-female attendance at a GSA luncheon in her honour; she noted that women geologists did not have much more in common than blue-eyed geologists! (Rossiter 1982).

Bascom's comments seem to imply that she did not consider her academic life at Bryn Mawr to be career-fulfilling. Her potential success at a prestigious women's college would not qualify her as a 'true geologist'. In order to be that 'regular fellow', Bascom's participation in the GSA and the USGS were vital to her own perceived success. Without her USGS fieldwork, Bascom would have felt that her professional accomplishments were lacking and incomplete.

However, it was through her work at Bryn Mawr that Bascom was also deservedly acclaimed and noted for her ability to infuse in other women 'her own enthusiasm and knowledge' (Eckel 1982, p. 37). It is in her mentoring of many of the early US women geologists that Bascom had her greatest and most lasting effects. Several of the prominent geologists of the early 20th century studied under Florence Bascom, including Ida Ogilvie, Eleanor Bliss (Knopf), Anna Jonas (Stose), Isabel Smith, Julia Gardner and Mary Porter. Other than Porter, the names were all included in the lists published in *American Men of Science*. (The *American Men of Science* rated Bascom as a four-star geologist, or one of the top 100 US geologists, in their very first issue of 1906.) Louise Kingsley, Katherine Fowler Billings (1902–97), petroleum geologist and micropalaeontologist Maria Stadnichenko, Dorothy Wyckoff and Anna Heitonen also rose to prominence. Bascom's teaching methods produced female geologists who could effectively compete in a discipline dominated by men (Eckel 1982).

Bascom's teaching pedagogy advanced her geological aspirations far beyond the impact that her own solitary life contributed. Since one usually teaches as one has been taught, it is probable that Bascom's students who pursued careers in academe held high expectations and set noble goals for their own geology students. Some of Bascom's instructional approaches – including her semi-radical notion at the time that women should conduct fieldwork in a similar manner to men – have become standard in the curriculum as society has moved toward the equal treatment of all students, regardless of gender. It is undeniable that the legacy of Florence Bascom continues in the United States today.

The authors wish to thank the reviewers of this paper, A. Bowden and Mary Creese, for their constructive suggestions and comments. A. Bowden was especially helpful in his annotations, as well as his suggestions for additional illustrative material. The authors also wish to acknowledge Karen V. Kukil, Associate Curator of Rare Books, and Susan Barker, Administrative Assistant of the Sophia Smith Collection and College Archives, for their timely and helpful assistance in procuring images of Florence Bascom from the Sophia Smith Collection, Smith College.

References

ARNOLD, L. B. 1983. The Wissahickon controversy: Florence Bascom vs. her students. *Earth Sciences History*, **2**, 130–142.

ARNOLD, L. B. 1984. *Four Lives in Science*. Schocken Books, New York.

ARNOLD, L. B. 1993. The Bascom–Goldschmidt-Porter correspondence: 1907 to 1922. *Earth Sciences History*, **12**, 196–223.

ARNOLD, L. B. 1999. Becoming a geologist: Florence Bascom in Wisconsin, 1874–1887. *Earth Sciences History*, **18**, 159–179.

ARNOLD, L. B. 2000. Becoming a geologist: Florence Bascom and Johns Hopkins, 1888–1895. *Earth Sciences History*, **19**, 2–25.

BASCOM, F. 1893. The structures, origin, and nomenclature of the acid volcanic rocks of South Mountain. *Journal of Geology*, **1**, 813–832.

BASCOM, F. 1902. The geology of the crystalline rocks of Cecil County. *In*: CLARK, W. B. (ed.) *Maryland Geological Survey*. Johns Hopkins University, Baltimore, 83–148.

BASCOM, F. 1925. The university in 1874–1887. *Wisconsin Magazine of History*, **8**, 300–308.

BASCOM, F. 1932. Letter to Professor Herman Fairchild. *In*: FAIRCHILD, H. (ed.) *The Geological Society of America 1888–1930*. Geological Society of America, New York.

BASCOM, F. & STOSE, G. W. 1932. *Description of the West Chester and Coatesville Quadrangles*. USGS Geologic Atlas of the United States, Folio 223.

BASCOM, F., WHERRY, E. T., STOSE, G. W. & JONAS, A. I. 1931. *Geology and Mineral Resources of the*

Quakerstown-Doylestown District, Pennsylvania and New Jersey. U.S. Geological Survey Bulletin, **828**, 62 pp.

COATES, M. S. 1986. Women geologists work toward equality. *Geotimes*, **31**, 11–14.

DOTT, JR, R. 2003. Two remarkable Wisconsin women of the 1920s. *Outcrop*, **39**, 1–4.

ECKEL, E. B. 1982. *The Geological Society of America Memoir 155: Life History of a Learned Society.* Geological Society of America, Boulder, Colorado.

HEMINWAY, C. E. 1946. Women in geology. *Geological Society of America Bulletin*, **57**, 1281.

HILD, A. 1999. Field research: a world of opportunities and challenges, *AWIS Magazine*, **28**, 10–18.

KNOPF, E. B. 1946. Memorial of Florence Bascom. *American Mineralogist*, **34**, 168–172.

OGILVIE, I. H. 1945. Florence Bascom: 1862–1945. *Science, New Series*, **102**, 320–321.

MILLER, B. 1935. Age of the schists of the South Valley Hills, PA. *Bulletin of the Geological Society of America*, **46**, 715–756.

REYNOLDS, M. D. 1999. Florence Bascom: geologist and teacher. *In*: REYNOLDS, M. D. (ed.) *American Women Scientists: 23 Inspiring Biographies, 1900–2000.* McFarland & Company, Inc., North Carolina, 13–18.

ROSSITER, M. 1981. Geology in nineteenth-century women's education in the United States. *Journal of Geological Education*, **29**, 228–232.

ROSSITER, M. 1982. *Women Scientists in America: Struggles and Strategies to 1940.* Johns Hopkins University Press, Baltimore.

SCHNEIDERMAN, J. S. 1997. A life of firsts: Florence Bascom. *GSA Today*, **7**, 8–9.

SMITH, I. F. 1981. *The Stone Lady: A Memoir of Florence Bascom.* Bryn Mawr College, Bryn Mawr, Pennsylvania.

The role of women in the history of geological studies in Ireland

BETTIE HIGGS[1] & PATRICK N. WYSE JACKSON[2]

[1]*Department of Geology, University College, Cork, Ireland (e-mail: b.higgs@ucc.ie)*

[2]*Department of Geology, Trinity College, Dublin 2, Ireland (e-mail: wysjcknp@tcd.ie)*

Abstract: During current research being carried out into the role of women in the history of geological study in Ireland, interesting social and cultural factors are emerging. A list of people who contributed to data gathering, and the unravelling of the complexity of Ireland's geology would characteristically contain only male names. Yet when one begins to look more closely, important roles were played by women. The story is one of women carrying out many and varied supporting roles, including stone-workers, illustrators, tutors, assistants, collaborators, wives, mothers, and later, curators, cartographers and technicians. From 1950 onwards, women begin to occupy professional roles as geologists, particularly in the Geological Survey of Ireland, but more slowly in academic circles. This paper concentrates on women now deceased, who paved the way for others, and only briefly indicates their legacy with selected examples leading to the present day.

Early geological activities

It is interesting to start at the very beginning of human settlement in Ireland and speculate who were the first women to take an interest in geology. It is clear from all accounts that they would have been Mesolithic. From about 9500 years ago, women were using different types of natural stone to do work to help them survive. Irish stone tools are different from those in the rest of Europe, and women would have been aware of the characteristics of these natural materials. Tools were made from a variety of lithologies: porcellanite from Rathlin Island; andesite (or 'Lambay Porphyry' as it is known colloquially) from Lambay Island; and flint nodules contained within the Chalk of County Antrim and in submerged outliers in the Irish Sea (Jackson 1991), the latter scooped up from the depths by glaciers and dumped along the southeastern coast. All would have been familiar to women in earliest times in Ireland, but it is not clear whether they sought out the rocks and shaped the tools, or simply used tools for tanning hides, preparing meat and the like. In any event, these were our first applied geologists. Later still, in Neolithic times, there is evidence from the southwest of the country that fossils were collected and placed within burial chambers (Wyse Jackson & Connolly 2002), and women probably played a part in the collection of these, to them, perplexing items.

In Ireland metal began to replace stone for some purposes around 2400 BC. We can speculate again about the role women played in labour-intensive activities such as copper mining, gold mining and the production of artefacts, but we do not doubt that they had a role. Were the impressive ornaments, made from copper, bronze and gold, all crafted by men? We should not simply assume so on flimsy evidence.

While the geological activities of women in prehistoric times are clouded by the expanse of time and the lack of a written record, by the 17th century some remote corners of Ireland, for example Allihies in County Cork, were thriving mining sites employing thousands of people. Allihies reached its peak in the 1850s when 1500 people worked there. The ore was crushed beside the mine and women and children were employed to further dress the stone and pick out the copper ore (Cowman & Reilly 1988). Historical and cartographic records, and observation of building materials by the authors, suggest that nearly every small village or scattered community had its own stone quarry, as well as a lime kiln for the burning of limestone for mortar and fertilizer. Women would have played their part in the exploitation of the geological resources at that time. On Valentia Island, County Kerry, information at the recently re-opened Valentia Slate Quarry records that, in the mid-1800s-'labour was, at times, so scarce, that women worked alongside the men' the going rate being 2s 6d for an 11-hour day.

The rise of institutional science in Ireland and the reluctant acceptance of women

To further study the role of women in the development of geology in recent times requires a slightly broadened vocabulary. During the past few hundred years people engaged in geological activity have been referred to as 'naturalists', 'natural historians', 'natural scientists', 'mineralogists', 'chemists', 'curators', 'tutors' and 'illustrators', as well as 'geologists', and they have been active both in the arts and scientific fields. For example, the 7th century cleric Augustin Hiberniae, who has been called 'Ireland's first naturalist', suggested that Ireland was once joined to mainland Britain and the Continent (Moriarty 2002). In addition, people interested in science were often referred to as 'polymaths', as their broad focus encompassed a number of the scientific disciplines as we view them today.

Until the second half of the 17th century, Ireland lagged behind its near neighbours in Europe in terms of scientific output, research and discovery. In the late 1600s this situation began to change, due to a number of factors which promoted geological and scientific study. Firstly, a good network of roads was available which allowed gentlemen and scientists ease of travel to more remote parts of the island, and secondly, a number of learned societies was established. The first such society was the Dublin Philosophical Society, later known as the Dublin Society for the Improving of Natural Knowledge, Mathematics and Mechanicks (Hoppen 1970), which began its work in 1683, some 83 years after the foundation of the Royal Society. Much of the energy for the establishment of the Dublin body came from William Molyneux, a natural historian and mathematician, who excelled in astronomy and optics (Simms 1982; Gabby 2002).

Later the Dublin Society and the Royal Irish Academy were founded, in 1731 and 1785 respectively. In Trinity College, Dublin, (TCD), a geological collection was assembled in its museum which had been established in 1777 (Wyse Jackson 1992). It was within the walls of these institutions that the new and interesting geological topics were debated by members. As we will see, the Dublin Society encouraged the geological and artistic work of some women, and later TCD opened up its museum to the general public and so enhanced the geological education of women. It was not until 1949 that women were admitted as members of the Royal Irish Academy.

Susanna Drury, artist, and the Giant's Causeway

For one of the first definite historical accounts of an Irish women influencing the development of the geological debate we must turn out attention to the famous Giant's Causeway, in County Antrim. The area was very isolated in the early 1700s, and only confused drawings and accounts were available. In 1740, Susanna Drury, a Dublin artist, was the recipient of a premium worth £25, which was awarded to her by the Dublin Society for her canvasses of the east and west prospects of the Giant's Causeway (Anglesea & Preston 1980). These paintings show accurate details of the structure of the basalt columns (Fig. 1), as well as the adjacent cliffs and bays of the Causeway (Lyle 1996). She produced three pairs of paintings: one pair is now in the Ulster Museum, Belfast, one in the possession of the Knight of Glin, Limerick, while the whereabouts of the third pair is unknown (Anglesea & Preston 1980).

Why were her paintings so important? They were subsequently engraved by François Vivarès, a Frenchman who worked in London, and it was the

Fig. 1. West Prospect of the Giant's Causeway (*c.*1739) by the artist Susanna Drury. (Photograph copyright of National Museums Northern Ireland and reproduced with kind permission of the Trustees.)

prints produced from these engravings that were widely circulated in scientific circles in Europe. These did much to advertise the unusual nature of the Giant's Causeway. Prior to the 1740s little was known of its geological structure; only a few papers on the area had been published, namely those by Richard Bulkeley, Bishop Samuel Foley of Meath and William Molyneux (Wyse Jackson 2000).

Soon various scientific men were embroiled in a fierce dispute relating to the origin of basalt and other igneous rocks, and much of the argument centred on the north Antrim coastline. The protagonists were divided into the Neptunists, or 'Watermen', who argued that basalt was a precipitate from seawater, and the Vulcanists, or 'Firemen', who said that basalt was a product of volcanic activity and was analogous with modern lava. The Neptunists were led by Abraham Werner (1749–1817) and supported by the Reverend William Richardson (1740–1820) and the influential mineralogist Richard Kirwan (Herries Davies 1981; Wyse Jackson 1998a). In 1799, Richardson discovered fossil ammonites in what he presumed was basaltic rock at Portrush on the Antrim coast, and this he felt supported the Neptunist cause. The French scientist Nicolas Desmarest (1725–1815) saw the prints of the Drury paintings and was able to compare the jointing Drury had accurately portrayed in the basalts with similar structures he had observed in the Auvergne (Taylor 1969), where lava flows are associated with obvious volcanic craters (Lyle 1996). The Reverend William Hamilton (1755–97), a Fellow of TCD and scientist, favoured the Vulcanist stance and did much to promote it through his book *Letters Concerning the Northern Coast of the County of Antrim* (Hamilton 1786) before being murdered by local people in his Donegal parish in 1797 (Wyse Jackson 2002).

We now know that Richardson had in fact mistaken several metres of baked Liassic mudstone, now essentially hornfels, containing impressions of the ammonite *Paltechioceras*, for basalt, and had not recognized the contact between these rocks and the underlying basaltic sill. In terms of the history and development of geological thought the Giant's Causeway and the Antrim coast is an important location where some fundamental ideas in the science were thrashed out. Unwittingly, Susanna Drury played a role in promoting research on the Giant's Causeway, but she did not carry out that research herself. This was to be typical of the role women took in the development of geology in Ireland over the next two centuries.

Popular education for women in Ireland

The literature on scientific developments has familiar themes. Quotations refer to male scientists as 'child prodigies' with 'natural ability'. Women are associated with quotations which include 'it was fashionable', 'she was encouraged to study by ... a man', 'she was inspired by ... a man'. The natural sciences, particularly botany, were considered appropriate subjects for women to study (Alic 1986). Women were certainly encouraged into the natural sciences as wives, assistants and carers, and they were also encouraged to take up art, where the desire for accuracy led some women to become outstanding scientific illustrators.

By the early 1800s there was a growing scientific and economic interest in geology. There had been a gold rush in County Wicklow in 1795. An independent geological branch of the Ordnance Survey of Ireland began mapping around 1820, although their collection of minerals and fossils was described as random and haphazard. Richard Griffith (1784–1878) was appointed Mining Engineer to the Dublin Society in 1812, and he also delivered lectures on geology to the general public. In 1820 he faced audiences of over 400 men and women (Herries Davies 1980). Against this backdrop, Jane Marcet (née Haldimand) (1769–1858) published a series of highly influential books entitled *Conversations* on various aspects of science. Marcet's writings did much to promote science among women (Alic 1986; Bahar 2001) and were aimed by her 'more particularly for the female sex' (Alic 1986). These books may well have been read by the wealthier women in Ireland.

Publication by women in the 19th century did come at a price however, and many published their books anonymously. Mary Somerville (nee Fairfax) (1780–1872), who together with her second husband, William, studied geology, published under her own name an important book *Physical Geography* (Somerville 1848) in which she explained the geological ideas of Charles Lyell and Roderick Murchison in a manner understandable to the educated reader. However, she was denounced from the House of Commons and York Minster for accepting and preaching on the antiquity of the Earth. Charles Lyell wrote to his future wife, Mary Horner, in 1831:

... had our friend Mrs Somerville been married to La Place or some mathematician, we should never have heard of her work. She would have merged it with her husband's and passed it off as his. (Alic 1986)

Nevertheless, Mary was lauded elsewhere for her efforts. She was elected an Honorary Member of the Royal Irish Academy in 1834, and in 1869 received a medal from the Royal Geographical Society. The Oxford College, Somerville, was named in her memory (Creese 2004).

In 1844 Robert Kane, chemist and educationalist, published his influential and detailed account of Ireland's natural resources and sources of industrial power (Wheeler 1944; Davis 2002). One year later, in 1845, the Geological Survey of Ireland was established, and the Museum of Economic Geology was opened in Dublin with Kane as its Director. Historical accounts of Kane usually omit to say that his wife, Katherine Sophia Baily of Newbury, Berkshire, was the anonymous author of *The Irish Flora* (1833) and the first woman admitted to the Botanical Society of Edinburgh, shortly after its founding in 1836 (Alic 1986). This illustrates the point that the role of women as intellectual wives is often forgotten in the accounts of eminent Irish geologists and scientists.

The struggle of women to be accepted by learned societies and academia in Ireland is a long and complex story. A woman, Queen Elizabeth I, was the founder of its first university, TCD, in 1592, but for over 300 years women were excluded from this institution (Parkes 2004). In contrast, the Royal Institution in London, founded in 1799 by Count Rumford, depended on subscriptions from men and women. It attracted fashionable London ladies, who flocked to Albemarle Street to attend lectures there. In Ireland, a campaign began in Cork to establish a similar institute for the 'diffusion of scientific knowledge by holding lectures'. In 1802 the Royal Cork Institution was founded by the Reverend Thomas Dix Hincks, a local nonconformist cleric, and it soon became an important venue for the promotion of cultural activities in the city (Pettit 1976; Davis 1985). The Cork establishment employed four professors, including one of natural history, who gave lectures to both sexes and brought in subscription money. All went well until Edmund Davy (1785–1857), Professor of Chemistry and Secretary of the Institution, left to take up a professorship at the Royal Dublin Society. Edmund was a cousin of Humphrey Davy and had come to Ireland in 1813. Subsequently, the Royal Cork Institution went into decline and in 1830 Parliament withdrew its grant. By 1838 teaching had ceased. Its grand premises are now the Crawford Art Gallery and are home to lectures of the Cork Historical and Archaeological Society. The teaching role was to be taken over by Queen's College, Cork, which opened in 1845. In the interim, the British Association for the Advancement of Science (BAAS) meeting in Cork in 1843 would have been a welcome event, although it was considered to be a financial failure on account of the low number of fee-paying attendees.

In 1831 the BAAS struggled with the question of admitting women (Alic 1986). In March 1832 the President-elect, the Reverend William Buckland, wrote to Roderick Murchison, 'Everybody whom I spoke to on the subject agreed that if the meeting is to be of scientific utility, ladies ought not attend the reading of papers', it being disruptive to a 'serious' philosophical union of working men'. This led to a decision by the officers that women could only attend the social functions. Charles Babbage objected to the exclusion of women, and consequently women were invited to occasional evening lectures as 'associates' of members. By 1833 large numbers of women were attending the meetings (Alic 1986), and in 1834 they actually outnumbered the men. To try to remedy this imbalance, the officers drew up strict rules designed to curb the attendance of women. Famously, at the 1835 meeting in Dublin, the women disregarded the rule that barred them from seasonal meetings and effectively gate-crashed the lectures. As a result, at the meeting held in Liverpool in 1837, women were officially invited to the Geology and Natural History Sections, where they were confined to special galleries. Women persisted in ignoring the rules of the BAAS and were eventually invited to all divisions. Alic maintains that, for the most part, the women attending the lectures did so out of genuine interest, but were not taken seriously by the men. However, it is certain that women interested in science benefited from these lectures, even though the first women full-member was not admitted until 1853. The association stated that there was no rule preventing women from going forward as officers, but in 1876 the council ruled against a women nominee on the grounds that they had never had a woman officer. However, the BAAS was well ahead of the Geological Society of London, which did not admit women members until the early 20th century.

Snapshots of scientific contribution made by women in Ireland in the mid-1800s

In Ireland in the 1800s, opportunities for the education of women in science had begun. Mary Ball (1812–98) is well known as a naturalist, but we are told she had an advantage through her brother's interest in natural history (Chesney 2002). Her observations had been obscured until recently as they had been published by her brother Robert, the Director of the Dublin University Museum (Hutchinson 1982). Even as late as the 1850s women did not publish under their own name in scientific journals. However it seems the close-knit and supportive community of naturalists in Ireland at the time encouraged women in their scientific endeavours.

The lives of Mary, 3rd Countess of Rosse (1813–85) (Fig. 2) and Mary Ward (1827–69) are examples that further highlight the role that women played in the natural and physical sciences

Fig. 2. A painting of Mary, 3rd Countess of Rosse, around the time of her marriage. (Photographed with permission of Alison, 7th Countess of Rosse.)

in Ireland. Mary Rosse was an influential scientist and pioneer photographer of the mid-1800s. Not only was she a collaborator in the erection at Birr Castle of the largest telescope in the world (it remained the largest until 1918), but she also researched and experimented with hazardous chemical processes to produce a photographic record of the telescope. These photographs were exhibited in London and gained great publicity in scientific circles. Although she worked together with her husband William Parsons, 3rd Earl of Rosse (1800–67) in their scientific research, it is William who is credited for the important work of building and using the Birr Telescope. The light-gathering power of the telescope allowed the observation of planets and objects never before seen in space, which were important steps in our understanding of planetary geology. Many new discoveries and observations were made, such as the existence of spiral nebulae, crucial to our understanding of the Earth's origin. Many well-known scientists visited the Birr estate to view and use the telescope, and Mary was part of the scientific debate. She was considered an unusual woman for her time because of her interest in science. Unfortunately, at that time, it was not acceptable for a married woman to be corresponding with male scientists and so Mary had to correspond through her husband.

Through all of this, Mary Rosse gave birth to 11 children, with only four surviving to adulthood (Davison 1989). Her children were educated at home and Mary took an active interest in this aspect of their lives. She was helped by tutors such as Robert Stawell Ball (1840–1913), nephew of Mary Ball, and later Astronomer Royal of Ireland. Mary Rosse's scientific background would have undoubtedly influenced the education of her children. In her cabinet, preserved in Birr Castle, are some of Mary's prized possessions. These include minerals and fossils that she collected herself. The fossils were mounted by her in a systematic way onto hand-coloured card (Fig. 3). A wooden box, dated August 1848 and purchased by Mary Rosse, has been recently discovered. It contains over 100 mineral specimens (Fig. 4) and was purchased from James Tennant (1808–81), one-time Professor of Mineralogy and Geology at King's College, London, and well-known dealer in geological specimens at 149 Strand Street, London. It was probably purchased as a teaching set (Higgs 2003).

Mary Rosse's eldest son, Laurence Parsons, 4th Earl of Rosse (1840–1908), took the first photographs of the Moon, using Mary's photographic equipment. He devised a way of taking accurate temperatures of the Moon's surface and his research on heat radiated from the Moon in the infra-red range is well known. Again, this information was important to the understanding of the geology of our own planet. His account of observations made at Birr, published by the Royal Society, contain an excellent engraving of the Great Nebula Orion, M42, as seen through the telescope. When widowed in 1867, Mary remained at Birr Castle only long enough to oversee the tuition of her youngest son, Charles Parsons (1854–1931). Charles went on to invent and develop the steam turbine, which he famously demonstrated in his craft *Turbinia* during the Navy Review held at Spithead for Queen Victoria's Diamond Jubilee in 1897. The present Earl of Rosse, and Alison, 7th Countess of Rosse, have recently had archives catalogued by Anthony Malcolmson and made available, and have highlighted Mary's achievements in the Birr Castle Historic Science Centre.

Mary Ward was a frequent visitor to Birr Castle, being a cousin of the Earl. Like Mary Rosse, she did not attend school or university, but was educated at home by a governess (Kavanagh 1997; Harry 2002). In 1845 her parents bought her a fine microscope which she used (and demonstrated) all her life. She chronicled the building of the giant telescope at Birr Castle in sketches. She met many of the eminent men of science of that day and became well known as an artist, naturalist, astronomer and microscopist, although she never received formal honours. At that time, women could not become members of learned societies or institutions, gain degrees or diplomas, or even use libraries during their lifetime. It was very difficult for them to become recognized in scientific fields. Nevertheless, Mary was the first to write and have published a book on the microscope, in spite of the fact that it

Fig. 3. Fossils collected by Mary, 3rd Countess of Rosse, and mounted on to hand-coloured card. (Photographed with permission of Alison, 7th Countess of Rosse.)

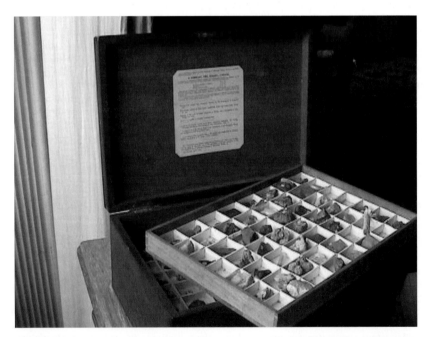

Fig. 4. A recently discovered wooden box, dated August 1848, containing over 100 mineral specimens. It was purchased by Mary Rosse, probably as a teaching set, from James Tennant one-time Professor of Mineralogy and Geology at King's College, London, and well-known dealer in geological specimens (Photographed with permission of Alison, 7th Countess of Rosse.)

was difficult to find publishers for a book written by a woman. She had to have the book, *Sketches with the Microscope* (Ward 1857), published privately by Shields of Parsonstown (now known as Birr) in 1857. A total of 250 copies were printed, and the following year an imprint was issued by Groomsbridge of London under the title *The World of Wonders Revealed by the Microscope*. She went on to write three books and several articles on scientific subjects, as well as performing the duties of a wife and mother, looking after eight children. She illustrated her own books and those of others, including a paper in the *Transactions of the Royal Society of Edinburgh* published in 1864. She made the original drawings of Newton's telescope and the Birr telescope (and perhaps even the famous illustrations of the spiral galaxy). She was well respected by contemporaries, such as the geologist Richard Owen, founder of the British Museum (Natural History) in London, who asked Mary to send him copies of her drawings. However, at the end of it all, Mary is perhaps best remembered for being the first person killed in a steam-carriage accident. She was only 42 years old at the time.

The role of women as amateur geologists in natural history societies in Ireland

The principal geological society in the 19th century was the Geological Society of Dublin, founded in 1831, which later became the Royal Geological Society of Ireland in 1864. Unfortunately women were excluded from this society (J. Archer pers. comm. 2006). However, a small number of women played significant roles as amateur geologists in various natural history societies in Ireland, and here we focus on the field clubs of Belfast and Dublin by way of example.

In Belfast, Sydney Mary Thompson (1847–1923) was, for many years, one of the most active and much appreciated members of the Belfast Naturalists' Field Club (founded in 1863) and served on the committee. Inspired by Percy Fry Kendall (1856–1936), Professor of Geology at the University of Leeds, Thompson, together with Mary Andrews and a few others, carried out research on the local glacial deposits, especially collecting and naming of erratics with a view to determining the direction of ice-flow (Wyse Jackson 1997). Their reports were published in the Club's *Proceedings* (Praeger 1949) and found their way into reports published by the BAAS. This avenue of research was quite in vogue at the time, and the Belfast group made valuable contributions to the BAAS committee that reported on the subject. Its importance can be measured in that, during the period 1874–1908, the BAAS granted the considerable sum of £157 16s. 6d. to support this work (Howarth 1931). Thompson departed Ireland on her marriage to the Swiss artist Rudolphe Christen, after which she was known as Madam Christen. She died in Llandudno on 16 July 1923.

Miss Mary K. Andrews, daughter of Thomas Andrews, Vice-President and Professor of Chemistry at Queen's University, Belfast, was also a leading light of the Belfast Club. Not only did she work on erratics but she also contributed geological photographs of the Ballintoy district to the collection assembled by the BAAS (Anon. 1909).

The Dublin Naturalists' Field Club was established in 1886 but its strengths have largely lain in the field of botany. However, over the years between the 1950s and 1970s, the two geological mainstays were two women associated with TCD, who between them did much to promote the subject amongst fellow members. Barbara Miller (née Thompson) who had been a graduate student in geology at TCD served as President of the Club in 1965/6 and led 24 geological excursions between 1957 and 1979, while Veronica Conroy Burns (1914–98) led 19 trips over a 27 year period until 1986 (Wyse Jackson 1987). Although Burns (Fig. 5) was considered 'self-educated', as she did not hold a degree in geology, in 1960 she became a research assistant to Robert George Spencer Hudson (1895–1965) and later, in 1964, became curator of the TCD Geological Museum until she retired in 1980 (Wyse Jackson 1998*b*, *c*). She reorganized the mineral collection and catalogued over 20000 specimens. She was an avid collector of fossils and was responsible for accumulating a large suite from the Waulsortian mudmound at Feltrim in northern County Dublin; these were utilized by Hudson *et al.* (1966). She was particularly fond of graptolites and had a penchant for finding specimens in successions that other less-patient palaeontologists passed over. Although she published only two geological papers, these are significant in that they provide the precise age of the Balbriggan Outlier situated in Counties Dublin and Meath (Rickards *et al.* 1973; Burns & Rickards 1993). These Dublin women served as a link between the academic institutions and the activities of amateur geologists.

Geology and the universities: first steps to providing a formal geological education for women

In the 1800s it was certainly necessary for women with aspirations in geology to move in circles that supported natural history and inventiveness, and to have some financial backing. However, it was the change in the formal education of women, and in particular the movement to admit women to

Fig. 5. 'The Rivals': Veronica Conroy Burns (1914–98), an active member of the Dublin Naturalists Field Club, is pictured here showing her dislike of modern extraction techniques (Geological Museum, TCD).

degree examinations, that was to be a crucial development in Ireland, as elsewhere in Europe. There are detailed accounts in the literature of the struggle of women, and their male academic supporters, to gain entry to academia, and of the procrastination of committees within the universities (Parkes 2004; Sheffield 2004; Burek 2007).

Trinity College, Dublin (TCD)

In TCD the first-ever motion at the Academic Council to admit women to the University was moved on 10 March 1880 by the Professor of Geology, the Reverend Samuel Haughton (1821–97). Haughton proposed that:

In the opinion of the Council, the time has come when the Degrees in Arts of the University should be opened to women, by examination, on the same terms as men. (Thomson 2004)

Haughton had through his work as a geologist and in his involvement with the Royal Zoological Society of Ireland no doubt come across many women interested in natural history, and was patently aware of the popularity of the meetings of the BAAS to women audiences. Such a motion was radical, but Haughton, who was on the liberal side in College matters, a legacy of his Quaker background, was no stranger to controversy, having already been involved in a set-to with the Chancellor of the University (Wyse Jackson 1994).

The TCD Provost, Humphrey Lloyd (1800–81), was opposed to the admission of women, but died soon after Haughton's motion, and was succeeded by John Hewitt Jellett (1817–88), who was not so opposed. He coincidentally is commemorated in the mineral name 'Jellettite', a yellowish green garnet. A committee of ten men was appointed to look into the matter, and this body included Sir Robert Kane, mineralogist, who was an opponent of the admission of women. Nothing was reported in the Council minutes for the next 10 years! However we are told that Haughton's name weaves in and out of this issue almost until his death in 1897.

Advances in school education for girls in Ireland led to an increased demand for university places. The Central Association of Irish School Mistresses (CAISM) was founded in 1881, with the first President being Mrs Dora Jellett, wife of the Provost. In 1891 the Board of TCD was forced to re-open the debate on the admission of women due to the campaigning of CAISM, led by Alice Oldham (1850–1907) of Alexandra College, the premier girls' school in Dublin. Unfortunately, Provost Jellett died in 1888, and the new Provost, George Salmon (1819–1904), was vehemently opposed to the admission of women. For the next 16 years he did his utmost to prevent change, even though his daughter was a member of CAISM (Thomson 2004). So it was that, in 1892, the Royal University was the only university in Ireland to offer degrees to women. It was an examining university only, and tuition had to be organized by CAISM. In England, Oxford and Cambridge had opened their doors to women for exams and lectures (see Burek 2007) but not for the award of degrees. CAISM still had a number of allies on the Board of TCD, including Haughton, and in March 1893 a straightforward motion was tabled, seeking that the university would confer degrees on women; surprisingly, it was passed. Unfortunately, procrastination and delay due to legal difficulties followed, mainly due to the efforts of Salmon (Thomson 2004).

In 1892, the tercentenary of TCD inspired a petition and a campaign of letter writing. The views of Professor Thomas McKenny Hughes (1832–1917) of the Woodwardian Museum in Cambridge were sought; his mother was from Dublin, and so it may have been perceived that he would have approved of the campaign in Ireland. He received the following letter, dated 10 October 1892, from Eleanor Mildred (Nora) Sidgwick (1845–1936), head of Newnham College and sister of Arthur Balfour, later Prime Minister (1902–05):

Dear Prof Hughes

An effort is being made to get the classes of Trinity College, Dublin, opened to women, and in view of this, I have been asked to collect information as to the working of mixed classes of men and women here, which those who are promoting the movement in Dublin may lay before the authorities.

I enclose a set of the questions they would like to have answered. They are especially anxious for information about such classes as those referred to in questions 6 and 7 – in which the students are not merely attending lectures but take an active part. We have, I think, very few of such classes in Cambridge, but it struck me that the work done in your geological excursions and museum is something of the same kind, and I am therefore especially desirous to be able to report your opinion on the subject to my correspondents.

If you can kindly send me answers to the questions, I shall be very much obliged

Believe me,

Yours sincerely, Eleanor Mildred Sidgwick (of Hill Side, Chesterton Road, Cambridge)

The tone and content of Hughes' response, dated 17 October 1892, was no doubt welcomed by the campaigners:

My Dear Mrs Sidgwick

I have consulted my colleagues in the Woodwardian Museum and some other persons of discretion before replying to your letter respecting the effect of instructing men and women together –

I will now answer your questions to the best of my knowledge.

Q1 *Do you lecture to a mixed class of men and women? And if so on what subject?* Yes – on Geology

Q2 *Do you find difficulty from the mixture of the sexes?* No

Q3 *Do you think male students object to the presence of women?* Those men who do object are much benefited by the restraint which the presence of women lays upon them.

Q4 *Do you think the presence of women interferes with the successful teaching of men?* No

Q5 *Do you think a chaperon necessary or desirable?* Unnecessary and undesirable – (as she must be either a bored or unsympathetic part of the audience or take a book to read in either case being a damper.) (she cannot be looking about all the time to see whether looks are exchanged and nothing else is possible from the point of view of the lecturer)

Q6 *In the mixed classes you teach, are questions put and answered by lecturer and students? Or is work done by the students, such as translation, working of mathematical problems, etc?* In some mixed classes questions are put by lecturer and by students – sometimes work is done involving the same amount of presentiment as in the case of oral translation or writing of problems on the board before the rest of the class.

Q7 *What is your opinion of the efforts of such lectures or classes being given to students of both sexes together?* In my department I have not found that lectures being given to students of both sexes together makes any difference except the formation of a stronger public opinion in favour of order and propriety.

Unfortunately, in Dublin, Provost Salmon took no notice of the petition and campaign and it is recorded that he 'presided with dignity over the tercentenary'.

The breakthrough for the admission of women to TCD came between 1896 and 1901, when six anti-admission members of the Academic Council died, including Sir Robert Kane, and were replaced by pro-admission members. The original proposal of Haughton was finally carried in 1902. The campaign had taken over 20 years, and some of those who fought hardest, including Haughton, never lived to see the admission of women, although ironically Salmon did just before he died.

The women who availed themselves of these new opportunities included Beatrice Hammond, who received a junior moderatorship in Natural Science in 1908, and Georgina Fisher (née Revington), who, in 1910, was the first-ever woman to be awarded a Gold Medal in Natural Sciences. In contrast, the Universities of Oxford and Cambridge resisted conferring degrees on women until 1920 and 1948 respectively (Kölbl-Ebert 2001), despite having established women's colleges much earlier than the first admission of women to university in Ireland (see Burek 2007). It was for this reason, that between 1904 and 1907, many Oxbridge women travelled to Dublin to be conferred with University of Dublin degrees.

In agreeing to confer degrees on Oxbridge students, TCD had thought that a few Irish women would apply, but much to their embarrassment 750 women applied in the first 3 years! TCD was accused of selling degrees; the fees were £10 3s. for a BA degree, and £9 16s 6d. for an MA (Parkes 2004, p. 88). The money paid in was used to set up Trinity Hall, a residence for women students situated 4 km south of the College in a leafy suburb of Dublin (it now accommodates students of both sexes). Most of the Oxbridge women came to Dublin for one night only, travelling on the Holyhead mailboat, and they were referred to as the 'Steamboat Ladies' (Parkes 2004, pp. 87–112) (Fig. 6). In 1905 the degree of DSc was awarded to five pioneering Oxbridge women scientists, including Gertrude Lilian Elles (1872–1960) the palaeontologist from Newnham College (see Burek 2007). Many years later she was to be the first woman to be conferred with a Sc.D by Cambridge, and the first woman to serve on the council of the Geological Society (1923–27) (Rickards 1999). It is said that Irish women were grateful to these pioneering women for showing what could be attained, but by 1907 TCD decided to discontinue the practice of conferring degrees earned in other Universities.

By the 1920s the study of science at TCD had became more popular for women. Evelyn Fisher (née Ross) was one of the first women to venture down a mine, working in colliery medicine. She reported that she was horrified by the hardship and misery of the mining community.

A schoolbook that proved highly popular and did much for the dissemination of information on geology was Elenor Butler's *Structural Geography of Ireland* (Butler 1924). This was published by the Educational Company of Ireland and was used in secondary schools for many years. Contrary to what its title suggests, the book was really an introduction to the geology of Ireland and it benefited from the inclusion of maps, diagrams and photographs supplied by Grenville Arthur James Cole (1859–1924), who was Director of the Geological Survey of Ireland. This unassuming text would have been familiar to a generation of girls and boys, for whom it was probably their first and possibly only exposure to geology. Undoubtedly it encouraged some to enter university to study the subject.

Queen's College, Cork

Queen's College, Cork, had been founded in 1845, and interestingly one of its priorities was to provide the College with a geological museum. Within 19 years, over 8000 rocks, minerals and fossils had been acquired. The first President of the College was the infamous (though usually referred to as renowned) Robert Kane. His presidency was a troubled time, given that he was double-jobbing at the Museum of Economic Geology in Dublin, causing him to take lengthy absences from Cork (Murphy 1995). Women had been frequent attendees at lectures and demonstrations on natural history in Cork since pre-College days, and so it was noticeable that women were conspicuously absent from the new Queen's College, which remained the exclusive domain of men.

There was no doubt that Cork women were actively pursuing geological interests. For example, a rare form of quartz was discovered at Mallow, County Cork, in 1876 by Miss E. Cotter and was named 'cotterite' after her by the Queen's Professor Robert Harkness (1816–78), the Professor of Natural History at the College between 1853 and his death. In the same year, the Council of Queen's College, Belfast, received a request from Mary Edith Pechey to attend medical courses. She was turned down. This incident prompted the issue to be discussed by the council of Queen's College, Cork, who agreed with Belfast that 'the idea was inexpedient' (Murphy 1995, p. 128). However, a group called 'the Ladies Association for the Promotion of Higher Education of Women in Cork' organized lectures, which were given by College staff and other experts. This increased the pressure on the

Fig. 6. Oxbridge women who came to Ireland to be conferred with their degree by Trinity College, Dublin. Women usually stayed for one night only, travelling on the Holyhead mailboat, and were referred to as the 'Steamboat Ladies' (Parkes 2004, pp. 87–112).

College to facilitate women, and by 1878 the College allowed the Ladies Association to use the Aula Maxima – the main College building – as a lecture venue. Later, the women arranged for an application to be submitted to the College Council for admission to courses of lectures in arts. In the Council meeting of 1883 it was stated:

College has no objection to ladies attending courses of lectures delivered by the arts professors ... on paying the usual fees. ... [but] Council will not require any professor to alter his course ... divide his classes ... deliver any additional course ... in consequence of admission of any lady to the college.

In additon, there were to be no special classes to prepare them for the Royal University examination (Murphy 1995). In a report in the academic year 1886/7, it is stated that:

... a separate cloakroom having been provided ... we were able to admit women last session [1985/6].

Although it had taken 10 years, women could now study for a degree in Queen's College, Cork, but they had to be examined by the Royal University. These first women received high honours grades, and the President remarked that the presence of the ladies had improved discipline! However this was not the start of a landslide, as only 30 women registered in the first 12 years. Of these, only six were Roman Catholics. Why was this? It is possible that the fees prohibited some women matriculating, but more importantly the Roman Catholic authorities pressurized women students and their families not to attend what was often referred to as the 'Godless' college. At this time, one Cork mother indignantly complained that Bishop O'Callaghan and other clergy had intimidated two sets of parents 'by terrifying them with excommunication' into withdrawing two girls from the arts courses in 1906' (Murphy 1995, p. 173).

The Association of Women Graduates, formed in 1902, helped promote women's issues. Life was not easy, and some professors refused to let women into their classes (even up to 1940). However, from 1904 onwards, under the presidency of Sir Bertram Windle, the situation for women generally improved in Cork. In 1909 the formidable Countess Aberdeen (Ishbel Maria née Marjoribanks), wife of the Lord-Lieutenant of Ireland, addressed the College and asked that girls be eligible to apply for scholarships. Windle started a students club, but the difference in rules for men and women caused some controversy.

Women began to be appointed to the college staff: the first was appointed in the area of Education, but the second, Miss Peggy Bowen-Colthurst, was appointed Demonstrator in Mineralogy in 1907. The university archives contain a letter from Marcus Hartog, Professor of Natural History at the time, to the Council of Queen's College, Cork, requesting permission to employ

Bowen-Colthurst at the rate of 1s an hour, with total payments not to exceed £20, to assist him in the arrangement of the mineral collection. It was his wish to display a good collection to the public and he suggested that an application be made for a grant of £25 from lapsed balances. He subsequently sent a note informing the Council that he had made the appointment, adding that she had obtained a first-class honours in geology at 'Greats' at Oxford, and requested confirmation that the appointment was approved. President Windle has annotated the note to say that the appointment is approved. It is interesting to note that Lady Windle, the President's wife, was a collector of minerals (her collection was later donated to the Geology Department), and it seems that she occasionally exerted some influence over her husband on whom was appointed to the College staff (Murphy 1995).

Soon afterwards, the Windles met Issaac Swain (1874–1963) in Portrush, during an archaeological flint-discovering trip, and Sir Bertram observed that 'it was an appropriate place to find a geologist'. Swain was at that time employed by the Royal College of Science for Ireland, where he was an assistant to Grenville Cole, the Professor of Geology and Mineralogy (Wyse Jackson 1989). Windle (or perhaps Lady Windle) was obviously impressed by the young man, and in 1910 he was appointed Professor of Geology and Geography a post which he held until his retirement in 1944.

Queen's College, Cork, holds the distinction of being the first university in the United Kingdom and Ireland to appoint a woman to a chair. This momentous event in the furtherance of women's education took place in 1910. Cork was not such an educational backwater. The Windles had high hopes that the number of female students attending the College would increase, but it was not until the establishment of the National University of Ireland in 1908 that the numbers of women attending Queen's College, Cork, began to rise. This change occurred after a deal brokered with the Roman Catholic church; this saw the building of the Honan Chapel on the campus, together with the College adopting an increased Roman Catholic ethos. By 1934 there were 296 women students, out of a total of 906 students, on the books of the College. In 1986, one century after female students had been admitted to courses in Cork, they outnumbered men on the campus, and this too was reflected in the gender balance in the student intake in the Department of Geology.

In sharp contrast the number of women in senior staff positions did not increase. Part of this can be attributed to the fact that professors could remain in a post for as long as they liked, even up to the age of 81 years – so not many positions became vacant. The principal of gender equality enshrined in the 1908 Act (1909 Charter) was a 'dead letter' according to Murphy (1995). The numbers of women employed in the subprofessorial grades rose, but they had to resign their positions on marriage. John A. Murphy, the College historian, damningly wrote:

Even in the more enlightened age of the 1980s and 1990s the men in power had no intention of sharing it with women, while deeming it expedient to pretend the contrary. (Murphy 1995)

The roles of women in the Geological Survey of Ireland

The 20th century saw interesting roles for women associated with the Geological Survey of Ireland (GSI). The history of the GSI typically describes 'Victorian days when an officer – and his family, would disappear off into the field, perhaps for years on end'. Thus, a common role was that of the supportive wife, but mothers could occasionally play their part.

William Bourke Wright, who was born in Dublin in 1876, joined the British Geological Survey in 1904. However, in 1910, Wright's mother became ill, causing him to move back to Ireland, where for the next 10 years he was a dynamo in Irish geology, particularly in the Kenmare and Killarney regions. During this period he published the important book *The Quaternary Ice Age* (W. B. Wright 1914) for which he is primarily remembered. In the same year as the publication of this book, he married Mabel Crawford MacDowall in London; she was one of Grenville Cole's students at the Royal College of Science for Ireland. Herries Davis (1995) reports that Mrs Wright 'soon after her marriage ... was deploying her talents in the Survey's interest through study of fossils from the Leinster Coalfield', a subject upon which she published a paper in the *Proceedings of the Royal Irish Academy* (M. C. Wright 1920). It is possible that this was the first geological paper published by a woman in Ireland. Recently her material has been re-examined and shown to be a rare and important lagerstätte (Orr et al. 1996). Unfortunately for Irish geology William Wright was appointed District Officer with the British Geological Survey in 1921 and given responsibility for the South Lancashire Coalfield, and so Mabel Wright left Ireland.

It is interesting to note that Blanche Vernon, who belonged to a wealthy north Dublin family, was also one of Cole's students. Eventually she was to marry Cole and assist him by producing fine drawings of thin section petrography, some of which illustrated his geological papers (Wyse Jackson 1989).

During its first 70 years of existence, the GSI apparently employed no women in an official capacity, although many of its early published maps were hand-coloured by women. After 1920 a number of women were employed mainly as cartographers, draughtsmen and technicians. One of the early workers was Eileen Barnes, who worked part-time on the Survey's collections in the National Museum of Ireland during 1920, and thereafter as an occasional draughtsman and modelmaker until about 1938 (Herries Davies 1995). Most of the women worked only for a short time, typically between 1 and 6 years, some because the marriage bar forced their resignation.

The GSI played an important part in the advancement of women as professional geologists in Ireland. It was in September 1951 when the first woman was appointed as a full-time Geologist. She was Dylis Penelope Lindsey Jones (b. 1929), a graduate in geology from the University of Manchester. The Senior Geologist (then in temporary charge of the Survey), Thomas John Duffy, wanted Jones to work in ground adjacent to that being mapped by Mike O'Meara, in Carrickmacross and Ardee. Such were the times, that he thought it necessary to ask Mrs O'Meara if she would have any objection to that plan (Herries Davies 1995). Jones and O'Meara mapped for several seasons in County Monaghan, before O'Meara was sent to oversee extensive drilling operations in the Leinster coalfield. Jones took over these responsibilities when O'Meara's health failed, and she carried out extensive logging of the core material. She must have been an unusual figure in the coal industry at that time.

By 1959 there were two women, out of a total of seven geologists, working for the GSI. One of the women was Jones and the other was Jean Margaret Wilkinson, who was appointed on 4 February 1959. Wilkinson was born in Batley, Yorkshire, in 1937, and she had also studied at the University of Manchester. After her appointment she worked with various colleagues on the production of maps of the drift deposits of Counties Carlow, Kildare and Limerick, but she only stayed with the Geological Survey for 10 months, leaving on 16 December 1959. Jones resigned her position in 1963 (Herries Davies 1995).

Following the departure of Dilys Jones from the GSI it was 8 years before another female geologist was appointed to its staff. Loreto Philomena Catherine Farrell (a geography graduate of University College, Dublin) was appointed Assistant Geologist in the Quaternary section in 1971. She transferred to the Geological Survey Mineral Exploration section in 1992 and is currently a Senior Geologist with responsibility for overseeing the rehabilitation of lands contaminated during late 20th-century extraction of lead and zinc ores in the Silvermines district of County Tipperary. To date, she is the longest serving female professional geologist in the Geological Survey.

Jean Barbara Archer (a geology graduate of TCD, and the first to be conferred with a PhD in the Geology Department, University College, Galway) was appointed Geologist in the bedrock mapping section in 1976. Archer, in addition to her field mapping and editorial work, assembled from scattered localities the materials that now constitute the Survey's archives. She resigned in 1997.

Bettie Matheson (later Higgs) was appointed to a temporary position in 1977 and became a permanent Geologist in 1978 (just after the abolition in 1976 of the directive which forced female civil servants to resign on their marriage; she was to become the first married woman, and mother, to work as a geologist). Having previously worked in the Computer Unit of the British Geological Survey, she began the drive to digitize field data and produce automated maps – a task ably carried on by Mary Gormley (later Carter) and others in the 1980s. The Geological Survey has continued its record. The most senior (high-ranking) position held by a female geologist in Ireland at present is in the GSI. Eibhlin Doyle was promoted to Principal Geologist in 2005, with responsibility for marine surveying.

The roles of women who studied or worked outside of Ireland

In 1933 Doris Livesey Reynolds (1899–1985) was appointed lecturer in geology at Durham University. She had lived for a while in the Belfast area and carried out research on the northern end of the Newry–Slieve Croob Granodiorite and the adjacent Triassic sandstones. Much of this detailed work was written up in a series of publications in the 1920s and 1930s (Reynolds 1928, 1931, 1937). It seems that Doris loved Ireland and made many trips to study its geology in the company of her husband, Arthur Holmes (1890–1965).

On the fringes of the geosciences, Kathleen Lonsdale (née Yardley) (1903–71) of Newbridge, County Kildare, went on to make her name in X-ray crystallography. She was the 10th child of the town's postmaster and his wife. The family moved to Essex in 1908, and there she attended the county high school for boys – the only girl to do so – and studied physics, chemistry and mathematics before entering Bedford College for women at the age of 16 years. She came first in her year in 1922. Indeed, she gained the highest score in 10 years in the final examination in London University. She was immediately offered a research place by William Bragg (one of the examiners) and began work on X-ray diffraction of crystals of organic molecules. In 1923 she transferred with Bragg to the Royal Institution, but her marriage to Thomas Lonsdale resulted in a move to Leeds, where he was working. By the time she had three children, she considered giving up science to become a good wife and mother, but Thomas would hear none of it. So she progressed in her career, studying the benzene ring. Bragg offered her £200 to get help at home so that she could work with him. She calculated it would cost her £277, so he came up with £300. During World War 2, Kathleen was sent to prison as a result of her pacifist stance, but was released in time to attend the 1943 meeting at the Dublin Institute for Advanced Studies. She became Professor of Chemistry in 1949 at University College, London. She broke new ground for women, being the first women to be elected a Fellow of the Royal Society in 1945 (some 285 years after its foundation), in 1968, and later became the first women president of the BAAS (since then a further four women have served in this position).

Gwyneth Evans (née Jones) (1908–2003) was born in Wales but came to Belfast on her marriage in 1931. She had relatives in Londonderry and eventually became a key person in the political and cultural life of Ulster for over 50 years (Catherine Byron pers. comm.). She was the wife of Emyr Estyn Evans (1905–89), a professor of geography in Queen's University, Belfast, from 1945 to 1968. She participated in her husband's field research in a very hands-on way, living in a remote cottage in the Mourne Mountains with Estyn and their four small sons in the late 1940s, with no water or electricity, as part of his 'action-research' into the interaction of geology and folk and farming traditions. Gwyneth oversaw the publication of Estyn's books and papers, and was instrumental in the production of the posthumous *Ireland and the Atlantic Heritage: Selected Writings* (Evans 1996), for which she wrote a substantial biographical memoir of her husband, reflecting the breadth of their scholarship and interests.

Patricia Yates (1932–60) contributed to Irish geology during her short life. While studying for a PhD at London University she carried out fieldwork in the Connaught Coalfield of County Leitrim (Herries Davies 1995). Here she studied the Lower and Upper *Eumorphoceras* stages of the Namurian, and described a number of new goniatite and bivalve species. Unfortunately, she died on 7 August, only 5 days before the examination of her PhD thesis, and while her first paper (Yates 1961) was in press. Her work was considered so important that her collections are housed in the Murchison Museum, Imperial College, the British Geological Survey Museum, and the Natural History Museum in London. Mrs Judith Creighton undertook the task of sorting the author's manuscripts and cataloguing her extensive collections. Her thesis was published posthumously (Yates 1962). The editors of the Palaeontological Association stated 'since it is a major contribution to Carboniferous stratigraphical palaeontology, it has been edited for publication' by Dr W. H. C. Ramsbottom (BGS, Leeds) and Dr Gywn Thomas (Department of Geology, Imperial College, London).

In 1969 Audrey A. Jackson was the first female field geologist to be appointed to the British Geological Survey; she had completed a PhD at TCD a year earlier. It is interesting to note that, following her third year of study as an undergraduate at TCD, the staff debated whether females should be allowed to proceed to the fourth and honours year. This was a prerequisite for post-graduate study, and although one member of staff objected, Audrey proceeded, a step that led to a fulfilling career as a field geologist (A. A. Jackson 1994).

Slow progress in academia

There has never been a female professor of geology in Ireland. The first female lecturer in geology appointed to an Irish university was Gillian Clare Lewarne (now Sheehan) who was appointed to the faculty of TCD, in 1959 (Wyse Jackson 1994). She had studied under Frank Hodson and together they published an important paper on the Upper Carboniferous geology of southern County. Clare (Hodson & Lewarne 1961). In 1964 she reported on the presence in the same area of the resting trace of the rare starfish *Asteriacites* (Lewarne 1964). She resigned her lectureship in 1963 on marriage and moved to Limerick, and it was not for another 35 years that another female lecturer was employed in a permanent position in a geology department in Ireland. In 1998 Catherine Coxon was deployed in the Geology Department, TCD, when the academic staff of the Environmental Sciences Unit were dispersed around the campus. In 1999 Kate Moore was appointed to a full-time permanent academic position, as a geologist in the Department of Earth and Ocean Sciences, in the National University of Ireland, Galway. Bettie Higgs received a half-time contract of indefinite duration in the Geology Department, University College, Cork (previously Queen's College, Cork) in 2003, after working there in a temporary capacity since 1987. None of these are senior positions, and the women are not graduates of Irish universities. The Geology Department of University College,

Dublin, has the largest number of staff of any such unit in Ireland, but has never at any time employed a woman geologist in a permanent position. During the past few decades the universities in Ireland have produced large numbers of female graduates who frequently make up greater than 50% of their year, and many of these move on to post-graduate work and geological careers. Those who have progressed to academic careers are those who have left Ireland.

Conclusions and the future

In the 19th century, women generally did not seek to make their own names scientifically; they were not normally admitted into the inner academic scientific circles where stimulation and competition would have driven them on. They probably were responsible for far more illustrative and artistic work of a scientific nature than we know about. Men were polymaths, and so it is likely that women were polymaths too. The mid-1800s saw women being admitted to some organizations and societies that fuelled their desire for geological information, and it was a period when popular literature became available to them. The late 1800s was a period marked by significant struggles that eventually saw women being admitted to universities in Ireland. In the 20th century, formal education began to change the status of women, but families with limited income sent only their sons to university: 'How things would have been different if the wealthy had put their daughters through TCD instead of their sons!'

In the past, society frowned on women taking an interest in science. To this day the legacy is still apparent. Comments made to the senior author by well-intentioned people include:

in 1969 in England, by a Professor of Geology:

You can study geology, but you cannot work as a geologist

in 1970s Ireland:

A university place is wasted on a woman.

When there is a shortage of jobs women should stay at home.

in 1980s Ireland:

Women don't make good scientists.

These relatively recent comments reflect society, and while they would no longer be acceptable, they convey a mood that has not quite disappeared. Although women are doing serious geological work in Ireland, at the turn of the century, and by that we mean 6 years ago, the number of full-time, permanent, women academics in geology departments in Ireland was two.

In recent decades, at least 50% of geology undergraduate and postgraduate positions have been filled by women, and women have obtained as many awards for achievement as men, but this is not reflected in academic positions. Women geologists have made major contributions to research into the geology of Ireland, and to industry, exploration and development. Nevertheless, there is still a gender bias in Ireland, in favour of male geologists. These issues were highlighted at the annual Irish Geological Research Meeting in 2005 (Higgs *et al.* 2005), and caused a little stir among the delegates, with equal measures of admission, denial and defensiveness. Hopefully this volume will cause a little more consternation. This is long overdue.

The authors would like to thank Cynthia Burek for her continued support, in particular for uncovering the letters to and from T. McKenny Hughes. The authors would also like to thank Jean Archer, and other reviewers, for comments which significantly improved the paper.

References

ALIC, M. 1986. *Hypatia's Heritage: A History of Women in Science From Antiquity to the Late Nineteenth Century*. Women's Press, London & Beacon Press, Boston.

ANGLESEA, M. & PRESTON, J. 1980. 'A philosophical landscape': Susanna Drury and the Giant's Causeway. *Art History*, **3**, 252–273.

ANON. 1909. *Report of the 78th Meeting of the British Association for the Advancement of Science, Dublin 1908*. John Murray, London.

BAHAR, S. 2001. Jane Marcet and the limits to public science. *British Journal for the History of Science*, **34**, 29–49.

BUREK, C. V. 2007. The role of women in geological higher education – Bedford College, London (Catherine Raisin) and Newnham College, Cambridge, UK. *In:* BUREK, C. V. & HIGGS, B. (eds) *The Role of Women in the History of Geology*. Geological Society, London, Special Publications, **281**, 9–38.

BURNS, V. & RICKARDS, R. B. 1993. Silurian graptolite faunas of the Balbriggan Inlier, counties Dublin and Meath, and their evolutionary, stratigraphical and structural significance. *Proceedings of the Yorkshire Geological Society*, **49**, 283–291.

BUTLER, E. 1924. *Structural Geography of Ireland*. The Educational Company of Ireland, Dublin.

CHESNEY, H. 2002. Mary Ball. *In:* MOLLAN, R. C., DAVIS, W. J. & FINUCANE, B. (eds) *Irish Innovators in Science and Technology*. Royal Irish Academy, Dublin, 110–111.

COWMAN, D. & REILLY, T. A. 1988. *The Abandoned Mines of West Carbery: Promoters, Adventurers and Miners*. Geological Survey of Ireland, Dublin.

CREESE, M. R. S. 2004. Somerville, Mary (1780–1872). *Oxford Dictionary of National Biography*, Vol. 51. Oxford University Press, Oxford, 617–619.

DAVIS, W. J. 1895. Royal Cork Institution. *In*: MOLLAN, R. C., DAVIS, W. J. & FINUCANE, B. (eds) *Some People and Places in Irish Science and Technology*. Royal Irish Academy, Dublin, 104–105.

DAVIS, W. J. 2002. Robert Kane. *In*: MOLLAN, R. C., DAVIS, W. J. & FINUCANE, B. (eds) *Irish Innovators in Science and Technology*. Royal Irish Academy, Dublin, 100–101.

DAVISON, D. H. 1989. *Impressions of an Irish Countess: The Photographs of Mary, Countess of Rosse 1813–1885*. Birr Scientific Heritage Foundation, Birr, Co. Offaly.

EVANS, E. E. 1996. *Ireland and the Atlantic Heritage: Selected Writings*. Lilliput Press, Dublin.

GABBY, A. 2002. William Molyneux. *In*: MOLLAN, R. C., DAVIS, W. J. & FINUCANE, B. (eds) *Irish Innovators in Science and Technology*. Royal Irish Academy, Dublin, 23–24.

HAMILTON, W. 1786. *Letters concerning the Northern Coast of the County of Antrim. Containing a natural history of its Basaltes: with an account of such circumstances as are worthy of notice respecting the antiquities, manners and customs of that country*. L. White, Dublin.

HARRY, O. 2002. Mary Ward, Microscopist, astronomer, naturalist, artist, 1827–1869. *In*: MOLLAN, R. C., DAVIS, W. J. & FINUCANE, B. (eds) *Irish Innovators in Science and Technology*. Royal Irish Academy, Dublin, 140–141.

HERRIES DAVIES, G. L. 1980. Richard Griffith – his life and character. *In*: HERRIES DAVIES, G. L. & MOLLAN, R. C. (eds) *Richard Griffith 1784–1878*. Royal Dublin Society, Dublin, 1–31.

HERRIES DAVIES, G. L. 1981. The history of the Earth sciences. *In*: SMITH, D. G. (ed.) *Cambridge Encyclopaedia of Earth Sciences*. Cambridge University Press, Cambridge, 12–23.

HERRIES DAVIES, G. L. 1995. *North from the Hook: 150 years of the Geological Survey of Ireland*. Geological Survey of Ireland, Dublin.

HIGGS, B. 2003. Mary Countess of Rosse (1813–1885). *Journal of the Cork Geological Association*, **6**, 22.

HIGGS, B., BUREK, C. V. & WYSE JACKSON, P. N. 2005. Is there gender bias in the geological sciences in Ireland? *Irish Journal of Earth Sciences*, **23**, 132–133.

HODSON, F. & LEWARNE, G. C. 1961. A mid-Carboniferous (Namurian) basin in parts of the counties of Limerick and Clare, Ireland. *Quarterly Journal of the Geological Society of London*, **117**, 307–333.

HOPPEN, K. T. 1970. *The Common Scientist in the Seventeenth Century: A Study of the Dublin Philosophical Society, 1683–1708*. Routledge & Kegan Paul, London.

HOWARTH, O. R. J. 1931. *The British Association for the Advancement of Science: A Retrospect 1831–1931*. British Association for the Advancement of Science, London.

HUDSON, R. G. S., CLARKE, M. J. & SEVASTOPULO, G. D. 1966. A detailed account of the fauna and age of a Waulsortian knoll reef limestone and associated shales, Feltrim, Co. Dublin, Ireland. *Scientific Transactions of the Royal Dublin Society*, **2**(A), 251–272.

HUTCHINSON, G. E. 1982. The harp that once . . . a note on the discovery of stridulation in the corixid water bugs. *Irish Naturalists' Journal*, **20**, 457–466.

JACKSON, A. A. 1994. Recollections 1964–1968. *In*: WYSE JACKSON, P. N. (ed.) *In Marble Halls: Geology in Trinity College, Dublin*. Department of Geology, Trinity College, Dublin, 106–107.

JACKSON, J. S. 1991. The geology and raw materials of the stone age. *In*: RYAN, M. (ed.) *The Illustrated Archaeology of Ireland*. Country House, Dublin, 33–35.

KAVANAGH, I. 1997. Mistress of the microscope. *In*: ANON. (ed.) *Stars, Shells and Bluebells: Women Scientists and Pioneers*. Women in Technology and Science, Dublin, 56–65.

KÖLBL-EBERT, M. 2001. On the origin of women geologists by means of social selection: German and British comparison, *Episodes*, **24**(3), 182–193.

LEWARNE, G. C. 1964. Starfish traces from the Namurian of County Clare, Ireland. *Palaeontology*, **7**, 508–513.

LYLE, P. 1996. *A Geological Excursion Guide to the Causeway Coast*. Environment and Heritage Service, Belfast.

MORIARTY, C. 2002. Augustin Hiberniae. *In*: MOLLAN, R. C., DAVIS, W. J. & FINUCANE, B. (eds) *Irish Innovators in Science and Technology*. Royal Irish Academy, Dublin, 17.

MURPHY, J. A. 1995. *The College: A History of Queen's/University College Cork, 1845–1995*. Cork University Press, Cork.

ORR, P. J., BRIGGS, D. E. G. & PARKES, M. A. 1996. The 'Castlecomer Fauna': a new Konservat Lagerstätte from the Upper Carboniferous of Ireland. *Irish Journal of Earth Sciences*, **15**, 93–106.

PARKES, S. M. (ed.) 2004. *A Danger to the Men? A History of Women in Trinity College Dublin, 1904–2004*. Lilliput Press, Dublin.

PETTIT, S. F. 1976. The Royal Cork Institution: a reflection on the cultural life of a city. *Journal of the Cork Historical and Archaeological Society*, **81**, 70–90.

PRAEGER, R. L. 1949. *Some Irish Naturalists*. Dundalgan Press, Dundalk.

REYNOLDS, D. L. 1928. The petrography of the Triassic Sandstone of north-east Ireland. *Geological Magazine*, **65**, 448–473.

REYNOLDS, D. L. 1931. The dykes of the Ards Peninsula. *Geological Magazine*, **68**, 97–111.

REYNOLDS, D. L. 1937. Contact phenomena indicating a Tertiary age for the Gabbros of the Slieve Gullion District. *Proceedings of the Geologists' Association*, **48**, 247–275.

RICKARDS, R. B. 1999. A century of graptolite research in Cambridge. *The Geological Curator*, **7**, 71–76.

RICKARDS, R. B., BURNS, V. & ARCHER, J. B. 1973. The Silurian sequence at Balbriggan, Co. Dublin. *Proceedings of the Royal Irish Academy*, **73B**, 303–316.

SHEFFIELD, S. L. 2004, *Women in Science*. ABC Clio Press, California & Oxford.

SIMMS, J. G. 1982. *William Molyneux of Dublin*. Irish Academic Press, Dublin.

SOMERVILLE, M. 1848. *Physical Geography*. John Murray, London.

TAYLOR, K. 1969. Nicolas Desmarest and geology in the eighteenth century. *In*: SCHNEER, C. J. (ed.)

Towards a History of Geology. MIT Press, Cambridge, Massachusetts, 339–316.

THOMSON, L. 2004. The campaign for admission. *In*: PARKES, S. M. (ed.). *A Danger to the Men? A History of Women in Trinity College Dublin 1904–2004.* Lilliput Press, Dublin, 19–54.

WARD, M. 1857. *Sketches with a Microscope.* Shields, Parsonstwon.

WHEELER, T. S. 1944. Sir Robert Kane, his life and work. *In*: ANON. (ed.) *The Natural Resources of Ireland.* Royal Dublin Society, Dublin, 5–42.

WRIGHT, M. C. 1920. Limnestheria: a new conchostracan genus from the Kilkenny coal-measures. *Proceedings of the Royal Irish Academy,* **35B**, 187–204.

WRIGHT, W. B. 1914. *The Quaternary Ice Age.* Macmillan, London.

WYSE JACKSON, P. N. 1987. *Statistical Survey of Geological and Geomorphological Excursions Undertaken by the Dublin Naturalists' Field Club, 1886–1986.* Dublin Naturalists' Field Club, Dublin.

WYSE JACKSON, P. N. 1989. On rocks and bicycles: a biobibliography of Grenville Arthur James Cole (1859–1924) fifth Director of the Geological Survey of Ireland. *Geological Survey of Ireland Bulletin,* **4**, 151–163.

WYSE JACKSON, P. N. 1992. The geological collections of Trinity College, Dublin. The Geological Curator, **5**, 263–274.

WYSE JACKSON, P. N. (ed.) 1994. *In Marble Halls: Geology in Trinity College, Dublin.* Department of Geology, Trinity College, Dublin.

WYSE JACKSON, P. N. 1997. Fluctuations in fortune: three hundred years of Irish geology. *In*: WILSON FOSTER, J. & CHESNEY, H. (eds) *Nature in Ireland.* Lilliput Press, Dublin, 91–114.

WYSE JACKSON, P. N. 1998*a*. Richard Kirwan (1733–1812): Chemist and Geologist. *Irish Chemical News* **12**, 36–39.

WYSE JACKSON, P. N. 1998*b*. Veronica Burns 1914–1998, Curator Geological Museum, Trinity College, Dublin 1964–1980. *The Geological Curator* **6**(9), 339–340.

WYSE JACKSON, P. N. 1998*c*. Veronica Conroy Burns (1914–1998). *Irish Naturalists' Journal,* **26**, 1–4.

WYSE JACKSON, P. N. 2000. Tumultuous times: geology in Ireland and the debate on the nature of basalt and other rocks of northeast Ireland between 1740 and 1816. *In*: WYSE JACKSON, P. N. (ed.) *Science and Engineering in Ireland in 1798: A Time of Revolution.* Royal Irish Academy, Dublin, 35–49.

WYSE JACKSON, P. N. 2002. William Hamilton. *In*: MOLLAN, R. C., DAVIS, W. J. & FINUCANE, B. (eds) *Irish Innovators in Science and Technology.* Royal Irish Academy, Dublin, 41.

WYSE JACKSON, P. N. & CONNOLLY, M. 2002. Fossils as Neolithic funereal adornments in County Kerry, southwest Ireland. *Geology Today* **18**(4), 139–143.

YATES, P. J. 1961. New Namurian goniatites of the genus *Eumorphoceras. Palaeontology,* **4**, 54–58.

YATES, P. J. 1962. The palaeontology of the Namurian rocks of Slieve Anierin, Co. Leitrim, Eire. *Palaeontology,* **5**, 355–443.

The role of British and German women in early 19th-century geology: a comparative assessment

M. KÖLBL-EBERT

Jura-Museum Eichstätt, Willibaldsburg, 85072 Eichstätt, Germany
(e-mail: Koelbl-Ebert@jura-museum.de)

Abstract: The history of the geosciences has largely been interpreted as a history of male scientists, but the inclusion of their social frame into historical research makes clear how women in various roles have participated in and shaped the history of geosciences.

The beginning of geological research in a modern sense occurred around 1800. In Germany, the early professionalization of geology and a rigid female gender model, idealizing female household duties and motherhood in a climate that was hostile to intellectual women, effectively precluded the collaboration of women, whereas in the largely non-professional culture of natural sciences in the United Kingdom, women were not excluded from participation.

In the United Kingdom at that time, wives, daughters and sisters, or even non-related female acquaintances, were an integral part of the infrastructure of British geology. They were often encouraged by leading scientists. As a result, there have been many female contributors, especially to palaeontology, in the early 19th century in the United Kingdom, forming a framework of assistants, secretaries, collectors, painters and field geologists to the leading figures in the geological sciences, thereby adding to and shaping their work.

Problems, however, arose, where women aspired to work on their own research programmes as independent female geologists.

Recent investigations of female participation in the history of geological research have made evident that numerous women, especially in the United Kingdom, played a not inconsiderable role in early 19th-century geology (Kölbl-Ebert 2001, 2002; and references therein). Most of them worked as assistants to male relatives, as field geologists, fossil collectors, taxonomists and draughtswomen, but there also existed fully mature women geologists with an independent research interest. A number of names come to mind: Mary Anning (see, e.g. Lang 1939, 1950; Taylor & Torrens 1987; Torrens 1995), Mary Buckland née Morland (Kölbl-Ebert 1997a), Charlotte Murchison (Kölbl-Ebert 1997b), Barbara Marchioness of Hastings (Edwards 1970; Kölbl-Ebert 2004) and Etheldred Benett (Torrens 1985; Torrens et al. 2000), to mention just a few.

On the other hand, in Germany, at that time, the influence of women on the development of the geological sciences had been negligible. There happened to be a few wealthy female collectors of minerals, fossils and other curiosities, such as Caroline Luise von Baden (1723–83) but with no particular scientific expertise and living earlier, in the 18th not the 19th century, well before geology developed an independent profile (Kölbl-Ebert 2001, p. 184).

This paper attempts to explain, why these differences occurred in these two countries.

Germany in the early 1800s

Around 1800, a time critical for the development of modern geosciences, the territory which today is Germany consisted of numerous small, more or less independent states, and among these, from a purely geological point of view, Saxony was among the most important. Saxony was rather well off and it owed much of its wealth to silver mining.

Consequently, late in the 18th century, a school of mines, the Bergakademie Freiberg, was established, forming an important nucleus in which geology developed from a technical, mineralogical tradition, and which produced surveyors, engineers, mineralogists and other specialists for the Saxonian and international mining industry, thereby encouraging early professionalization of the geosciences in Saxony and neighbouring countries. The male domination through both mining and formal training precluded the emergence of women geologists until the early 20th century. In Germany, professionalization of science in general became established towards the end of the 18th century (see, e.g. Weber-Kellermann 1998, p. 46).

Around the same time, the German states were shaken by news of the revolution in France and then even more by the Napoleonic Wars, which altered the political landscape quite drastically. Possibly as a result of this political unrest and insecurity, social patterns started to change too, causing German family life to retreat more and more into the private sphere, creating the ideal of a home, which provided a comfortable, secure nest to which to retire from the frightening political storms that swept Europe (e.g., Weber-Kellermann 1998, p. 46)

Both these processes promoted a rigid division of labour in the wealthier families, the man venturing out into the public, whereas his wife stayed at home, attending to her household duties and the children. Geology in Germany, however, was shaped in the public and professional sphere of the universities and museums, a sphere that was closed to women.

A German woman's world at the beginning of the 19th century was very much limited to her house. Of course, a statement like this is rather simplistic, but it seems to be true for the social circles that are relevant for this study. In Germany at that time, geology was practised by university professors and curators, and, to a lesser extent, as a by-product to their professional interests, by high-school teachers, apothecaries and clergymen. All of these professions belonged to a middle, not too wealthy social level (see Weber-Kellermann 1998, p. 49). In the United Kingdom, however, we are talking about gentlemen geologists and their relations, nearly all of them being decidedly upper class. But even, if we compare this social level directly with its counterpart in Germany, we still have the German ideal of womanhood described in this paper (see, e.g., Weber-Kellermann 1998, pp. 23 & 26), and which famous German poet, naturalist and politician Johann Wolfgang von Goethe (1749–1832) defined like this:

To serve, the woman has to learn early, according to her destiny; because through serving only, she will finally achieve to rule, to achieve the proper power, which is hers in the house ... so that she will forget herself completely and may live only in others! (Goethe 1798)

Goethe's contemporary, the writer and political revolutionary Friedrich Schiller (1759–1805), also described the role model for husband and wife in one of his ballads:

The man must go forth into the hostile life, must effect and endeavour and plant and create, trick, grab, must bet and dare, to hunt his luck ... and at home the demure housewife attends to her duties, the mother of the children, and rules wisely within the range of the house. (Schiller 1799)

It is rather difficult to translate these lines from the original German into English, since the connotations of many of the relevant words can hardly be condensed into a single English expression; for example, the German expression *züchtig* is translated here as 'demure', but it also carries the following meanings: modest, restrained, frugal, without personal needs, disciplined, even asexual. And the word *Hausfrau*, here translated literally as 'housewife', also strongly implied industry, thriftiness, devotion and never-ending painstaking diligence and an enormous workload, since a typical German household at that time had far fewer servants than was the custom in British households of comparable standing (see, e.g., Weber-Kellermann 1998, pp. 49 & 62).

In very wealthy social circles with a more cosmopolitan experience, women might hold fashionable soirees in the drawing-rooms, where they acted as hostess to men of culture, creating a meeting place of artists, musicians and men of letters, as well as businessmen and politicians with a taste for the arts. However, these soirees were not comparable in quality to the famous French salon culture and the custom tended to peter out towards the early 19th century. In any case, it was the exception rather than the rule in female social life; discussing the arts and admiring the (usually male) artist, was a leisured, private and recreational activity, and part of a woman's task of making private life agreeable to husband, brothers and father. The same women would not have been able to discuss economics or science with the same freedom, since nobody would have considered them to be serious partners for the discussion of subjects which belonged to the public domain – and men's life.

Consequently, the very few German women with connections to geology were hired collectors or professional dealers in fossils and minerals. They met male geologists in the public sphere of the business relationship, but none of them achieved anything remotely similar to the scientific acceptance experienced by Mary Anning (1799–1847).

German and British women's view of each other

In 1803, Johanna Schopenhauer, wife of a wealthy German tradesman, travelled to England, and what she reported home about English women was far from favourable. She noticed that breast-feeding among the wealthier mothers in England was much less common than in Germany (Schopenhauer 1982, p. 97), and she complained that the English girls learned fine embroidery, the arts of fashioning paper flowers, shaping little boxes from golden paper and creating tiny vessels from egg shells, and a thousand other nonsensical knickknacks, but that they were completely ignorant of what was really needed in a household (Schopenhauer 1982, p. 169).

At that time, the German housewife was accustomed to doing everything herself, with the aid of her elder daughters and maybe a maid-servant or two. This incluced: cooking, baking bread, making pickles, preserves and wines, drying apples and other fruit, washing and mending clothes and linen, cleaning the house, serving dinner and so on, while any number of small children were clinging to her skirts (Weber-Kellermann 1998, pp. 49–52, 62). With respect to other, more intellectual accomplishments, the general opinion was that an educated woman was unable to secure herself a husband and, if she did, she would neglect her household duties completely (Leisner 1998, p. 27).

The different gender models in Germany and the United Kingdom is highlighted by an anecdote from Mary Somerville's autobiographical notes. Somerville (1780–1872), a British scientific writer, astronomer and mathematician, accompanied Sir James McIntosh on a journey up the Rhine valley. (It was especially British women, on their Grand Tours, travelling along the Rhine Valley to Italy, who introduced contemporary men and women in Germany to the sight of travelling females (Pelz 1991, pp. 177–178).) The following incident happened near Bonn:

A professor of his acquaintance in one of the towns invited us to dine, & I was astonished to see the lady of the house going about with a great bunch of keys dangling at her side assisting in serving up the dinner, & doing all the Duty of Carving, the husband taking no part whatever in it. I was annoyed that we had given so much trouble by accepting the invitation. In my younger days in Scotland a lady might make pastry, jelly, or direct in the Kitchen, but she took no part in cooking or serving up the dinner, & never rose from the table till the ladies went to the Drawing room. (First draft of Mary Somerville's autobiography, p. 157; Bodleian Library, Somerville Collection, Department c355, MSALL-2.)

This short account conveys the indignation of a British lady of her house on encountering a typical, demure German housewife, whose badge of honour was indeed the big bunch of keys (Fig. 1) that gave her power over the kitchen and the larders, the attics and the cellars, the washing house and the linen closet. In the United Kingdom, however, the Industrial Revolution, which had started in the 18th century and gained much momentum around the turn of the century

... was changing the whole social pattern of Britain. The new middle class was largely an urban class contemptuous of the domestic arts and skills which gentlewomen, hitherto, had been expected to have at their finger-tips. In a town house, a woman had no dairy or herb-garden in which to display a personal interest; fashion now bade her stay out of her kitchen and still-room. The arts of baking and preserving, pickling and wine-making were surrendered to housekeeper, cook and butler. (Turner 1962, p. 100)

And the great bunch of keys was equivalent to a badge of office to the housekeeper, an upper servant, but definitely a servant, while a proper lady of the house gave her orders but never set foot below stairs (Turner 1962, pp. 117 & 133).

A lady above a certain social level, even when she was too poor to afford a large number of domestics, at least made an effort to keep up public appearance. This is wonderfully illustrated in the novel '*Cranford*' by Elizabeth Gaskell (1810–65), which she published in 1851–53:

I imagine that a few of the gentlefolk of Cranford were poor, and had some difficulty in making both ends meet; but they were like the Spartans, and concealed their smart under a smiling face ... When Mrs. Forrester, for instance, gave a party in her baby-house of a dwelling, and the little maiden disturbed the ladies on the sofa by a request that she might get the tea-tray out from underneath, every one took this novel proceeding as the most natural thing in the world, and talked on about household forms and ceremonies as if we all believed that our hostess had a regular servants' hall, second table, with housekeeper and steward, instead of the one little charity-school maiden, whose short ruddy arms could never have been strong enough to carry the tray upstairs, if she had not been assisted in private by her mistress, who now sat in state, pretending not to know what cakes were sent up, though she knew, and we knew, and she knew that we knew, and we knew that she knew that we knew, she had been busy all the morning making tea-bread and sponge-cakes. (Gaskell 1851–53, pp. 17–18).

In contrast to a German housewife, a lady in a British house would never have served dinner personally, and would even pretend not to know what the kitchen would bring to the table. Her place was a public presence at soirees and dinner parties.

No wonder the words used to describe the ideal British lady of the house – that we find in novels as in letters and other historical sources – are so different from those used to describe her German counterpart.

From a fashionable, accomplished lady of the house, honourable British society expected 'the courteous bearing of a cultivated woman', to which she 'added a brightness of conversation, an intelligence, and a range of knowledge which gave her a peculiar charm, and enabled her to please people of the most varied tastes and acquirements' (Geikie 1875, vol. 2, p. 339). This is a contemporary description of Charlotte Murchison, who was active as geological amanuensis to her husband and other geologists (Kölbl-Ebert 1997*b*).

British geology in the early 1800s

During the early 19th century, there still existed a largely non-professional culture of natural sciences in the United Kingdom. Outside industrial geosciences, geology remained a private interest for people with sufficient money, time and leisure to study, travel and publish. Geology was fashionable and accepted topic for small-talk at dinner parties as well as for presentations at scientific societies.

In these informal, non-professional surroundings, women were welcomed – within the restrictions of the social order – as fellow enthusiasts, often

Fig. 1. German housewife with typical bunch of keys (illustration by Julius Oldach 1804–34).

working alongside husbands or brothers, and also for non-relatives.

The geology that was emerging in the United Kingdom around the turn of the 19th century was furthered also by opportunities offered by the Industrial Revolution, which, in the United Kingdom, started decades earlier than in Germany (Weber-Kellermann 1998, p. 46). The intensive building of canals, and later of railway tracks, opened new outcrops all over the country, in addition to the many established small and local quarries and the natural outcrops along the coast. And at that time, every outcrop – accessible on private walks – could hold unknown fossil treasures. The French Revolution had brought a fashion in female clothes that, for a few decades, allowed fashionable and wealthy women to move around in comparatively comfortable dresses (Fig. 2). Thus fashion contributed to their 'mobility'.

Another aspect is the fact that, within the non-professional culture of British science at that time, the geological interest of the men was also part of their private life, their private studies and cabinets, and their private summer travels, and on all these occasions, women could participate, even though they were not allowed to enter the rooms of the Geological Society of London.

Geological education was available to women (see Burek 2007) by means of books and private tuition, and also through public lectures, which were often delivered by leading scientists of the day, for example, in the Royal Institution or at the annual meetings of the British Association for the Advancement of Sciences (Phillips 1990, pp. 200–207).

It is interesting to note that, during the years 1860 to 1862, the Geological Society of London undertook considerable efforts to open its meetings to women. At that time, Leonard Horner was president and the action was also very much approved of by his son-in-law, Charles Lyell. The meetings were even moved, with quite some inconvenience for the Society's officers, from Somerset House to Burlington House, where the available rooms were deemed more suitable for ladies.

Unfortunately, only a couple of women appeared, among them Lady Lyell and her sisters. Because of this lack of interest, the attempt was

Fig. 2. Fashionable bride and bridegroom from an advertisement (around 1810).

discontinued in 1863, and the men returned to their usual business at Somerset House.

When, in 1889, a revision of the Society's bye-laws was discussed and the admittance of women became an issue, there was no longer a majority to be gained among the fellows. This was a time, when the first university-trained women endeavoured to become fellows of the Geological Society, and so the men possibly saw them as unwelcome competitors for scarce jobs (Woodward 1908, p. 242–244). The point, however, is that by the year 1860, Leonard Horner's unsuccessful attempt to open the meetings for a mixed audience came too late. The window of opportunity for scientific ladies (see Fig. 3) had already started to close. During the 19th century, several factors favoured female participation in geology. In Britain, these factors came together between about 1795 and 1840, forming a unique, local window of opportunity in the gap between the early onset of industrialization, which freed women of higher social standing from their traditional household duties, and the late professionalization of geology, which required an informal workforce to help the typical gentleman geologist. A comparatively liberal gender model, which did not rigidly discriminate between a private, family sphere of life for women and a public sphere for men, together with a rather uncomplicated fashion in clothes allowed women to meet these demands. Informal geological education for these women was available through many different sources, even though universities did not open their doors to women until the 1870s.

In Germany, there was no such gap and professionalization occurred much earlier, while the onset of industrialization and its consequences on household economics happened decades later. Consequently, there was no requirement for a female workforce in geology. Also, a rigid female gender model precluded female participation in the public, male sphere of science, while the general social climate was hostile to intellectual women.

It is possible to cite numerous examples of British women as diligent gatherers of information, as assistants to their husband and as efficient guardians of a collection, and these have been described in more detail elsewhere (see Kölbl-Ebert 2002 and references therein). Many of them were connected to some extent with William Buckland.

Dr William Buckland (1784–1856), Canon of Christ Church, Oxford, was a great popularizer of geology, becoming Reader in Mineralogy at Oxford in 1813, and Reader in Geology in 1819. His main interest was palaeontology, but he was interested not only in anatomy but more especially in reconstructions of ancient ecosystems and food chains. His wife, Mary (1797–1857), née Morland, had well before her marriage made herself a name as a scientific draughtswoman, who had contributed to works of Conybeare, Cuvier and Buckland. In around 1819, she also mapped for Buckland the distribution of certain pebbles in the surroundings of Oxford. In 1825, she married William Buckland and continued to be a practising geologist, accompanying her husband on field trips, making observations, illustrations and models of fossils, and assisting in the collection of specimens and the preparation of papers – as well as giving birth to nine children. This sometimes proved to be quite a heavy workload, as Charlotte Murchison wrote to Mary Somerville: 'Mrs. Buckland not very well, quite overworked assisting at the birth of his essay – too bad with so many births of another description to endure' (fragment of a letter from Charlotte Murchison to Mary Somerville, dated 6 May [1833]; Bodlein Library, Somerville Collection. Folder MSM-5, SC Box 21).

In Mary, Buckland had a most supportive wife-assistant, who also continued to work

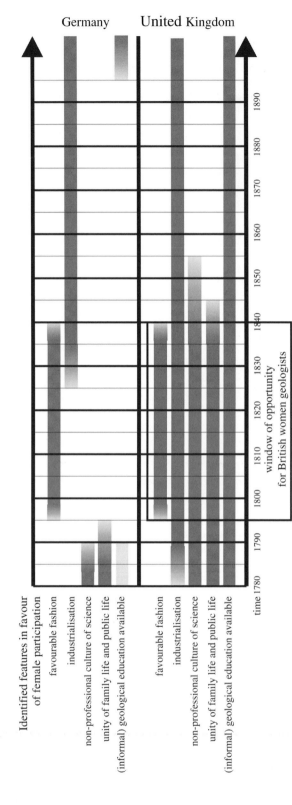

Fig. 3. Factors favourable to female participation in Geology between 1780 and 1900.

scientifically on her own, after her husband's mental health declined in his last years of life.

It is quite possible that Buckland's own favourable experience with his wife allowed him to think it natural to fall back on many more women of his acquaintance as his assistants, and what is more, he did not hesitate to quote these women and their findings in his letters, as well as in his publications, thus giving an example to the next generation of young gentlemen geologists.

However, even in this comparatively favourable situation, women tended to run into trouble as soon as they tried to be independent scientists. Whereas women featured quite commonly as geological help-mates to gentleman geologists, a true female equivalent to the gentleman geologist, that is, a 'lady geologist' following an independent research programme, publishing her results and presenting them to the contemporary scientific community, was an extremely rare phenomenon.

Such independent 'lady geologists' were, for example, Etheldred Benett (1775–1845) (see Torrens 1985; Torrens et al. 2000) and Barbara, Marchioness of Hastings (1810–58) (see Kölbl-Ebert 2004), who both corresponded widely with male colleagues. Their stratigraphical work and their collection were indeed recognized by these, but nevertheless, because of their sex, they stood separated from the geological community. They could not become members of important scientific societies, such as the Geological Society of London, and without chaperone they had difficulties in obtaining books and journals from libraries, thus being reduced to petitioners, whereas male colleagues simply had to order.

One of the less well-known examples of these few 'lady geologists' was Lady Eliza Maria Gordon-Cumming, née Campbell (1798–1842). Her interest in fossils was, however, of rather short duration, owing to her early, tragic death. In around 1839, she began to collect fossil fish of Devonian age from the Old Red quarries in Morayshire. She paid workers to do the excavations and to collect the fossils, which were found in concretionary nodules. She presented fossil fishes to other collectors of her social class, such as the Earl of Enniskillen, and also to the Geological Society and other collections and museums. Being rather class-conscious, it was not so easy to deal with her, but she was delighted to make drawings of her fish, an occupation which her daughter Ann Seymour had to follow upon her mother's request.

Louis Agassiz spoke highly of Lady Gordon-Cumming's drawings (see Andrews 1982, p. 31), but after a while, Lady Eliza became tired of drawing only what she saw, and endeavoured to reconstruct complete fishes from fragments using assumptions of symmetry.

Only 2 or 3 years after she started her collection, Lady Eliza died a few weeks after the birth of what was 'her thirteenth living child' (Gordon-Cumming 1904, p. 43). A few days before her death, she dictated a letter to Roderick Murchison, complaining that some of her drawings had been copied and circulated without her consents, and adding:

The further I excavate the more I am convinced that a world of treasure lies before us & I could now make out a set distinct from what Agassiz saw which would make a very beautiful little work & which work I mean to bring out myself if nothing better offers_ I think it would bind in very prettily at the end of the 'Silurian System' if the Author of the said Silurian would only add a few lines of letter-press with each.

I have several *entirely* new fishes & very beautiful specimens & I am convinced that by perseverance we shall hit upon some perfect specimens both of Pterichthys & Cocosteus

Tell me distinctly from your experience of the Russian specimens Has the Pterichthys one or two fins on the tail – Does the said tail conclude in an earwiglike manner & has it winged-like horns?

Has the Cocosteus more than one pair of decided paddles (I think it has) & have you ever seen the fins or the hip of the tail complete? I send you facsimiles of a Pterichthys & a Cocosteus which are the most distinct which we have as yet brought to light on those prints, but I am breathless to be at work again whereas for a short time at least I must expect to be perfectly powerless. (Geological Society, London, Murchison Papers, M/G11/1a-b).

We do not know what Murchison thought about her Ladyship's idea to intrude into his book '*Silurian System*', but we do know that her daughter Ann Seymour felt quite uncomfortable with the idea of fossil reconstruction and, after her mother's death, tried – with all due filial respect – to withdraw the circulated interpretative drawings. She wrote to Murchison:

My dear Mr. Murchison

it has occurred to me that you may shew some of my drawings now in your hands to geologists as having been faithfully copied from specimens like some of those you saw & so I write you this one line & will not long encroach upon your valuable time only I wish to remind you that with all our quarryings we have never yet hit upon a Pterichthys with *two fins* on the tail

It was my dear mothers persuasion that there *were two* & that we should ere long find a perfect one ... I should be sorry that those half fanciful creatures should be shewn as authentic specimens to scientific people – all the other bits of anatomy in the Pterichthys indicated in those drawings are I think established on pretty good authority – Then the *cocosteus* is *put together* from several specimens we have never yet found one so perfect as that drawing seems to assert without the necessary explanation ... It was much against my own inclination that I painted the *restored creatures* but now how thankful I am that I did any thing to amuse our darling angel during her last illness. (Altyre 8 June 1842; Geological Society, London, Murchison Papers, M/G11/3a-b)

Interestingly, the whole story is not so much whether or not Eliza Gordon-Cumming was right in her assumption of a symmetrical tail fin (she was not), but that until today she is disqualified as having drawn fanciful creatures when she should have drawn only observable facts. While nobody ever bullied Buckland and Owen for having depicted *Iguanodon* with its thumb spike stuck to the nose, or Hugh Miller, who reconstructed at least one Old Red fish by merging two different genera. (Andrews 1982, fig. 14). It seems that Lady Eliza Gordon-Cumming violated the scientific code for women.

British women at that time were allowed, or even desired, to assist and to contribute observations and data. They certainly also talked to their husbands or non-related mentors about their work, but a public display of female creativity was beyond the invisible boundary of social conduct. So it is no wonder that a self-conscious and well-educated lady geologist like Etheldred Benett (1775–1845) complained 'that scientific people in general have a very low opinion of the abilities of [the female] sex'. (letter to Samuel Woodward 12 April 1836, quoted after Torrens *et al.* 2000, p. 64).

Nevertheless, apart from those few women who aspired to their own 'fame', at the same time the more ordinary female geological assistants probably experienced scientific and social acceptance. Geological research opened to them a kind of intellectual freedom, being able to share their interests and being an accepted and necessary part of the British scientific culture during the early 19th century, before the professionalization of geology set in.

Conclusion

During the first few decades of the 19th century, there are several factors in the United Kingdom that allowed the emergence of female geologists (Fig. 3): There was a gap between the early onset of industrialization, which together with its social implications, freed wealthier women from their traditional household duties, and the late professionalization of geology, which required an informal workforce to help the typical gentleman geologist. A fairly liberal gender model for women, together with a rather uncomplicated fashion in clothes, allowed women to address these demands. They were often encouraged by leading scientists, such as William Buckland, who himself had a supportive, intelligent wife, and therefore knew of the benefits and did not hesitate to 'employ' all sorts of other females to help him, thus setting an example for the next generation: his pupils. Consequently, it was only natural to allow a certain, informal geological education for these women.

In Germany, on the other hand, there was no such gap between industrialization and professionalization. The latter occurred much earlier, because geology developed from the ore-mining industry, on which some of the small German states, notably Saxony, depended very strongly, while the onset of industrialization and its consequences on household economics happened decades later than in the United Kingdom. There was no requirement for a female workforce in geology, since geology was mainly practised by professionals with all the resources related to their position.

The female gender model in Germany was much more rigid than in the United Kingdom, idealizing female household duties and motherhood in a climate that was hostile to intellectual women. This gender model was promoted and stabilized by the political and social situation in Germany around the Napoleonic Wars.

I wish to thank the staff of the Bodleian Library (Oxford) and the library of the Geological Society of London for their kind help. I am also greatly indebted to Cynthia Burek for improving my English. This work was financed by the Deutsche Forschungsgemeinschaft (DFG).

References

ANDREWS, S. M. 1982. *The Discovery of Fossil Fishes in Scotland up to 1845 with Checklists of Agassiz's Figured Specimens.* Royal Scottish Museum Studies, Royal Scottish Museum, Edinburgh.

Bodleian Library, University of Oxford. Somerville Collection.

BUREK, C. V. 2007. The role of women in higher education—Bedford College, London (Catherine Raisin) and Newnham College, Cambridge. *In:* BUREK, C. V. & HIGGS, B. (eds) *The Role of Women in the History of Geology.* Geological Society, London, Special Publications, **281**, 9–38.

EDWARDS, N. 1970. The Hastings Collection (fossil vertebrates): history of additions made by the Marchioness of Hastings between 1845-1851 from the Upper Eocene Beds at Hordle Cliff, Hampshire. *Journal of the Society for the Bibliography of Natural History*, **5**, 340–343.

GASKELL, E. C. 1851–53. *Cranford.* 1993 edn, Wordsworth Classics, Ware.

GEIKIE, A.1875. *Life of Sir Roderick I. Murchison, Bart., K.C.B., F.R.S.; Sometime Director-General of the Geological Survey of the United Kingdom. Based on his Journals and Letters; with Notices of his Scientific Contemporaries and a Sketch of the Rise and Growth of Palaezoic Geology in Britain.* John Murray, London.

Geological Society, London. Murchison papers.

GOETHE, J. W. VON 1798. Hermann und Dorothea. World Wide Web Address: http://www.von-goethe.net/werke/werke/hermann_und_dorothea.html.

GORDON-CUMMING, C. F. 1904. *Memories.* William Blackwood & Sons, Edinburgh & London.

KÖLBL-EBERT, M. 1997a. Mary Buckland née Morland 1797–1857. *Earth Science History*, **16**, 33–38.

KÖLBL-EBERT, M. 1997b. Charlotte Murchison née Hugonin 1788–1869. *Earth Science History*, **16**, 39–43.

KÖLBL-EBERT, M. 2001. On the origin of women geologists by means of social selection: German and British comparison. *Episodes*, **24**, 182–193.

KÖLBL-EBERT, M. 2002. British geology in the early 19th century: a conglomerate with a female matrix. *Earth Sciences History*, **21**, 3–25.

KÖLBL-EBERT, M. 2004. Barbara Marchioness of Hastings (1810–1858): fossil collector and 'lady-geologist'. *Earth Sciences History*, **23**, 75–87.

LANG, W. D. 1939. Mary Anning and the pioneer geologists of Lyme. *Proceedings of the Dorset Natural History and Archaeological Society*, **60**, 142–164.

LANG, W. D. 1950. More about Mary Anning including a newly-found letter. *Proceedings of the Dorset Natural History and Archaeological Society*, **71**, 184–188.

LEISNER, B. 1998. 'Unabhängig sein ist mein heißester Wunsch' – Malwida von Meysenburg. Econ, München.

PELZ, A. 1991. Reisen Frauen anders? Von Entdeckerinnen und reisenden Frauenzimmern. *In*: BAUSINGER, H., BEYRER, K. & KORFF, G. (eds) *Reisekultur. Von der Pilgerfahrt zum modernen Tourismus*. Verlag C.H. Beck, München, 174–178.

PHILLIPS, P. 1990. *The Scientific Lady. A Social History of Women's Scientific Interests 1520–1918*. St Martin's Press, New York.

SCHILLER, F. 1799. Das Lied von der Glocke. World Wide Web Address: http://www.historisches-wuerttemberg.de/kultur/dichter/schiller/glocke.htm.

SCHOPENHAUER, J. 1982. *Reise nach England*. Rütten & Loening, Berlin.

TAYLOR, M. A. & TORRENS, H. S. 1987. Saleswoman to a new science: Mary Anning and the fossil fish *squaloraja* from the Lias of Lyme Regis. *Proceedings of the Dorset Natural History and Archaeological Society*, **108**, 135–148.

TORRENS, H. S. 1985. Women in Geology, 2. Etheldred Benett. *Open Earth*, **21**, 12–13.

TORRENS, H. S. 1995. Mary Anning (1799–1847) of Lyme; 'the greatest fossilist the world ever knew'. *British Journal for the History of Science*, **28**, 257–284.

TORRENS, H. S., BENAMY, E., DAESCHLER, E. B., SPAMER, E. E. & BOGAN, A. E. 2000. Etheldred Benett of Wiltshire, England, the first lady geologist. Her fossil collection in the Academy of Natural Sciences of Philadelphia, and the rediscovery of 'lost' specimens of Jurassic Trigoniidae (Mollusca: Bivalvia) with their soft anatomy preserved. *Proceedings of the Academy of Natural Sciences of Philadelphia*, **150**, 59–123.

TURNER, E. S. 1962. *What the Butler saw – Two Hundred and Fifty Years of the Servant Problem*. Reprinted 2001. Penguin, London.

WEBER-KELLERMANN, I. 1998. *Frauenleben im 19. Jahrhundert: Empire und Romantik, Biedermeier, Gründerzeit*. 4th edn. C.H. Beck, München.

WOODWARD, H. B. 1908. *The History of the Geological Society of London*. Longmans, Green & Co., London.

Invincible but mostly invisible: Australian women's contribution to geology and palaeontology

S. TURNER

Monash University School of Geosciences, Victoria 3800, Australia;
Queensland Museum, 122 Gerler Road, Hendra, Queensland 4011, Australia
(e-mail: sue.turner@qm.qld.gov.au)

Abstract: Women have played a significant role in Australian geoscience, and especially in palaeontology. 'Australian' women gained degrees by the early 20th century and began to contribute intensively. Australian-born young women already immured to the rigours of climate and culture, collected and illustrated fossils, enrolled in the first university courses, thrived in the field, in some instances outnumbering and out-achieving men. Where women palaeontologists made their mark they often energetically concentrated on a taxonomic group, making them their own, as Isabel Cookson did with palynology, Joan Crockford with bryozoans, Irene Crespin especially with foraminifans, Dorothy Hill with corals, Ida Brown with brachiopods, Nell Ludbrook with molluscs, Elizabeth Ripper with stromatoporoids, Kathleen Sherrard with graptolites, and Mary Wade, initially with foraminiferans and then the Ediacaran fauna. Brown, Crespin, Hill, Ludbrook, Wade and their contemporaries did alpha taxonomy, classical geology and biostratigraphical studies that laid the foundations for making maps and work that became recognized nationally and internationally. Some achieved greatness; some – Hill, Cookson, Ludbrook and Phillips Ross – by leaving the country, either to gain their higher degree or to work. Many – for example, Hosking, Johnston, Prendergast, Richards, Ripper, Sullivan and Vincent – are or have been mere shadowy figures with a few publications and then oblivion or even tragedy. Women in geosciences spanning the 20th century in Australia contributed some hundreds of scientific papers, maps and textbooks.

Abbreviations: AAP, Association of Australasian Palaeontologists; AAS, Australian Academy of Science; AMDEL, Australian Mineral Development Laboratories; ANZAAS, Australian and New Zealand Association for the Advancement of Science; ARC, Australian Research Council; BAAS, British Association for the Advancement of Science; BMNH, British Museum (Natural History), now The Natural History Museum; BMR, Bureau of Mineral Resources, now Geoscience Australia; CSIRO, Commonwealth Scientific and Industrial Research Organization; FGS, Fellow of the Geological Society, London; GA, Geological Association, London; GSA, Geological Society of Australia; GSL, Geological Society, London; GS, Geological Survey (GSNZ, New Zealand; GSQ, Queensland; GSSA, South Australia; GST, Tasmania; GSV, Victoria; GSWA, Western Australia); IGCP, International Geological Correlation Programme (now International Geoscience Programme); IUGS, International Union of Geological Sciences; MBE, Member of the order of the British Empire; NSW, New South Wales; OBE, Officer of the British Empire; PIRSA, Primary Industries Research, South Australia; QLD, Queensland; SA, South Australia; U, University (ANU, Australian National; CU, Cambridge, UK; MU, Melbourne; MUGS, MU Geology Section; SU, Sydney; UA, Adelaide; UMA, MU Archives; UN, University of Newcastle; UNE, New England; UNSW, New South Wales; UQ, Queensland; UT, Tasmania; UWA, Western Australia); UNESCO, United Nations Educational, Scientific and Cultural Organization.

Geologists, whether men or women, start off in Australia with the same daunting and exhilarating task. This unforgiving continent is not easy to work in, but if you can survive fieldwork in some of the most extreme of conditions, then the rewards from the continent, which always seems to have been a far-flung outpost on earth, are paramount (Brill 1953; Seddon 1996).

Geological research in Australia has always been swayed towards the practical. Even universities have shied away from a fully academic, pure research philosophy, and instead applied research to solve Australia's problems; the needs of primary production have always been at the forefront. Not only that but Australian academic life, when it did get underway in the late 19th century, was dominated for good or bad by the British presence (e.g., Home 1988). Only after World War II was there a real push to gain academic independence post-the Murray Commission (in 1957; Murray 1958), which had been set up by Prime Minster Robert Menzies to fulfil a promise to returned servicemen and which provided a new funding base for universities (Murray 1958), when it became the norm to undertake doctoral (PhD) training at 'home' in Australia (see, e.g., Branagan 1973, appendix V). Beryl Scott (Nashar; Fig. 7b), principally a petrologist and mineralogist, and also a popularizer of Australian geology, was the first to gain a geological PhD in Australia in 1952 in Sam Carey's department in Tasmania (p. 13); Professor June R. Phillips Ross (Fig. 7c) did so in 1959 at SU but the DSc award had been available earlier, gained, for instance, by Georgina Sweet (see below).

Some women in Australia, and especially Australian-born women, have taken to all aspects of geoscience with great verve and a few have risen to the highest ranks. They participated in the scientific endeavour by writing papers, books, producing maps and running research programmes, fostering students and founding societies and journals (see Appendix). Not all were able to make it through the layers of constraint (see also Hooker 2004) and here I shall examine those who succeeded to make a significant contribution along the way, considering also the brick walls that are still hard for women to climb or burrow through! The principal of these constraints is still lack of job opportunities and lack of dedicated funding, especially for palaeontology.

Only when the first, 'sandstone', universities – SU, MU and later UA – opened in the mid–late 1800s (UQ and UWA opened with teaching in UT in the early 1900s and others later in the 20th century), could women achieve a more professional approach to science. Still only a few made their studies in geology or palaeontology, probably, thought Kelly et al. (1993), because of the association of the earth sciences with mining and engineering, which were basically a masculine domain; Schools of Mines, such as Charters Towers, a boom gold-mining town in northern Queensland in the early 20th century, were essentially male-oriented. Of those women who succeeded in gaining degrees right up to the present, many decided to use their knowledge in helping others or in practical work, such as using fossils to date rocks or making geological maps, but they entered often in the teaching profession, thus providing role models for the next generation. The few who entered into a research career did so usually by forsaking marriage and/or children. This was the case at least until the last few decades, when women have gained more control over their domestic situations, especially the vexed problems of child and parent care.

The work of these particular 'invincible' women is commemorated here. Some, like Dorothy Hill, are well known as her career spanned 60 years and she achieved so many 'firsts' (e.g., Runnegar & Jell 1983; Campbell & Jell 1998); others are virtually unknown, even if their work stands as their record (see Tables 1–3; see also the Appendix, which is available online at http://www.geolsoc.org.uk/sup18265 and as a hard copy, obtainable from the Society Library; further updates can be provided by the author on request), and many are unsung. This is an attempt to recount their stories before they slip into oblivion. I concentrate on work done before 1970, when the plate tectonics revolution began in earnest, and on the main figures, such as Hill, Brown, Cookson, Crespin, Joplin, Ludbrook and Wade, providing an introduction to and seeking answers for the more obscure. Essentially, most of the subjects of this paper are no longer living, but their students and colleagues often are and it is to them that we turn for enlightenment in the face of lack of documentary information.

Sources

As judged by Falk (2000), finding data on the lives of women scientists is a hard road. Fortunately, many of the Australian geologists have gained obituarists and in recent times in Australia some redress has been made (e.g., Branagan 1973; Darragh et al. 1976; Dettmann 1988; Cooper & Branagan 1994; Allen 1997). Entries are appearing in the Australian Dictionary of Biography (e.g., Dettmann 1993), and the institution of such events as International Women's Day (e.g., McKay 1997) and the Women's Scientific Network with its WISENET newsletter (e.g., Guy 2004) are providing useful sources of information.

A deeper analysis of MU women graduates (Kelly 1993) and work on early scientists by Moyal (e.g., 1976, 1993b) have discussed the difficulties and triumphs of many Australian women who persevered and managed to follow their desire to understand and investigate their physical surroundings. Earlier studies of the history of geology in Australia would leave the impression that women were conspicuous by their absence (e.g., Bryan 1954). Unlike Brown (1946), Vallance (1978) in his history mentioned not one woman, likewise bicentennial texts in 1988. This has been rectified with recorded memories (e.g., Teichert 1991), studies by Rich et al. (1982), Turner & Wade (1986), Turner (e.g., 1994, 1996, 1998a, b), and Vickers-Rich (e.g., in Vickers-Rich & Archbold 1991).

Recent years have seen an upsurge in interest in women's history in general in Australia, with that of scientists being featured. Ann Moyal (e.g., 1993a & b, 1994) and Nessy Allen (e.g., 1994, 1997) have led the field while later studies include those of historians Jane Carey (2001a & b) and Claire Hooker (2004). Histories of universities (e.g., Branagan 1973, Kelly 1993, Waterhouse 2004), surveys (e.g., Johns 1976), and museums (e.g.,

Mather 1986) have helped put women's scientific work in context. Historians, however, often do not always consider the actual scientific work done and its context, and in this paper I try to consider the significance of the work of Australian women geologists, especially palaeontologists both within Australia and on the world stage.

Useful sources apart from the *Australian Dictionary of Biography*, the AAS and Bright Sparcs websites include, from 1953 to 1956, the nine *News Bulletins* of the GSA, later, from the 1970s onward named *The Australian Geologist* (or *TAG*). Also helpful are state and divisional newsletters of the GSA Inc. and eventually, after the founding of the AAP (*cf.* Cooper & Branagan 1994), its newsletter *nomen nudum*. In recent years, an amateur association has been united by its own newsletter, *The Fossil Collector*, which lists the work of amateurs and some professional palaeontologists in each state. These, plus newspapers and societies and associations, are the main repositories for obituaries and memorials. Each state has a Royal Society and most have a naturalists' association and, importantly, there is the Linnean Society of New South Wales. One of the most comprehensive sources is the SU Department of Geology centenary volume (Branagan 1973), which gives graduate class lists and other data compiled by departmental graduate and secretary, Molly Breckenbridge, herself a geology graduate. Sadly, the second book (Branagan 1994) does not distinguish women from men [this was not a decision by the Author; Branagan, pers. comm. 2006]. Based on these and correspondence from contemporaries, birth and death dates are given for the women discussed in this paper when they are known (see, e.g., Table 1).

Many of the illustrations used here show images of women doing geology in the field (e.g., Figs 1–5) but sadly, in most cases, we do not know who they are. This is also the case for most of the men. It highlights the need for better documentation of photos, even today (Branagan pers. comm. 2006). What is fascinating is the evolution of field dress, particularly the millinery (a topic emphasized at the symposium), which rapidly became oriented to hard bushwork by the 1920s, with jodpurs or moleskins and strong high boots and gaiters instead of long skirts, to combat spiny plants and snakes etc. (compare changes in Figs 1–5).

History

19th century

By the 1830s and 1840s white settlers were finding leisured time to pursue knowledge for its own sake. Scientific societies began as soon as the budding colonies had developed more than subsistence living (e.g., Moyal 1976). Women were allowed to attend some meetings and encouraged to participate in nature study. Many were accomplished illustrators. There were artists and a few collectors in most of the states as they developed but little is known about them (e.g., Turner 1998*b*; Vickers-Rich & Archbold 1991); a 'Mrs Brown' apparently first recorded galena mining at Geraldton, Western Australia (Archbold 1981).

Most women who made contributions did so as amateurs in this century, with a sprinkling of women in scientific societies before 1900 joining philosophical, naturalist and geographical associations (cf. Branagan 1972, Mather 1986); records of their activities are few but can be gleaned by looking through newspapers, membership lists and annual reports. Vickers-Rich & Archbold (1991) noted perhaps the earliest woman to investigate and draw fossils: the wife of Dr E. C. Hobson. Margaret Hobson (neé or previously married to Adamson; see also Kenyon 1930, and Vickers-Rich & Archbold 1991, and Table 1) was a keen naturalist who collected the first Silurian trilobites near Melbourne and illustrated fossil marsupial remains in her husband's papers. Whether she was a native-born Australian like her husband is also not clear; her birthplace is given as Walbrook, which might be in England.

Unlike the situation in the northern hemisphere (e.g., Torrens 1995; McCall 1999), women do not enter the academic realm in geology until the late 19th to the early 20th centuries. As noted above, only a handful of women in the first 120 years of Australian history contributed to geological or palaeontological knowledge, mainly as scientific illustrators (often unacknowledged by their male colleagues) or collectors. These few were often the daughters of scientists, such as young Georgina Sweet (1875?–1946; P.C.M. 1946), who graduated with BSc at MU in 1896 (Kelly 1993; Hooker 2004). Daughter of mine manager and geologist George Sweet, who joined Edgeworth David on the Funafuti expedition (Branagan 2005), as a child she had happily collected fossils at Mansfield, Victoria, accompanying her father (Turner 1998*b*). She was the first woman DSc (1904) and rose to be an assistant professor in zoology at MU, and even acting professor in 1916. She had to retire early in 1924 because of ill health related to stress (Kelly 1993).

As Kelly *et al.* (1993) showed, up to about the 1920s, few women had the opportunity or stamina to contradict convention. Unfortunately, because of their 'invisibility' (Moyal 1993*a*), it is especially difficult to discover much about their lives or their ways of combating difficulties that arose. Catherine

Table 1. *Australian women in geology and palaeontology from c. 1850*

Main name	Other name	Dates	Birth place	Univ.	Degree	Highest position	Research interests	Links	Obit. biog	Comments
Louisa ATKINSON	Mrs James Calvert	1834–72				Illustrator naturalist	Fossil collecting	d. Cavan NSW	Y	Noted artist
Margaret HOBSON	Mrs Admson	1837–18??	Walbrook England?			Illustrator (unpald?)	Collected Silurian trilobites	Married Dr Edmund Charles Hobson 1814–48		Drawings for husbands papers; *Diprotodon* jaw
Fancy COHEN		1887–1975	Grafton NSW	SU	BA 1908 BSc 1909 MA1911	Junior demonstrator, Headmistress		TWED	Y	Early lecturer, SU Geol. Dept.
Isabel Clifton COOKSON	Cookie	1893–1973	M	MU grad. 1916	PhD Cantab. DSc 1932	Leader Pollen Research Unit MU, Senior Lecturer	Palaeobotany Palynology	A.C. Seward W.H. Lang A. Eisenack	Y	Festschrift
Adele V. VINCENT		1896?–19??		MU grad.	BSc 1917 MSc 1921		Palaeozoic plants			
Irene CRESPIN		1896–1980	M?	MU grad.	BA 1919 DSc 1960	BMR Commonwealth Palaeontologist	Micropalaeontology forams	F. Chapman	Y	Festschrift FAAS, AC, OBE, Pres. GSA
Dorothy HILL	Dotty	1987–97	B	UQ grad. 1929	PhD Cantab. 1932	Professor Personal Chair	Palaeozoic corals archaeocyathids	Gt Barrier Reef	Y	Festschrift CBE FRS Pres. & FAAS Pres. GSA
Catherine DRUMMOND	neé Smith	18??–19??		SU	BSc 1911	Demonstrator lecture to 1919		SU Geol. Dept.		TWED
Kathleen SHERRARD	neé McInerny	1898–1975		MU grad.	BSc 1918 Msc 1921	Lecturer	Building stones later graptolites	SU	Y	
Dorothy K. POWELL	Dip	1898?–19??		SU	BSc 1919	Acting Lecturer	Taught palaeoptology 1920–34	SU Geol. Dept.		
Ida BROWNE	neé Brown	1900–76		SU	BSc Hons 1922	Senior Lecturer SU Geol. Dept.	Petrology then palaeontology Palaeozoic brachiopods	TWED 2nd wife Prof. W. R. Browne	Y	Pres. GSA
Florence Mabel QUODLING	Flo	190?–19??		SU	BSc Hons 1922	Lecturer SU Geol. Dept.	Mineralogy	TWED		

Name	Nickname	Dates	Location	Degree 1	Degree 2	Position	Field	Associates	Y	Notes
Dorothy CARROLL		190?–19??	WA?	UWA grad. BA	BSc 1920s PhD Lond 1936		Mineralogy, sedimentology, soils	Research Fellow 1939		
Elizabeth Arnold RIPPER	Betty	1909–200?	M	MU grad. BSc 1931	MSc 1932 PhD Cantab. 1936		Stromatoporoids	Married Stanley Holmes FGS	Y	Lived in London from 1940s
Germaine Ann JOPLIN	Joppy	c. 1910–89		SU grad.	MSc PhD Cantab. 1936	Lecturer SU geol. & ANU	Petrology, wrote several textbooks	TWED Alma Culey		University House library room
Kathleen Laura PRENDERGAST		191??–54	WA	UWA BSc 1933	PhD Cantab. 1939	Curator UWA	Palaeontology		Y	MD 1944
Joyce Gilbert TOMLINSON		1916–81		MU grad.		Assistant Commonwealth Palaeontologist	Palaeontology	BMR Crespin		
Lucy F.V. HOSKING	Mrs Hanrahan	1919–??	WA	UWA	BSc		Palaeontology (brachipods)			
Joan Marion CROCKFORD		1919–	Hunter's Hill NSW	SU grad.	DSc 1951	Research fellow	Palaeontology (bryozoans)	TWED Brown	Y	Lecturer in Education, Kuringai Coll.
Beryl SCOTT	Mrs Nashar	1923–	S	SU grad.	PhD Tas	Foundation	Palaeontology	UT Carey	Y	Retired 1980
Suzanne LAWLESS	Susanne (or Sue) DUIGAN	1924–93		MU grad. BSc	1952	Chair, UN	Sedimentology	Newcastle Egypt		
					PhD Cantab.	Lecturer	Palaeontology Quaternary brown coals	Harald Godwin Cookson		
Kathleen PIKE		1926?–	Fiji	MU grad.	BSc MSc 1953		Quaternary plants	Godwin Cookson		Married Ross McWhae; lives in Canada
June R. Phillips	J.R.P. ROSS	193?–	Taree NSW	SU	BSc PhD 1959	Professorship in USA Zoo Dept.	Palaeozoic bryozoans	Married Charles Ross USA		
Lorna N. MEDWELL	Mrs P. Burke	193?–??		MU	MSc	Research asst?	Palaeobotany	Cookson		Later grew apples then disappeared
Mary J. WADE		1932–2005	SA	UA BSc	PhD 1959	Temporary lecturer UA, Senior Curator	Precambrian Ediacara nautiloids, Cretaceous reptiles	M.F. Glaessner Queensland Museum	Y	GSA Honour Symposium

Abbreviations: AC, Companion of the Order of Australia; B, Brisbane; FAAS, Fellow of the Australian Academy of Science; M, Melbourne; MD, Doctor of Medicine; OA, OBE, Order of the British Empire; S, Sydney; TWED, Edgeworth David. (See also abbreviation list in text.)

Fig. 1. Welshman Tannatt Edgeworth David (affectionately known as TWED), first Professor of Geology at Sydney University encouraged many Australian women to be geologists and palaeontologists. He is depicted here with his students and on field trips, showing the number and field dress code of early women around the turn of the 19th century (photos courtesy of David Branagan and modified from Branagan 1973 & 2005, SU archives).

Fig. 2. Edgeworth David: (**a**)–(**c**) in the field with his students, Fanny Cohen, centre laughing in **c**; (**d**) identifying a specimen (photos courtesy of David Branagan and modified from Branagan 1973 & 2005, SU archives).

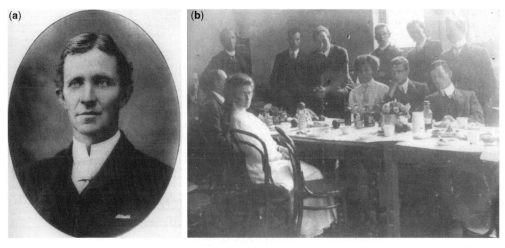

Fig. 3. Edgeworth David: (**a**) portrait; (**b**) with colleagues in the SU geology department 1910, including two women who became the first geology lecturers: Fanny Cohen (centre) and Catherine Drummond-Smith (nearest and left), who were employed as demonstrators (photo modified from Branagan 2005, p. 219, SU archives).

Smith (Fig. 3) at SU, with prizes in Geology II (in 1909, equal with Frank Debenham) and Geology III (in 1910), was appointed demonstrator in 1911 by Edgeworth David, a post which she held until 1919 (Branagan 1973, 2005). She apparently did not publish and married physiologist Frank Cotton (Branagan pers. comm. 2006) and 'disappears'. What is clear is that only a very few of the female population had any access to education at any level when Australia was a relatively unexplored country. The few who do surface were clearly determined. They include Louisa Atkinson (1834–72), an artist, botanist and 'collector of curiosities', including fossils near her home in New South Wales (Lawson 1995). The only serious geological thinker in this early era was another self-educated amateur-deemed eccentric, Georgina King (1845–1932). Fostered by her friendship with Frederick McCoy and George Bennett around the turn of the century, she explored geological ideas and became involved in the debate regarding the age of productive coal (Branagan 1982). King was encouraged by these men and advised by them not to marry; she had a strong belief in her own, essentially unorthodox, ideas, later claiming they had been stolen or plagiarized by the scientific men of Sydney (Branagan 1982; Kelly 1993).

By the end of the 19th century young women were already immune to the rigours of the Australian climate and cultural scene. They enrolled in the first university courses from the 1880s onward, thriving even in the field, in some instances outnumbering men. These included: Jessie Leonora Little (b. 1867), who, in 1885, was the first woman to gain a BSc from MU (Kelly 1993), and Edith Emily Dornwell, first female science graduate (December 1885) at UA; Georgina Sweet herself; Daisy Symonds and Sarah Taylor in 1892; and women each year for the remaining years of the century at SU (Branagan 1973, pp. 109–110). Some, such as the Sutherlands (sisters?), studied geology under the first professor, Arthur Liversidge (1846–1927; e.g., Branagan & Vallance 2001), and possibly under (Mr. W. J.) Stephens, in the 1880s, prior to Edgeworth David's appointment (see below), but they probably did only a 1-year course.

As with Louis Agassiz in the United States (e.g., Warner 1978), certain men had an important influence. Although, in general, young women might not have been expected to be attracted to a life of smashing rocks with hammers, or tramping the bush in heat, with snakes, flies and dust, geology proved to be a very popular subject with female students from the late 19th century onward. Women flocked to learn from newly appointed professors, such as the charismatic Tannatt Edgeworth David at SU (Branagan 1973, 2005; Turner 1998b; see (Figs 1–3). The women seen in these photos must include those in the graduate class lists for 1892–99 (Branagan 1973): the above-mentioned Daisy Symonds and Sarah Taylor, Agnes E. L. Bennett, Kate E. Hogg, Lizzie Dalmas, Millicent Elliott, Lucy Flavelle, Hortense H. Montefiore, Mabel A. Noakes, Marion Horton, Isabella E. Langley, Bertha Symonds, Annie H. Beaumont, Sarah O. Brennan, Agnes M. H. Davis, Marian Harris and Alice I. Newsham, as well as Sarah M. Roseby, who majored in geology in the faculties of arts or science, having had no less than 3 years of instruction. Several went on to further studies gaining second degrees but most appear to have married.

Fig. 4. Fieldwork in the 1920s to 1930s: (**a**) SU breakfast at field camp, 1930s (courtesy of David Branagan, SU Archives); (**b**) Joan Crockford and (**c**) Ida Brown in a hired car; note the solar topées; (**d**) 'Betty Basnett' and Joan Colditz in snake-proof boots and jodhpurs at Wellington Caves, 1939 (from Facer 1996, SU Archives); (**e**) 'Betty' Ripper and colleagues on a MUGS field trip in Victoria, 1929 (Sherbon Hills archive, NLA MS8564 17/11/2, see Sherratt & McCarthy 1992; with permission of copyright holder National Library of Australia).

Fig. 5. (**a**) 'Twenties girls': Dorothy 'Dip' Powell (centre, with toothy grin) (**b**) other SU women geology staff; (**c**) students on an excursion to Mt Kosiusko *c.* 1930 with W. R. Browne and Leo Cotton (photos courtesy of D. Branagan, SU archives).

Early 20th century

Here I interpret 'Australian' broadly, some women being émigrés, but in general most early to mid-20th century women were home-grown. Their success did depend on support of family and mentors (usually male until the later 20th century). With degrees in geology and often 'glittering' prizes (e.g., Table 2), most decided or were forced to become teachers or wives. Despite professional appointments appearing in the early 1900s, jobs were scarce until the post World War II boom time, and even then the Marriage Rule precluded many from working until post-1968 (Kelly 1993).

In the first half of the 20th century, women were more successful in geology than in other sciences; looking at the SU graduation lists (Branagan 1973), they pour into Edgeworth David's classes; notable are, in 1910, Constance MacLean (who married palaeobotanist A. B. Walkom), and, in 1912, Olga Pauss (1890/91?–1948), first curator of the departmental museum (1913–15) and the first wife of Professor W. R. Browne. Women were appointed to academic staff at universities (MU, SU and the budding UT), although few reached full and permanent lecturer status (e.g., Summers 1947), and to government institutions (surveys and museums), with few becoming principal scientific officer. As noted, Edgeworth David

Table 2. *Sydney University Prize for women in geology (1892–1967) (from Branagan 1973)*

Date	Recipient
1892	Eleanor M. Whitfield (shd)
1893	Florence .J. Murray
1894	Edith H. Hansard prox.acc.
1897	Florence M. Rutherford
1898	Marjorie Jarrett (shd)
1899	Constance Mackness (shd)
1904	Dorothy Taylor
1905	Birdie K. Brodziak
1913	Dorothy Law
1914	Edna Sayce (shd)
	Vera M. Doubleday prox. acc.
1916	*Dorothy Powell* (1918) (shd)
	Marie Bentivoglio prox. acc. (photo in Bygott & Cable 1985)
1917	Mary M. Bingham (1918[*]) (shd)
	Margaret H. O'Dwyer
1918	*Ida A. Brown* (1919[*] shd, 1922)
1922	Annie E. Richardson
1923	Kathleen E. Kelly
1926	*Germaine A. Joplin* (1927 # 2nd and 1930)
1927	Mary J. Spruson
1928	Minnie Snyder
1929	*Dorothy M. York*
1932	Marjorie H. Taylor (shd)
1933	Rita H. Harradene (married John Cornforth, a Nobel prize winner, tutored Dcm. in org. chemistry – to Joan, with 'Coop' Marjorie Cooper)
1935	*Elizabeth M. Basnett*
1936	*Joan M. Crockford*
1937	Jean S. Murie (went to teach in the Hunter Valley, killed by a hockey ball blow to temple according to Joan Beattie)
1939	Betty Walker
1941	Beryl E. Brookes
1942	Ruth E. Meadows (shd)
	Beryl Scott
1948	Shirley Wiles
1951	Elizabeth M. Cardwallader and Judith M. Wright (shd)
1952	Ann R. Mutton
1953	Barbara G. Briggs
1956	Robin M. Mackay and Anne D. Thomson (shd)
1962	Rochelle Rodier
1967	Ann Vandyke

(Continued)

Table 2. *Continued*

Date	Recipient
Slade Prize	
1937	*Joan M. Crockford*
1943	*Beryl Scott*
Edgeworth David Prize for Palaeontology (donated by Dr & *Mrs A. A. Pain*, geol. grad. 1912)	
1935	*Joyce Gilbert-Tomlinson*
1938	*Joan Crockford*
1944	*Beryl Scott*
1969	Ann Vandyke
Sheila Mitchell Swain Memorial Prize (for 3rd year fieldwork)	
1950	1 women
University Medalist	
1940	*Joan Crockford*
1946	*Beryl Scott*
Science Research Scholarship (no fees for teaching but could assist: $1000)	
1919	*Dorothy K. Powell*
1920–21	Marie Kent
1920	May Bligh
1922	*Ida Brown*
1940–41	*Joan Crockford*
Linnean Macleay Fellow	
Ida Brown	
Joan Crockford for her work on Palaeozoic bryozoa – 1943–1946	
Postgraduate Travel	
June Phillips 1957	
1851 Exhibition	
Mary Bentivoglia 1916	
American Association of University Women Fellowship	
June Phillips 1958	

BA exam (from 1907 prev. Physiography/Natural History); *, Head of Dept Prize; shd, shared; italic, career or publication.

(e.g., Fig. 3) was a major influence for women in geology. Before World War II, 30 (out of 75) women won the university prize for geology; post-war, up to 1972, there were 8 out of 22. In the early years women consistently won most of the prizes (Table 2). Edgeworth David was also the first professor to appoint women to academic staff in geology (e.g., Branagan 1973; Hooker 2004). Margaret Deer (BSc 1908, employed 1908), Fanny Cohen MBE (employed 1909–11; Radi 1988; Fig. 3b), the first woman to publish that I have found (on azurite, 1911; see in Supplementary Publication) and Catherine Drummond Smith (employed 1911–19; Fig. 3b) began a long tradition of female staff presence there until the retirement of Ida Brown/e when she married in 1950. These young women were often demonstrators and curators (Branagan 1973; Turner 1998b), such as Dorothy Powell, and later Germaine Joplin (*c.* 1910–89) and Dorothy Crosby, who went on to be lecturers but often only 'Acting'.

Professor Sam Carey (pers. comm. to ST 1994) noted especially the palaeontological teaching he received from 'Dip' Powell (Fig. 5). Eva M. Dolan, a 1908 graduate at UT(?) was one of a few early part-time lecturers teaching geology in Tasmania (Banks pers. comm. 2006) but no more is known of her.

Under Professor Ernest W. Skeats (1875–1953), Kathleen McInerny (1898–1975; Quodling 1976) was the first woman geology graduate at MU in 1918. She won scholarships, gained her MSc in 1921, and was among the first to be employed, with an assistant lectureship in 1920, lecturing on building stones (e.g., McInerny 1929) and road metal to engineering and architectural students (Summers 1947; Kelly 1993; Hooker 2004).

Elsewhere only a few became academics, especially after seeking higher education overseas, notably Dorothy Hill (UQ), Isabel Cookson (MU), Germaine Joplin (SU) and Nell Ludbrook (Senior Palaeontologist, Department of Mines, South

Australia). In Western Australia, Lucy Hosking (NB. not Hoskins or Hoskings, cf. Branagan & Vallance 2001; Hooker 2004), Kathleen Laura Prendergast and at MU Adele Vincent began down the research road but then more or less disappear after publishing a few papers. However, Hosking was the first woman to be prominent in ANZAAS Section C in Western Australia in 1932. Prendergast succeeded her in 1935 (Branagan & Vallance 2001). Miss Prendergast was acknowledged by Dorothy Hill in a 1939 paper on Western Australian corals in the Wade collection, possibly when departmental curator. Both Cookson and Irene Crespin were some of the first paid government scientific staff (Carey 2001a), with Crespin eventually attaining the status of Commonwealth Palaeontologist (see below). In the 1960s Hill and Nashar (Facer 1996; Allen 2001) were among the first women full professors in Australia in any subject. Crespin, Ludbrook (e.g., Alley 1996), Hill (e.g., Runnegar & Jell 1983) and Cookson (e.g., Andrews 1980) had spectacular, internationally applauded research careers and published prolifically.

Nearly all women noted here joined their respective state Royal and other societies, and Brown (see Fig. 6), Hill, Crespin, Cookson, Crockford, Ludbrook, Joplin, Quodling and Sherrard were founding, prominent and active members of scientific associations (e.g., Dean, in Cooper & Branagan 1994) and were frequently the first women to hold high office in them; Hill was even a force in the foundation of societies (e.g., Cooper & Branagan 1994).

The first wave of women taught in high schools (e.g., Turner 1998b), as well as in academe, teaching and mentoring the next generation of women students, such as Ida Brown, Joan Crockford,

THE LAST MONTHLY MEETING HELD IN THE LINNEAN HALL, ELIZABETH BAY, 25TH JUNE, 1924.
Left row: A. S. Le Souef, G. A. Waterhouse, C. Anderson, R. Greig Smith, May M. Williams, Ida A. Brown, Jessie K. Steel, Sarah Hynes, J. H. Campbell.
Seated at table: W. Welch, T. Iredale, A. Musgrave, E. Cheel, A. G. Hamilton, A. B. Walkom, R. H. Cambage, J. J. Fletcher, D. G. Stead, W. W. Froggatt, E. G. Jacobs.
Right row: E. le G. Troughton, G. M. Goldfinch, A. J. Nicholson, P. D. F. Murray, I. M. Mackerras, Mrs. Mackerras, Eleanor E. Chase, Marjorie I. Collins, R. T. Baker, W. L. Wearne.

Fig. 6. Young women attending the Linnean Society of New South Wales, 1924; young Ida Brown marked with an asterisk (from annual report and history of society).

Beryl Scott (Nashar) and Kathleen Sherrard. Nashar (Fig. 7a) was taught palaeontology at SU (1942–44) by Dr Ida Browne, and Crockford elected to take honours. She later took leave and spent some time at Cambridge University, but returned and took her PhD in petrology, this despite winning the Edgeworth David Prize for Palaeontology in Geology III at SU during her time there. Max Banks (pers. comm. 2006) noted that hers was the first PhD in geology at UT and may well have been the first overall. Nashar wrote (pers. comm. to ST 1994):

Teaching of palaeontology in those days (50 years ago!) was based on classification and range. We described, sketched and labelled the specimens for prac. work. We really learned our fossils! However, when I taught palaeontology and several other branches of Geology in my very early days at the Newcastle University College.... I did try to make palaeontology more useful for the study of stratigraphy.

Later 20th century

Full professionalism of science did not come until after World War II and women were affected in various ways (e.g., Allen 1994; Jane Carey, see Hooker 2004, p. 32). Until the 'modern', post-plate tectonic era, as noted above, Hill and Nashar were among the first women in Australia to attain professorial level. Even so Hill had to wait until she was in her early 50s before achieving a full personal

Fig. 7. The first PhDs in geology: (**a**) Beryl Scott (Nashar), SU graduate; (**b**) Dr Beryl Scott (Nashar), UT; (**c**) June Phillips, SU; (**d**) Professor June Phillips Ross with husband Charles 'Chuck' Ross (photos a–b courtesy Australian Academy of Science website; photos c–d courtesy of Professor Ross).

professorship. Nevertheless, she did not get offered Head of Department status, even though her application was accompanied by references from most of Australia's top geologists, as well as from overseas. These included: Bryan (UQ), Bulman (Cambridge), Denmead (GSQ), Fisher (BMR), Hills (MU), Mackerras (CSIRO), Raggatt (BMR), Captain Thomas (her war service chief), Traves (BMR), and Wells (Cornell) (UQ Archives, S130 Staffing Geology Full Time, dated 25 March 1959).

Early SU PhD graduate, June R. Phillips Ross (b. 1931; Fig. 7c); (e.g., Turner 1998b) opted for the overseas option by gaining an Edgeworth David Travelling scholarship. June wrote (pers. comm. to ST 2005):

I was born in 1931 in Taree, NSW. My parents were always supportive in my activities, both in my learning and also my sports activities. My mother was an outstanding pianist and I gained all my understanding and love of music from her. I enjoyed all my primary and high school classes and did quite well. In high school, I was captain of my class. I also was captain of my grass hockey team and led the girls swimming team. My high school Head Mistress (and English teacher) encouraged me to apply for a scholarship for university. I won a four-year scholarship to the University of Sydney. I took Geology as a fourth subject and was captivated, particularly by the subject and the field work!

In my honours year, I had a field area near Queanbeyan, NSW and, as I did not have a car, I located myself by renting a room with board at a farm in the middle of the field area so I could explore the geology of the area on foot. For my PhD field work, which covered a much greater area of NSW, I still did not have a car, so passenger and goods trains were my means of transportation. When I did need to go farther afield, I arranged to go to a farm house in that particular area and rented a room with board. I finished my PhD thesis in four years and then went overseas. As I was absent from Australia, it took the University of Sydney two years to contact me that I had successfully completed the thesis requirements. Subsequently, based on my accumulated published studies, I was awarded a DSc (in 1974). Leaving Sydney to go to New Haven, Connecticut involved support of an AAUW (American Association of University Women) fellowship to be held at Yale University in New Haven. The Department of Geology, University of Sydney, partly supported my travel to the U.S. Yale gave me a fellowship for a second year and then I succeeded in obtaining U.S. National Science Foundation Grants which included a half-time stipend.

Phillips attended the graduate school lectures given by fellow Australian and SU graduate, Sam Warren Carey at Yale in 1959–60 and decided to make her career in the United States. She became a full professor of Western Washington University in 1970, having gained a post in biology (she was precluded from getting employment with her husband Charles Ross in the geology department), and is now Emeritus Professor. She widened her palaeontological scope accordingly, and most of her massive publication record (now over five decades) is on non-Australian geology, although she did at first follow down some of her predecessor Crockford's research path (e.g., Ross 1957, 1963); her early Australian publications are recorded in the Appendix.

Other women did not gain post-graduate work. Some still opted for marriage and then the overwhelming problems of child care slowed or stopped them entirely (e.g., Crockford and A. Kemp, pers. comm. to ST 1991). One MU graduate, Judy Sullivan made great headway with her MSc studies on gyracanth fishes under Professor Sherbon Hills in the 1950s, even travelling to the United Kingdom to view type material, a research option still difficult when most original palaeontological specimens are in northern hemisphere museums. Sadly, her marriage broke down and parental responsibilities became paramount and so the thesis did not materialize (letters in Sherbon Hills archive, Sherratt & McCarthy 1992). The salary cap for women (e.g., Kelly 1993; Hooker 2004) affected palaeontologists even in industry as late as the 1960s. Dr Marjorie Apthorpe at Woodside Petroleum in Western Australia had problems with GSWA Director, J. H. Lord (Johns 1976), who did not want women working in the field. Dr Kath Grey (pers. comm. to ST 2004) came from England in the 1970s to work for the GSWA in Perth; she also had to negotiate her way into fieldwork with male colleagues latitude by latitude! Even in the late 20th to the early 21st century, however, few women palaeontologists are in secure professional positions and many who gained headway earlier are approaching retirement, have changed work emphasis or have retired (e.g., Rich 1999; Guy 2004). SU Geology Department at least has had one event to honour their women graduates (Facer 1996).

Dr Nell Ludbrook (1907–95) fared better. This Australian had an atypical career path; Nell ('never Nelly', according to Alley 1996) Hooper Ludbrook (neé Woods, 1907–95; Fig. 8) succeeded with a broad spectrum of research (Lindsay 1985). As with Crespin (see below), she achieved a position of eminence and travelled widely, personally collecting comparative material throughout the world. She became a prominent member of the scientific community and was always an active member of the GSA, holding office at all levels (Cooper & Branagan 1994). She was elected Federal President in 1968 (Fig. 8d), the first woman to hold this position, and Honorary Member in 1976. She also actively participated in the affairs of the Royal Society of South Australia and was its first woman president in 1961/2. In 1963 the Society awarded her its highest honour, the Sir Joseph Verco Medal. In 'retirement' from 1967 to 1980, she edited the *Handbooks of the Flora and Fauna of South Australia*. In 1981 she was made a Member of the Most Excellent Order of the British Empire 'for service to science'.

Fig. 8. Dr Nell Ludbrook MBE, South Australian all-round geologist and palaeontologist: (**a**) sampling in Eromanga basin sediments, 1963; (**b**) at her desk in the Department of Mines and Energy Core Library, 1985; (**c**) examining Caenozoic sediments along the River Murray, South Australia; (**d**), as one of the presidents of the GSA, taken at Mt Coot-tha, Brisbane in 1971 by one-time Director of the GSQ, A. K. Denmead (photos a & c modified from Alley 1996; photo b from Lindsay 1996 and photos courtesy of Primary Industries and Resources South Australia; photo d modified from TAG Newsletter, 58, p. 12, courtesy of the GSA).

Nashar also thought her case typical and, as a geologist, she felt that she did not meet with any difficulty in her working life in academia (Allen 1997). She recounted (pers. comm. to ST 1994):

Interestingly, the only job I ever applied for was the Chair of Geology at the University of Newcastle to which I was appointed as Foundation Professor in 1965. All other positions were offered to me. Again, the little I did in industry was by invitation. Having taken on a job, I worked very hard so as not to let the 'side' down and to show that a woman could do it as well as a man.

The key to success was to have the right qualifications and to be in the right place at the right time. In her case it was her doctorate that put her in the front running, although, in retrospect, overall Nashar's career track is most unusual.

Disciplines

Women have played a significant role in Australian geoscience, especially in palaeontology. Crespin, Cookson, Crockford, Hill, Ripper and Wade became pioneers or were foremost in their disciplines and their work is internationally or nationally known. These and others have acted as role models and mentors for the next generation.

Micropalaeontology

Irene Crespin (1896–1980; Casey 1976; Crespin 1972, 1975; Fig. 9) graduated in 1919 with a BA from MU, and started in 1927 as assistant to Frederick Chapman (Benson 1944) at the then National Museum of Victoria (Turner 1998a;

Fig. 9. Dr Irene Crespin, then Assistant Commonwealth Palaeontologist working at the National Museum of Victoria: (**a**) with W. Baragwanath, Secretary of Mines for Victoria, probably visiting a *Cooksonia* plant site, c. 1927 (cf. Douglas & Holmes 2006); (**b**) leaving for collect samples at Lakes Entrance in Gippsland, 1942; (**c**) Crespin, then Commonwealth Palaeontologist, on Dr Harold Raggatt's appointment to the new BMR in January 1941(with, from left to right, Recka Johnston, Dr W. G. Woolnough, Dr Raggatt, Mr 'Happy' Hawkins and Irene's assistant Joyce Gilbert-Tomlinson); (**d**) Crespin's photo of her aeroplane and crew on an overseas trip to Java, Indonesia, 1939; (**e**) Crespin at her microscope after the discovery of oil at the Rough Range (from media?); (all photos courtesy of Crespin archive via Dr H. J. Harrington).

Carey, in Rasmussen 2001). She began to learn micropalaeontology from Chapman, then Chief Government palaeontologist, undertaking work for the GSV, taking part in at least one collecting trip with W. Baragwanath Jnr, Secretary for Mines, Director of the Geological Survey and Chief Mining Surveyor (Fig. 9a; see also Douglas & Holmes 2006). She took over as Commonwealth Palaeontologist in 1935 when still in Melbourne, moving to Canberra in the early 1940s when the BMR in Canberra began under Sir Harald Raggatt (Fig. 9c) and held the top job until 1961, remaining well into her retirement. At the fledgling BMR, she was assisted by Joyce Gilbert-Tomlinson (see below).

In 40 years of research, mainly in the area of oil exploration, Crespin completed 87 papers as a single author, 23 as a joint author and more than 100 open-file reports and notes. Her work, particularly on the use of Foraminifera in oil-bearing shales (e.g., Fig. 9b) helped make micropaleontology a respectable science in Australia, contributing as it did to mineral wealth. Her role in the immediate post-war burst of development and recognition of the first oil show at the Rough Range in Western Australia were duly noted in the media (Fig. 9d & e).

Crespin gained her knowledge of foraminiferans from Chapman, then virtually the only authority on the organisms in Australia. Chapman's style of taxonomy did not translate into a more modern era and this may be why her DSc was first turned down in 1944 (it was finally awarded in 1960). Crespin and Chapman's first joint paper appeared in 1928 and, in 1935, they recognized the first rocks of Eocene age. She succeeded Chapman as Commonwealth Palaeontologist but at around half his salary. Despite this, Crespin did much pioneering taxonomy and was honoured with taxa in turn, such as *Crespinina kingscotensis* Wade.

The Assistant Commonwealth Palaeontologist, Joyce Gilbert-Tomlinson (1916–81; Dickins 1982; Fig. 9c), an MU graduate, became a senior palaeontologist in turn but did not publish in the same way, perhaps as a reaction to her efficient boss. Gilbert-Tomlinson's research on Cambrian gastropods from the Northern Territory began in 1953 (GSA 1953–56, vol. 1). She was at first active in the GSA with talks (e.g., on Proterozoic algal structures from the Sleisbeck area; GSA 1953–56, vol. 2) and made contributions to the Canberra ANZAAS (White 1954). Nevertheless, as years past, little publication ensued.

Like Cookson (see below), with whom she worked, Crespin focused much of her research on locating fossils associated with oil and coal. Crespin did like to spend time in the field (e.g., Fig. 10a) but more often collated material brought

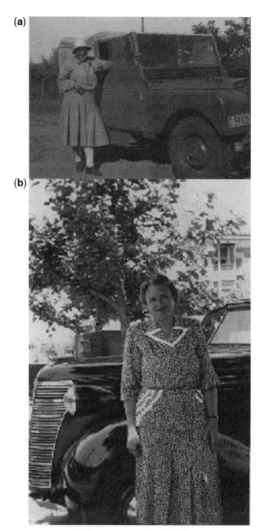

Fig. 10. Irene Crespin (**a**) in the field at Bullara in the Carnarvon Basin, WA, 1950; (**b**) with her favourite car 'Bluebird' outside the Ainslie Hotel where she lived in Canberra (photos courtesy of Crespin archive, via Dr H. J. Harrington).

or sent by the field parties (e.g., Wilkinson 1996). In the newly established *Geological Society of Australia News Bulletin* (GSA 1953–56, vol. 1) a report for the Australian Capital Territory stated:

Miss Crespin continues the task of sorting and correlating the masses of microfossils accumulated by workers in the Mesozoic and Tertiary and her reports on the Nelson Bore and the Adelaide–Aldinga area will be published shortly.

Her research took her all over Australia (even down a 400-m descent in a mine shaft), up through Indonesia, where she was probably the first woman geologist to visit the region (Fig. 9d), and to the Philippines (Crespin 1954). Crespin recalled in her memoirs:

An event of some importance affecting the Palaeontological, especially the Micropalaeontological approach to oil-field investigation took place in early 1939. Approval was given by the Minister of the Department of the Interior for me to visit Java and Sumatra to discuss problems relating to foraminiferal assemblages of the larger and smaller forms for correlating some of the Tertiary deposits in the Netherlands East Indies with Papua and New Guinea. I was to visit Dr. Tan Sin Hok, micropalaeontologist to the Geological Survey of the Netherlands East Indies [later murdered during the war], with headquarters of the Nederlandische Koninklijke Petroleum Maatschappj (N.K.P.M.) at Palembang, Sumatra. I had been hoping for some time that I could talk to these authorities. Whilst in Java I was the guest of the Netherlands East Indies Government; and in Palembang, a guest of the N.K.P.M. It was agreed that I should fly to Java and Sumatra on April 10th for six weeks. It was an historic flight for me as I had not previously been in an aircraft. I travelled in a Dutch KNILM two-engined Lockheed Hudson, SE. 14, with a crew of four and was the only passenger between Sydney and Sourabaya, East Java.

Crespin noted down in her memoirs in amazement how, at Palembang, the staff of the two major oil companies actually completely ignored one another. Her host, Dr Thalmann of NKPM, and Mr Evers of the Shell Company were rivals and she was not permitted to visit the Shell Laboratories at Plagio. Nevertheless, Irene adapted well to these exotic locations. In southwestern Timor she adapted to the local food: 'I noticed that the crew was eating fried rice. I soon became accustomed to that dish and now I enjoy it whenever the opportunity presents itself'. Even the presence of big predators in eastern Sumatra did not faze her: 'Whilst I was in Palembang, one of the boys working on one of the oilfields was taken by a tiger'.

One important outcome of her visit came from her work at the Geological Museum in Java on Dr Oostingh's collection of Pliocene molluscs. Crespin was struck by the modernity of the scientific instruments; she was envious of a wonderful camera which took photographs of thin sections of rocks and fossils, a gift from an appreciative mining company. Returning to Canberra she reported on this instrument and eventually the BMR obtained one very similar to that in Bandoeng.

Crespin was well respected internationally within her specialities, although she may not always have accepted younger women trying to enter her field (see comment by Mary Wade below). Ludbrook appreciated her help when she went to work in Canberra during the war years (noted in Goggs 1994) but at that stage she was assisting in mineralogical and economic work. Irene stayed on in retirement, living in a nearby hotel (Fig. 10), often working on projects and being consulted by the BMR. She was awarded the Clarke Medal from the Royal Society of New South Wales, and an Order of the British Empire in 1959.

Ludbrook later tackled foraminiferan stratigraphy in her work for South Australia and in later life served as Australian news correspondent for *Micropalaeontology* from 1962 to 1966. In the years after World War II she lived in London and worked at the BMNH, honing her palaeontological research. Following the death of her husband, and on her return to Australia, Ludbrook had joined the South Australian Department of Mines in 1952 and decided to concentrate on developing the application of micropalaeontology to subsurface stratigraphy. She sought to change the prevailing view in the GSSA that palaeontology had only limited applied or economic use. A cooperative research programme to collect cuttings from water bores at regular intervals was extended and samples were submitted regularly for micropalaeontological examination (Alley 1996). She was appointed Palaeontologist in 1957 and, until her retirement in 1967 as Senior Palaeontologist, was engaged principally in the study of Foraminifera to elucidate subsurface stratigraphy of sedimentary basins, particularly in the search for petroleum and groundwater. In this way, Ludbrook contributed greatly to the understanding of the stratigraphy of the Murray, Eucla and Great Artesian basins.

Mary Wade was one of Martin F. Glaessner's first students after he joined the UA department in the 1950s (Fig. 11a) and he put her to work on Foraminifera. She recounted (pers. comm. to ST 2004) how she came to loggerheads with Crespin over identifications when she was undertaking her PhD. Fellow student, Professor Brian McGowran, recounted (pers. comm. to ST 2005):

In dissecting and reconstructing the anatomy of foraminiferal shells, Mary was quite superb. She had a secure grasp of the relationship between morphology, taxonomy and evolution. She could not draw very well in an aesthetic sense but she could accurately depict dissected anatomy in drawings and plasticine models. She had an excellent feel for rock relationships and correlation and age determination, and of all the research in the Cenozoic in southern Australia in the 1950s, I would rate her work on the Cenozoic planktonics as preeminent. Martin always had the highest regard for her overall intelligence and general capabilities in geohistory and biohistory. She was not a narrow specialist in her mindset as were many micropalaeontologists in those days and I think that Martin saw in her the qualities that were emphasized in the holistic palaeobiology of the Vienna of his own formative years.

Mary published her first paper on foraminiferans from Kingscote, Kangaroo Island, in 1955. Subsequent papers followed on southern Australian Tertiary foraminiferans and material from the Parr Collection of Antarctica, one with A. N. Carter of the GSSA and her first joint paper with Glaessner.

In this early work she named two new genera and five species of foraminiferans.

Invertebrate palaeontology

Elizabeth (Betty) Ripper (Figs 4e) was a brilliant student at MU in the 1920s who first worked on graptolites discovered during her honours work. Ripper, like the much better known Dorothy Hill, graduated in the late 1920s and went to Newnham College, Cambridge, taking Australian material with her for her doctoral studies. Like Hill she was also guided by Gertude Elles (see below). She changed, however, to Devonian stromatoporoids for her PhD work and went on to produce some nine scientific papers over six years, with pioneering alpha taxonomy (Turner & Webby 2004; Turner 2006a). In 1936, Ripper donated stromatoporoid material, including figured specimens, to the Sedgwick Museum in Cambridge and the BMNH (Cleevely 1983). She then came back home but, being unable to find a permanent job, gave up her palaeontology. Returning to Britain she married a geologist, had children and kept her geological interests alive via her husband, who was an FGS and by joining the GA in London. However, as late as 1966, Stearn cited Ripper for her recognition of pores in stromatoporoid pillars, and many of her named taxa remain (Turner 2006a).

Kathleen McInerny, moved to Sydney after her marriage to Howard Sherrard, and by 1953 was considered part of the SU department. She continued her geology and changed direction, eventually becoming a keen graptolite worker (Darragh et al.

Fig. 11. (a) The Palaeontology Club, UA c. 1950s–60s; Mary Wade at top far right (photo from UA Geology archives, ex Franklin 1985); (b) Joan Crockford and Ida Brown on the south coast of New South Wales; (c) Crockford in a gas mask (b,c courtesy of Joan Beattie); (d) 'Mrs' Ida Browne (photo from Anon. 1951, a newspaper article about the ANZAAS meeting in town aimed at women, where she was noted as 'Mrs W. R. Browne, *wife* of a scientist' - my italics); (e) Crockford-Beattie, 1951 communication at Brisbane ANZAAS (photo modified from Cooper & Branagan 1994, courtesy of GSA, also possibly from newspaper).

1976; D. F. B. 1975; Kelly 1993). Sherrard refined Ordovician to Silurian graptolite zones (e.g., Chalmers in GSA in 1953–56, vol. 1), writing a series of papers providing much useful stratigraphic information (e.g., Sherrard 1942; Quodling 1976); she also tackled tentaculites and other invertebrates. Joan Crockford Beattie noted (pers. comm. to ST 2005) that when they knew each other, Sherrard was always totally focused on graptolites and, when on a joint excursion, would not even look at Joan's fenestellid bryozoans. Sherrard's link, through her husband, the Chief Engineer of the NSW Department of Main Roads, helped the SU geology department to get permission for excursion bus stops on busy main roads, equipping the party with special signs (Branagan pers. comm. 2006).

Sherrard had been granted leave from MU to go overseas in 1929 to do research for a doctorate in Cambridge. Instead, in 1928, she had married and, in Sydney, had involved herself in social causes, education and motherhood, busily working for many associations (Hooker 2004). Sherrard (e.g., 1944) wrote several articles for magazines and journals, advocating that women become more involved in science. She maintained her own scientific life via SU, which resulted in regular papers, about 15, on aspects of palaeontology. A provocative conversationalist and correspondent, her contributions to scientific debate were welcomed, not least for her wonderful voice (Branagan pers. comm. 2006). Sadly, I have found no picture of McInerny/Sherrard.

Most of the other women who published at this time are known from one or a few papers describing alpha taxonomy. Contemporaries were Prendergast (listed in the *Journal of the Royal Society of Western Australia*, vol. 17, p. viii; obituary Prider (R.T.P.) 1954) and Lucy F. V. Hosking, who, in the early 1930s, collected and wrote four important papers on Palaeozoic brachiopods, naming several taxa from the remote Kimberleys of Western Australia. Mentions are brief, as in Clarke and Wells (1931) regarding a Royal Society field excursion to see the Bullsbrook Marl: 'Misses Carroll, Prendergast and Bowley, with Miss Hosking have been mapping'. Lucy, then Mrs Hanrahan, did become an early member of the GSA in 1954 (GSA 1953–56, no. 3(1)).

Ida Brown (Figs 4c, 6, 11b & d), most fascinating perhaps of all the SU geological women, graduated in the early 1920s. She began as a petrologist, examining unusual rock types, unique occurrences in Australia and laying down her reputation as a field geologist. She was noted for her research, which encompassed the geology, especially igneous, of much of the southern area of New South Wales, displaying a perceptive feel for field relations; this work gained her a DSc in 1932. Mr D. J. Mahoney, Director of the National Museum of Victoria, wrote to W. R. Browne in August 1932 (perhaps only slightly patronizingly), regarding the thesis of the young geologist:

> What amazed me most was the immense amount of good field work and mapping carried out by Dr Ida Brown in rough country and her sound interpretation of such a variety of problems.

She had a Linnean Macleay Fellowship from 1927 and then became assistant lecturer in palaeontology at SU in 1934, replacing W. S. Dun, who died of cancer that year. She then began research mainly on Middle Palaeozoic brachiopods and, for 3 years, she conducted a full research programme on top of her teaching. Brown's ability to move across the disciplines and do such good work later in palaeontology is laudable, although her teaching was very much old-style taxonomy (see above mention by Nashar). Her studies were typical of the time: meticulous, detailed and local. Despite some attempts to work with colleague and friend Dorothy Hill, most of her papers were also published locally and only one internationally: in 1948, in the *Journal of Paleontology*. Ida would gather fossils from a particular area, then spend hours cleaning, dissolving away unwanted rock, breaking some apart to look at the internal structure and using a microscope to measure fine detail in the fossil.

Brown passed on her many research attributes to her students, including the two Joans – Crockford (now Beattie, see below) and Johnston – the latter writing one pivotal paper in 1941, after research in the Baradilla district of New South Wales. Beattie (pers. comm. 2005) reckons her friend Joan Johnston was earmarked by Brown for great things. However, Johnston married fellow student Daniel Moye and gave up her research. Later the family died tragically (see Probert 1975).

Brown rose to be senior lecturer through the 1930s and 1940s and might have become the most brilliant of this mid-century group (Hooker 2004). However, despite a publishing span of 40 years and respected contributions to Siluro-Devonian geology, she hardly left her university or New South Wales and did not attain the international recognition or distinction of Hill, Crespin or Cookson. She did become the first female president of both the Royal and the Linnean Societies of New South Wales, and then, following her marriage to W. R. Browne, spent most of her time in teaching and the routine work of the department. She did this long into her retirement, being lauded for her help with the centenary volume (Branagan 1973). She died in 1975 with subsequent tributes from colleagues (e.g., Anon. 1976; Branagan 1977). Cain (2000) and Hooker (2004) have recent provided outlines of her career and life, and the influence of her work continues (see General geology and mapping below).

Joan Marion Crockford (later Mrs George Beattie) (Figs 4b, 11b, c & e), in the 1940s to 1950s, maintained a career as a geologist and invertebrate palaeontologist and specialist on bryozoans and gained a high professional reputation, which has been maintained. She undertook basic and pioneering taxonomic research on bryozoan faunas at SU, supported primarily by scholarships and fellowships. Her subsequent marriage and mothering of four children slowed her down but she kept up her research for a while. In later life she went on to new challenges in teaching at graduate level.

Crockford was born in January 1919 at Hunter's Hill, not far from the centre of Sydney. Developing an interest in rocks and fossils at an early age, she was taught geology at high school by Marjorie Collins (Mrs Shiel) (see Fig. 6 right) and went to SU in 1938 to read Geology in the department created by Edgeworth David, graduating with a BSc honours I in 1939 and the university medal in 1940. Under the tutelage of Ida Brown, Crockford began work on Palaeozoic brachiopods, doing honours on Silurian specimens from Yass and the Ordovician of Bowan Park. She spent a long time etching the faunas with acid: engaging with 'the really exciting Bryozoa in these collections' [which Phillips Ross (1961) later amplified and used]. Brown and Professor L. A. Cotton put her to work on this latter group because of their stratigraphic interest and because little was known of them in Australia. Supported by a Science Research Scholarship in 1940, full-time 'demonstrating' in 1941, a Linnean Macleay Fellowship for 3 years and a Commonwealth Research Grant in early 1946, she pursued her research, gaining her MSc in 1942. Linnean Macleay Fellowships were supposed to preclude teaching but, as there were staff shortages during the war, she assisted with practical classes: palaeontology to years I–III and crystallography, mineralogy and blowpipe analysis. She collected most of the New South Wales material she described and was also sent material, first from Queensland, Tasmania and WA. Wartime restrictions on travel precluded interstate travel.

In mid-1946, after her marriage to naval lieutenant George Beattie, who became a mining engineer after his discharge in 1945, Crockford-Beattie went to live in Cobar, New South Wales, which was then very remote. She managed to work on her DSc thesis, which was awarded by SU in 1951 and published, while bringing up her children and being virtually unpaid. Some later contract work for the BMR through the auspices of Dr Norman Fisher, then Assistant Director (Johns 1976), allowed her to look at western Australia Permian Bryozoa sent in by the field parties then working in the far northwest Kimberley region. Her monographic work in the resultant 1957 *BMR Bulletin* (for which she received the sum of £500) covered material from the west Kimberley; she recorded 79 species, 42 of which were new. Three new genera were named and a new family erected in the Cyclostomata. In her active palaeontological years (1940–56), Crockford erected over 80 species of bryozoan.

Australia's most notable palaeontologist of the 20th century, Dorothy Hill (1907–97; Fig. 12) got 'hooked' as a young girl after seeing fossil corals when visiting friends, but only began serious palaeontological study during her honours degree e.g., Turner 1996; Gregory 2005). The geological and palaeontological work of Hill, and her national contributions, for example to the Great Barrier Reef Committee, have been extensively recognized and discussed elsewhere (e.g., Runnegar & Jell 1983; Gregory 1987; Jell 1997; McKay 1997; Campbell & Jell 1998; Simpson 1999; Bowen & Bowen 2002); she also often represented Australia overseas, for example at the 100th anniversary of the GSL (e.g., Flett 1937). Here I shall make reference to certain aspects only.

Winning an overseas travel award, Hill took Queensland material with her to Cambridge University in 1930 (Fig. 12), and Gertrude Elles (see Burek 2007) allowed her to incorporate a comparative study into her thesis on the Lower Carboniferous corals of Scotland, thus setting her life's taxonomic work (Jell 1997). A long and distinguished career followed in Queensland, after her return in the late 1930s, resulting in more than 100 research papers. She built up a research library second to none (Hill & Willadsen 1980; Anon. 1985; Sherratt 1994), now subsumed in a larger Physical Sciences collection, which still retains her name, and a school of invertebrate palaeontologists. By advising her students to take their higher degrees in Australia, she almost single-handedly decolonized Australian palaeontology. Many of the guests at her 80th birthday celebration her former, students, were heads of their respective institutions. In later life, with colleague John Wells (Fig. 12g), she prepared the definitive treatise on the coelenterates (e.g., Sherratt 1994). She also tidied up the taxonomy of the unusual Archaeocyatha, using Australian and Antarctic material, producing a *Cambridge Biological Review* and a definitive chapter (Hill 1967) in *The Fossil Record*. Her body of work and influence on students constitutes one of the most comprehensive of all Australian palaeontology.

In the same vein, former Director General of the South Australian Department of Mines and Energy, R. Keith Johns (in Lindsay 1985), concluded that Ludbrook's work had a profound influence on and made a major contribution to the knowledge of

Fig. 12. Dorothy Hill (a) as a young graduate of the University of Queensland (courtesy of UQ archives); (b) undergraduate sportswoman, Hill the rowing cox, University of Queensland, 1927 (Fryer Library, UQ library pamphlet UQL 98, p. 17); (c) H.C. Richards, her teacher and mentor (UQ archives); (d) at Munduberra, Queensland, sitting with notebook, 1920s (photo courtesy of J.S. Jell); (e) in field with horse; (f) with friends in the Lake District c. 1932; (g) Newnham College girl at a picnic party at Thetford c. 1932 (from the Newnham College Roll Letter 1998, courtesy Betty Ripper); (h) Professor John W. Wells (Cornell, University, Ithaca) and Hill in Brisbane 1954 (photo by and courtesy of J.S. Jell); (i) portrait of Hill (UQ Library pamphlet); (j) Hill on the Marco Polo, Great Barrier Reef Committee trip c. 1980s (photo by and courtesy of Patricia Mather).

South Australian and Australian palaeontology and stratigraphy with her 'indefatigable' dedication and productivity, and because of her guidance and ready counsel to her peers. Ludbrook, like Crespin and Hill, was always mindful of the important relationship between basic taxonomy and its application to biostratigraphy, economic or otherwise. While an undergraduate at UA, and despite the lack of a formal course in palaeontology, young Nell Woods became interested in the study of Pliocene molluscs in the departmental museum collections (Sir Joseph Verco and Professor Ralph Tate classic collections from South Australia and Victoria). She continued these studies after graduating as a BA (in 1928) and MA (in 1930), and during a period as teacher at Mount Barker High School. Woods was awarded the Tate Medal of the UA for her first paper in 1931.

For most of the century, and before the current age of political equality, to get on in geology or palaeontology most women gave up thoughts of a professional career because of the Marriage Bar that was in force in the public service and elsewhere until 1966 (Kelly 1993). Woods married Wallis Verco Ludbrook in 1935 and went to live in Canberra, but she was able to maintain an interest in Tertiary molluscs after commonwealth palaeontologist Crespin transferred from Melbourne to Canberra. From 1942 to 1949, she worked as assistant geologist at the BMR, primarily on statistics of strategic minerals. Only later, in 1950, by going overseas, coupled with the subsequent death of her husband, could she further her early studies of Tertiary molluscs. Ludbrook was attached to Imperial College of Science and Technology while being accommodated as a visiting research worker in the BMNH. She gained her PhD from the University of London in 1952 for a study of Pliocene molluscs underlying the Adelaide Plains, as well as her DIC. Her return to Australia and her launch into micropalaeontological work have been noted above. From 1967 she was Consultant in Palaeontology to the South Australian Department of Mines and Energy (Johns & Cooper 1995). In this capacity she was able to describe Tertiary molluscan faunas, using them in biostratigraphy and stratigraphic sequences (Taylor 1985). Over her research career Ludbrook erected many tens of taxa and over 20 are named for her (Lindsay 1985), including a mollusc *Ludbrookia* Chavan, 1951.

In 1947 Mary Wade had chosen to study in the geology department of Sir Douglas Mawson (e.g., Cooper 2000) coming under his thrall. She took to geology with gusto and majored in geology and zoology in 1951 completing a first class BSc with honours in micropalaeontology by 1954. As noted above, she continued to work on Tertiary Foraminifera under Glaessner, gaining her doctorate in 1959. Her first-class honours degree led to employment as a full-time senior demonstrator. In the late 1950s, when Glaessner took up the momentous discovery made by Reg Sprigg in 1947 (e.g., Turner & Vickers-Rich 2004) and decided to pursue Precambrian fossils, he was awarded US research money and, as a consequence, Wade too changed fields. In her postdoctorate life, she concentrated on the Ediacara fauna, then the oldest known macrofossils, from the Flinders Ranges of South Australia. Wade and Glaessner continued as a team (B. McGowran pers. comm. to ST 2005):

Martin's drive, awareness of the big picture, and grip on the background in several languages; Mary's tenacious enquiring into field context, taphonomic pathways, and anatomical reconstruction.

Mary produced two major papers with Glaessner and at least nine single-authored works, culminating in her contribution on the fossil Scyphozoa to the *Traité de Zoologie* (Wade 1994). One of the early forms, which Glaessner and Wade identified as a marine worm, *Spriggina? ovata*, in 1966, was subsequently named *Marywadea* by Glaessner (1984). Their joint work on what have been regarded as the most momentous fossils of the 20th century, the Precambrian Ediacara fauna, contributed significant papers over a decade, most in international journals and a book from Glaessner (1984), for which Mary helped to design the cover. This oeuvre sparked a new revolution in thinking about the earliest metazoans that continues most recently with an international research project (UNESCO-IUGS, IGCP 493) and the recognition of the Ediacaran period on the Geological Timescale (e.g., Turner & Vickers-Rich 2004). After several attempts, and working as a temporary lecturer over many years, Wade realized that she was not going to gain promotion at UA and so moved to the Queensland Museum as Curator of Geology in 1971. In her last three decades of work she became equally famous both for her field collecting and other abilities as she moved into new phyla: she discovered new dinosaurs (Fig. 13) and marine reptiles; a new order of nautiloids; giant squid; and a GSA special symposium was held in her honour in July 1998 (Rozefelds & Turner 1998; Cook 2005, Turner 2005).

And what of those who remain almost invisible?: Edith Margaret Smith (1913–67), born in Forth, northern Tasmania, saw World War II service as an aircraftwoman and then decided to gain an education. After graduating in Geology at UT in 1949, in Sam Carey's department, the first woman to do so, she was working as a research assistant on Tasmanian Permian Bryozoa in late 1952 (GSA 1953–56, vol. 1(1), p. 12). Max Banks (pers. comm. to ST 2006) noted that

Fig. 13. Dr Mary Wade: (**a**) at the time of her move to Queensland; (**b**) with new bones from western Queensland, (**c**) photograph c. 1995; (**d**) in the field c. 1972, north Queensland; (**e**) working on the Lark Quarry dinosaur trackway site c. 1977 (photos courtesy of the Queensland Museum).

she did work as a geologist after graduation until her death in 1967. Sadly she did not publish her research but her work did trigger Banks' own interest in the history of geology in Tasmania and inspired another student, Catherine Reid, who completed her PhD on Permian bryozoans and went to work in Canada. 'Edie' Smith enjoyed the Tasmanian wilderness of her youth and went on to become the first woman president of the Australian Speleology Society (Poulter & Goede 2001, with photos).

Vertebrate palaeontology

Here in Australia, contrary to the image portrayed of Australian University life by Monty Python *et al.*, women are prominent in vertebrate palaeontology and, if postgrads are included in the census, they actually form the majority. Women are active in all areas of vertebrate palaeontology including fish (Sue Turner, Anne Kemp), amphibians (Anne Warren), marine reptiles (Mary Wade), dinosaurs and birds (Pat Rich), and mammals (Sue Hand, Jeanette Muirhead) all of which are in positions above that of postgrad in universities and museums throughout the country. Our [i.e. Australia's] sole coprolite researcher (Caroline Northwood) is also female and continues her research undeturd (yes, I stole that one from John Long, who no doubt stole it from someone else). (Quoted from vertebrate palaeontologist and ABC TV presenter, Paul Willis, pers. comm. to ST 1996)

Rich *et al.* (1982) and Turner (e.g. 1986; Turner & Wade 1986) and Vickers-Rich & Archbold (1991) have emphasised the role of women in vertebrate palaeontology in Australia, especially over recent decades.

As a supreme example, Mary Wade was pivotal in recent years in putting Australian dinosaurs on the world stage. Excavation and subsequent description of the 95-million-year-old Dinosaur Trackways at Lark Quarry near Winton began in the mid-1970s. Under the fierce western Queensland sun, and with the aid of a group mainly of UQ volunteers and a contingent from the army, she and her co-worker, Dr R. A. 'Tony' Thulborn (then UQ), cleared out and cast hundreds of chicken-sized to elephantine dinosaur footprints from the red Winton Formation rock (Fig. 13). This is one of the world's few evidences of running dinosaurs, which Wade and Thulborn identified and described as the only known dinosaur stampede (Thulborn & Wade 1984; Thulborn 1990). The find, with over 3000-exposed dinosaur footprints of three distinct taxa (formerly displayed in the Queensland Museum foyer in Brisbane), is unique and its reputation is now international and featured widely in textbooks. The trackways are a major generator of tourist income to the Winton Shire district.

Palaeobotany/Palynology

The earliest Australian research palaeobotanist is Adele V. Vincent (1896?–19??), who graduated with a BSc from MU in 1917 and made her MSc (1919) study on the important Silurian–Devonian floras in Victoria, published through the GSV in 1925 (Douglas & Holmes 2006). No more is known of her.

By contrast, the scientific work of Isabel Clifton Cookson (1893–1973; Fig. 14) is well known and recognized throughout the palaeobotanical world (Baker 1973; Dettmann 1988, 1993); a recent review was made by Turner & Dettmann (2005). Her published record spans more than five decades – some 86 papers with 19 co-authors – 'a splendid example of active co-operation' (Baker 1973), but also of determination, as much of her work was done on soft money or self-funded; 50 joint papers and 35 papers as sole author. She became the pioneer and 'doyen' of palaeobotany and palynology in Australia. However, Douglas & Holmes (2006) regard her as a plant taxonomist, as she 'never ventured into the field'. This is at odds with some memories of her, and notebooks etc. will have to be checked to verify this point of view.

Starting out with training in botany at MU (Fig. 14c), she worked for a brief time at the National Museum of Victoria and began to look at fossil plants collected by GSV men. Cookson described her first fossil in 1926, when she identified specimens from Walhalla and Rhyll (the latter in a bore core), as belonging to the northern hemisphere Early Devonian species (now probably referable to *Baragwanathia*; Douglas & Holmes 2006). Soon after this, Cookson left for Britain and became a postgraduate, first at Cambridge and then with Professor W. H. Lang at Manchester University. She found her palaeobotanical mentor in Lang. On arrival she discovered that the fungal specimens she had brought to work on were too damaged for study, which meant that she turned to her Australian fossil plant material, beginning with the mid-Palaeozoic Baragwanathia flora from Victoria. Cookson worked with Lang on plants from Victoria from 1926 to 1935, together re-identifying the Rhyll specimen as *cf. Thursophyton* (Lang & Cookson 1927). In addition, they also described a number of fossil specimens from the Jamieson district, which they could not relate to any northern hemisphere pre-Carboniferous plant. Eight years later, similar specimens were to be described as *Baragwanathia longifolia*; they used material collected by survey trips in the late 1920s (Fig. 9c) to set a Silurian age. Graptolites associated with the plants were submitted to Elles at Cambridge (see Lang & Cookson 1935), who placed their age 'beyond doubt' as Early Ludlow (Late Silurian; Douglas & Holmes 2006). Although this age has been disputed over the years, their work revolutionized thinking on the evolution of land floras and led to her DSc in 1932 (Fig. 14); she was only the fourth Australian woman to gain this degree. Lang (with Croft in 1942) later honoured her by naming the genus *Cooksonia* for Early Devonian plants from the Welsh Borderland.

Cookson returned to MU as a lecturer in botany in 1932. Through her stimulating lecturing and demonstrating at Melbourne, she encouraged many senior students and collaborators, including Kathleen McWhae (née Pike; Fig. 14d), Lorna Medwell and Mary Dettmann. During the 1940s Cookson began concentrating on microscopic fossil plant remains and the fruit of this labour was the establishment in 1949 of the Brown Coal Pollen Research Unit at MU (Fig. 14). She was appointed leader of the Pollen Research Unit, where she undertook palaeobotanical studies with pollen analysis of Victoria's coal for CSIRO. This she headed with Pike, and later Mary Dettmann, as research assistants. In this era she established the usefulness of plant microfossils for biostratigraphy and correlation. She also laid the foundations for tracing the history and evolution of Australia's present-day vegetation, putting some students to work on modern pollen.

Although Cookson was a respected research leader, she only achieved senior lecturer status in the botany department and, after her official retirement in 1959, maintained her research mainly by self-funding especially for her overseas trips, which extended her international profile. These led to productive partnerships with Leflandre in Belgium, Eisenack in Tübingen, and Manum in

Fig. 14. Isabel Cookson: (**a**) portrait (photo courtesy Mary Dettmann); (**b**) Cookson in graduation robes; (**c**) Cookson and an unidentified woman in her office in the Pollen unit at MU in 1954; (**d**) with her student and assistant Kathleen Pike (centre) and fellow botany/geology students at Melbourne University in 1947 around the time of their graduation (photo courtesy Professor Trevor Clifford); (**e**) as a student (top row, right) in a group photograph of MU 1st-year biology women, *c.* 1921 (photos b; c & e courtesy of MU Archives).

Sweden (e.g., Dettmann 1988; Turner & Dettmann 2005). Her later work with Alfred Eisenack (1891–1982), who named 'Chitinozoa' (Sarjeant 1999), led to a new research paradigm.

Cookson's student, Pike, was born in Suva, Fiji, in 1927 and moved to Australia when she was 10 years old (pers. comm. to ST 1994). She graduated with a BSc from MU in 1948. From 1948 to 1950 and 1951 to 1955, she worked in the botany school under the supervision of Isabel Cookson, funded by the State Electricity Commission of Victoria. Dr A. B. Edwards (Geology) was responsible for the funding and directed the research on the palynology and palaeobotany of the Yallourn Brown Coal Deposits. Pike spent 18 months at Cambridge University in 1950–51, working on the palynology of interglacial deposits at Clacton-on-Sea under the supervision of Professor Harry Godwin at 'Whitstead', the School of Botany at Cambridge University, which resulted in one paper in the *Quarterly Journal of the Geological Society*; at this time she met her husband-to-be. She graduated with MSc in 1954 with a thesis on living pollen grains of the Myrtaceae. Marriage to oil geologist Dr Ross McWhae of Perth, Western Australia, in 1955, a move to Canada and three children effectively ended her scientific life. When the children were older she did try to obtain palynological work in Calgary; a senior university man offered $3.00 per hour to prepare his samples, which Kathleen rightly deemed beneath contempt!

Medwell carried out work on the Jurassic and Cretaceous floras of Victoria (Runneymede Formation) and then concentrated 'on growing apples under the name of Mrs P. Burke' (according to B. Glenister, GSA 1953–56, vol. 1(1)) later 'disappearing off the face of the earth' (H.T. Clifford, pers. comm.).

Dr Suzanne (Sue) Lawless Duigan (1924–93) was another pioneer palynologist. She did her BSc and MSc at MU, and then won a CSIRO Scholarship to study in England, also under Godwin, where she received her PhD Cantab. in 1954. On returning to MU, Sue worked with Cookson on the brown coals and, *c*. 1960/61, was appointed as a Lecturer in Botany. She specialized in Quaternary flora (Hart 2007). Colleagues Jack Douglas and Mary Dettmann, in their memorial, said:

In the late 40's and 50's she contributed greatly to knowledge of the Tertiary flora of eastern Australia and collaborated with Isabel Cookson on several papers.

In her early years at Cambridge she undertook some excellent research on pollen morphology and the Quaternary palynology of Lake Mountain and Macquarie Island, working closely with Cookson. Kershaw (1997, n.d.) noted that she was a perfectionist who was unable to 'reveal most of her research in print'. A special issue followed a symposium dedicated to her work (Kershaw 1997). Another Cookson colleague, Lucy Cranwell, a New Zealander, contributed significantly to the debate on the southern Gondwanaland continental configuration with her studies describing plants from Antarctica.

'She learned palynology in Sweden, and applied it in N.Z. where she was botanist in the Auckland War Memorial Museum, and was so prominent that she was made an FRSNZ at a very early age. She married an American serviceman during the War, used the name Lucy Cranwell Smith, lived in Tucson where her husband was an archaeologist of independent means, and continued to do palynology of southern regions and was *the* authority on *Nothofagus*'. (Harrington pers. comm. 2000)

This palaeobotanical sorority, where Cookson collaborated with her colleagues and students, was outstandingly successful, producing 30 palynological papers in 17 years. Several of these women were also pioneers in the foundation of the Palynological and Palaeobotanical Association of Australasia in 1991. Cookson herself was honoured with a *Festschrift* late in life (Glover & Playford 1973).

Other aspects of geology

General geology and mapping. Ida Brown's mapping skills need to be emphasised as probably her most major achievement. She excelled in her mapping studies of New South Wales and her Taemas map is still in use (e.g., Branagan & Packham 1967; Basden 1999). Before World War II, very little of Australia's geology had been comprehensively studied, so her detailed mapping and stratigraphy made a substantive contribution. Her maps of Yass and Good Hope link her palaeontological knowledge with the stratigraphy of the region. Her students including E. M. 'Betty' Basnett and Joan Colditz (Fig. 4d), and Phillips followed in her footsteps.

Hill initiated research mapping in Queensland with students (such as G. Tweedale) and brought many maps to fruition in co-operation with GSQ. She was Queenland's representative on The Tectonic Map of Australia Committee (Fisher 1962). She also commenced joint sedimentary petrography and tectonic studies with R. Gradwell in 1952.

Igneous and sedimentary rocks. Women were prominent in the fields of sedimentary and igneous petrology and mineralogy at first; Brown's early work is noted above. Again, little is known of most of these women, and obituaries and memorials seem conspicuous by their absence.

Germaine Anne Joplin (*c*. 1910–89) is the equivalent of Dorothy Hill in this respect, and she dominates in the field until the late 1970s. Joplin

Fig. 15. (a) Germaine Anne Joplin in her DSc robes, 1950 (modified from Waterhouse 2004); (b) the Joplin Room in the library of University House, ANU, Canberra (photo by ST).

was forthright and appeared to be ambitious; a university medalist in 1930, she received her PhD at Cambridge University in 1936. She resigned her assistant lectureship in petrology in 1941 to take a full-time Linnean Macleay Research Fellowship, spending 5 years researching the Cooma district. She co-authored more than 50 research papers and wrote definitive texts. In the late 1920s, she began to complete her first major work on the petrology of plutonic and associated rocks in the Hartley district in NSW (Joplin 1931, 1933, 1935). She was awarded a DSc by SU in 1950 (Fig. 15). Her long-standing career in the department there, which might have led to greater rewards, was cut short for personal reasons.

Joplin joined the new Geophysics Department at the prestigious ANU in 1952 and began to work on metamorphic rocks from the Mt Isa-Cloncurry area (Townley, GSA 1953–56, no. 1); Carey was there on sabbatical leave, 'developing his views on continental drift' and Germaine began planning some experiments on the alteration of rocks. She began a co-operation with E. K. Carter and J. K. Burnett on the occurrence of soluble salts in the Mount Isa slates and on the salt content of adjacent rocks, as well as the chemical metamorphism of a group of pelitic schists (GSA 1953–58, no. 6 (1)).

She entered into the work of the fledgling GSA, becoming divisional chair in Canberra in 1955 and running the Standing Committee on Collection & Recording of Chemical Analyses of Australian Rocks from 1964 to 1969. Joplin was also foundation member of the Specialist Group in Geochemistry, Mineralogy and Petrology (Cooper & Branagan 1994). An application from Joplin to become an academician in the AAS was rejected (papers in Glaessner Archive, Barr Smith Library), and she did not reach professorial level, despite her move to ANU and writing several major textbooks on petrography and geochronology (e.g. Joplin 1964, 1968; Joplin et al. 1972). Her body of scientific papers and books spans 1931 to 1971. Joplin was awarded the W. R. Browne Medal and became a Member of the Order of Australia (General Division). Sadly, although internationally recognized for her often-pioneering work, she did not achieve the high academic position she deserved, although she seems to have mentored many and is remembered with great affection and honoured at Canberra's University House (Fig. 15).

Alma J. Culey, a contemporary collaborator at SU during the 1930s, wrote two of five papers with Joplin between 1933 and 1938. Culey entered SU in 1929, taking her BSc in 1931 and gaining an MSc in 1934. Her work covered general geology, sedimentology and mineralogy, mainly in the New South Wales coal measures. But what became of her post-1938?

Coal studies became prominent in the 1950s to 1960s. Many young women at SU for example, were assistants in the Coal research laboratory. Noreen Allen (later Morris), a 1949 SU graduate, moved to UN and published on coal palaeobotany. Scott was another of the post World War II achievers. Born in Maryville, a suburb of Newcastle, New South Wales, during the depression, she was influenced by her science mistress at Newcastle Girls' High School, a former graduate of SU, and had done geology at school, coming first in the state at the leaving certificate. She won prizes as an undergraduate and, after her pass degree, continued as a demonstrator. This meant deferring her honours course for 1 year, until 1946, but she still gained a first-class honours and won the university medal as well as a research scholarship. Her early research addressed the geology of the Stanhope district in the Hunter Valley. This was later extended to embrace the mineralogy, geochemistry and genetic relations of the Carboniferous and Permian andesitic associations of eastern New South Wales, and the conditions of formation of secondary minerals in these andesitic and basic rocks. A few years later, in 1954, after residence in Spain and Egypt, she returned to Australia. When the former 'well known' Miss Scott, as Dr B. Nashar, was appointed to UN College, School of Mining Engineering and Applied Geology, as lecturer in geology, it was reported as the most significant event in the new year (GSA 1953–56, no. 3(1), p. 5). She continued in her chosen field of igneous petrology, but turned more to sedimentary petrography, publishing books on geology in New South Wales (e.g., Nashar 1967). However, she turned more and more to organizational work, for which she became soundly respected (e.g., Allen 2001).

Others are less well known and often did not get as far as publication. Basnett (1941, 1942), however, pursued metamorphic studies on dynamic and contact metamorphism and hybridization of rocks in New South Wales. Colditz (1948) tackled the petrology of the Silurian volcanic sequence in the Wellington district of New South Wales. On the other hand, Miss Margaret Sando at UA, began investigating granitic and metamorphic rocks in South Australia in 1954 and was made a demonstrator the following year (GSA 1953–56, no. 2(2), p. 11) but only published a little.

Mineralogy/Economic geology. Another hard-rock specialist, Florence Mabel ('Florrie' or 'Flo') Quodling, began at SU in 1924, going on to become a lecturer in 1945 (Branagan 1973; Hooker 2004), providing academic training to many in the post-war boom time, primarily as a fine teacher through her laboratories. Interestingly, she was deaf, but this impediment did not indispose her for a teaching role. She became a respected mineralogist, penning at least nine scientific papers between 1936 and 1964, most with collaborators or students and some in overseas journals.

Miss Dorothy Carroll, an early graduate with a BA and BSc at UWA, took up mineralogy possibly for honours, publishing two papers on the mineralogy of fine sands in the early 1930s. She may have continued this work, for she gave a lecture on sedimentology-heavy mineral separation at SU in around 1960 (Branagan pers. comm. 2006).

During and after the war, economic needs dominated. Both Crockford (1942) and Crespin (1946) tackled the problems of diatoms and diatomite (diatomaceous earth) during the war years. They studied them to see if Australian diatomites had any value to the army as a filter medium for water in the more tropical parts of Australia and in Papua New Guinea. Prior to this investigation, all diatomites suitable for filtration were imported from the United States. Ludbrook helped to compile maps and statistics on this and several other important minerals for major reports. Another UA graduate, Sylvia Whitehead, still a GSA member in the South Australian Division, was a petrologist who spent many years before retirement at AMDEL in South Australia.

Other important Australian work. In their biostratigraphic work and contributions to overall geological mapping projects, several of these Australians became members of important committees, such as the Australian Stratigraphical Nomenclature Committee and Tectonic Map of Australia Committee. Women represented their country overseas at symposia and on councils, and have also served on the national IGCP Committee post-1972 (Turner 2006b). Many women have also been the recorders, curators and librarians and compilers of bibliography and indices: Hill compiled one part of the *Lexique stratigraphique international* and recorded state nomenclature and bibliography; Edie Smith did even more work on the Australian contributions to the *Lexicon* and became a noted editor for their GST (Poulter & Goede 2001). The work of Ngaire Teesdale-Smith (neé Dolling), as another geological bibliographer, was noted by O'Neil (1998).

Notably they have pioneered and been major contributors to geoheritage protection in Australia, with E. Maude McBriar, a UA graduate being active for most of her working life (Joyce, in Cooper & Branagan 1994) and she still contributes to the GSA Geological Monuments committee. Her pioneering contribution on geological monuments

Table 3. *Chronology of Australian Women who have published in geology, especially palaeontology*

Name	Publishing span	Subject	University
Fanny Cohen[†]?	1911	Mineralogy	SU
Marie/Mary Bentivoglio[†]	1921	Mineralogy	SU (see Radi 1988)
Ida A. Brown[†]	1923–63	Mineralogy, then stratigraphy/structure, palaeontology	SU
Irene Crespin[†]	1923–74	Micropalaeontology	MU
Adele V. Vincent[†]?	1925	Important Silurian–Devonian floras in Victoria	MU
Isabel C. Cookson[†]	1926–82?	Earlier botanical papers, palaeobotany	MU
Cecily Briggs (neé Moxton)[†]	1929	Petrology	UQ
Kathleen McInerny (Mrs Sherrard)[†]	1929–?	Building stones	MU
Dorothy Hill[†]	1930–72	Stratigraphy, palaeontology	UQ
Lucy F. V. Hosking[†]?	1931–34	Stratigraphy, palaeontology	UWA
Germaine A. Joplin	1931–72	Petrology	SU
Nell Hooper (neé Woods) Ludbrook[†]	1931–87?	Mineralogy, palaeontology	UA
Elizabeth 'Betty' A. Ripper[†]	1932–38	Palaeontology	MU
Dorothy Carroll[†]?	1933–34;	Mineralology	SU
Alma G. Culey[†]	1933–38/9	Sedimentology/mineralogy/igneous petrology	SU
Kathleen Laura Prendergast[†]	1935–51	Palaeontology	UWA
Florence M. Quodling[†]	1936–64	Mineralogy	SU
Kathleen Sherrard	1936–52?	Now in palaeontology	SU
Marjorie J. Whitehouse[†]	1937–38	Mineralogy	UQ
Joan M. Crockford (Mrs Beattie)[†]	1940–57	Palaeontology	SU
Joan Johnston (Mrs Moye)[†]	1941	Palaeontology	SU
Elizabeth (Betty) M. Basnett[¶]	1941–46	Palaeontology?	MU
M. Joan Colditz[†]	1943–48	Geology, petrology	SU
Beryl Scott/Nashar[†]	1948–67	Geology/palaeontology	SU
E. Betty Dodds[†]	1949	Palaeontology	UQ
Suzanne Lawless Duigan[†]	1951–66	Palaeobotany	MU
Kathleen Pike (Mrs McWhae)	1953–	Palaeobotany	MU
Joyce Gilbert-Tomlinson[†]	1954–68	Palaeontology	MU
Lorna Medwell[†]	1954	Palaeobotany	MU
Mary Julia Wade[†]	1955–	Still posthumous mss to publish; palaeontology	UA
Cochrane, Nevis (neé Ould)[†]	1956–60	Geology	UQ
Margaret Sando	1956	Petrology	UA
June R. P. Phillips (-Ross)	1956	Palaeontology	SU
Edith Margaret Smith[†]	1957–66	Stratigraphy	
Bever(e)y Houston[†]	1965	Geology	UQ

[†] Australian-born
Abbreviations: MU, Melbourne; SU, Sydney University; UA, University of Adelaide; UQ, University of Queensland; UWA, University of Western Australia.

History of geology. Interestingly, another discipline where Australian women have contributed significantly is in the history of their science. Dorothy Hill made a point of this throughout her later career and in retirement (e.g. departmental and bibliographic work; Hill & Willadsen 1980; Hill 1981). Ida Brown also, when given the opportunity to make her (1946) presidential address, addressed, the history of her discipline. Irene Crespin's (1972, 1975) contributions to BMR history are far more enlightening than the official version (Wilkinson 1996), certainly on the role and work of palaeontologists in that organization. More recent studies have been mentioned above, with at least one woman (ST) now elected to the International Union of Geological Sciences History of Geological Sciences Commission (INHIGEO).

Popularization and general books on geology. Women get involved in promoting geology and palaeontology at all levels, contributing to local and international associations, participating in BAAS and ANZAAS and writing general reviews and texts (e.g., Hill, Joplin & Gilbert-Tomlinson; see White 1954). Hill's part in creating the Queensland Palaeontographical Society, which produced booklets on fossils from each geological period led on to the formation of the professional palaeontological association in the region (Runnegar 1975).

Conclusions

Many Australian women geoscientists did not and would not consider that they had impediments put in the way of their scientific work for being women. They wished to be known for their achievements and these are clearly manifold. Nevertheless, even today, very few women reach high university or other academic positions, or places of influence in scientific societies, principally because of the loss of continuity in their career path if they decide to leave to marry, to follow partners/spouses, or to spend time in raising a family; little has changed in that regard.

Women have played a significant role in Australian geoscience, especially in palaeontology. Here I have interpreted 'Australian' broadly, as there have been émigrés, but in general, for most of the early to mid-20th century, the geologists were homegrown. Following on the crest of the 19th-century craze for all things natural (Moyal 1986), young women enrolled in the first university courses, thriving in the field, in some instances outnumbering men (a trend that is back today; e.g., Facer 1996). The research university of ANU did not come into being until 1954, but, post-war, the newly formed BMR was almost equivalent. Only in the 1960s did research possibilities open up, with the creation of many 'new' universities; however, the last decade has seen a reversal of both funds and attitudes. The current trend is for research to be restricted to the 'Big 8' universities – 'Go8 (Monash, MU, ANU, UNSW, UA, UWA, UQ, SU)' – and, because of lack of jobs and funds in universities, museums and surveys, pure research in geoscience, especially palaeontology, has stultified. All this has affected women's employment and research chances, with many women now only existing on 'soft money'.

Original specimens are the key to any taxonomic work and its significance. Pioneer studies on Australian fossil faunas and floras, such as those by Ripper, Hill, Cookson and the SU women, are the cornerstone of palaeontological studies, even if such work is no longer appreciated. (In the current political climate, alpha taxonomy is virtually unsupported in Australia, with no taxonomic palaeontologist, male or female, being appointed to any Australian institution since the mid-1990s e.g., Rich 1999). The sort of major monographic contributions made by Australian women in the past (e.g., Hill to Raymond C. Moore's *Treatise on Invertebrate Paleontology*; Ludbrook on molluscs, especially scaphopods; Wade to the *Traité de Zoologie*) are less likely, unless Mather's recent call (in Archer *et al.* 2006) for an Australian Institute of Taxonomy is answered.

The success of earlier women depended on support from family and mentors (usually male until the later 20th century) and to gain academic status meant forgoing marriage and/or children. The 1930s was an exceptional decade, with over 80 publications by graduate women, mainly in palaeontology. Hill and Cookson, both in the Britain at the same time and influenced by their mentors, were the first to publish in overseas journals in 1935: Cookson in the prestigious *Philosophical Transactions of the Royal Society*, with her ground-breaking work on the earliest terrestrial plants (Cookson 1935), and Hill in the *Geological Magazine*, followed the next year by a work in the *Philosophical Transactions* on British Silurian rugose corals (Hill 1935, 1936). By contrast, most Australian women published in local journals until late in the 20th century, which might account for their lack of recognition (invisibility) on the world stage. Overseas colleagues still claim that Australian serials are 'obscure'.

During the war years (World Wars I and II) and in the post World War II boom, women were involved in war work, but new studies accumulated, some of economic significance (e.g., Hill's,

Cookson's and Crespin's involvement in the search for oil and coal, Crockford on diatomites, and Ludbrook's mineral statistical studies). The post-war era was bountiful, especially during the exploration for natural resources, and Cookson, Hill, Ludbrook and Crespin especially undertook consultancy work. Coal resources were important in Australia, and many young women (not considered here in detail) worked in the unit at SU, for instance (Branagan 1973). Others, such as Brown and Crockford, were assisting with biostratigraphical studies that enabled correlation across Australia and the region, and underpinned the push to map Australia. The principals had between 40–50 publishing years each.

Women were appointed early to academic staff at universities and in government institutions (surveys and museums), although few reached full lecturer status until after World War II. Concurring with the earlier findings of Moyal and Hooker, I too have to stress how easily it is for a woman to be effaced from the history and consciousness of her discipline, even when she has been the first in a discipline or has led scientific research programmes (*cf.* Turner & Cadée 2005). Until the 'modern', post-plate tectonic era, Hill and Nashar were among the first and only women in Australia to attain professorial level. Others, such as Basnett and Colditz, who were the *tour de force* behind the beginning of the geology department at UNE, hardly rate a mention in Voisey's memoirs (Harrington *et al.* 1991). Young Joan Crockford was one who succeeded in having a good early scientific career in the post World War II era at a time of optimism in Australia, when the nation was in an industrial revolution. In a different culture (one that encouraged women to maintain and develop their intellectual skills even after marriage and children), Crockford-Beattie might have been able to continue her research and gain a viable geological job. Nevertheless, women such as Cookson, Hill, Ripper, Crockford and Wade, were foremost in their disciplines and acted as role models and mentors for the next generation.

As in the 19th century, when for most people, science was an avocation to be enjoyed for itself, usually with no thought of reward, in Australia we are returning to this paradigm in the 21st century, with few jobs for palaeontologists. As Rich noted in 1999, most of the people then employed have retired or been forced out of their jobs, so there are few jobs to aspire to but plenty of volunteer work to do! Already the bibliography of scientific papers by women in geosciences spanning the 20th century comprises many hundreds of contributions (See the Appendix which is available online at http://www.geolsoc.org.uk/).

Bruce Waterhouse (the only other holder of the Dorothy Hill Professorship of Palaeontology at UQ, quoted on the GSWZ website) summed up the situation well:

It is a strange thing, that whereas Australia has been fortunate in having had a number of senior women geologists, at least one or two per state born near the turn of the century, it was very much later, as far as I know, when women such as Jean Luke or Heather Leed took a career in geology in New Zealand – and then, for only a few years, after the second world war. Yet the population of Queensland is the same, roughly, as New Zealand, and South Australia and Western Australia populations are smaller. Women also were remarkably prominent in the geological sciences in Canada, Britain, let alone the United States and Russia, as may be seen from perusing Bill Sarjeant's mammoth 10-volume bibliography. South Africa had the remarkable Edna Plumstead. We seem to fall closer to the geological fraternity of India, of all places – long established geological survey, museums, university departments, and few if any women, until well into the second part of this century. It goes without saying that we have therefore been the poorer. But why has it been so? It cannot be that Australia provided much more incentive or encouragement for women geologists – the Australian heroines such as Hill, Joplin, Browne, Ludbrooke[sic] etc. **had such hard times before and after graduation**. Dorothy was forty before she found permanent employment. Were they more profoundly moved to explore the planet earth and wonder about rocks and fossils, and so more staunchly pursued their devotion to science, no matter what the difficulties? How strange. Yet that was not only our gain: it was theirs as well, I am sure that Dorothy would say, not for the honours, positions, salaries, but for the thrill and challenge and fulfilment to be found in the earth sciences.

Emulating, and perhaps improving on Professor G. Philip in his foreword to the Sydney department centenary volume (Branagan 1973)...

> 'Let us now praise famous women
> And our mothers that begat us'.

This essay has evolved from a database of women geologists in Australia begun and continued with assistance from many people: the substance was presented at the Women's Day at the GSL in 2005. I happily thank the organizer Cynthia Burek for her invitation and the financial support from the GSL. I am indebted to many for encouragement and original pictures, notably D. Branagan and L. Harrington, both of whom acted as suitably critical reviewers over the years; Marjorie Apthorpe, M. Banks, A. Bartholomai, Joan Beattie, Cynthia Burek, S. Carey, O. Chalmers, T. Clifford, B. Cooper, Mary Dettmann, the late M. Dickens, Margaret Eva, J. Jell, R. Jenkins, Patricia Mather, B. McGowran, Kathleen McWhae, Ann Moyal, Beryl Nashar, Sue Parfrey, the late Betty Ripper, June Phillips Ross, the late J. Shergold, Patricia Vickers-Rich, Liz Truswell, the late Mary Wade, and B. Webby all gave freely of their time, personal photos and stories. I am indebted to Nessy Allen, who provided a final polish. The thoughts expressed here are, however, my own. In addition, I thank the archivists and librarians in the Queensland Museum, the Barr Smith Library, UA, the UA, MU and

SU archives and the National Library of Australia, who gave help and allowed access to documents within their care. Photographic acknowledgements are to the MU, UQ and SU archives, the National Library of Australia, South Australia Mines Department PIRSA, and University of Melbourne Photographic Services, the personnel of which were all wonderfully helpful. Thanks to Annemarie Williamson and to Mineral Resources, PIRSA, for photographic files. Financial support came from an ARC Discovery grant no. DP0453155 to Patricia Vickers-Rich, which aided work in 2004. A contribution to travelling was given by Bradford University Women into Science, Engineering and Technology Resource Centre. My 'scientific wife', Judy Bracefield, gave all other kinds of support.

References

ALLEN, N. 1994. Achievement in science: the careers of two Australian women chemists. *Historical Records of Australian Science*, **10**, 129–141.

ALLEN, N. 1997. Test tubes and white jackets: the careers of two Australian women scientists. *Journal of Australian Studies*, **52**, 126–137.

ALLEN, N. 2001. Interview with Professor Beryl Nashar. In: *Interviews with Australian Scientists*. Australian Academy of Science. World Wide Web Address: http://www.science.org.au/scientists/mashar.htm

ALLEY, N. F. 1996. Obituary. Nelly Hooper Ludbrook MBE MA PhD DIC FGS 14.vi.1907–9.v.1995. *Transactions of the Royal Society of South Australia*, **120**, 74–77.

ANDREWS, H. N. 1980. *The Fossil Hunters in Search of Ancient Plants*. Cornell University Press, Ithaca, 421 pp.

ANON. 1951. Wives of scientists have many interests. *Brisbane Telegraph*, 22 May, 17.

ANON. 1953. Federal Territories Reports. *Geological Society of Australia News Bulletin* **1**(1), 3.

ANON. 1976. Ida Alison Browne. *The Australian Geologist, Newsletter*, No. 12, (December), 13.

ANON. 1985. Another honour for geologist. *University News*, (Queensland) 20 Nov. 4.

ARCHBOLD, N. W. 1981. Western Australian geology: an historical review to the year 1870. *Journal of the Royal Society of Western Australia*, **63**, 119–128.

ARCHER, M., MATHER, P. & TALBOT, F. 2006. Rescuing the key to nature's databank. *Australasian Science*, **27**, 35–37.

BAKER, G. 1973. *Dr Isabel Clifton Cookson*. Geological Society of Australia, Special Publication, **4**.

BASDEN, A. 1999. Emsian (Early Devonian) microvertebrates from the Buchan and Taemas areas of southeastern Australia. *Records of the Western Australian Museum*, (Supplement), **57**, 15–21.

BASNETT, E. M. 1941. Studies in metamorphism and assimilation in the Wellington District, New South Wales. I. Hybridisation in the Wuuluman Creek Intrusion. *Journal and Proceedings of the Royal Society of New South Wales*, **73**, 161–189.

BASNETT, E. M. 1942. Studies in metamorphism and assimilation in the Wellington District, New South Wales. II. The dynamic and contact metamorphism of a group of ultrabasic rocks. *Journal and Proceedings of the Royal Society of New South Wales*, **76**, 55–81.

BENSON, W. N. 1944. Frederick Chapman obit. *Transactions of the Royal Society of New Zealand*, **74**, 302–303.

BOWEN, J. & BOWEN, M. 2002. *The Great Barrier Reef: History, Science, Heritage*. Cambridge University Press, Cambridge, 454 pp.

BRANAGAN, D. F. 1972. Words, actions, people: 150 years of the scientific societies in Australia. *Journal & Proceedings of the Royal Society of New South Wales*, **104**, 123–141.

BRANAGAN, D. F. (ed.) 1973. *Rocks, Fossils, Profs. Geological Sciences in the University of Sydney 1866–1973*. Science Press, Sydney, 184 pp.

BRANAGAN, D. F. 1977. Ida Alison Browne. *Journal & Proceedings of the Royal Society of New South Wales* **110**, 75–77.

BRANAGAN, D. F. 1982. Georgina King: geological prophet or lost? *University of Sydney Archives Office of Registrar, Record* No. 2, 4–9.

BRANAGAN, D. F. 1994. *More Rocks, Fossils, Profs: The Department of Geology and Geophysics University of Sydney, 1972–1992*. Science Press, Sydney.

BRANAGAN, D. F. 2005. *T. W. Edgeworth David. A Life. Geologist, Adventurer, Soldier and 'Knight in the Old Brown Hat'*. National Library of Australia, Canberra, 648 pp.

BRANAGAN, D. F. & PACKHAM, G. H. 1967. *Field Geology of New South Wales*. 2nd edn. Science Press, Sydney, 1970, 191 pp.

BRANAGAN, D. F. & VALLANCE, T. 2001. Earth sciences. In: MACLEOD, R. (ed.) *The Commonwealth of Science. ANZAAS and the Scientific Enterprise in Australasia 1888–1988*. Oxford University Press, Melbourne, 130–146.

BRILL, K. G. 1953. Impressions of geology in Australia. *Geological Society of Australia News Bulletin*, **1**, (3), 19–20.

BROWN, I. A. 1946. An outline of the history of palaeontology in Australia. *Proceedings of the Linnean Society of New South Wales*, **71**, v–xviii.

BRYAN, H. C. 1954. Samuel Stutchbury and some of those who followed him. *Queensland Government Mining Journal*, **55**(634), 1–8.

BUREK, C. V. 2007. The role of women in geological higher education – Bedford College, London (Catherine Raisin) and Newnham College, Cambridge, UK. In: BUREK, C. V. & HIGGS, B. (eds) *The Role of Women in the History of Geology*. Geological Society, London, Special Publications, **281**, 9–38.

BYGOTT, U. & CABLE, K. J. 1985. *Pioneer Women Graduates of the University of Sydney 1881–1921*. Sydney University Monographs, **1**, 55 pp.

CAIN, ALEXANDRA 2000. Biography. Ida Alison Browne, Palaeontologist 1900–1975, Sydney. Private publication, 8 pp.

CAMPBELL, K. S. W. & JELL, J. S. 1998. Dorothy Hill 1907–1997. *Historical Records of Australian Science*, **12**(2), 205–228.

CAREY, J. 2001*a*. Engendering scientific pursuits: Australian women and science 1880–1960. *Limina*, **7**, 10–25.

CAREY, J. 2001*b*. No place for a woman? Intersections of class, modernity and colonialism in the gendering of Australian science, 1885–1940. *Lilith*, **10**, 153–172.

CASEY, J. N. 1976. Forward. *In*: BELFORD, D. J. & SHEIBNEROVA, V. (eds) *The Crespin Volume: Essays in Honour of Irene Crespin*. Bureau of Mineral Resources, Geology and Geophysics, Bulletin, **192**, viii–ix.

CLARKE, E. DE C. & WELLS, J. E. 1931. Excursion to Bullsbrook. *Journal of the Royal Society of Western Australia*, **17**, xx.

CLEEVELY, R. J. 1983. *World Palaeontological Collections*. British Museum (Natural History) and Mansell Publishing Ltd, London, 365 pp.

COLDITZ, M. J. 1948. Thepetrology of the Silurian volcanic sequence at Wellington, New South Wales. *Journal and Proceedings of the Royal Society of New South Wales* 81 (for 1947), 180–197.

COOK, A. 2005. Dr Mary Wade. *TAG*, **137**, 45.

COOKSON, I. C. 1935. On plant remains from the Silurian of Victoria, Australia, that extend and connect floras hitherto described. *Philosophical Transactions of the Royal Society B*, **25**, 127–148.

COOPER, B. J. & BRANAGAN, D. F. (eds) 1994. *Rock me Hard ... Rock me Soft ... A History of the Geological Society of Australia*. Geological Society of Australia Inc., Sydney, 194 pp.

COOPER, J. (ed.) 2000. *Records and Reminiscences. Geosciences at the University of Adelaide 1875–2000*. University of Adelaide Department of Geology & Geophysics, Adelaide, 254 pp.

CRESPIN, I. 1946. Diatomite. Bureau of Mineral Resources Survey, Australia. Summary Report **12**, 5–14.

CRESPIN, I. 1954. 8th Pan-Pacific Science Congress, Manila. *Geological Society of Australia, News Bulletin*, **2**(1), 5–6.

CRESPIN, I. 1972. *Recollections of Growth of Commonwealth Interests in Geological sciences*. BMR Record, 1967/157. Reprinted in *Records of the Australian Academy of Science* **2**(2), 29–46.

CRESPIN, I. 1975. *Ramblings of a Micropalaeontologist*. BMR Record, 1975/83, 61 pp.

CROCKFORD, J. M. 1942b. Notes on the Structure of Some Diatomites from Australia and New Zealand with Reference to their use in Filtration. University of Sydney Department of Geology.

DARRAGH, T. A., KNIGHT, J. L. & TATTAM, C. M. 1976. History of Geology in Victoria. *In*: DOUGLAS, J. G. & FERGUSON, J. A. (eds) *The Geology of Victoria*. Geological Society of Australia, Special Publications, **5**, 1–9.

DETTMANN, M, 1988. Isabel Cookson. *In*: RADI, H. (ed.) *200 Australian Women: A Redress Anthology*. Women's Redress Press Inc., Broadway, New South Wales, 182–183.

DETTMANN, M, 1993. Cookson. *In*: RITCHIE, J. (ed.) *Australian Dictionary of Biography*, **13**(1940–80), 491–492.

D. F. B. 1975. Kathleen Sherrard. *The Australian Geologist*, No. 8, (December), 11.

DICKINS, J. M., 1982. Joyce Gilbert Tomlinson. nomen nudum, **12** (3).

DOUGLAS, J. & HOLMES, F. 2006. The Baragwanathia story: an update. *The Fossil Collector*, **77** (Jan.), 9–26.

FACER, R. A. 1996. Professional profiles of the women graduates honoured by this symposium. *In*: FACER, R. A. (ed.) *Geology and the Community. Honouring Sydney Women Graduates*. The 9th Edgeworth David Day Symposium, 6 September, University of Sydney. Edgeworth David Society & Earth Resources Foundation, Sydney 6 September, xi–xv.

FALK, D. 2000. Careers in science offer women an unusual bonus: immortality. *Nature*, **407**, 833.

FISHER, N. H. (ed.) 1962. *Geological Notes in Explanation of the Tectonic Map of Australia. Tectonic Map Committee of the Geological Society of Australia*. Commonwealth Bureau of Mineral Resources, Authority of Hon. HW Spooner, Minister for National Development, 72 pp.

FLETT, J. S. 1937. *The First Hundred Years of the Geological Survey of Great Britain*. HMSO, London, 280 pp.

FRANKLIN, B. (ed.) 1985. Historic know your geologists from the University of Adelaide. *The Australian Geologist*, **58**, 15–16.

GEOLOGICAL SOCIETY OF AUSTRALIA (GSA). 1953–56. *Geological Society of Australia News Bulletins*, **1–4**.

GLAESSNER, M. F. 1984. *The Dawn of Animal Life: A Biohistorical Study*. Cambridge University Press. Cambridge, 244 pp.

GLOVER, J. E. & PLAYFORD, G. (eds) 1973. *Mesozoic and Cainozoic Palynology Essays in Honour of Isabel Cookson*. Geological Society of Australia, Special Publications, **4**, Sydney.

GOGGS, M. R. 1994. Now in Retirement – Dr Nell Ludbrook. ABC Radio interview.

GREGORY, H. 1987. Dorothy Hill CBE PhD DSc Hon. LlD, FRS, FAA. Research Professor of Geology 1959–1972. *In*: *Vivant Professores: Distinguished Members of the University of Queensland 1910–1940*. Fryer Memorial Library Occasional Publication, **7**, 51–58.

GREGORY, H. 2005. Dorothy Hill (1907–1997). *In*: *Great Queensland Women*. Office for Women, State of Queensland.

GUY, A. 2004. Scientific women successful despite the odds. *WISENET Journal*, **67**, 32–34.

HARRINGTON, H. J., YEATES, A. J., BRANAGAN, D. F & MCNALLY, G. H. 1991. *Sixty Years on the Rocks. The Memoirs of Professor Alan H. Voisey*. Geological Society of Australia Inc. with Earth Sciences History Group, Sydney, 124 pp.

HART, J. 2007. The role of women in British Quaternary Science. *In*: BUREK, C. V. & HIGGS, B. (eds) *The Role of Women in the History of Geology*. Geological Society, London, Special Publications, **281**, 83–95.

HILL, D. 1935. British terminology for rugose corals. *Geological Magazine*, **72**, 481–519.

HILL, D. 1936. The British Silurian rugose corals with acanthine septa. *Philosophical Transactions of the Royal Society of London B*, **226**, 189–217.

HILL, D. 1967. Phylum Archaeocyatha Vologdin 1937. *In*: HARLAND, ET AL. (eds) *The Fossil Record*. Geological Society, London, Special Publications, **2**, 341–345.

HILL, D. 1981. *The First Fifty Years of the Department of Geology of the University of Queensland*. Department of Geology, University of Queensland, Papers, **10**, 1–68.

HILL, D. & WILLADSEN, C. 1980. *Bibliography of Australian Geological Serials and of Other Australian Periodicals that include Geological Papers*. Department of Geology, University of Queensland, Papers, **9**, 76 pp.

HOME, R. W. (ed.) 1988. *Australian Science in the Making*. Australian Academy of Science, Cambridge University Press, Cambridge, 413 pp.

HOOKER, C. 2004. *Irresistible Forces: Australian Women in Science*. Melbourne University Press, Melbourne, 288 pp.

JELL, P. A. 1997 Dorothy Hill (1907–97). *Nature*, **388**, 234.

JOHNS, R. K. (ed.) 1976. *History and Role of Government Geological Surveys in Australia*. A. B. James Government Printer, South Australia, 1–111.

JOHNS, R. K. & COOPER, B. J. 1995. Obituary: Nelly Hooper Ludbrook (1907–1995). *The Australian Geologist*, **95**(June), 48–49.

JOPLIN, G. A. 1931. Petrology of the Hartley district. i. The plutonic and associated rocks. *Proceedings of the Linnean Society of New South Wales*, **56**, 16–59.

JOPLIN, G. A. 1933. The petrology of the Hartley district. ii. The metamorphosed gabbros and associated hybrid and contaminated rocks. *Proceedings of the Linnean Society of New South Wales*, **58**, 152–158.

JOPLIN, G.A. 1935. The petrology of the Hartley district. iii. The contact metamorphism of the Upper Devonian (Lambian) Series. *Proceedings of the Linnean Society of New South Wales*, **60**, 16–50.

JOPLIN, G. A. 1964. *A Petrography of Australian Igneous Rocks*. Angus & Robertson, Sydney.

JOPLIN, G. A. 1968. *A Petrography of Australian Metamorphic Rocks*. Angus & Robertson, Sydney.

JOPLIN, G.A., RICHARDS, J. R. & JOPLIN, C. A. 1972. *Finding the Age of Rocks*. Angus & Robertson, Sydney.

KELLY, F. (ed.) 1993. *On the Edge of Discovery. Australian Women in Science*. The Text Publishing Company, University of Melbourne.

KENYON, A. S. 1930. Our first Victorian naturalist: Dr. Edmund Charles Hobson. *Victorian Naturalist*, **67** (Oct.) 94.

KERSHAW, A. P. n.d. *Dr Suzanne (Sue) Lawless Duigan, B.Sc., M.Sc., Ph.d. Cantab: 1924–1943*. World Wide Web Address: http://www.ctie.monash.edu.au/hargrave/duigan_suzanne_bio.html

KERSHAW, A. P. (ed.) 1997. Australian Palaeoclimates: Refinement of Estimates from Palaeobotanical Data. Proceedings of the Palynological and Palaeobotanical Association of Australasia conference, Monash University, November 1995. A Special Issue Dedicated to the Memory of Suzanne (Sue) Lawless Duigan (1924–1993). *Australian Journal of Botany*. **45** (3), 1–617.

LANG, W. H. & COOKSON, I. C. 1927. On some Early Palaeozoic plants from Victoria Australia. *Memoirs and Proceedings of the Manchester Literary and Philosophical Society*, **71**, 45–51.

LANG, W. H. & COOKSON, I. C. 1935. On a flora, including vascular land plants, associated with Monograptus, in rocks of Silurian age from Victoria, Australia. *Philosophical Transactions of the Royal Society London B*, **224**, 421–449.

LAWSON, E. 1995. *The Natural Art of Louisa Atkinson*. State Library of New South Wales Press, Sydney, 144 pp.

LINDSAY, J. M. (ed.) 1985. *Stratigraphy, Palaeontology and Malacology: Papers in Honour of Dr Nell Ludbrook*. Department of Mines & Energy, South Australia, Special Publications, **5**, 1–387.

MCCALL, J. 1999. Mary Anning & her times: the discovery of British palaeontology 1820–1850. *Geoscientist*, **9**(10), 4–6.

MCINERNY, K. 1929. The building stones of Victoria: part 2. The igneous rocks. *Proceedings of the Royal Society of Victoria*, **41**, 121–159.

MCKAY, J. (comp.) 1997. *Brilliant Careers: Women Collectors and Illustrators in Queensland*. Queensland Museum, S. Brisbane.

MATHER, P. (ed.) 1986. *A Time for a Museum. History of the Queensland Museum 1862–1986*. Memoirs of the Queensland Museum, S. Brisbane.

MOYAL, A. M. (ed.) 1976. *Scientists in Nineteeth Century Australia. A Documentary History*. Cassell, Australia.

MOYAL, A. 1986. *'A bright and Savage Land'. Scientists in Colonial Australia*. Collins, Sydney, 192 pp.

MOYAL, A. 1993a. Invisible participants. women in science in Australia. *Prometheus II*, **2**, 175–187.

MOYAL, A. 1993b. Invisible heritage. *Australasian Science*, **1993** (Summer), 2–3.

MOYAL, A. (comp.) 1994. *Portraits in Science*. Australian National Archives, National Library of Australia, 203 pp.

MURRAY, SIR K. A. H. 1958. *Report of the Committee on Australian Universities*. Australian Government Printer, Canberra.

NASHAR, B. 1967. *Geology of the Sydney Basin*. The Jacaranda Press, Brisbane.

O'NEIL, B. 1998. *Above and Below: The SA Department of Mines & Energy 1944–1994*. Department of Mines and Energy, Adelaide, Special Publications, **10**, 655 pp.

P. C. M. 1946. The late Dr. Georgiana Sweet, O. B. E. *The Victorian Naturalist*, **62**, 211–212.

POULTER, N. & GOEDE, A. 2001. Who was Edie Smith? *In: A Cave Odyssey*, 23rd Australian Speleological Federation 23rd Biennial Conference, Bathurst, NSW, 28 December 2000–3rd January 2001. Privately printed, 24 pp.

R. T. P. [Prider] 1954. The late Dr K. L. Prendergast. *Geological Society of Australia News Bulletin*, **2**(3), 19.

PROBERT, D. 1975. Obituary: Daniel Moye. *The Australian Geologist*, **5**, 15.

QUODLING, F. 1976. Kathleen Sherrard. *Journal & Proceedings of the Royal Society of New South Wales*, **109**, 168.

RADI, H. 1988. Fanny Cohen. *In*: RADI, H. (ed.) *200 Australian Women. A Redress Anthology*. Women's Redress Press Inc., Broadway NSW, 156–157.

RASMUSSEN, C. (ed.) (with 46 specialists) 2001. *A Museum for the People. A History of Museum Victoria and Its Predecessors*. Museum Victoria, Melbourne, Scribe Publications Pty Ltd, 423 pp.

RICH, P. V., LONG, J., WARREN, A., RICH, T. R. & MOLNAR, R. 1982. An all too brief and superficial history of Australian vertebrate palaeontology. *In*: RICH, P. V. & THOMSON, E. M. (eds) *The Fossil Vertebrate Record of Australasia*. Monash University Printing, Clayton, 2–26.

RICH, T. H. 1999. Australia: vertebrate palaeontology. *In*: SINGER, R. (ed.) *Encyclopedia of Paleontology*. 2 vols. Fitzroy Dearborn Publishers, Chicago, 1700 pp.

Ross, J. P. R. 1957. Geology of the Queanbeyan District, NSW, Australia. *Journal of the Royal Society of New South Wales*, **89**, 116–112.

Ross, J. P. R. 1961. Liscombea, a new Silurian tabulate coral genus from New South Wales, Australia. *Journal of Paleontology*, **3**, 1017–1019.

Ross, J. P. R. 1963. Lower Permian Bryozoa from Western Australia. *Palaeontology*, **6**, 70–82.

Rozefelds, A. & Turner, S. 1998. Dr Mary Wade – Collector and facilitator – Queensland Field Work 1971–1992. 14th Australian Geological Convention, Townsville, July 1998. *Geological Society of Australia, Abstracts*, **49**, 384.

Runnegar, B. 1975. The message of Alcheringa. *Alcheringa* **1**, 1–2.

Runnegar, B. & Jell, J. S. 1983. Dorothy Hill, C. B. E., Ph D, DSc, L. L. D., F. R. S., F. A. A., F. G. S. *Memoirs of the Association of Australasian Paleontologists*, **1**, 9–15.

Sarjeant, W. A. S. 1999. Eisenack, Alfred. In: Singer, R. (ed.) *Encyclopedia of Paleontology*. Fitzroy Dearborn Publishers. Chicago, **1** (A–L), 401–402.

Seddon, G. 1996. Thinking like a geologist: the culture of geology. Mawson Lecture 1996. *Australian Journal of Earth Sciences*, **43**, 487–495.

Sherrard, K. 1942. Upper Ordovician graptolite horizons in the Yass-Jerrawa District, NSW. *Journal and Proceedings of the Royal Society of NSW*, **76**, 252–257.

Sherrard, K. 1944. 'Are You Planning a Science Career?' *Australian Women's Digest*, November, 15–16.

Sherratt, T. 1994. An eye on the past. Finding life in ancient corals – Dorothy Hill. *Australasian Science*, **1994** (Summer), 64.

Sherratt, T. & McCarthy, G. 1992. *A Guide to the Records of Edwin Sherbon Hills. Deposited with the National Library of Australia NLA*. Australian Science Archive Project, Melbourne University. 149 pp.

Simpson, A. 1999. The work and type collections of the Australian palaeontologist, Professor Dorothy Hill (1907–1997). *The Geological Curator*, **7**(2), 51–69.

Stearn, C. W. 1966. The microstructure of stromatoporoids. *Palaeontology*, **9**, 74–124.

Summers, H. H. 1947. Clarke Memorial Lecture: Teachers of Geology in Australian Universities. *Journal of the Royal Society of NSW*, **81**, 122–146.

Taylor, D. 1985. Review of 'Stratigraphy, Palaeontology and Malacology: papers in honour of Dr Nell Ludbrook'. *The Australian Geologist. Newsletter*, **56**, 28.

Teichert, C. 1991. Palaeontology in Australia 50 years ago. *nomen nudum* **20**, 2–4.

Thulborn, R. A. 1990. *Dinosaur Tracks*. Chapman & Hall, London.

Thulborn, R. A. & Wade, M. 1984. Winton dinosaur footprints. *Memoirs of the Queensland Museum*, **21**.

Torrens, H. 1995. Mary Anning (1799–1847) of Lyme: 'the greatest fossilist the world ever knew'. *British Journal for the History of Science*, **28**, 257–284.

Turner, S. 1986. Vertebrate palaeontology in Queensland. *Earth Science History*, **5**, 50–65.

Turner, S. 1994. Women in palaeontology in Australia. In: Branagan, D. F. & McNally, G. H. (eds) *Useful and Curious Geological Enquiries Beyond the World. Pacific-Asia Historical Themes*. 19th International INHIGEO Symposium, Sydney, Australia, 4–8 July. Conference Publications, Springwood, 248–250.

Turner, S. 1996. Women hooked on fossils. *Australasian Science*, **1996**(May), 61–62.

Turner, S. 1998a. Paleontology in Australia. In: Good, G. A. (ed.) *Sciences of the Earth. An Encyclopedia of Events, People, and Phenomena*. Garland Press, New York, 662–666.

Turner, S. 1998b. Women in paleontology in Australia. In: Good, G. A. (ed.) *Sciences of the Earth. An Encyclopedia of Events, People, and Phenomena*. Garland Press, New York, 848–852.

Turner, S. 2005. Obituary Dr Mary Wade. *The Queensland Geologist, Geological Society of Australia, Queensland Division Newsletter*, **105**(4), 4.

Turner, S. 2006a. Dr Elizabeth Arnold Ripper 1909–2004: early 20th century Victorian palaeontologist. In memoriam. *Proceedings of the Royal Society of Victoria*, **117**(2), xlix–liv.

Turner, S. 2006b. Rocky road to success. A new history of the International Geoscience Programme (IGCP). In: Petitjean, P., Zharov, V., Glaser, G., Richardson, J., de Paridac, B. & Archibold, G. (eds). *Sixty Years of Science at UNESCO 1945–2005*. UNESCO, Paris, 297–314.

Turner, S. & Cadée, G. C. 2005. Dr Margaretha Brongersma-Sanders (1905–1996), Dutch scientist: an annotated bibliography of her work to celebrate 100 years since her birth. *Zoologische Mededlingen Leiden*, **80**, 183–204.

Turner, S. & Dettmann, M. 2005. *Dr Isabel Clifton Cookson (1893–1973) – micropalaeontologist, palynologist and palaeobotanist of world acclaim*. For International Botanical Association, USA. World Wide Web Address: http://www.botany.org/newsite/awards/detail/cookson2.php.

Turner, S. & Vickers-Rich, P. 2004. Reg Sprigg, Martin F. Glaessner, Mary Wade and the discovery of the Ediacaran fauna. IGCP 493: *Rise and Fall of the Vendian Biota, Workshop*, Prato, Italy, 30–31, August 2004. IGCP493 Abstracts, 1 p.

Turner, S. & Wade, M. 1986. The records in the rocks. In: Mather, P. (ed.) *A Time for a Museum. The History of the Queensland Museum 1862–1986*. Queensland Museum, Brisbane, 128–149.

Turner, S. & Webby, B. 2004. Dr Elizabeth Arnold Ripper 1909–2004. *The Australian Geologist*, No. 133, (December), 45.

Vallance, T. G. 1978. Pioneers and leaders – a record of Australian palaeontology in the nineteenth century. *Alcheringa*, **2**, 243–250.

Vickers-Rich, P. & Archbold, N. 1991. Squatters, priests and professors: a brief history of vertebrate palaeontology in *Terra Australis*. In: Vickers-Rich, P., Monaghan, J. M., Baird, R. F. & Rich, T. H. (eds) *Vertebrate Palaeontology of Australasia*. Pioneer Design Studios in co-operation with Monash University Publications Committee, Melbourne, 1–43.

Vickers-Rich, P., Monaghan, J. M., Baird, R. F. & Rich, T. H. (eds) 1991. *Vertebrate Palaeontology of Australasia*. Pioneer Design Studios/Monash University Publication Committee, Melbourne.

WADE, M. 1994. Fossil Scyphozoa. *In*: GRASSÉ, P. (ed.) *Traité de Zoologie*. Masson et Cie, Paris.

WATERHOUSE, J. 2004. *University House as They Experienced It. A History 1954–2004*. Australian National University, Canberra, 375 pp.

WARNER, D. (curator) 1978. *Women's Science in Nineteenth Century America*. (Designer Deborah Bretzfelder, Smithsonian Institution Exhibition, September 1978–February 1979.) National Museum of History & Technology, Smithsonian Institution, Washington DC, 16 pp.

WHITE, H. L. (ed.) 1954. *Canberra, A Nation's Capital*. 30th Congress, ANZAAS, Canberia, 13–20 January. Halstead Press, Sydney.

WILKINSON, R. 1996. *Rocks to Riches. The Story of Australia's National Geological Survey*. Australian Geological Survey Organization, Allen & Unwin, St Leonards, NSW, 446 pp.

Rediscovering and conserving the Lower Palaeozoic 'treasures' of Ethel Woods (neé Skeat) and Margaret Crosfield in northeast Wales

C. V. BUREK[1] & J. A. MALPAS[2]

[1]*University of Chester, Parkgate Road, Chester, CH1 4BJ, UK*
(e-mail: c.burek@chester.ac.uk)

[2]*NEWRIGS, Millennium EcoCentre, Borras Airfield, Tarmac,*
Holt Road, Wrexham, LL13 9SE, UK

Abstract: This paper explores, within a historical context, the importance of geoconservation of not only sites but also artefacts, collections and specimens as well as letters and original documents. It sets out the search and finding of sites in northeast Wales and materials thought lost then found and the subsequent nomination of Regionally Important Geological/Geomorphological Sites (RIGS) conservation status of the sites to safeguard them for the future. It is important to note that RIGS can be designated for their historical value alone, which is in contrast to Sites of Special Scientific Interest (SSSIs), which are protected solely for their national scientific and research value.

The role of Ethel Woods (neé Skeat) and Margaret Crosfield in developing an understanding of the geological history of northeast Wales had been lost over time. This paper contains biographical sketches of the two women, followed by their Lower Palaeozoic lithological, structural and graptolite research and places it in an historical context. This case study illustrates how female curiosity, perseverance and attention to detail unearthed previously forgotten treasures.

The importance of conserving their sites, specimens and sketch field notebooks in our electronic and throw-away age is vital. The role of the North East Wales Regionally Important Geological/Geomorphological Sites (NEWRIGS) in conserving this information is put forward as an example of good practice.

During the early 19th century, a scientific revolution occurred which fundamentally changed the thinking about the age of the Earth and geological time. Gradually the concept of the stratigraphic column developed (Oldroyd 1996). However, there were problems with the Lower Palaeozoic, caused by the pride and stubbornness of two eminent men within the field of geology: the Reverend Adam Sedgwick and Sir Roderick Murchison (Rudwick 1985; Secord 1986). The division between the Upper Cambrian and Lower Silurian Periods was jealously fought over as each man sought to impose his ideas over the other's. Following the death of both individuals and, after one of the most acrimonious episodes in the history of geology, a solution was needed.

In 1879, Lapworth's Ordovician Period had been created to solve the stratigraphical dilemma left by Sedgwick and Murchison. In Kuhn's terms (Kuhn 1962), this paradigm precedes normal science, but a scientific revolution needs consensus building. Thus, Lapworth needed further stratigraphic and palaeontological evidence to support his creation. So, in the early 1890s, Professors McKenny Hughes and Marr at the University of Cambridge, together with Professor Lapworth at Mason Science College in Birmingham, (which became the University of Birmingham in March 1900), encouraged a small group of women at Newnham College, Cambridge, to investigate the Silurian and Ordovician rocks of North Wales and the borders in order to support Lapworth's creation and solve the so-called 'Silurian' problem. Their names were Gertrude Elles, Ethel Shakespear (neé Wood), Ethel Woods (neé Skeat) and Margaret Chorley Crosfield (Burek 2007). It was considered suitable and appropriate work for females and required great patience and perseverance.

The classic work of Elles and Wood on British graptolite identification, classification and taxonomy is well known through their monograph on graptolites (under the editorship of Lapworth), which remains a research source nearly 100 years after its publication (Elles & Wood 1901–18). However, the contribution of the other two women in developing and understanding the Silurian and Ordovician stratigraphy of northeast Wales had been lost over time. It is their research we wish to evaluate following a biographical context.

Ethel Gertrude Skeat (1865–1939)

Ethel Gertrude Skeat was born on 14 May 1865, the third daughter of Professor William Walter Skeat (Elrington and Bosworth Professor of Anglo-Saxon at the University of Cambridge) and Bertha (neé Jones). Her sisters were Bertha and Margaret; her brothers were Walter and Arthur.

From: BUREK, C. V. & HIGGS, B. (eds) *The Role of Women in the History of Geology*. Geological Society, London, Special Publications, **281**, 203–226. DOI: 10.1144/SP281.12
0305-8719/07/$15.00 © The Geological Society of London 2007.

She was educated from the age of 12 years (1878) at Bateman House, a private school in Cambridge, until she left in 1885 at the age of 20 years. Professors Marr and McKenny Hughes, colleagues of her father, probably kindled her interest in geology during this time. In 1885 she travelled south to St Leonards on Sea, East Sussex, to study at a private boarding school for a year until she was 21 years old. At the age of 25 years (1891), she went to Newnham College, Cambridge, at the same time as Gertrude Elles and Ethel Wood. There, she also met both her life-long friend and collaborator, Margaret Crosfield, (Fig. 1) and her future husband, Henry Woods. In 1894 she became the Arthur Hugh Clough Scholar and completed the Natural Science Tripos Certificate part 1, gaining a Class 1 at the age of 29 years. In 1893, while she was a student, she joined the Geologists' Association (GA). Also during this time, she collaborated with Margaret Crosfield on their first paper on Welsh stratigraphy in the Carmarthen area, which was published in the *Quarterly Journal of the Geological Society* (Woods & Crosfield 1896). Between 1895 and 1897, she was elected to a Bathurst research studentship, which was awarded from time to time for proficiency in Natural Science. This was not restricted to Newnham students but was an open competition for women. In 1910 it was worth £70 and was normally awarded for a year (Morley 2004). However, Ethel received it 2 years running. During this time she went to the University of Munich, Germany, to work under Professor K. A. von Zittel, the first female student to do so (Kölbl-Ebert pers. comm. 2005). Thus she was a scholar of Newnham College and later an Associate from 1896 until her marriage in 1910.

In Munich she investigated the palaeontology of Jurassic Neocomian glacial boulders from Denmark with Dr Victor Madsen, and this led to an important paper in *Danmarks Geologiske Undersogelse* (Skeat & Madsen 1898; Skeat 1904). She also travelled to Paris, Geneva and Lausanne as a geological research worker.

While in Munich, she must have paved the way for, and met if not overlapped with, the most prolific female Scottish geologist of the 19th century, Maria Ogilvie, who received her doctorate in 1900 (Burek 2005; Wachtler & Burek 2007). Although no documentary evidence has so far been found they were roughly the same age and both were British.

By 1898, Ethel, now 35 years old, was employed as a science schoolmistress and boarding at 68 Stanwell Road, Penarth, Glamorgan, Wales, with landlady Mrs Emma Perry (1901 Census). Here she met and boarded with Beatrice Elizabeth Clay, the headmistress of Seconder Public School, Penarth, who was to become her life-long friend; it was Beatrice Clay who became the executor of her will and to whom she left all her jewellery and clothes on her death (Crosfield 1939).

From South Wales she then moved north and across the border into England, following her friend Beatrice Clay, who had left in March 1903 to take up the position of headmistress of Queen's School, Chester. Miss Clay held this position until she retired in 1925. Her subjects were English literature and scripture (Queens' School staff register). Meanwhile, in January 1904, Ethel had been given the position of second mistress, head of science and geography, and in charge of the sixth form at Queen's School, Chester, at a wage of £130 (Queen's School 1910). Her special subjects are listed as botany, geology, chemistry, zoology, French and German (Queen's School staff register). These last two were probably a result of her sojourn overseas. However, the subjects she taught and was responsible for are listed as botany, geology, nature study, geography, chemistry and zoology. The experience she gained from teaching geography shows itself later in life in her publications (Woods 1922).

Ethel Skeat carried on her geological work with Margaret Crosfield during this time and together they went on GA field trips. Figure 1 shows the two women sitting together near Oswestry, in the North Wales borderlands, in 1908. This was very close to their geological study area at the time.

During her 6 years in Chester, Ethel entered into the life of the city by joining the Chester Society of Natural Science, Literature and Arts between 1904 and 1909 and gave at least three talks to the Society, each illustrated with lantern slides. These talks were as follows:

- 1905: 'Jurassic shorelines; or a fragment of world history';
- 1909: 'Bernese Oberland';
- 1910: 'Life's failures'.

Fig. 1. Margaret Crosfield and Ethel Skeat sitting together in 1908 near Oswestry, in the North Wales borderlands on a Geologists' Association fieldtrip (courtesy of Geologists' Association).

Interestingly her affiliations for her talk on 14 April 1905 are listed under the geological section of the society as 'Assoc. Newnham Coll., Mem. Geol. Association' but, by 31 March 1910, she has become 'Miss E. G. Skeat, DSc' and her affiliations are not mentioned (*Chester Society of Natural Science, Literature and Arts Annual Report, 1909–1910*). In May 1905, Ethel Skeat had been awarded a DSc by Trinity College, Dublin, in 'recognition of her contribution to geological research'. This was a common practice at that time because the University of Cambridge did not award women higher degrees, and this tradition continued until 1948 (Higgs *et al.* 2005; Higgs & Wyse-Jackson 2007).

As further evidence of her social interaction, on 7 April 1905 she gave a demonstration to schoolchildren at the Grosvenor Museum on recent exploration in Greenland, using materials lent to her by the Danish authorities (Queen's School 1905). This was a sign of things to come.

Ethel Skeat's research continued while she was teaching at Chester and, in recognition of her achievements, she was awarded the Murchison Fund in 1908 by the Geological Society of London. It amounted to £25.6s.4d. She was the first English woman to receive this fund and the eighth woman to receive any kind of funding from the Geological Society. The citation by Archibald Geikie in 1908 reads:

... the Murchison Geological Fund as a mark of appreciation of your geological work, especially among the glacial deposits of Denmark and the Lower Palaeozoic rocks of Wales. It is with much gratification that we hail in you another woman who is worthily placed on the roll of those who have gained our awards.

In this, he was referring to former Murchison Funds awarded to Jane MacDonald in 1898 (£28.14s.3d) and Elizabeth Gray in 1903 (£22.15s.10d). Interestingly, both these women were from Scotland. In her obituary, which appeared in the roll call of Newnham College on her death in 1939, Gertrude Elles erroneously states that Ethel was the first recipient of the Fund (she probably meant 'English' recipient) and Volume 1 of the *Newnham College Register 1* for the year 1891 also states that it was the Murchison Medal that she received. Original sources show this to be incorrect (Geikie 1908).

Ethel left Chester on 17 June 1910, when she returned to the place of her birth, Cambridge, to marry Professor Henry Woods, the famous palaeontologist. She was 45 years old. As a good and popular teacher 'with what admirable results is testified not only by the excellent School reports, but – better far – by the enthusiasm she has evoked from her pupils', she was wished well by all her friends and colleagues. 'She has always entered, heart and soul, in the interests, the work and the play of the School, and she will be sorely missed: but we must all unite in wishing that our loss may be her gain' (Queen's School 1910). Ethel had met Henry Woods in Cambridge during her student days. He had entered St John's College in 1887 and became a demonstrator in geology there in 1892.

In 1911, a few months after her marriage, she became a lecturer at the Cambridge Training College for Women, where she remained for 2 years. She returned in 1919, after World War 1, and was Registrar and Honorary Secretary until 1937, a year before her death.

During the war her knowledge of German found her working in the postal censorship code department. This was also true of Maria Ogilvie Gordon, who had been in Munich with her during their student days (Burek 2005).

After the war she carried on her research in North Wales with Margaret Crosfield. The period of her life which is the subject of this study revolves around the time she moved to Chester and worked with her friend and colleague Margaret Crosfield. This culminated in the publication of their paper in the *Quarterly Journal of the Geological Society* in 1925, which she read on 7 May 1924 as Mrs Woods and which was entitled 'The Silurian rocks of the central part of the Clwydian range' (Woods & Crosfield 1925).

Margaret Chorley Crosfield (1859–1952)

Margaret Chorley Crosfield was born on 7 September 1859 in Reigate, Surrey, where she lived all her life. She was educated at The Mount School, York, a Quaker school founded in 1785 and one of the first schools to send girls to university. Later she went to Newnham College, Cambridge. She entered Newnham in 1879 at the age of 20 years but left a year later. Her studies there were interrupted by ill health (Johnston 1953) and she returned to complete her studies 10 years later, in 1890. This meant she was significantly older than some of the friends she made there. With the permission of the authorities she only took geology as a subject, not the usual natural tripos.

At Cambridge she made some life-long friends: Ethel Skeat (later Mrs Woods), who was 6 years younger, and Mary Johnston, who was 16 years younger. She researched on the Welsh borders and Carmarthen and also published papers with both of them.

She was also a contemporary of Gertrude Elles, who was 13 years younger, and Ethel Wood (later to become Dame Ethel Shakespear), who was 12 years younger. Figure 2 shows the Dublin British Association participants in 1908, with Gertrude Elles in the front row and Margaret Crosfield two rows behind. Professors Marr and McKenny Hughes

Fig. 2. The Dublin British Association participants in 1908 with Gertrude Elles in the front row and Margaret Crosfield two rows behind (courtesy of Geologists' Association).

influenced all these young women in their studies. Later, it is interesting to note that, in her will, Crosfield mentions both Gertrude Elles and Mary Johnston (Crosfield 1939).

She was an active member of the GA, which she joined in 1892; 17 years later she was among the first group of women to be elected Fellows of the Geological Society of London. She became a GA Council member in 1918 and the GA librarian from 1919 until 1923 (Newnham College Register 1878). Figure 3 shows her on a GA field trip in 1912 to Leith Hill with Professor Lapworth. She was also a member of the Palaeontological Society from 1907 to 1932 (Elles 1953; Johnston 1953).

Margaret Crosfield published three important papers. The first, which was on Carmarthen with Ethel Skeat (Woods & Crosfield 1896), was regarded as a very fine piece of work and of sufficient depth and clarity that it formed the basis of the geological map produced by the British Geological Survey for the area. In 1914 she published a work on the Wenlock limestone of Shropshire, with her friend Miss Mary Johnston and then, in 1925, she published her second paper with Ethel Skeat (now Mrs. Woods) on the geology of the Silurian rocks of the Clwydian Range (Woods & Crosfield 1925) and the subject of this paper.

Margaret had a great interest in education. She served on the Reigate Borough Council Education Committee (Newnham College Register, 1878; Johnston 1953) and was a school governor of the Reigate Girls County School (Burek 2007). Margaret was also a great promoter of women's suffrage. Indeed some of her field notes are written on the back of suffragette notepaper (Fig. 4).

She travelled widely and attended many meetings. Her training at Cambridge ensured that she kept meticulous notes, field notebooks and records of specimen locations. This was remarked upon in her obituary by Mary Johnston:

> ... but by far the most outstanding of her many characteristics was her wonderful accuracy and industry, as shown in her voluminous, perfectly and artistically-kept note-books, and her large collection of specimens, all of which are place-marked, thus ensuring them of permanent value.

These can still be seen at the British Geological Survey in Keyworth and at the Grosvenor Museum in Chester. It was her distinctive labels that led to the discovery of her previously lost specimens in the museum (Fig. 5). She also presented specimens from the Silurian of Wenlock Edge to the geology department of the University of Liverpool, in 1919 and to the Natural History Museum in London between 1906 and 1933. These were mostly invertebrates, as shown in a letter dated 1933 (Fig. 6). She maintained her interest in palaeontology to the end of her life and presented

Fig. 3. Margaret Crosfield on a Geologists' Association fieldtrip to Leith Hill with Professor Lapworth (courtesy of Geologists' Association).

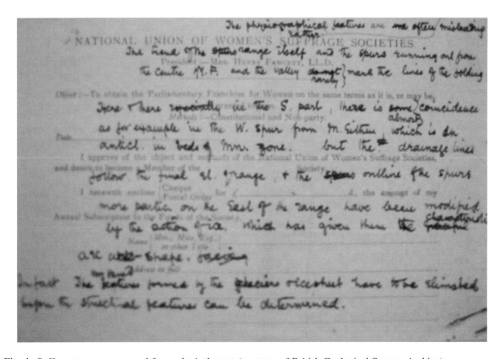

Fig. 4. Suffragette notepaper used for geological notes (courtesy of British Geological Survey: Archive).

Fig. 5. Distinctive labels on the specimens helped identify the lost specimens in the Grosvenor Museum Chester (courtesy of Grosvenor Museum, Chester).

Palaeozoic invertebrate fossils to the Museum in 1937 at the age of 78 years!

Margaret was keen to share her knowledge and led two field excursions to Reigate, Surrey, for the GA in 1899 and 1911. She also lectured on many subjects to her local Holmesdale Natural History Society, as well as to other local societies. She exhibited material at the GA in 1919, where she

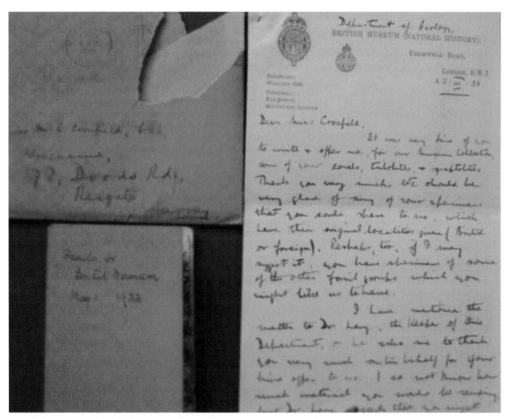

Fig. 6. Letter from the British Museum (Natural History) accepting specimens from Margaret Crosfield (courtesy of British Geological Survey: Woods & Crosfield Archive).

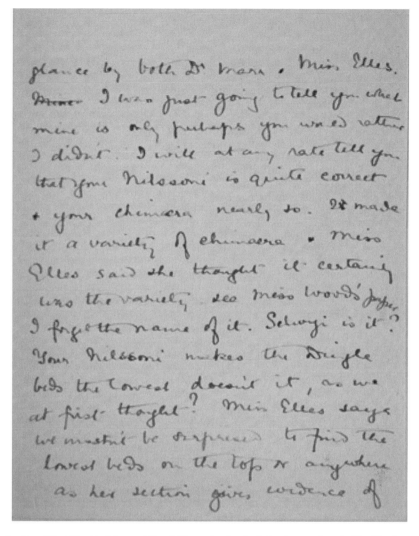

Fig. 7. Letter from Ethel Skeat to Margaret Crosfield, mentioning the help given to them in identification by Gertrude Elles (courtesy of British Geological Survey: Woods & Crosfield Archive).

displayed 'Worked flakes from the fields near Savernake Forest and from the neighbouring Wiltshire Down' (Wright 1919).

Although the second paper Margaret wrote with Ethel Skeat was published in 1925, work for this research started much earlier. They began work for the paper by visiting sites between 1906 and 1909, when Ethel was teaching at the Queen's School, Chester, and again in 1911 and in 1922, after Ethel had married. There is evidence that they had help with some of the identification of the graptolites from Gertrude Elles and Ethel Wood, who were part of the old girl network from Newnham College, from a letter (Fig. 7) found in the archives of British Geological Survey.

Obviously Margaret Crosfield, Gertrude Elles, Ethel Skeat (Mrs Woods) and Ethel Wood (Dame Shakespear) all knew each other as students at Cambridge and collaborated with each other while they were alive. Even death did not separate this tight network. When Ethel Skeat died in January 1939, Gertrude wrote her obituary for the *Proceedings of the Geologists' Association* and the *Newnham College Roll Letter* (Elles 1940) where she also refers to 'her friend Margaret Crosfield'. Perhaps they were the first female old girls, geological network! This then is an exposé of their work.

Discovery

The present research stems from an unrelated investigation into the early history of the new University of Chester. In 1872, the Reverend Charles Kingsley, when he was Canon of Chester Cathedral, gave six lectures to the young men of Chester with titles such as 'The lime in the mortar' and 'The slates on the roof'. These were published under the title *Town Geology*, or 'the poor man's science' as Kingsley described it. In the preface to this book he stressed the importance of studying the natural sciences and so these talks were reported in the annual report of the Chester Society of Natural Science, Literature, and Arts. This publication also contained the *Annual Report of the Curator and Librarian for 1925/6*, dated 27 May 1926, from Alfred Newstead F.R.S. [sic], the curator and librarian of the Grosvenor Museum, Chester. Alfred Newstead stated that

'Mrs. Ethel Woods and Miss Margaret Crosfield had donated their collection of graptolites to the Grosvenor Museum in Chester and a copy of their reprint Silurian rocks' etc. It is a 'valuable addition' (Newstead 1926, p. 9).

Initially these samples could not be located. However, recent recataloguing of the Grosvenor Museum fossils revealed a collection of graptolites, many of which were identified as collected from sites listed in Woods & Crosfield (1925).

Figure 5 illustrates one of the fossils from the Grosvenor Museum, Chester, with its distinctive label, while Figure 8 illustrates the final map from the publication used to locate the sites where their specimens were found.

Further work showed that the major part of the Woods and Crosfield collection was lodged with the British Geological Survey, Keyworth. These had been donated by Professor Henry Woods of the Sedgwick Museum, Cambridge, following the death of his wife Ethel Woods in 1939. A further collection of specimens and field notebooks was donated/purchased in 1952 following the death of Margaret Crosfield.

The Woods & Crosfield (1925) paper includes a map and list of the localities visited for their research. The details in Margaret Crosfield's field notebooks lead to the rediscovery by the authors of many of their field localities (Fig. 9). In several cases, these localities were overgrown and difficult to identify, as there are many small quarries in the Clwydian Range and no grid references. Subsequently, nine sites have been notified as Regionally Important Geological/Geomorphological Sites (RIGS) under scientific and historical criteria. This is the only conservation status that specifically uses historical scientific importance as a designation criterion. The sites are those that represent the stratigraphy of Woods & Crosfield (1925) and also have graptolite specimens in the Grosvenor Museum, Chester. Figure 10 illustrates the sketches by Margaret Crosfield of the fossils they found in the Clwydian Range.

The geological setting of Woods and Crosfield

In 1879, Lapworth clarified the Cambrian–Silurian boundary 'problem' by introducing the Ordovician Period to solve the stratigraphic dilemma left by Sedgwick and Murchison over the 'upper Cambrian–lower Silurian' beds.

Prior to 1885, there had been little work done on the Silurian stratigraphy of the Clwydian Range. Professor McKenny Hughes had produced a careful lithological description but Woods & Crosfield (1925) thought that the structure was 'less simple than he describes it'.

Woods and Crosfield had difficulty in determining the stratigraphy because of the 'bewildering similarity of the strata' (Woods & Crosfield 1925). This was due to repetition of the strata at various horizons. In their paper, Woods & Crosfield (1925) based their stratigraphy on the Glyn Dyfrdwy, Nant-y-bache and Vivod Groups of Wills & Smith (1922) (Figs 11 & 12). In contrast, they described the succession in terms of the graptolites found rather than the lithology (Fig. 12). Although Woods and Crosfield used graptolites for their stratigraphy, they do state that the graptolites were generally ill preserved and allowed recognition of only one horizon in the lower beds. Figure 13 is a copy of the detailed graptolite table from the field notebooks (Woods & Crosfield 1925).

The general sequence in the Moel Famau area described by Woods & Crosfield (1925) consists of: a basal unit of uncleaved mudstones with flaggy and gritty beds, and cleaved mudstones with finely laminated siltbands; the *Monograptus nilssoni* Zone (the Slab Horizon of Wills & Smith 1892), which Woods and Crosfield divided into the Y Fronwerth Beds; 'hard banded grey flags' found in the Coed Ceunant area and the Bwlch pen-Barras Beds; and 'blue, grey slaty rock' which is well exposed in a quarry south of Moel y Gaer and Old Mold Road quarry. The Rhiwisg Beds, overlying the 'Slab Horizon' in the Moel Famau area are fine-grained, blue-grey flags or shales with bands of whitish silty material. The Y Fronwerth, Bwlch-pen-Barras and lower Rhiwisg Beds are now placed in the Nantglyn Flags Formation, Lower *M. nilssoni* Biozone (Fig. 11).

Woods and Crosfield divided the overlying units into six beds.

1. The Teiran or Passage Beds, present above Teiran Farm, west of Pen Barras, are described as having an intermediate lithology between

Fig. 8. The Woods and Crosfield map of the central part of the Clwydian Range with the localities and some of their lost specimens (Woods & Crosfield 1925) (courtesy of British Geological Survey: Archive).

those under and overlying units and consist of 'grey, sandy flags, unevenly bedded, and occasionally containing decomposing bands similar to those of the "Whetstone" below'. Certain horizons are fossiliferous containing abundant fragments of brachiopods and crinoids.

2. Overlying the Passage Beds is the *M. scanicus* Zone consisting of the Penmachno Beds; hard, pale grey fossiliferous 'flaggy' sandstone beds exposed at their maximum thickness at Penmachno in the north of the area, near Moel Arthur.

3. The Plâs-newydd flags, grey sandstones and shales with 'contorted beds', which are also present in the Penmachno area and represent the *M. scanicus* horizon, are a series of flags and shales containing the 'true *M. scanicus* fauna'.

4. The *M. tumescens* Zone, overlying the *M. scanicus* Zone, consists of the Eithinen mudstones 'coarse twisted rock, lenticular with large brachiopods in masses' and the Cefn Goleu flags 'clear blue-grey flags banded with yellowish or greenish sandy layers'.

5. The *M. leintwardinensis* Zone is represented by the Bryn-y-saeson flags and shales, which

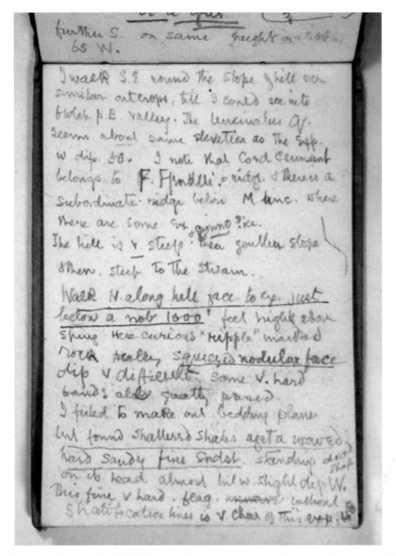

Fig. 9. Detail from the Margaret Crosfield field notebooks, showing her attention to detail (courtesy of British Geological Survey: Woods & Crosfield Archive).

are only present in the Moel Famau area and consist of 'pale-grey micaceaous banded flags and sandy shales'.

6. The upper Rhiwisg Beds and the overlying beds are now placed in the Elwy Formation.

Woods & Crosfield (1925) describe some lateral variation in some of these beds with others, for example the Bwlch-pen-barras flags and shales, which are only present in the Moel Famau area (Fig. 12).

Woods and Crosfield summarized their detailed findings as having been subjected to:

...great compressional and torsional forces which, in this case, have thrown beds into a succession of anticlines and synclines, crossing the present range in directions varying from north-north-west to south-south-east... permitting rocks to move with some individual freedom. (Woods & Crosfield 1925)

In the discussion that followed the presentation of the paper, Wedd, who later wrote the Wrexham memoir (Wedd et al. 1927), congratulated the authors on 'the completion of a decidedly arduous undertaking, and on the interesting results obtained (Woods & Crosfield 1925). It is clear from the discussion that the origin of the now named 'disturbed beds', cleavage, crushing, shearing, folding and faulting of the beds was far from being understood.

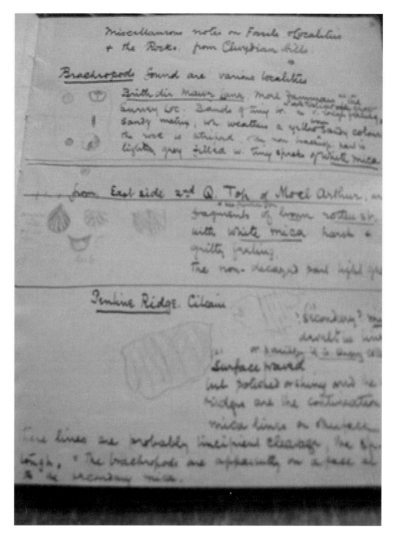

Fig. 10. Details of the fossil sketches in Margaret Crosfield's field notebooks (courtesy of British Geological Survey: Woods & Crosfield Archive).

Woods & Crosfield (1925) also describe, in some detail, the faults, cleavage and rock structures of the Clwydian Range. Figure 8 was the final map they produced. Figures 14 and 15 show the evolution of their ideas on structures during the course of 1923 from March through to June.

The more than 30 notebooks in the British Geological Survey's collection reveal that Woods and Crosfield were excellent field geologists. They observed and questioned constantly. For example on 16 June 1907 an entry in the notebooks states:

Then EGS after tea rode away and I walked by the road to Caerwys Church which is a mile from Caerwys station on the limestone hills as a Welsh service was progressing I did not stay – but returned via foot path to C. Station, enjoying the view down the Wheeler Valley into the Vale of Clwyd. And the high W mts afar off. How has the Wheeler managed to cut through the C. range while the other stream have not and we only have the high Bwlchs with deep consequent streams from them. Is the Alyn, a subsequent or a consequent from an older range? Did the Wheeler once flow out to the sea independently of the Clwyd and as the Alyn? Close to the C. Station are great depths of a sandy deposit 60? Ft thick – glacial, presumably. The dip should be noted of these sands in the valley so as to find out the direction of the currents. I noted that the Wheeler valley across the range is a broad and well worn one, not a narrow gorge. (Crosfield 1907)

In preparing the 1925 paper, they shared the work, as is evident from these field notebooks. Often Ethel is taking the dip measurements while Margaret is writing. Margaret Crosfield's field

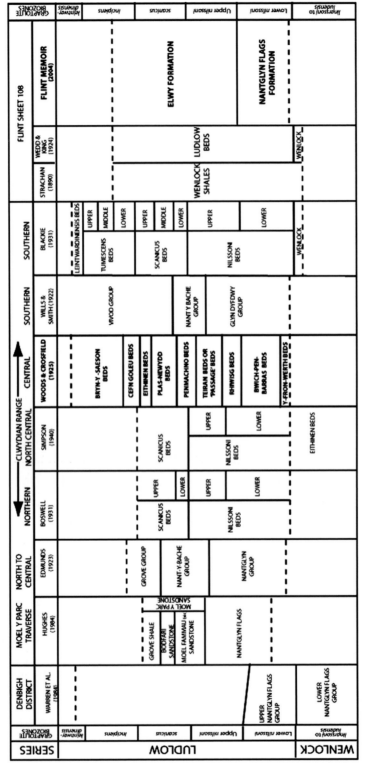

Fig. 11. Current and former classifications of Silurian rocks of the Clwydian Range and adjacent areas (adapted from Davies et al. 2004).

WILLS & SMITH			MOEL EITHINEN AREA	MOEL FAMMAU AREA	MOEL DYWYLL AREA	MOEL ARTHUR AREA	BRYN GOLEU AREA
VIVOD GROUP	Bryn-y-sason flags and shales	*Monograptus leintwardinensis* Zone		Pale-grey micaceous banded flags and sandy shales. These flags weather to a rich brown and golden yellow			
	Cefn Golen flags	*Monograptus tumescens* Zone	Clear blue-grey flags banded with yellowish sandy layers. Sometimes large nodular masses in the bedding planes.	Clear blue-grey flags banded with yellowish or greenish sandy layers. Bedding planes sometimes curved owing to an inclination to nodular structure.		Flaggy beds, containing sandy layers; variegated surfaces, with coloured circles and ovals; speckly fragments	Hard flags splitting irregularly; dark with lighter bands; sometimes lenticular or nodular
	Eithinen mudstones		Dark mudstones in thick blocks with irregular surface; many cavities with broken brachiopods and crinoids.	Coarse twisted rock, lenticular, with large brachiopods in masses	Hard irregularly-bedded mudstone, full of crinoids and brachiopods		
NANT-Y-BACHE GROUP	Pennachno flags	*Monograptus scanicus* Zone	Bluish-grey banded flags	Grey flags, with light brownish layers	Thin sandstones and thick shales; sandstones broken up into lenticular blocks and shales twisted round them	Grey flags with pale sandy layers, splitting into biscuity fragments	Grey flags with yellowish or brownish sandy layers; sometimes lenticular or nodular
	Plas-newydd sandstones		Thin sandstones and thick shales, violently contorted; sandstones broken up into lenticular blocks. Massively-bedded tough greenish sandstone, with finely banded partings of sandy shales. Sandstone bands up to 1 foot thick.	Contorted beds, probably representing thin sandstones and shales. Hard pale-grey sandstones in bands about 2 feet thick; narrow bands of uneven flags, with shaly parings.	Hard pale-grey bands of sandstone up to 4 feet thick, forming features; narrow, bands of shales interbedded	Thick contorted shales twisted round occasional blocks of sandstone. Fine-grained thickly-bedded sandstones (4 to 8 feet) with crushed shales cleaved at high angle	Some contorted shales beds
	Teirarn or Passage Beds			Bluish-grey bedded grits about 1 foot thick, with shales cleaved at a steep angle.	Bluish-grey bedded grits about 1 foot thick; sandstones often cleaved at a steep angle.	Thinly-bedded sandstones and flaggy shales; sandstones often current-bedded; gingerbread-coloured rottenstone containing small brachiopods	Grit-bands, with shales cleaved at high angle
	[Moel Fammau Sandstones of Hughes]			Hard blue-grey gritty micaceous flagstones, with fine pale-grey bands; unevenly bedded, sparsely fossiliferous. Some grit bands.	Grey flags, either thickly bedded or showing fine pale-grey laminae; nodular lumps of decomposed sandy material; some crinoid fragments	Tough flags showing bright sandy bands alternating with more shaly beds, current-bedded; some brown decomposing rock containg mica and brachiopods	
	Passage Beds, intermediate as regards both lithology and fauna		Banded flags much decomposed and nodular, surfaces showing brightly coloured ovals and circles	Fine-grained blue-grey flags or shales, with bands of whitish silty material. Some surfaces show circles or ovals of brown or grey. Blue-grey slaty rock, with spaced bands of lighter material alternating; sometimes well-cleaved. Hard, banded, grey flags weathering yellowinsh or reddish-brown, with lighter and darker lines alternating.	Grey flags, with whitish or yellowish bands; surfaces showing decomposing ovals and	Flags marked by pale stripes and of 'whetstone' appearance. Tough grey mudstones, with occasional lines of bedding, weathering to a buff, sandy colour; cleavage generally poor and irregular.	Flaggy beds with decomposing surfaces showing differently coloured circles concentric. Hard pale-grey flags, weathering reddish-yellow; occasional lighter bands, often lenticular in shale;
GLYN DYFRDWY GROUP	Yfron-Bwlch-pen-werth Barras flags	*Monograptus nilssoni* Zone	[SLAB HORIZON (of L J Wills & B Smith)]	Tough dark-grey mudstones, with spaced bands of silty material resembling music-ruled paper. Cleavage variable; sometimes poor, and sometimes very well developed.			
	Rhiwig Beds						
	Ngf, Nantglyn Flags						

Fig. 12. Biozone and lithological classification of the central part of the Clwydian Range (Woods & Crosfield 1925). **Key:** Permo-Triassic KnS, Kinnerton Sandstone Fm. Carboniferous Erb, Erbistock Fm; Gwp, Gwespyr Fm; Hsh, Holywell Shale Fm; Cfs, Cefn-y-fedw Fm; ICFS, Lower tongue Cfs; MrF, Minera Fm; CML, Cefn Mawr Fm; LoL, Loggerheads Fm; LeL, Leete Fm; LmL, Llanarmon Fm; Lfs, Llyn-y-frân Sandstone; Fol, Foel Fm; BB, Basement Beds. Silurian EY, Elwy Fm; Dd, Disturbed Beds; sa, thickly bedded sandstone; Ngf, Nantglyn Flags.

Fig. 13. The table of graptolite localities produced by Woods and Crosfield (courtesy of British Geological Survey: Woods & Crosfield Archive).

notebooks (1907/8) contain meticulous detail, illustrated with sketches, descriptions and evidence of their careful sampling (Fig. 16). There are also frequent personal details about their accommodation and state of health (Fig. 17). A quote from the field notes of 7 August 1908:

7th visit Stayed at Miss A. J. Jones Tairan Farm Llanbedr. At 30/- a week. This is a v. conv loc. But the sitting room is poor. Ethel's eyes bad.

Arrived Aug 7th 1908 Friday from Oswestry G. Ass. Ex. (Fig. 1). On Saturday Aug 8 1908 worked in the crushed beds by wall at foot of hill slope just above Tairan (sic) (Crosfield 1908)

However, it is the personal details of life in Wales at that time that make them invaluable. A quote from Sunday 9 June 1907 illustrates this point:

After tea Ethel started on her bicycle about 4.15. I rested till about 5.30 and then went to Church at Nannerch the rector Mr Watkins Williams is 89. A fine old man, He has just ordered a brick grave to be made for himself. (Crosfield 1907)

The descriptions and interpretations presented in Woods & Crosfield (1925) form the basis of the structure described in the Flint memoir (Davies

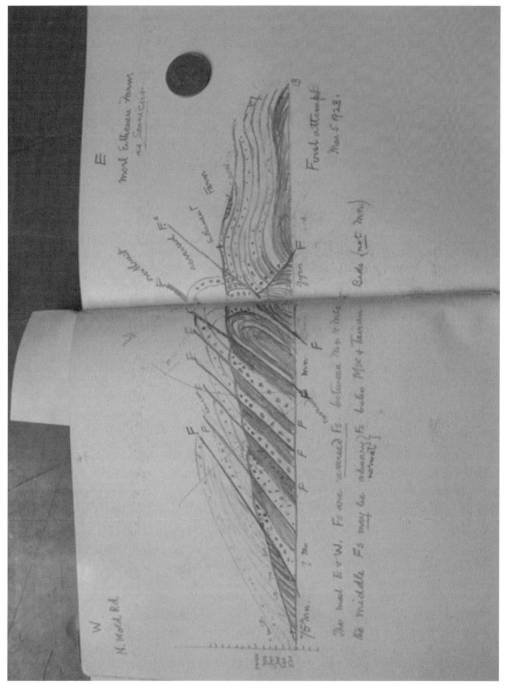

Fig. 14. The first attempt at the structure of the Clwydian Range March 1923 (courtesy of British Geological Survey: Woods & Crosfield Archive).

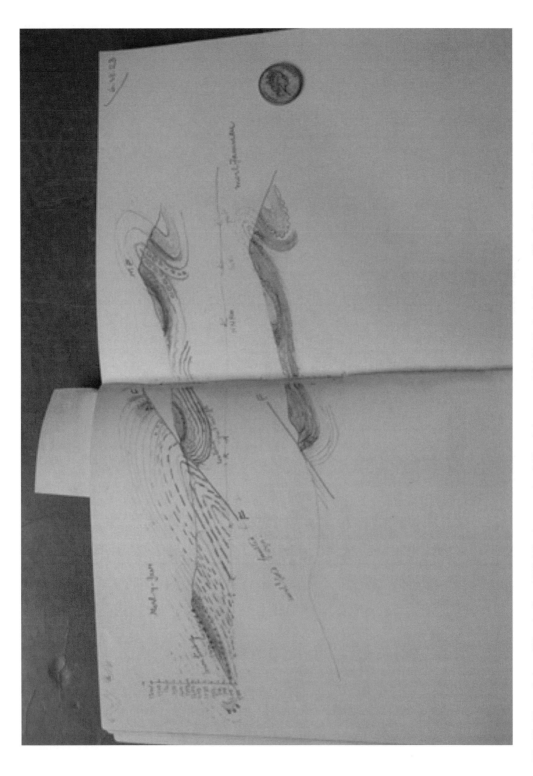

Fig. 15. The evolution of ideas by June 1923 following extensive fieldwork (courtesy of British Geological Survey: Woods & Crosfield Archive).

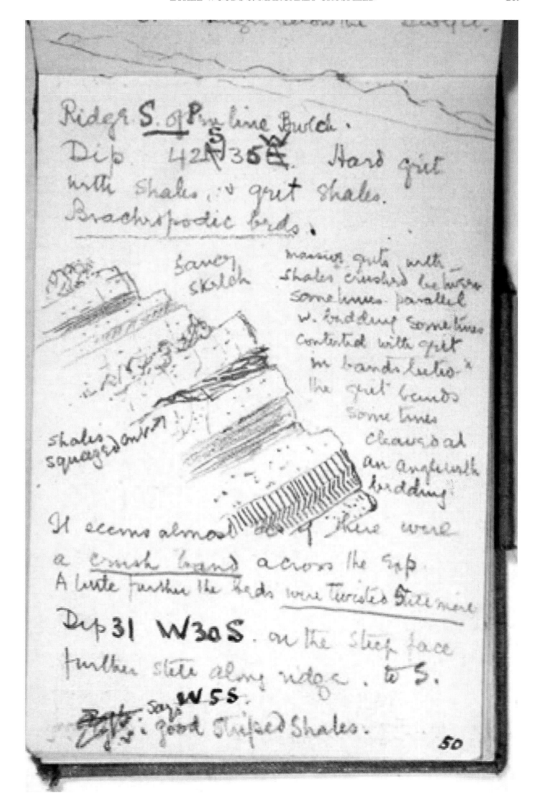

Fig. 16. Detailed geological observations from the field notebooks (courtesy of British Geological Survey: Woods & Crosfield Archive).

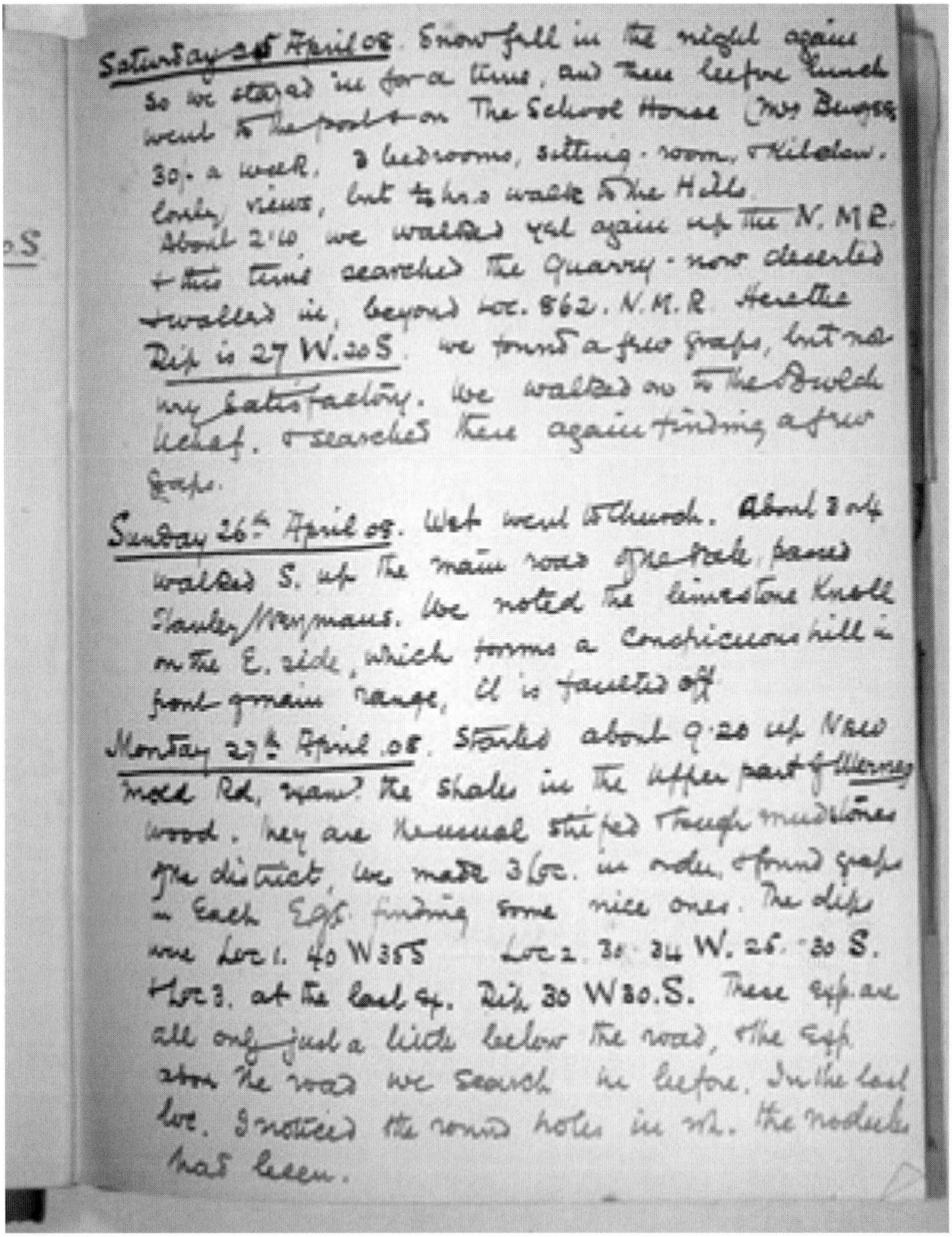

Fig. 17. Details from the field notebook of personal details bringing the fieldwork alive (courtesy of British Geological Survey: Woods & Crosfield Archive).

et al. 2004) and Flint Sheet 108 (Figs 18 & 19). Interpretation of the processes that caused the complex structure of the Silurian strata in the Clwydian Range had to wait until the theory of plate tectonics.

Current geological context

The area that Ethel Woods (neé Skeat) and Margaret Crosfield researched was centred on Moel Famau, in the Clwydian Range, Flintshire,

North Wales, and their work forms the basis of the current understanding of the geology of the Clwydian Range. This area is now an Area of Outstanding Natural Beauty (AONB) and is crossed by Offa's Dyke Path, a long-distance walk.

The Clwydian Range consists of Silurian strata that are exposed on the eastern side of the Vale of Clwyd. The Clwydian Range is bounded by two major faults: to the east by the Alyn Valley Fault, which separates the Silurian from the downthrown Lower Carboniferous limestone, and to the west by the Vale of Clwyd Fault, with the downthrown Permo–Triassic sandstones; thus the Clwydian Range is a prominent horst block (Fig. 18).

During much of the Late Pre-cambrian and Early Palaeozoic, Wales was influenced by crustal subsidence and sediment accumulation and is known as the Welsh Basin. This elongated, intracratonic basin trends northeast–southwest and, together with the rest of southern Britain, formed part of the supercontinent of Gondwana until earliest Ordovician times. It was located on the southern side of the Iapetus Ocean (Cocks & Fortey 1982). In the early part of the Ordovician, the southern part of Britain was subject to rifting and the microcontinent of Eastern Avalonia broke away from Gondwana (Cocks & Fortey 1982). It was during the Ludlow Stage (Upper Silurian), when this area was part of the Iapetus Ocean, that the mudstones of the Clwydian Range were deposited (Davies et al. 2004). The sediments were subject to a period of compressional deformation during the Late Caledonian (Acadian) orogeny as Eastern Avalonia collided with Laurentia (Soper & Woodcock 1990).

There is an estimated 1000 m of Silurian, Ludlow Stage rocks in the district that were deposited in the northern part of the Welsh Basin in the Denbigh Trough (Cummns 1959). They comprise a basinal sequence of turbidites and hemipelagites, as well as slumped and destratified beds known as the 'disturbed beds', which are thought to be derived from the Irish Sea Platform, which was to the west and NW of the district (Davies et al. 2004).

In NE Wales, the principal outcrop of Silurian strata runs the length of the Clwydian Range and comprises only two formations: the Nantglyn Flags Formation and the overlying Elwy Formation (Figs 11 & 19). There are discrete successions of sandstone turbidites and 'disturbed beds' in both formations but they are most common in the Elwy Formation (Davies et al. 2004).

The Silurian strata in the Clwydian Range are composed mainly of four facies.

1. Ribbon-banded mudstones (Warren et al. 1984): rhythmically interbedded silty mudstones and sandstones with subordinate thin calcareous siltstones.
2. Laminated muddy siltstones (Warren et al. 1984): irregular alternations of thinly bedded silty mudstone, siltstone and fine-grained sandstones with very subordinate beds of muddy siltstone (hemipelagite). These are also blue-grey but are coarser grained, relatively carbon-rich and yield graptolites.
3. Sandstones: both thin and thick-bedded facies.
4. Disturbed beds: with complex slump folds, disrupted bedding, internal slide planes and units where all primary sedimentary structures have been destroyed. These beds have been interpreted to represent slope failure, downslope mass movement and folding and destratification due to load or seismically induced diapirism (Eva & Maltman 1994).

The ribbon-banded mudstone facies of the Nantglyn Formation reflect hemipelagic deposition in an anoxic environment on a mud-dominated slope apron interrupted periodically by turbidity current (Davies et al. 1997). Turbiditic siltstones and sandstones increase in frequency in the upper part of the Nantglyn Formation and, together with the presence of a sequence of thin-bedded sandstone facies, are interpreted to reflect a change in sedimentation pattern that resulted in the deposition of the Elwy Formation (Davies et al. 2004).

The Elwy Formation composed primarily of striped silty mudstone facies. These are thinly bedded turbidite mudstones, siltstones and sandstones with minor laminated hemipelagites, a lower proportion than the Nantglyn Formation (Davies et al. 2004). There are also laterally impersistent disturbed beds and units of thin- and thick-bedded sandstones. The depositional environment is interpreted as recording a major increase in the grade and volume of sediment supplied to the slope apron, resulting in thick-bedded sandstones and disturbed beds. The continued presence throughout the Elwy Formation of thin beds of laminated hemipelagites suggests that anoxic bottom conditions prevailed at this time (Davies et al. 2004).

Importance of geoconservation and RIGS

Geoconservation is about conserving natural diversity; it is the active management of geological resources (Prosser 2002; Gray 2004). Geoconservation is proactive and can allow change and investigation, which may include the destruction or removal of material, for example, fossils; in this respect, geoconservation mirrors archaeology. This is in contrast to biological conservation, which constantly seeks to enrich, enhance and enlarge species numbers and habitat area. Biological conservation cannot allow the destruction or removal of material.

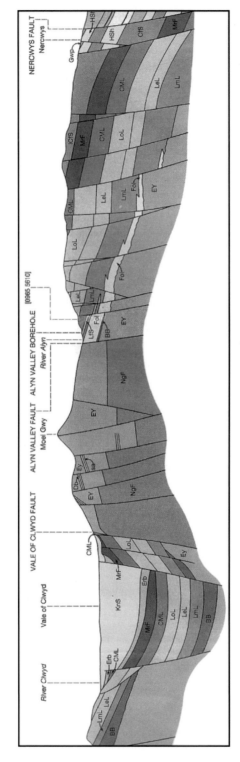

Fig. 18. Cross-section of Clwydian Range from British Geological Survey Sheet 108 Flint. IPR/73-47C (British Geological Survey © NERC; all rights reserved). **Key:** Permo-Triassic KnS, Kinnerton Sandstone Fm. Carboniferous Erb, Erbistock Fm; Gwp, Gwespyr Fm; Hsh, Holywell Shale Fm; Cfs, Cefn-y-fedw Fm; lCFS, Lower tongue Cfs; MrF, Minera Fm; CML, Cefn Mawr Fm; LoL, Loggerheads Fm; LeL, Leete Fm; LmL, Llanarmon Fm; Lfs, Llyn-y-frân Sandstone; Fol, Foel Fm; BB, Basement Beds. Silurian EY, Elwy Fm; Dd, Disturbed Beds; sa, thickly bedded sandstone; Ngf, Nantglyn Flags.

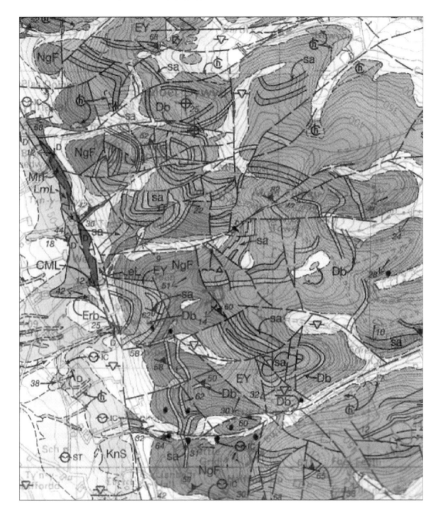

Fig. 19. Detail of the Moel Famau area from British Geological Survey Sheet 108 Flint. IPR/73-47C (British Geological Survey © NERC; all rights reserved). **Key**: Permo-Triassic KnS, Kinnerton Sandstone Fm. Carboniferous Erb, Erbistock Fm; Gwp, Gwespyr Fm; Hsh, Holywell Shale Fm; Cfs, Cefn-y-fedw Fm; lCFS, Lower tongue Cfs; MrF, Minera Fm; CML, Cefn Mawr Fm; LoL, Loggerheads Fm; LeL, Leete Fm; LmL, Llanarmon Fm; Lfs, Llyn-y-frân Sandstone; Fol, Foel Fm; BB, basement beds. Silurian EY, Elwy Fm; Dd, disturbed beds; sa, thickly bedded sandstone; Ngf, Nantglyn Flags.

The term 'geodiversity' was first coined by Sharples in 1993 in a Tasmanian forestry document (Sharples 1993) and did not find wide acceptance in the UK until 2001. Geodiversity is:

... the variety of rocks, fossils, minerals and natural processes. It is a powerful term that compliments biodiversity. (Prosser 2002; Gray 2004)

However, the importance of geodiversity and its benefits to people are often overlooked (Larwood & Durham 2005). The same is true of geoconservation.

RIGS are Regionally Important Geological/Geomorphological Sites (nowadays more commonly referred to as geodiversity sites, to sit alongside biodiversity). RIGS are non-statutory sites that have a measure of protection granted by designation within the planning systems of local authorities. They are much broader in their conservation designations than SSSIs and can be designated on the basis of one of four criteria:

- scientific
- education
- aesthetic
- historical.

North East Wales RIGS Group Site Record

General	North East Wales
Site Name: Coed Ceunant Old Quarry	**File Number**: D0030
RIGS Number: 80	**Surveyed by**: Dr J. Malpas, Dr C. Burek, P. Appleton
Grid Reference: SJ 151 602	**Date of visit**: 02/02/05
RIGS category: Scientific & Historical	**Date Registered**: Owner: 31/03/06
Earth Science Category: Lower Palaeozoic Stratigraphy	Planning Authority: 31/03/06
Site Nature: ED - disused quarries, pits and cuttings	**Documentation prepared by**: Dr J. Malpas
Unitary Authority: Denbighshire County Council	**Documentation last revised**: 29/09/05
1:50,000: Sheet 116, Denbigh and Colwyn Bay	**Photographic record**: yes
1:25,000: Explorer 265, Clwydian Range	**BGS sheet**: Flint 108

RIGS Statement of Interest:

Coed Ceunant Old Quarry is of regional geological importance because it is an important fossil locality that was described by pioneering women geologists during the late 19th and early 20th centuries. Ethel Woods and Margaret Crosfield, investigated the rocks and fossils of the Silurian Period that were deposited 430 to 425 million years ago. The most important fossils are graptolites, extinct, tiny drifting, marine animals that evolved relatively quickly during the Silurian. Graptolites from this and associated quarries were used to describe, quantify and understand the order in which these rocks were deposited, their relative age and the nature of the faults. Woods and Crosfield's research in this and other areas of North Wales resulted in their definitive work on British Silurian graptolites published between 1901 and 1918.

The graptolites that Woods and Crosfield found here (named by Woods and Crosfield as 'Coed Ceunant quarry on 900 ft contour line') placed the rocks in the Y Fron-werth Flags at the base of the graptolite *Monograptus nilssoni* Zone, which is now attributed to Nantglyn Flags of Ludlow age.

This site is one of a network of sites in the Clwydian Range that were used by Ethel Woods and Margaret Crosfield in their research. The specimens they used for their analysis are in the Grosvenor Museum, Chester. Their work was fundamental to the development of British Silurian stratigraphy that underpins modern understanding of the Silurian System of Period.

Fig. 20. Site documentation by NEWRIGS for Coed Ceunant Old Quarry, one of the network of Woods and Crosfield RIGS which now has protection.

This is the importance of RIGS for the present paper because it can conserve sites for their historical importance alone.

This is a measure of the geoconservation protection that SSSI status cannot award because 'historical' is not a recognized criterion within the Geological Conservation Review (GCR) or SSSI system. All of the sites described fit into the historical category, although some have also been designated as scientific sites. Examples of the designation documentation are attached (Fig. 20). NEWRIGS has nominated nine sites for designation to Denbighshire County Council and has submitted the necessary documentation. These sites will now be safeguarded for future generations.

Discussion

The importance of conserving sites, samples and specimens for the next generation of geologists cannot be underestimated (Burek 2003). Important sites, specimens and samples are being lost annually, both by natural and human erosion (Burek 2001; Lenham 2006). Some examples have been lost and then found, for example Hutton's drawings in the archives of the University of Edinburgh Library (Craig et al. 1978), and the specimens recovered, or should we say recognized, as those of Ethel Skeat and Margaret Crosfield at the Grosvenor Museum, Chester. But this is the exception rather than the rule. In order for research into the history of geology to be sustainable, it is necessary to recognize the work of former generations of geologists and conserve their legacy. If those geologists happen to be one of the few women working in this discipline, it is even more important because their work is more likely to have been lost. Reasons put forward for this include name change job, change or loss of job, lack or difficulty of professionalism due to educational deprivation and, finally, society's perception of the role of women. The RIGS designated here are part of a network of sites and were a direct result of a RIGS audit of North Wales supported by the Welsh Assembly Government. RIGS recognize the importance of historical sites and seek to conserve them, alongside other criteria for designation.

The choice of Ethel Skeat and Margaret Crosfield's nine sites in northeast Wales marks a significant change in geoconservation. These sites have been designated as a package of sites honouring the work of two female geologists who contributed significantly to our understanding of the structure and stratigraphy of the Clwydian Range. Their work was part of the consensus building and further understanding of the Silurian and Ordovician period in this part of Wales. As stated in the NEWRIGS site documentation Statement of Interest:

This site forms part of a network of sites in North East Wales that demonstrate the development of British stratigraphy that underpins modern understanding of the Silurian system of rocks. (Malpas et al. 2005)

Conclusions

So what are the rediscovered treasures of Woods and Crosfield?

- The field notebooks and diaries of Margaret Crosfield, with their detailed descriptions, measurements and drawings with their ideas and statements of thought;
- The specimens relating to the 1925 Woods and Crosfield paper found in the Grosvenor Museum, Chester, and the British Geological Survey archives in Keyworth, Nottinghamshire;
- The network of sites relating to the 1925 paper;
- The establishment of a series of RIGS designated sites on important historic criteria by Denbighshire County Council;
- The letters written by these women and photographs that establish the networking which took place following their departure from Cambridge;
- The importance of a RIGS audit to geoconservation in order to safeguard materials and sites for future generations.

All the above are the treasures mentioned in the title of this paper and are important in the recognition of the melding together of the history of geology and current research.

The authors wish to gratefully acknowledge the help given to us in the preparation of this manuscript by the Queens' School, Chester; Marjorie Carreck of the Geologists' Association, for access to photographic archive; M. Howe and P. Shepherd of the British Geological Survey; colleagues at the University of Chester; Kate Riddington, Keeper of Natural History at the Grosvenor Museum, Chester; the NEWRIGS group for support and encouragement, especially P. Appleton and R. Roberts of the Countryside Council for Wales. This work came about as a result of a geodiversity audit undertaken by NEWRIGS for the Welsh Assembly Government and was funded by the Aggregate Levy Sustainability Fund for Wales. Without this work, these sites and samples would not have been rediscovered. We gratefully acknowledge receipt of this Welsh Assembly Government Aggregates Levy Sustainability Fund Grant.

References

BUREK, C. V. 2001. The uses and abuses of RIGS sites. *In*: ADDISON, K. (ed.) *Geoconservation in Action*. Proceedings of the 3rd UKRIGS annual conference, Newton Rigg, Cumbria.

BUREK, C. V. 2003. Time to take responsibility for collections. *Earth Heritage*, **20**, 22–23.

BUREK, C. V. 2005. Dame Maria Matilda Ogilvie Gordon. *Teaching Earth Science*, **30**, 42–44.

BUREK, C. V. 2007. The role of women in geological higher education – Bedford College, London (Catherine Raisin) and Newnham College, Cambridge, UK. In: BUREK, C. V. & HIGGS, B. (eds) *The Role of Women in the History of Geology*. Geological Society, London, Special Publications, **281**, 9–38.

COCKS, I. R. M. & FORTEY, R. A. 1982. Faunal evidence for oceanic separations in the Palaeozoic of Britain. *Journal of the Geological Society*, London, **139**, 167–480.

CRAIG, G. Y., MCINTYRE, D. B & WATERSON, C. D. 1978. *James Hutton's 'Theory of the Earth': the lost drawings*. Scottish Academic Press, Edinburgh.

CROSFIELD, M. C. 1907. 'Field notebook 2'. British Geological Survey, Archive material, biostratigraphy collections.

CROSFIELD, M. C. 1908. 'Field notebook 3'. British Geological Survey, Archive material, biostratigraphy collections.

CROSFIELD, M. C. 1939. Will. Probate Office, London, England.

CUMMINS, W. A. 1959. The Lower Nantglyn Flags; mid Salopian basin facies in Wales. *Liverpool and Manchester Geological Journal*, **2**, 159–167.

DAVIES, J. R., WILSON, D. & WILLIAMSON, I. T. 2004. *Geology of the Country around Flint*. British Geological Survey Memoirs, Sheet 108.

DAVIES, J. R., FLETCHER, C. J. N., WATERS, R. A., WILSON, D., WOODAHALL, D. G. & ZALASIEWICZ, J. A. 1997. *Geology of the Country around Llanilar and Rhayader*. British Geological Survey Memoirs, Sheets 178 & 179.

ELLES, G. L. 1940. Ethel Gertrude Woods D.Cs (1891–1894). *Newnham College Roll Letter*, **1940**, 62–64.

ELLES, G. L. 1953. Obituary: Margaret Chorley Crosfield. *Proceedings of the Geological Society*, **1502**, 131–132.

ELLES, G. L. & WOOD, E. M. R. 1901–18. *A Monograph on British Graptolites*. Parts 1–11. Palaeontological Society, London, Monograph.

EVA, E. J. & MALTMAN, A. J. 1994. Slump-fold and palaeoslope orientations in Upper Silurian rocks, north Wales. *Geological Magazine*, **13**, 685–691.

GEIKIE, A. 1908. Award of the Murchison Geological Fund. *Proceedings of the Geological Society*, **64**, 48–49.

GRAY, M. 2004. *Geodiversity Valuing and Conserving Abiotic Nature*. John Wiley & Sons Ltd, Chichester.

HIGGS, B. & WYSE-JACKSON, P. 2007. The role of women in the history of geological studies in Ireland. In: BUREK, C. V. & HIGGS, B. (eds) *The Role of Women in the History of Geology*. Geological Society, London, Special Publications, **281**, 137–154.

HIGGS, B., BUREK, C. V. & WYSE-JACKSON, P. 2005. *Is there Gender Bias in the Geological Sciences?* Poster presented at the Irish Geological Research Meeting, 18–20 February 2005, Trinity College, Dublin, Ireland.

JOHNSTON, M. 1953. Margaret Chorley Crosfield Obituary. *Annual Report of the Council, Geologists' Association*, **64**, 62–63.

KINGSLEY, C. 1873. *Town Geology*. D. Appleton & Company, New York.

KUHN, T. S. 1962. The structure of scientific revolutions. In: NEURATH, O. (ed.) *International Encyclopaedia of Unified Science*, University of Chicago Press, Chicago, **2**, 1–174.

LARWOOD, J & DURHAM, E. 2005. *Involving People in Geodiversity*. [Workshop held at the Earth Heritage: World Heritage conference, 7–10 September 2004, Warehand, Dorset, organized by the Joint Nature Conservation Committee.] Report by English Nature.

LENHAM, J. 2006. Climbers lobby for rock face respect. *Earth Heritage*, **25**, 5.

MALPAS, J. A. & BUREK, C. V. 2006. A ton of new RIGS! *Earth Heritage*, **26**, 12–13.

MALPAS, J. A., BUREK, C. & APPLETON, P. 2005. *Coed Ceunant Old Quarry Site Documentation*. NEWRIGS D0030, RIGS documentation, Wrexham, UK.

MORLEY, L. 2004. Interrogating Doctoral Assessment. *International Journal of Educational Research*, **41**, 91–97.

NEWNHAM COLLEGE 1878. *Register (1871–1923)*, **1**, 65.

NEWSTEAD, A. 1926. *Annual Report of the Curator and Librarian for 1925–26*. Grosvenor Museum, Chester.

OLDROYD, D. 1996. *Thinking about the Earth: A History of Ideas in Geology*. Athlone Press, London.

PROSSER, C. P. 2002. Terms of endearment. *Earth Heritage*, **17**, 12–13.

QUEEN'S SCHOOL. 1905. *Have Mynde*. Queen's School, Chester, 6 May.

QUEEN'S, SCHOOL. 1910. *Have Mynde*. Queen's School, Chester, 6 July.

RUDWICK, M. J. S. 1985. *The Great Devonian Controversy*. University of Chicago Press, Chicago.

SHARPLES, C. 1993. *A Methodology for the Identification of Significant Landforms and Geological Sites for Geoconservation Purposes*. Forestry Commission, Tasmania.

SECORD, J. E. 1986. *Controversy in Victorian Geology: The Cambrian-Silurian Dispute*. Princeton University Press, New Jersey.

SKEAT, E. G. 1904. The Jurassic rocks of east Greenland. *Proceedings of the Geologists' Association*, **18**, 336–350.

SKEAT, E. G. & MADSEN, V. 1898. The Jurassic, Neocomian and Gault boulders found in Denmark. *Danmarks Geologiske Undersogelse*, **2**(8), 1–213.

SOPER, N. J. & WOODCOCK, N. H. 1990. Silurian collision and sediment dispersal patterns in southern Britain. *Geological Magazine*, **127**, 527–542.

WACHTLER, M. & BUREK, C. V. 2007. Maria Matilda Ogilvie Gordon (1864–1939): Scottish researcher in the Alps. In: BUREK, C. V. & HIGGS, B. (eds) *The Role of Women in the History of Geology*. Geological Society, London, Special Publications, **281**, 305–317.

WARREN, P. T., PRICE, D., NUTT, M. J. C. & SMITH, E. G. 1984. *Geology of the Country around Rhyl and Denbigh*. British Geological Survey Memoirs, Sheets 95 and 107.

WEDD, C. B., SMITH, B. & WILLS, L. J. 1927. *Geology of the Country around Wrexham*. British Geological Survey Memoirs, Sheet 121.

WILLS, L. J. & SMITH, M. A. 1922. The Lower Palaeozoic rocks of the Llangollen district with special reference to the tectonics. *Quarterly Journal of the Geological Society, London*, **78**, 176–226.

WOODS, E. G. 1923. *The Principles of Geography, Physical and Human*. Clarendon Press, Oxford.

WOODS, E. G. & CROSFIELD, M. C. 1896. The geology of the neighbourhood of Carmarthen. *Quarterly Journal of the Geological Society, London*, **52**, 523–541.

WOODS, E. G. & CROSFIELD, M. C. 1925. The Silurian rocks of the central part of the Clwydian Range. *Quarterly Journal of the Geological Society, London*, **81**, 170–194.

WRIGHT, 1919. Report of the session Ordinary Meeting Friday 7th November 1919 List of Exhibits – Miss M. C. Crosfield. *Proceedings of the Geologists' Association*, **30**(4), 208.

Marie Stopes and the Fern Ledges of Saint John, New Brunswick

H. J. FALCON-LANG[1] & R. F. MILLER[2]

[1] *Department of Earth Sciences, University of Bristol, Bristol, BS8 1RJ, UK*
(e-mail: howard.falcon-lang@bris.ac.uk)

[2] *Natural Sciences Department, New Brunswick Museum, 277 Douglas Avenue, Saint John, New Brunswick, E2K 1E5, Canada*

Abstract: Marie Stopes (1880–1958) is best known for her controversial writings on sex, marriage and birth control, but during her 20s and 30s she carved out a successful career as a palaeobotanist. Here, we discuss her work on the Fern Ledges of Saint John, New Brunswick. The age of these fossil beds had long been shrouded in controversy. The eminent 19th-century geologist, Sir William Dawson, had argued that they were Devonian and represented remains of the oldest known terrestrial ecosystem. In 1910, Stopes was commissioned by the Geological Survey of Canada to reassess the taxonomy and age of this fossil flora. We provide the first detailed chronology of this 18-month long research project and highlight some of the most important aspects of the study. Her outstanding monograph, characterized by precise observation and interpretation, cut through decades of muddled thinking to prove that the beds were, in fact, Pennsylvanian. In addition, her palaeoecological inferences were well ahead of their time and also had biostratigraphical implications. Although she continued to intermittently publish geological works until the mid-1930s, the Fern Ledges project, which coincided with her disastrous first marriage to Reginald Gates, marked the beginning of the end of her palaeobotanical career and the start of her more extraordinary and enduring contribution to society.

Marie Carmichael Stopes (1880–1958) was one of the most flamboyant and influential figures of the 20th century (Briant 1962; Hall 1977; Rose 1992). She is, of course, most famous for her controversial sex manual *Married Love* (Stopes 1918a), her pioneering work on birth control outlined in *Wise Parenthood* (Stopes 1918b), and numerous subsequent writings; the clinics that bear her name stand as her most significant and enduring legacy (Rose 1992). In 1999, readers of *The Guardian* newspaper (25 January) voted her 'Woman of the Millennium' for her massive contribution to Western society.

Less well known is the fact that Stopes initially trained as a geologist and pursued a highly successful scientific career in her 20s and 30s (Chaloner 1959, 1995, 2005). Her BSc (honours) degrees in botany (first class) and geology (third class) from University College London (UCL) in 1902, were followed by a PhD from Munich in 1904, and a DSc from the University of London in 1905 (Fig. 1; Rose 1992). Aged only 23 years, Stopes was appointed junior demonstrator at the University of Manchester where, among many other things, she carried out research on Mesozoic plant compressions (Watson 2005). Her early output was prolific, and by the time her popular palaeobotany textbook *Ancient Plants* (Stopes 1910) hit the bookshelves, she had published more than 20 scientific papers (see Chaloner 2005 for a full bibliography).

In 1910 Stopes was approached by the Geological Survey of Canada (GSC) to help resolve an acrimonious and long-running debate concerning the age of a succession of rocks in southern New Brunswick – the so-called 'Fern Ledges' of Saint John (J. W. Dawson 1868). In this paper we provide, for the first time, a detailed chronology of Stopes's research on the Fern Ledges material and critique of her monograph (Stopes 1914). Specifically, we emphasise several aspects of her scientific approach that marked her out from her contemporaries and set the standard for all future palaeobotanical monographs.

The Fern Ledges controversy

The Fern Ledges is the name given to the most famous of a series of rocky shelves that crop out along the Bay of Fundy at Saint John, southern New Brunswick, Canada (19T 728963 5014433). These beds, hewn by the world's highest tides, comprise the Lancaster Formation (of present nomenclature) and are also exposed at several other sites within the municipal boundaries of Saint John (Fig. 2). The other sites mentioned in this paper are Lepreau (19T 696592 5000432), Duck Cove (19T 728325 5014150), Barrack Shore (19T 731157 5016428), and Little River (19T 733202 5017677).

From: BUREK, C. V. & HIGGS, B. (eds) *The Role of Women in the History of Geology*. Geological Society, London, Special Publications, **281**, 227–245. DOI: 10.1144/SP281.13
0305-8719/07/$15.00 © The Geological Society of London 2007.

Fig. 1. Marie Stopes (1880–1958), photographed 1904/5, at about the time that she joined the University of Manchester. (Reproduced with kind permission of her son, Harry Stopes-Roe; from Chaloner 1995, 2005.)

First fossil collectors

Abraham Gesner (1797–1864) was the first person to document fossils from these beds, in the late 1830s, at the Barrack Shore locality (Gesner 1840). Gesner was New Brunswick's provincial geologist, and later found fame for his work on kerosene distillation (Falcon-Lang 2006). Between 1860 and 1863, members of the Steinhammer Club, an amateur geology group based in Saint John, followed up Gesner's report and made a huge fossil collection from the nearby Fern Ledges locality (Miller & Buhay 1988). From 1862, under the influence of Sir William Dawson (1820–99), well-known principal of McGill University (Sheets-Pyenson 1996; Falcon-Lang & Calder 2005), the Steinhammer Club evolved into the more eclectic Natural History Society of New Brunswick. Such was their interest in the Fern Ledges fossils that they later incorporated some of them into their official emblem (Fig. 3).

The principal Steinhammer fossil collectors at Fern Ledges were C. Frederic Hartt (1840–78) and George F. Matthew (1837–1923), who had first met as teenagers. Both were in their early 20s at the time of the discoveries. Matthew (Fig. 4a) was a clerk at the Saint John Customs House and had developed an interest in geology as a young boy (Miller 1987; Miller & Buhay 1988, 1990). Hartt (Fig. 4b) was a teacher at a local girls' school run by his father, having recently moved to Saint John after studying geology at Acadia College, Nova Scotia. Hartt later became the first geology professor at Cornell University (Brice 1994) and, following in the footsteps of Louis Agassiz, went on to conduct a geological survey of Brazil; he died in Rio de Janeiro, aged 37 years (Lopes 1994).

Hartt and Matthew's fossil collections of 1860–63 included both plant and animal remains and were purchased as the first acquisition of the Natural History Society of New Brunswick. Dawson (Fig. 4c) described most of these specimens (J. W. Dawson 1861, 1862, 1863, 1868, 1871, 1880, 1881), although he and Hartt sent the arthropods to international experts for

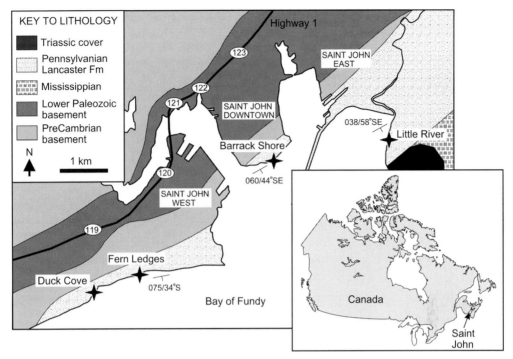

Fig. 2. Location details of the main sites mentioned in the text (Duck Cove, Fern Ledges, Barrack Shore, Little River), all within the municipal boundaries of Saint John, New Brunswick. Lepreau is positioned some 30 km to the west.

Fig. 3. The official emblem of the Natural History Society of New Brunswick showing fossil plants from the Fern Ledges. Beginning in 1861, members of this group collected more than 8000 specimens from the site. (Courtesy of the New Brunswick Museum.)

identification (Salter 1862; Scudder 1868a & b). Hartt also prepared a detailed stratigraphic section through the eight principal fossil beds (Hartt 1865); this was included in a geological report by Loring W. Bailey (1839–1925), professor of chemistry and natural sciences at the University of New Brunswick, and reprinted in the second edition of *Acadian Geology* (J. W. Dawson 1868). Of greatest interest to the wider geological community were the numerous fossil insects in the assemblage (J. W. Dawson 1868); one specimen famously preserved a stridulating organ (Scudder 1868a & b). Darwin (1871), and others, later picked up on this observation, imagining the primeval forests alive with chirping (Miller & Buhay 1988).

As part of his larger research on the geology of New Brunswick, Quebec and Maine, Dawson assigned the Fern Ledges succession to the Devonian System (J. W. Dawson 1861, 1862, 1863, 1868). In doing so, he effectively recognized it as the oldest known terrestrial ecosystem and duly caused a sensation in Victorian scientific circles (Miller & Buhay 1988). Others were more sceptical, however, noting that at least some of the fossil plants found on the same bedding surfaces as the insects were apparently Pennsylvanian in age

Fig. 4. Key figures in Fern Ledges research shortly after the first fossil discoveries: (**a**) C. Frederic Hartt in 1870; (**b**) George F. Matthew in 1870; (**c**) J. William Dawson in 1865. (a & b courtesy of the New Brunswick Museum.)

(Geinitz 1866; Hagen 1881). Nevertheless, Dawson remained dogmatic (J. W. Dawson 1882), and his influence was such that, until he died in 1899, there was little serious discussion about age dating (Falcon-Lang & Calder 2005).

Geological Survey of Canada mapping project

Controversy about the age of Fern Ledges really erupted at the end of the 19th century, when officers of the GSC started to remap the complex geology of the Bay of Fundy (Fig. 5). Initial studies by Hugh Fletcher (1848–1909) and Dr R. W. Ells seemed to concur with Sir William Dawson, apparently showing that the Fern Ledges was Devonian, based on inferred stratigraphic field relationships and their work on similar rocks in Nova Scotia (Fletcher 1900; Ells 1901). Matthew, who had recently started to redescribe the Fern Ledges fossil assemblage (formerly he had merely been a collector), was delighted that the views of his former mentor were being borne out by on-going research. In contrast, others in the GSC, such as palaeontologist Henry Ami (1858–1931), cogently argued that the rocks were, in fact, Pennsylvanian based on fossil content (Ami 1900a & b).

The director of the GSC was, at this time, George M. Dawson (1849–1901), Sir William

Fig. 5. Matthew with the officers of the Geological Survey of Canada discussing geology at West Bay, Parrsboro, on the Bay of Fundy, Nova Scotia. Left to right: G. F. Matthew, Hugh Fletcher, W. F. Ferrier, A. P. Low, and G. A. Young. (Courtesy of GSC, Ottawa, Archive Image No. 201753.)

Dawson's son (Smith 2002). Concerned that the publication of several map-sheets was being held back by the dispute, Dawson contacted international experts to garner their opinion concerning the fossils found in the Nova Scotian rocks mapped by Fletcher and Ells. Robert Kidston in the United Kingdom (Whiteaves 1899; G. M. Dawson 1900) and David White (1902, 1911) in the United States were asked to examine the fossil plants, while Woodward, Traquair and Jones were asked to consider the fishes and crustaceans; independently all concluded that the succession in question was undoubtedly Pennsylvanian. Kidston and White went on to compare their work to the Fern Ledges and also to confirm their age as Pennsylvanian. On receiving this intelligence, Dawson wrote to Matthew on 1 March 1900, stating:

Naturally [I would] be pleased if my father's view in regard to the Saint John plant beds [could] be maintained as correct in every particular; but one cannot afford to be dogmatic upon any scientific question. (NBM, George Frederic Matthew *fonds*, S104, F21).

George Dawson died suddenly the following year. Had he lived he would have probably brought closure to the Fern Ledges debacle, but as it was, infighting within the GSC intensified, even at times becoming manifest in personal attacks (Zaslow 1975). For example, in a letter to Matthew, dated 8 April 1907, Hugh Fletcher alleged that his [Fletcher's] opinion of the palaeobotanical interpretations had been called 'low, mean [and] contemptible' by others, and that his palaeontological sparring partner, Henry Ami, had once described him as a 'real geologist' (NBM, George Frederic Matthew *fonds* S104, F6). Encouraged by Matthew's ongoing research of the Fern Ledges assemblages (which supported Sir William Dawson's age determinations), and angered by Ami's comments, Fletcher vowed to maintain his belief in the Devonian age of the succession 'more strongly than ever!' A couple of years later, Fletcher died of a chill, caught in the field at Joggins, and so played no further part in the debate (Zaslow 1975).

When Reginald Walter Brock (1874–1935) became director of the GSC in 1907, he made a concerted effort to sort out the Fern Ledges problem once and for all. One of his key personnel in this work was William J. Wilson (*c*. 1852–1920), a former schoolteacher from Saint John and, from 1884 to 1889, recording secretary of the Natural History Society of New Brunswick. Wilson had been a GSC palaeobotanist since 1891 (Zaslow 1975) and, beginning in 1908, at Brock's instruction, he began collecting fossil plants from the Bay of Fundy region and southern New Brunswick 'with the object of definitely fixing the geological horizons of the formations found there' (Wilson 1909, p. 183). Familiar with the Fern Ledges from his Saint John days, he made several major collections, but failed to resolve the crucial question of the age of the site.

In early 1910, Toronto was chosen as the location for the 12th International Geological Congress (to be held in 1913). Brock decided to showcase the geology of eastern Canada in a major pre-conference field trip, and 'immediately began to direct the fieldwork for 1911 towards the districts and places to be covered by the tours' (Zaslow 1975, p. 301). But what was he to do about Fern Ledges? Previously Kidston (in Whiteaves 1899, p. 216) had advised George Dawson that:

A thorough revision of the [flora], especially in the light of subsequent collections and possible discovery of more perfectly preserved specimens seems most desirable, and also that a better series of figures be published.

But who could undertake such a work? Clearly given the recent history of conflict, it had to be an unbiased outsider.

Marie Stopes's commission

In 1910, Brock approached Marie Stopes to undertake the Fern Ledges project, which was to be completed in time for the arrival of the 12th International Geological Congress. It is not known why she was considered the best candidate. After all, she had no prior experience of working with Carboniferous compression floras, although she had studied anatomically preserved floras in coal balls (Stopes 1903, 1906; Stopes & Watson 1908) under the tutorage of Professor Francis W. Oliver (1864–1952) at UCL, the man who had discovered pteridosperms with D. H. Scott (Chaloner 2005).

It is possible that she had made contacts during her first trip to North America. In early 1909 she had traversed the entire continent by train, from Vancouver to New York, on the way home from a scientific expedition to Japan, and lectured widely to Canadian women's groups (Rose 1992). In fact, she had even visited McGill University in Montreal, where Dawson had formerly been principal (Sheets-Pyenson 1996), and this was probably the first occasion she saw the type specimens from Fern Ledges (see Stopes 1914, p. 12). Had she met Brock during this trip? We simply do not know. However, one way or another, she was evidently eager to work in Canada because that year she applied (unsuccessfully) for a job at the University of Toronto (Rose 1992).

Stopes was, of course, one of the rising stars of British geology, so perhaps she was the obvious choice for the job. Although aged only 30 years,

Fig. 6. A palaeobotanical meeting in Manchester, England, May 1910: seated (left to right), C. Grand'Eury, W. J. Jongmans, F. W. Oliver, M. C. Stopes, P. Bertrand, M. D. Zalessky; standing (left to right), F. O. Bower, unidentified, W. H. Lang, unidentified, unidentified, F. E. Weiss, unidentified. (Courtesy of C. Blanc and the Paris National Museum of National History Collections.)

she had recently accepted a prestigious lectureship at UCL, and had just published an influential textbook (Hall 1977; Rose 1992). She was also well known among the international palaeobotanical community. A photograph of a meeting at Manchester University, dated May 1910, shows her in the illustrious company of European palaeobotanists such as Bertrand, Grand'Eury, Zalessky and Jongmans, as well as members of the British palaeobotanical community (Fig. 6; Laveine 1995).

Chronology of Stopes's Fern Ledges research

No satisfactory historical account of Stopes's work on the Fern Ledges material exists at present. In producing the following chronology we have examined, in the British Library (BL), her diary for January–February 1911 (BL MS58745) and correspondence (BL MS58477-58480) as well as pertinent documents in the archives of the Natural History Museum, London (NHM uncatalogued) and the New Brunswick Museum, Saint John (NBM, George Frederic Matthew *fonds*). Stopes's Fern Ledges work (February 1911 to July 1912) coincided with the early part of her disastrous first marriage (see below), a turbulent period formative for her later writings on sex, marriage and birth control (Rose 1992).

McGill and GSC collections (spring 1911)

Following Brock's commission, Marie Stopes travelled to North America to begin her Fern Ledges research at the end of 1910. She arrived just before Christmas and, on 29 December, attended the 'botanical dinner' at the American Association for the Advancement of Science meeting in St Louis, Missouri. It was here that she met Reginald Ruggles Gates (1882–1962), a Canadian geneticist, with whom she fell instantly and passionately in love; they were engaged by 31 December! However, Stopes's busy schedule of meetings, and Gates's own research, were then to keep them apart for several months (Rose 1992).

Beginning in late February 1911, after extensively touring around New England, Stopes visited collections of Fern Ledges material in the GSC archives in Ottawa (including those made by Wilson in 1908), where she also enjoyed an energetic social life (Rose 1992). That year saw the GSC take on Alice Wilson (1881–1964), its first female professional employee, but there is no evidence that the two met. Stopes later went on to study Dawson's collection at McGill University, Montreal, which she had probably originally viewed in early 1909. She negotiated with both institutions and arranged for a significant part of these collections (including type specimens) to be shipped to the British Museum (Natural History) for further comparative study (Stopes 1914, pp. 12–13).

After briefly being reunited for a rapturous marriage ceremony in Montreal on 18 March 1911, Stopes and Gates boarded the *Empress of Britain* in Quebec, and sailed to Liverpool, England, arriving a day late on 1 April 1911 (Hall 1977; Rose 1992). Stopes informed her friends and colleagues, and subsequently her husband, that, in contrast to received tradition, she would retain her maiden name and be known as Dr Stopes-Gates (Hall 1977). Gates seems to have quietly accepted this embarrassment, but Henry Woodward, editor of *Geological Magazine*, queried the decision, enquiring as to by what surname any resulting children would be known (Rose 1992). For a few months, Stopes and Gates made their home in Hampstead Heath, London. During this time, Stopes worked at the British Museum (Natural History) London, where she not only undertook Fern Ledges research, but also compiled a catalogue of Mesozoic plants in the museum's collection (Stopes 1913a, 1915a).

Unable to find suitable employment in England, Gates returned to North America for the summer of 1911. Following a trip to Stockholm in June, Stopes also appears to have travelled back across the Atlantic and paid a visit to Saint John, New Brunswick, to undertake her second phase of Fern Ledges research (Stopes 1914, p. 13). Although the GSC Summary Report for 1911 (published in 1912) makes no mention of her work that year, the report focuses on fieldwork by GSC staff; hence Stopes's own work may have been omitted. In a letter to Brock, written in July 1912, she merely states that she 'spent one month in the field [in July/August 1911] collecting & also examining the stratigraphic evidence' (NHM; uncatalogued material). Ironically, Stopes's visit coincided with the 50th anniversary of the first discoveries at Fern Ledges by Matthew and Hartt in the summer of 1861.

Fieldwork in Saint John (Summer 1911)

During her time in Saint John, Stopes was entertained by an elderly Matthew, aged 74 years (Fig. 7). Beginning in the late 19th century, he had taken an active research interest in the Fern Ledges material and had generated 28 published papers on the flora and fauna, some in the *Transactions of the*

Fig. 7. George F. Matthew in 1921, a decade after Stopes's visit. He never accepted Stopes's results and, until his last day, worked to prove that Fern Ledges was Devonian. (Courtesy of the New Brunswick Museum.)

Royal Society of Canada and others in the more obscure *Bulletin of the Natural History Society of New Brunswick* (see Miller & Buhay 1990 for a full list). In one of these, published the year Stopes, received her GSC commission, he had infamously argued that the site was, in fact, Silurian (Matthew 1910), even older than the Devonian age suggested by Dawson!

Matthew was initially very helpful to Stopes.

He was kindness itself, and placed all he could at my disposal, as well as going with me on two excursions in the neighbourhood [of Saint John]. (Stopes 1914, pp. 12–13)

One of these excursions was certainly to the original Fern Ledges locality, where Stopes was disappointed to discover that the eight fossil beds described by Hartt (1865; in Dawson 1868) were 'practically worked out' (Stopes 1914, p. 9). In 2005, we re-identified Hartt's fossil beds in the field. As Stopes correctly notes, they had been so denuded by Natural History Society of New Brunswick geologists that new collections were practically impossible; the thin fossil-rich beds were locally recessed by up to 80 cm (Fig. 8)! At the time of Stopes's visit the shore bordering Fern Ledges had been converted into Seaside Park, a 'tourist trap' for American visitors, fitted out with a restaurant, dance hall and ornamental gardens.

Fig. 8. The Fern Ledges of Saint John, New Brunswick, showing the authors. Howard Falcon-Lang (right) and Randall Miller (left), in 2005. Note the deep recess above Falcon-Lang's knee, one of Hartt's fossil beds (no. 3) 'worked out' by geologists of the Natural History Society of New Brunswick.

The location of her second excursion with Matthew is unknown. However, Wilson (1915, p. 130) records that she visited nearly all the other outcrops of the Lancaster Formation at Lepreau, some distance west of Saint John, as well as Duck Cove and Little River within the municipality boundary of Saint John itself; any one of these trips may have constituted the second excursion with Matthew.

In addition, Stopes records how she specifically visited Duck Cove with William McIntosh (1867–1950), then the curator of the Natural History Society's museum in Saint John, who 'very kindly showed me a bed a little further on at Duck Cove which he had found rich in plant fossils' (Stopes 1914, p. 9). This was probably the same bed from which Matthew (1906) had recently published his unusual *Pseudobaiera* flora (*Rhacopteris busseana* Stur of present nomenclature). At Duck Cove, she

spent much time collecting, and found a series of plant-containing bands far more numerous than those reported by Prof. Hartt or Dr. Matthew. The beds ... are raised well above the tide, and thus are superior to the original Fern Ledges from the collector's point of view. (Fig. 9; Stopes 1914, pp. 9–10)

Presumably aided once again by McIntosh, Stopes also made studies of the extensive collections at the Natural History Society's museum in Saint John. Since the original work of Hartt and Matthew, this collection had grown from further discoveries by Wilson, McIntosh, A. Gordon Leavitt (1865–1945), Geoffrey Stead (1872–1943) and others. In fact just that year, Leavitt had donated a further 92 fossils (Anon. 1911), so that the collection now numbered more than 8000 specimens (Miller & Buhay 1988). Stopes would have visited McIntosh in the impressive new museum building on Union Street, acquired by the Society in 1906 (Fig. 10).

Mr. McIntosh ... gave me every facility for seeing the specimens in his keeping while I was in [Saint] John, and also lent me a number of the more interesting and important to bring to London for my comparative study. (Stopes 1914, p. 12)

Most of these original type specimens at the Natural History Society of New Brunswick, often thought lost by subsequent taxonomists and reviewers, have recently been rediscovered in Society collections

Fig. 9. The fossil sites around the time of Stopes's visit in 1911: (**a**) view east from Duck Cove towards Fern Ledges (distant right arrow) (Image X11456); (**b**) the classic Fern Ledges outcrop (Image X13087); and (**c**) Duck Cove outcrop at high tide. (Courtesy of the New Brunswick Museum, Image 1995-9-13 (2).)

Fig. 10. The Natural History Society of New Brunswick building at 72 Union Street, Saint John. Purchased in 1906, it served as the society's museum until the construction of the Provincial Museum in 1932. It was torn down in the 1980s to make way for a car park. (Courtesy of the New Brunswick Museum, Image No. 13305 (2)).

and are now catalogued with the New Brunswick Museum holdings (Miller 1995*a* & *b*).

Stopes also viewed Matthew's personal fossil collections in Saint John. Although she described her collaboration with Matthew in 1911 as a very positive experience, certain subtexts in her monograph may imply a degree of tension. Stopes commenced her work with 'an entirely open mind' (Stopes 1914, p. 125), but was clearly aware that the overwhelming consensus of opinion was that Matthew's dating was wrong. At times, she entered into vigorous arguments with Matthew about his taxonomy, and found him 'emphatic' (Stopes 1914, p. 48). Her frustration is quite evident in her monograph; 'I regret to add this to the many points on which we disagree' (Stopes 1914, p. 52), she wrote concerning yet another dispute about Fern Ledges taxa.

In contrast, Matthew's own writings show no obvious signs of friction with Stopes at the time of their meeting. A few weeks after her visit, he noted that:

Dr. Stopes-Gates, an authoress of note as a geologist, now engaged in work for the British Museum, has been studying the Paleozoic floras of this region, and especially those of the Little River group of which the first type series is in the museum of this Society. She has also carefully examined the set of these plants in the Museum of McGill University and the Devonian plants there, described by Sir Wm. Dawson. These plants are of especial value to science on account of their rarity, their great age and the important light they shed on the Palaeozoic floras. (Matthew 1911, p. 360)

However, when Stopes was to send a précis of her findings the following summer, openly criticizing Matthew's work, he was to become almost incandescent (see below).

In the course of her 1911 field season, Stopes also visited Joggins, the world famous Pennsylvanian fossil site a little further east along the Bay of Fundy in Nova Scotia (Fig. 10; Falcon-Lang & Calder 2004; Falcon-Lang *et al.* 2006). The purpose of her 2-day trip was to collect comparative material (Stopes 1914, pp. 50–51). That summer, Walter A. Bell (1889–1969), was just beginning his own palaeobotanical field studies at Joggins

(Bell 1913, 1914), but apparently the two did not meet (Falcon-Lang 2006).

> I understand that Mr. W. A. Bell of the Canadian Survey is re-examining the Joggins section, so that the detailed comparison of the [Fern Ledges] with those at Joggins must await his results. (Stopes 1914, p. 125)

This coincidence is of significance because, in later life, Bell was to build on Stopes's work (Bell 1943), eventually rising to become a world-renowned biostratigrapher (Zodrow 1995). One cannot seriously talk about missed opportunities when discussing an iconoclastic figure like Marie Stopes, but it is only natural to wonder what she might have achieved had she devoted her whole life to geology.

At the British Museum (Natural History) (autumn 1911 to summer 1912)

Back in England in the autumn of 1911, Stopes and Gates were once again reunited at their home in Hampstead Heath, London. At this point, and for the subsequent year, they appeared, at least on the surface, to be happily married. Stopes' Fern Ledges work at the British Museum (Natural History) focused on the specimens she had had shipped from Ottawa, Montreal and Saint John in 1911. In addition, she found small collections stored at the Museum itself, sent at one time or another by Sir William Dawson and members of the Natural History Society of New Brunswick, (Anon. 1913) and a larger collection, made later by Wilson. As she wrote (Stopes 1914, p. 13):

> I have been privileged to have opportunities and material for comparative study of these plants which have not been available for any of the other writers on the subject.

and consequently,

> I feel some confidence in the general results of the present work.

In addition, to having all the specimens to hand, the facilities offered to her at the British Museum (Natural History) were excellent. Dr Smith Woodward, keeper of the geological department, gave her 'every assistance', even providing the services of a photographer, a Mr. H. G. Herring, to document key specimens. She also had access to an unrivalled library of classic coal measure monographs and evidently perused the works of Andrä, Arber, Artis, Boulay, Brongniart, Corda, Crépin, Dawson, Ettinghausen, Geinitz, Germar, Goeppert, Grand'Eury, Jongmans, Kidston, Lesquereux, Nathorst, Potonié, Renier, Renault, Sauveur, Schimper, Schlotheim, Solm Laubach, Sternberg, Suckow, Stur, Unger, White and Zeiller, among others (Stopes 1914). Clearly, she had an excellent working knowledge of French and German, presumably having honed her linguistic skills during her year in Munich (1903–04).

Nevertheless, Stopes needed to be sure that her taxonomy was entirely up to date and accurate. In early 1912, she visited Monsieur René Zeiller (1847–1915), 'the distinguished palaeobotanist at the École des Mines' in Paris for 2 days (Stopes 1914, p. 12; NHM uncatalogued material). She took with her some of the more debatable type specimens from the various collections for comparison with those of Zeiller, and drew extensively on his expertise (Stopes 1914, pp. 20, 33–36, 40, 46, 57, 60–63, 91). Sadly, papers pertaining to this meeting were probably destroyed in the early 1980s, when the collections of the École des Mines were transferred to the University of Lyon (Laveine pers. comm. 2006). However, she obviously developed a close working relationship with Zeiller for, when he died a few years later, she wrote his glowing obituary in *The Times* (Stopes 1915b). In the course of her Fern Ledges work, Stopes also corresponded with a number of other leading palaeobotanists, including Robert Kidston (1852–1924) in Stirling, Scotland, Armand Renier (1876–1951) at the Corps des Mines in Belgium, Henri Potonié (1857–1913) at the Berg Akademie, Berlin, Germany, and Lignier in France (Andrews 1980), mainly concerning various taxonomic matters (Stopes 1914, pp. 7, 48, 81).

While Stopes was concluding her Fern Ledges work in the summer of 1912, she was visited in the British Museum (Natural History) by Wilhelmus Jongmans (1878–1957), a palaeobotanist at the Rijks Herbarium, University of Leiden, Netherlands (Stopes 1914). She had previously met Jongmans in Manchester in 1910 (Fig. 6). Now aged 35 years, he was working on fossils from borehole cores through the South Limburg coalfield (Wagner & van Amerom 1995). What was his reason for visiting the British Museum? Probably, he was either compiling the first part of the monumental *Fossilium Catalogus – Plantae* (Jongmans 1913) or writing, with Kidston, a massive monograph on *Calamites* (Kidston & Jongmans 1915–17). Given that Stopes mostly picked his brains about the taxonomy of calamiteans, the latter scenario is more probable, and perhaps it was Kidston who suggested that the two of them made contact. Stopes was evidently impressed by Jongmans, a man roughly her contemporary, writing that, with regard to taxonomy, 'his authority carries great weight' (Stopes 1914, p. 18).

Going public (autumn 1912)

With the Geological Congress only a year away, Brock at the GSC was anxious to know how

Stopes was getting on. In a letter dated 18 June 1912, he gave her a gentle nudge, remarking that Henri Potonié of Berlin was also interested in studying the Fern Ledges. 'Have you had any interesting results?' he enquired (BL MS58479). Brock did not have to wait long, for a few weeks later she sent him the final manuscript of her Fern Ledges work – more than 193 pages long! Apparently expecting a short note, which he could publish in the GSC Summary Report series, Brock appears to have been displeased with this gigantic tome. After consideration, he decided to hold it back for publication as a GSC Memoir, or to place it in some other series (letter from Brock to Stopes, dated 16 August 1912, BL MS58479).

The following month, Stopes attended the Dundee meeting of the British Association for the Advancement of Sciences (BAAS), a society of which she had been a member since 1903. On 10 September 1912, she gave her first public lecture about the Fern Ledges project, entitled 'Palaeobotany versus stratigraphy in New Brunswick'. Specifically, she spoke about her 'work in the field' at Saint John in 1911 and subsequent taxonomy studies in London and Paris, during the course of which she had identified 'a large proportion of well-known European types ... all [Pennsylvanian] and mostly typical of [the] Westphalian division in [the] Coal Measures' (Stopes 1912, p. 471). She also argued that assertions that the Fern Ledges were Devonian in age, based on stratigraphic studies, had overlooked the existence of a major overthrust in the region. Sketches of the inferred structure of the Saint John region are scribbled on the back of her geological map of New Brunswick preserved in the Natural History Museum, London (NHM uncatalogued material), together with other documents dating from 1912.

Before departing for the BAAS meeting, Stopes began to return the fossil collections that she had borrowed from Canadian institutions. Fossils were duly returned to the curator of the Peter Redpath Museum at McGill University, Montreal (letter from Adams to Stopes, dated 3 August 1912; BL MS58479), who was glad that they had 'proved to be of such interest'. Collections were also 'sent in' to Wilson at the GSC in Ottawa. This latter consignment included not only those specimens previously borrowed, but also new collections made by

'Miss M. C. Stopes ... [at] Duck cove, Lepreau, and east of [Saint] John harbour in the Little River group, and from Joggins, N.S.' (Wilson 1915, p. 130).

Wilson's reference to her as 'Miss' doubtless riled Stopes, who had earlier censured the Royal Society of London for failing to acknowledge her doctorate (Rose 1992).

Stopes also returned fossils to Matthew and McIntosh in Saint John, but additionally included a précis of her GSC manuscript, outlining the evidence for a Pennsylvanian age for the assemblage. McIntosh was out in the field for the whole summer of 1912, and so it was Matthew who first received this unwelcome intelligence. Absolutely furious with her results, Matthew fumed for some months before replying. When he finally put pen to paper on 25 October 1912, he sourly wrote:

You cannot think how much I *feel* your deflection from the causes of truth, to worship the fetish of a too rigid adherence to an old standard; we who depend on stratigraphy cannot look on the age of the St John plant beds other than we do. No bold reassessment that I can conceive of will make them [Pennsylvanian] so we will have to await the improvement of the standards – perhaps not in your day or mine, but in some time to come. (BL MS58479)

However, when McIntosh returned to Saint John, he was much more conciliatory and philosophical, even joking with Stopes that Matthew was getting a bit senile and set in his ways.

You fear we in [Saint] John 'will not like your work'. I think you are mistaken on this. There are a few of us who realise that dear old Dr. Matthew, clever old man as he is, stands alone in some of his ideas. And so it was not a great surprise to me that you should find our plants to be [Pennsylvanian]. I am glad our specimens were in your hands because that will settle the matter of their identity and you know, sooner or later, truth must prevail, and if your duty was an unpleasant one, you are too far away for us on this side to throw bricks at you! (letter from McIntosh to Stopes, dated 19 September 1912; BL MS58479)

To emphasise that there were really no hard feelings (on his part), he warmly invited her to join them at the Geological Congress the following summer, and sent her two photographs from her 1911 fieldwork as a memento.

Stopes must have realized that she could never win over Matthew, as implied by her fear that New Brunswick geologists would not like her work. Matthew had never accepted the opinion of any of the myriad dissenters about his 'Silurian-Devonian' age for Fern Ledges. His correspondence illustrates how he repeatedly tried to convert international palaeobotanists, without success. In a letter dated 14 November 1910, E. A. N. Arber (1870–1918), demonstrator in palaeobotany at the Sedgwick Museum, Cambridge, had written:

It will probably not surprise you to hear that in common with Kidston & others, I regard the 'Fern Ledges', with *Annularia*, *Alethopteris* & *Neuropteris*, as belonging to the Westphalian division of the [Pennsylvanian], & not to the Devonian. (NBM, George Frederic Matthew *fonds*, S104, F1)

Perhaps encouraged by receipt of the prestigious Murchison Medal of the Geological Society of London in 1917, Matthew spent his last years working on a paper to prove (once again) that the Fern Ledges were Devonian in age. Not long before he died, he had walked the entire shoreline from New Brunswick to Maine in order to demonstrate a connection with the unequivocally Devonian beds found there (Bailey 1923)!

Publication of GSC Memoir 41 (autumn 1912–14)

While these emotional events were unfolding in New Brunswick, Stopes's home life in London was itself far from serene. By the end of 1912, major cracks were developing in her marriage to Gates (apparently as a result of his impotence and her sexual frustration). By that time, Stopes had begun openly flirting with Alymer Maude, well-known biographer of Tolstoy, and, in February 1913, he had moved in as a 'lodger'. Using his city contacts, Maude even arranged for a grant to extend Stopes's tenure as lecturer in palaeobotany at UCL from 1913 to 1916 (Rose 1992).

Meanwhile in Ottawa, Stopes's Fern Ledges monograph (to be published as GSC Memoir 41) was slowly being prepared for publication. According to a note by McIntosh, pre-dating 30 September 1912:

Plates of a number of [the Saint John fossil specimens had] been prepared to illustrate a work soon to be published. (McIntosh 1913, p. 485)

However, publication was delayed by other works in press with the Government Printing Bureau.

I am doubtful whether we can get your [Saint] John paper out in time for the summer congress. (Letter from Brock to Stopes, dated 13 March 1913, BL MS58480)

Another delay was apparently caused because the quality of Stopes' plates did not meet Brock's high standards. Two months later, he requested that Stopes send him her original negatives so his staff could redo the 13 plates, but encouraged her that, at last, 'preparations are being made to send your manuscript on the flora at the Fern Ledges to press' (letter from G. A. Young, on behalf of Brock, to Stopes, dated 13 May 1913, BL MS58480). Ironically, photographs of certain specimens preserved at the Natural History Museum, London (unarchived material) show that Stopes's original prints were actually superior to those that eventually appeared in GSC Memoir 41!

As the publication of Memoir 41 was being held up and the Geological Congress rapidly approaching, an anonymous editor hashed together a brief précis of Stopes's work for the conference field guide. Apparently, Stopes was poorly known among North American geologists for her contribution (Stopes 1913b) was printed under the name Mary Stopes. Conference participants visited the Fern Ledges on the afternoon of 30 July 1913, with refreshments being served at Duck Cove, prior to an opportunity to study museum collections in the evening (McIntosh 1914). Interestingly, writing about the fieldtrip in the autumn of 1913, McIntosh noted that the age of the Fern Ledges remained controversial (McIntosh 1914), so, given his earlier private comments to Stopes, he was obviously being diplomatic for the sake of Matthew's feelings.

From its date of first submission in July 1912, GSC Memoir 41 took over 2 years to be finally published. In the GSC Summary Report for 1912 (published in 1914), Memoir 41 was listed as 'in press' as of 1 April 1914. The following day, Brock asked Stopes to put together a list of leading geologists to whom complimentary copies should be sent (letter from Brock to Stopes, dated 2 April 1914 BL MSS8480). Stopes left her husband a few weeks later and, as Britain declared war on Germany on 4 August 1914, she found herself living a destitute life in a tent on the Northumberland coast of England, part of the estate of Lord Grey, former governor-general of Canada. Following Gates' departure for Canada, Stopes returned to her London home in October (Briant 1962).

The GSC Summary Report for 1914 (published in 1915), which contains a list of GSC publications, records that Memoir 41 was finally 'Published October 28, 1914' (p. 190), and complimentary copies were swiftly mailed out. On receipt of their copies, warm letters of praise began to pour in from her appreciative colleagues, including Zeiller and Jongmans, who were now living at the heart of the European war zone (BL MS58480). David White in the United States described how his copy had been 'used and reused' (letter dated 30 March 1915; BL MS58480) while Robert Kidston even requested a second copy writing that:

I have read your memoir from beginning to end most carefully & can only heartily congratulate you on your work. The question [of the age of the Fern Ledges] is soundly settled now, once & for ever. (Letter dated 11 July 1915, BL MS58480)

Critique of the Stopes' Fern Ledges work

Stopes's monograph on the Fern Ledges flora, compiled from February 1911 to July 1912, remains today one of the great palaeobotanical monographs of all time. In a book review in *Science*, Bessey described Memoir 41 as of 'greatest value to students of Carboniferous plants' (Bessey 1914, p. 862). More recent revision of the Fern Ledges remains by Bob Wagner (Fig. 11), ironically a former student of Jongmans in the 1950s (Wagner & van Ameron 1995), has confirmed the

Fig. 11. Bob Wagner, a former of student of Jongmans, examining outcrops at Duck Cove in 2004. Wagner is undertaking a revision of Stopes's work.

extremely high quality of her taxonomic work (Wagner 2005a & b). In the second part of this paper, we highlight a few outstanding aspects of her monograph.

Careful observation

One of the characteristics that really marked out Stopes from her contemporaries was her commitment to precise observation. In the modern vernacular, she always tried to 'tell it like it is', letting the fossils speak for themselves. From the very start of her analysis, comparison of the taxonomic descriptions of Dawson and Matthew and the actual type specimens revealed massive problems with these earlier works. In some cases, descriptions were just simply inaccurate:

I cannot recognize the little flower-like structures shown by Dawson in his 1871 restoration. (Stopes 1914, p. 88)

But in others, figures had apparently been prepared with artistic composition or economy of page use in mind, rather than for the purpose of scientific accuracy. In one figure, two unlikely plant taxa were depicted in organic connection but, wrote Stopes:

... when Dr. Matthew kindly showed me his originals it became evident that [Dawson] had drawn his figures in the way they are from considerations of space, because the fructifications and the vegetative parts are on different slabs of matrix. (Stopes 1914, p. 31).

Dawson and Matthew could not be entirely blamed for these myriad problems. The Fern Ledges material had been metamorphosed (Falcon-Lang & Miller 2007) and was extremely difficult to work with. As Stopes (1914, p. 11) noted:

The plant impressions have been completely graphitised and most of them merely consist of a bright film or streak on the rock. This has made them difficult, not merely to study, but also to photograph. (Fig. 12).

However, if specimens were poorly preserved, then that is how any honest and impartial monograph should portray them. That honesty extended not only to verbal descriptions, but also to their illustrations. Of her own illustrations, Stopes wrote:

I wish here particularly to emphasize the fact that photographs reproduced in the plates are not in any way touched up or improved. The technical skill in what almost amounts to fakery, has reached such a pitch in some instances that the word photograph no longer conveys any guarantee that the object is impartially represented. (Stopes 1914, p. 11)

Matthew was particularly culpable in this regard, sometimes drawing outlines on his specimens to improve their impact (Miller 1987).

Precise interpretation

In addition to careful observation, Stopes was equally concerned with avoiding over-interpretation. By 1911, over 80 species of fossil plant had been

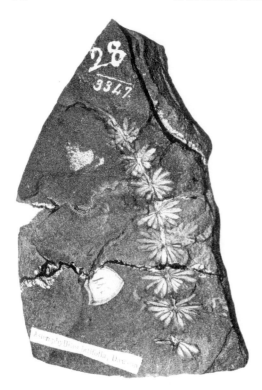

Fig. 12. A specimen of *Annularia stellata* showing the metamorphosed nature of the fossils. Specimens appear white because they are graphitized and highly reflective. (Reproduced from Stopes 1914, p. 148.)

described from Fern Ledges, but she was to cut down the list to just 40. Often new species had been distinguished on the basis of highly equivocal characters. For example, with regard to one of Matthew's putative species of *Asterophyllites*, Stopes pointed out that:

The fact that the nodes are indistinct is not a sufficient reason for forming a new species even were the specimen were a good one, but this is an exceedingly poor and indistinct fragment. (Stopes 1914, p. 98)

However, a much greater problem was Matthew and Dawson's tendency to make something out of nothing. Stopes (1914) is littered with comments like:

I cannot bring myself to identify specifically such incomplete fragments (p. 18),

or

In too many cases have the merest scraps of half-preserved debris been utilised for the types of new species (p. 113).

In her own work, Stopes was careful to clearly document specimens, providing both photographs and interpretative sketches (Fig. 13).

In easily her most stinging criticisms, she wrote:

There is a certain standard (which it is difficult to define, but which a *competent* palaeobotanist can recognize) that specimens must reach before they can be reliably determined. (Stopes, 1914, p. 113)

Did she then think Dawson and Matthew to be incompetent palaeobotanists? That certainly seems to have been her implication.

It is deplorable though none the less true that many palaeobotanists have slipped into the easy habit of naming as a new or distinct species imperfect or troublesome fragments. (Stopes, 1914, p. 115)

In fairness, Matthew's expertise was in invertebrate palaeontology, especially Cambrian trilobites, so it is hardly surprising that he had palaeobotanical shortcomings (Miller 1987). Dawson too was undoubtedly a gifted geologist; he described fossil plants in their sedimentary context and made quite revolutionary taphonomic and palaeoecologic inferences, well ahead of his time (Falcon-Lang & Calder 2005). Yet, in regard to plant taxonomy, Stopes' work clearly shows that, like Matthew, he was second-rate.

Part of Dawson's taxonomic naivety arose from his isolated position on the edge of the British Empire, such that it was difficult to access relevant literature (Sheets-Pyenson 1996).

At this early date [1860s] comparatively few figures of European and other American Palaeozoic fossil plants were available for his use, and so it is not surprising that Sir William [Dawson] made new species from most of the specimens. (Stopes 1914, p. 1)

In addition, Dawson was not familiar with the latest synonymy; Stopes writes:

Dawson ... compares [one specimen] to *Alethopteris serrula* of Lesquereux, which, as Schimper long ago pointed out is not an *Alethopteris* at all. (Stopes 1914, p. 96)

Matthew suffered from the same problems as Dawson, perhaps even more so, with only the resources of a small natural history society at his disposal. In addition, he was only a part-time geologist, his main employment being at the Customs House of Saint John (Miller 1987).

Palaeoecological insights

Stopes's secure understanding of Fern Ledges' taxonomy allowed her to make a number of quite brilliant insights into the palaeoecology of the plant assemblage. These insights had important implications for understanding biostratigraphy, which, after all, was the primary purpose of her GSC commission. Detailed analysis revealed that the Fern Ledges assemblage did not exactly

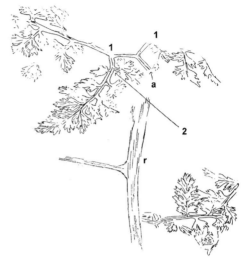

Fig. 13. *Diplothmema subfurcatum*, illustrating Stopes's meticulous approach to systematics, carefully documenting both photographs and interpretative sketches for the same specimens. (Reproduced from Stopes 1914, pp. 38 & 153.)

coincide with any of the floral biozones in the Westphalian of Europe, as defined by Zeiller,

... for species from the higher zones in considerable numbers of individuals are mixed ... [with those of] the lowest of the Westphalian zones. (Stopes 1914, p. 122)

Despite Matthew's (1906) protests to the contrary, there was no doubt that Fern Ledges represented a single plant assemblage deposited over a relatively short time period (Stopes 1914). So why was there this apparent mixture of taxa from different biozones?

In answering that question, Stopes showed the breadth of her palaeobotanical knowledge. Her own earlier work on the British Coal Measures had demonstrated that, in Pennsylvanian times,

... the highland flora differed from the [lowland] swamp flora, and the former included a number of forms generally considered to be ... [stratigraphically] higher ... even Permian. (Stopes 1914, p. 123)

What was the significance of this for the Fern Ledges? Taphonomic analysis showed that:

the flora did not grow in the place where it is now found. (Stopes 1914, p. 123)

The occurrence of a few 'Upper Pennsylvanian' or 'Permian' taxa (such as *Dicranophyllum*; Fig. 14) might therefore be explained if these

... fragments [had been] brought down from some inland of higher elevation [and then] mingled with those plants growing on the flats. (Stopes 1914, p. 123)

This insight pre-empted the classic paper by Moore *et al.* (1936) on the subject of 'Permian' plants in Pennsylvanian strata some 20 years later (see Falcon-Lang 2005 for discussion).

In addition, Stopes clearly recognized the importance of taphonomy in understanding plant assemblages, something that few palaeobotanists of her era understood: she wrote:

Fern Ledges represent plant debris [mixed together] from differing ecological situations. (Stopes 1914, p. 123)

For example, some assemblages pointed to the existence of

... groves of *Calamites* growing in almost 'pure formation' (in the language of the ecologist) just as modern *Equisetum* often does today. (Stopes 1914, p. 124)

Taken as a whole, the Fern Ledges assemblages was interesting because remains of lycopsids, a plant group typically abundant in Pennsylvanian swamps, were extremely rare. Hence, she inferred that, in general

... we are not dealing with the typical, mixed swamp-flora but one principally growing on dryland. (Stopes 1914, p. 124)

Understanding the opposition

A final strength of Stopes's Fern Ledges work was her sensitive understanding of her opponents reasoning. She tried to 'get into the head' of Dawson and Matthew and appreciate why they both so cogently

Fig. 14. *Dicranophyllum glabrum*, a putative coniferopsid. This unusual plant probably grew in the uplands and is more characteristic of Late Pennsylvanian, or Permian, rocks. (Reproduced from Stopes 1914, p. 160.)

believed in a Devonian age for Fern Ledges in order to effectively refute it. Their argument seemed to rest on several (mostly well-founded) observations, whose interpretation was debatable.

One pointer to the age of the rocks was their metamorphosed nature, showing a strong 'slaty cleavage'. When regional mapping also showed the apparent super-position of Mississippian conglomerates (J. W. Dawson 1862; Fletcher 1900; Ells 1901), together these data seemed to strongly imply a Devonian age. But, argued Stopes, as 'the region is so wooded ... stratigraphers have to see with an eye of faith what lies beneath the covered surfaces' and it is therefore possible that they 'have overlooked an important overthrust' (Stopes 1914, p. 126). This comment again emphasises Stopes's geological knowledge, for thrust tectonics had only recently been described from the Scottish Moine belt of Scotland (Peach *et al.* 1907). Localized thrusting would also explain the metamorphosed nature of the Fern Ledges relative to other Pennsylvanian outcrops nearby. Subsequently research has borne out this intuition and shown that the Fern Ledges belt occurs within a series of thrusted nappes (Nance 1987).

A second strand of Dawson and Matthew's argument related to the fossil assemblage itself. Clearly, the dominance-diversity characteristics of the flora (with its abundant cordaitaleans and very rare lycopsids) was strikingly different to typical Pennsylvanian swamp floras. In addition, Dawson recognized the presence of *Psilophyton*, a taxon known from undisputed Devonian successions elsewhere in eastern North America (J. W. Dawson 1862). However, Stopes's palaeoecologic inferences (above) removed the former difficulty: The Fern Ledges 'represent merely a different local *facies* of a portion of the [typical Pennsylvanian assemblage seen at] Joggins' (Stopes 1914, p. 125). In addition, her detailed taxonomic work dealt with the latter problem, showing that Dawson's specimens were probably not true *Psilophyton* at all (but indeterminate roots).

In her obituary, Chaloner (1959) wrote that:

Marie Stopes was rarely a detached and objective scientific worker. She made careful observations, came to her conclusions, and believed in them with an almost passionate fervour.

While we query Chaloner's inference that Stopes was not objective, her fervour is evident in her final arguments as she 'goes in for the kill' and emphatically defends the Pennsylvanian age of Fern Ledges.

It would be mere waste of time here to argue elaborately why this flora must be [Pennsylvanian], for every species is a typical [Pennsylvanian] one. (Stopes 1914, p. 119)

As a last 'warning shot' to those stratigraphers that might later marshal a response to her monograph, she closes:

If Geology is a science at all, Stratigraphy and Palaeontology must prove ultimately to be harmonious. (Stopes 1914, p. 126).

Conclusions

Stopes's research at the Fern Ledges of Saint John, New Brunswick, has stood the test of time, and her monograph is now recognized as a classic palaeobotanical work. In this paper, we have provided a detailed chronology of her work at Fern Ledges for the first time, and have highlighted many aspects of her monograph, which have set the standard for subsequent taxonomic studies. Particularly, we have been impressed by her careful observational skills and precise interpretation, which allowed the plant fossils to 'speak for themselves'. In addition, her palaeoecological inferences have proved to be well founded, with far-reaching implications for biostratigraphy. Stopes's Fern Ledges work coincided almost exactly with her disastrous first marriage to Reginald Gates, an event formative for her later writings on sex, marriage and birth control. It is, of course, inappropriate to talk of missed opportunities when discussing a person so influential as Marie Stopes, but it is only natural to wonder what she might have achieved had she devoted her whole life to geology. Similarly, in what kind of society might we live in today if she had.

We thank W. Chaloner for the photograph used in Figure 1 and for discussion about Marie Stopes. J.-P. Laveine kindly supplied the photograph used in Figure 6; Joan Watson and Chris Cleal helped identify some of these palaeobotanists. P. Kendrick alerted us to the existence of Stopes archives at the Natural History Museum, London. We also thank R. Wagner for discussion about Fern Ledges over several years. HJF-L gratefully acknowledges receipt of a NERC post-doctoral fellowship held at the University of Bristol, UK (NER/I/S/2001/00738), and a Matthew Fellowship (2005) from the New Brunswick Museum. RFM gratefully acknowledges receipt of a SSHRC – CURA grant (833-2003-1015) and the support of the New Brunswick Environmental Trust Fund. This manuscript benefited from the constructive reviews of Chris Cleal and Bill Chaloner.

References

AMI, H. M. 1900*a*. Notes bearing on the Devono-Carboniferous problem in Nova Scotia and New Brunswick. *Ottawa Naturalist*, **14**, 121–127.

AMI, H. M. 1900*b*. On the sub-divisions of the Carboniferous System in eastern Canada, with special reference to the position of the Union and Riversdale formations of Nova Scotia, referred to the Devonian System by some Canadian geologists. *Transactions of the Nova Scotian Institute of Science*, **10**, 162–178.

ANDREWS, H. 1980. *The Fossil Hunters: In Search of Fossil Plants*. Cornell University Press, Ithaca, NY, 421 pp.

ANON. 1911. Donations. *Natural History Society of New Brunswick Bulletin*, **29**, 378.

ANON. 1913. Donations. *Natural History Society of New Brunswick Bulletin*, **30**, 471.

BAILEY, L. W. 1923. George F. Matthew. *Royal Society of Canada, Proceedings and Transactions*, **17**, vii–x.

BELL, W. A. 1913. The Joggins Section. *In*: Anon. Excursion in Eastern Quebec and the Maritime Provinces. 12th International Geological Congress, Toronto, Excursion Guidebooks, **1**, 326–346.

BELL, W. A. 1914. Joggins Carboniferous Section, Nova Scotia. Geological Survey of Canada, *Summary Report*, 1912, 360–371.

BELL, W. A. 1943. *Carboniferous Rocks and Fossil Floras of Northern Nova Scotia*. Geological Survey of Canada, Memoirs, **238**, 120 pp.

BESSEY, C. E. 1914. Botanical notes. *Science*, **40**, 860–862.

BRIANT, K. 1962. *Marie Stopes: A Biography*. Hogarth Press, London, 286 pp.

BRICE, W. R. 1994. Charles Frederic Hartt (1840–1878): the early years. *Earth Sciences History*, **13**, 160–167.

CHALONER, W. G. 1959. Obituary: Marie Stopes. *Proceedings of the Geologists' Association*, **70**, 118–120.

CHALONER, W. G. 1995. Marie Stopes (1880–1958): The American connection. *In*: LYONS, P. C., MOREY, E. D. & WAGNER, R. H. (eds) *Historical Perspective of Early Twentieth Century Carboniferous Palaeobotany in North America*. Geological Society of America, Memoirs, **185**, 127–134.

CHALONER, W. G. 2005. The palaeobotanical work of Marie Stopes. *In*: BOWDEN, A. J., BUREK, C. V. & WILDING, R. (eds) *History of Palaeobotany: Selected Essays*. Geological Society, London, Special Publications, **241**, 127–135.

DARWIN, C. 1871. *The Descent of Man and Selection in Relation to Sex*. John Murray, London.

DAWSON, G. M. (ed.) 1900. Summary report on the operations of the Geological Survey of Canada for the year of 1899. *Geological Survey of Canada, Summary Report for 1899*, A1–A224.

DAWSON, J. W. 1861. On the Pre-Carboniferous flora of New Brunswick, Maine and Eastern Canada. *Canadian Naturalist*, **6**, 161–180.

DAWSON, J. W. 1862. On the flora of the Devonian period in northeastern America. *Quarterly Journal of the Geological Society, London*, **18**, 296–330.

DAWSON, J. W. 1863. Further observations on the Devonian plants of Maine, Gaspe, and New York. *Quarterly Journal of the Geological Society, London*, **19**, 458–469.

DAWSON, J. W. 1868. *Acadian Geology or The Geological Structure, Organic Remains, and Mineral Resources of Nova Scotia, New Brunswick and Prince Edward Island*. 2nd edn. MacMillan & Company, London, 694 pp.

DAWSON, J. W. 1871. The Fossil Land Plants of the Devonian and Upper Silurian Formations of Canada. *Geological Survey of Canada, Report*, **428**, 1–92 pp.

DAWSON, J. W. 1880. Revision of the land snails of the Palaeozoic era, with description of new species. *American Journal of Science*, **20**, 403–415.

DAWSON, J. W. 1881. Notes on New Erian (Devonian) plants. *Quarterly Journal of the Geological Society, London*, **37**, 299–308.

DAWSON, J. W. 1882. The fossil plants of the Erian (Devonian) and Upper Silurian formations of Canada. *Geological Survey of Canada, Report*, **429**, 95–142.

ELLS, R. W. 1901. The Devonian of the Acadian Provinces. *Canadian Record of Science*, **8**, 131–139.

FALCON-LANG, H. J. 2005. Earliest mountain forests. *Geology Today*, **21**, 181–184.

FALCON-LANG, H. J. 2006. A history of research at the Joggins Fossil Cliffs, Nova Scotia, Canada, the world's finest Pennsylvanian section. *Proceedings of the Geologists' Association*, **117**, 377–392.

FALCON-LANG, H. J. & CALDER, J. H. 2004. UNESCO World Heritage and the Joggins cliffs of Nova Scotia. *Geology Today*, **20**, 140–144.

FALCON-LANG, H. J. & CALDER, J. H. 2005. Sir William Dawson (1820–1899): a very modern palaeobotantist. *Atlantic Geology*, **41**, 103–114.

FALCON-LANG, H. J. & MILLER, R. F. 2007. Palaeoenvironments and palaeoecology of the Early Pennsylvanian Lancaster Formation ('Fern Ledges') of Saint John, New Brunswick, Canada. *Journal of the Geological Society, London*, **164**, in press.

FALCON-LANG, H. J., BENTON, M. J., BRADDY, S. J. & DAVIES, S. J. 2006. The Pennsylvanian tropical biome reconstructed from the Joggins Formation, of Nova Scotia, Canada. *Journal of the Geological Society, London*, **163**, 561–576.

FLETCHER, H. 1900. Geological nomenclature in Nova Scotia. *Transactions of the Nova Scotia Institute of Science*, **10**, 235–244.

GEINITZ, H. B. 1866. Report of meeting. Remark on Scudder's insects from Canada. Sitzungsberichte der Naturwissenschaftlichen Gellschraft Isis zu Dresden, **1–3**, 22.

GESNER, A. 1840. *Second Report on the Geological Survey of the Province of New Brunswick*. Henry Chubb, Saint John, 76 pp.

HAGEN, H. A. 1881. The Devonian insects of New Brunswick, with additional remarks upon a fern in the same slab with Plateparmera. *Bulletin of the Harvard Museum of Comparative Zoology*, **8**, 275–284.

HALL, R. 1977. *Marie Stopes: A Biography*. Andre Deutsch, London, 351 pp.

HARTT, C. F. 1865. Appendix L. On the Devonian plant locality of the 'Fern Ledges', Lancaster, New Brunswick. *In*: BAILEY, L. W. (ed.) *Observations on the Geology of Southern New Brunswick, Made Principally During the Summer of 1864 by Prof. L. W. Bailey, Messrs. Geo. F. Matthew, and C. F. Hartt.* Printed by order of the House of Assembly, Fredericton, 131–141.

JONGMANS, W. J. 1913. *Fossilium Catalogus. II: Plantae. Pars 1: Lycopodiales I*. W. Junk, Berlin, 52 pp.

KIDSTON, R. & JONGMANS, W. J. 1915–17. *Flore of the Carboniferous of the Netherlands and Adjacent Regions*. Vol. 1. *A Monograph of the Calamites of Western Europe*. 2 Vols. Medeelingen van Rijksopsporing van Delfstoffen, 207 pp. 158 pls.

LAVEINE, J.-P. 1995. Paul Bertrand: French Palaeobotanist. *In*: LYONS, P. C., MOREY, E. D. & WAGNER, R. H. (eds) *Historical Perspective of Early Twentieth Century Carboniferous Palaeobotany in North America*. Geological Society of America, Memoirs, **185**, 197–211.

LOPES, M. M. 1994. C. F. Hartt's contribution to Brazilian museums of natural history. *Earth Sciences History*, **13**, 174–179.

MCINTOSH, W. 1913. Curators report, Palaeontology. *Natural History Society of New Brunswick Bulletin*, **30**, 485.

MCINTOSH, W. 1914. Geological committee. *Natural History Society of New Brunswick Bulletin*, **31**, 80–81.

MATTHEW, G. F. 1906. New species and a new genus of Devonian plants. *Natural History Society of New Brunswick Bulletin*, **5**, 393–398.

MATTHEW, G. F. 1910. The geologic age of the Little River Group. *Transactions of the Royal Society of Canada, Section IV*, 67–75.

MATTHEW, G. F. 1911. Reports of committees: Geology (G. F. Matthew, Chairman). *Natural History Society of New Brunswick Bulletin*, **29**, 360.

MILLER, R. F. 1987. George Frederic Matthew: Victorian Science in Saint John. *New Brunswick Museum News*, August–September.

MILLER, R. F. 1995a. The status of *Belinuropsis wigudensis* Matthew, 1910, a Pennsylvanian merostome from New Brunswick, Canada. *Atlantic Geology*, **31**, 117–118.

MILLER, R. F. 1995b. The holotype of *Leaia silurica* Matthew, 1910, Pennsylvanian, Lancaster Formation, New Brunswick, *Atlantic Geology*, **31**, 23–24.

MILLER, R. F. & BUHAY, D. N. 1988. The Steinhammer Club: geology and a foundation for a natural history society in New Brunswick. *Geoscience Canada*, **15**, 221–226.

MILLER, R. F. & BUHAY, D. N. 1990. Life and letters of George Frederic Matthew: geologist and paleontologist. *New Brunswick Museum, Publications in Natural Science*, **8**, 1–89.

MOORE, R. C., ELIAS, M. K. & NEWELL, N. D. 1936. A 'Permian' flora from the Pennsylvanian rocks of Kansas. *Journal of Geology*, **44**, 1–31.

NANCE, R. D. 1987. Dextral transpression and Late Carboniferous sedimentation in the Fundy coastal zone of southern New Brunswick. *In*: BEAUMONT, C. & TANKARD, A. J. (eds) *Sedimentary Basins and Basin-forming Mechanism. Canadian Society of Petroleum Geologists, Memoirs*, **12**, 363–377.

PEACH, B. N., HORNE, J., GUNN, W., CLOUGH, C. T., HINXMAN, L. W. & TEALL, J. J. H. 1907. *The Geological Structure of the North-west Highlands of Scotland*. Memoirs of the Geological Survey of Great Britain, HMSO, Glasgow, 653 pp.

ROSE, J. 1992. *Marie Stopes and the Sexual Revolution*. Faber & Faber, London, 272 pp.

SALTER, J. W. 1862. On some fossil Crustacea from the Coal Measures and Devonian rocks of the British North America. *Quarterly Journal of the Geological Society*, London, **19**, 519–521.

SCUDDER, S. H. 1868a. The fossil insects of North America. *Geological Magazine, (Decade I)*, **5**, 172–177.

SCUDDER, S. H. 1868b. The insects of ancient America. *American Naturalist*, **1**, 625–631.

SHEETS-PYENSON, S. 1996. *John William Dawson: Hope, Faith and Science*. McGill-Queen's University Press, Montreal, 274 pp.

SMITH, C. H. 2002. George Mercer Dawson: pioneer explorer of western Canada. *GSA Today*, **12**, 16–17.

STOPES, M. C. 1903. On the leaf structure of *Cordaites*. *New Phytologist*, **2**, 92–98.

STOPES, M. C. 1906. A new fern from the Coal Measures: *Tubicaulis sutcliffi* sp. nov. *Memoirs and Proceedings of the Manchester Literary and Philosophical Society*, **50**, 1–34.

STOPES, M. C. 1910. *Ancient Plants*. Blackie, London.

STOPES, M. C. 1912. Palaeobotany versus stratigraphy in New Brunswick. *Report of the British Association for the Advancement of Science, Dundee, 4–11 September 1912*. John Murray, London, 471.

STOPES, M. C. 1913a. *Catalogue of the Mesozoic Plants in the British Museum (Natural History), Part I: Bibliography, Algae and Fungi*. British Museum (Natural History), London.

STOPES, M. C. 1913b. Fern Ledges. *In*: ANON. (ed.) *Excursion in Eastern Quebec and Maritime Provinces*. 12th International Geological Congress, Toronto. Excursion Guidebooks, **1**(2), 390–395.

STOPES, M. C. 1914. *The 'Fern Ledges' Carboniferous flora of St. John, New Brunswick*. Geological Survey of Canada, Memoirs, **41**, 142 pp.

STOPES, M. C. 1915a. *Catalogue of the Mesozoic Plants in the British Museum (Natural History), Part II: Lower Greensand (Aptian) Plants of Britain*. British Museum (Natural History), London.

STOPES, M. C. 1915b. René Zeiller. *The Times Literary Supplement*, 16 December.

STOPES, M. C. 1918a. *Married Love*. Firfield, London.

STOPES, M. C. 1918b. *Wise Parenthood*. Firfield, London.

STOPES, M. C. & WATSON, D. M. S. 1908. On the present distribution and origin of the calcareous concretions in coal seams known as 'coal balls'. *Philosophical Transactions of the Royal Society, London*, **B200**, 167–218.

WAGNER, R. H. 2005a. *Alethopteris lancifolia* Wagner, a rare element of the Lower Westphalian 'Fern Ledges' of Atlantic Canada. *Revista Española de Paleontología*, **20**, 15–19.

WAGNER, R. H. 2005b. *Dicranophyllum glabrum* (Dawson) Stopes, an unusual element of Lower Westphalian floras in Atlantic Canada. *Revista Española de Paleontología*, **20**, 7–13.

WAGNER, R. H. & VAN AMEROM, H. W. J. 1995. Wilhermus Josephus Jongmans (1878–1957): palaeobotanist, carboniferous stratigrapher, and floral biogeographer. *In*: LYONS, P. C., MOREY, E. D. & WAGNER, R. H. (eds) *Historical Perspective of Early Twentieth Century Carboniferous Palaeobotany in North America*. Geological Society of America, Memoirs, **185**, 75–90.

WATSON, J. 2005. One hundred and fifty years of palaeobotany at Manchester University. *In*: BOWDEN, A. J., BUREK, C. V. & WILDING, R. (eds) *History of Palaeobotany: Selected Essays*. Geological Society, London, Special Publications, **241**, 229–258.

WHITE, D. 1902. Stratigraphy versus palaeontology in Nova Scotia. *Science*, **16**, 232–235.

WHITE, D. 1911. 'A Carboniferous flora in the Silurian': discussion and correspondence. *Science*, **34**, 440–442.

WHITEAVES, J. F. 1899. The Devonian system in Canada. Address of the Vice President, Section E, American Association of Science, Ohio. *American Geology*, **24**, 210–240.

WILSON, W. J. 1909. Summary report dealing with the fieldwork in connexion with the collection of palaeontological material from the Devonian and Lower Carboniferous of New Brunswick. *Geological Survey of Canada, Summary Report for 1908*, 183–185.

WILSON, W. J. 1915. Palaeobotany. *Geological Survey of Canada, Summary Report for 1914*, 130–133.

ZASLOW, M. 1975. *Reading the Rocks: The Story of the Geological Survey of Canada, 1842–1972*. Macmillan, Toronto, 599 pp.

ZODROW, E. L. 1995. Walter Andrew Bell (1889–1969); Canadian palaeobotanist and earth scientist. *Geological Society of America, Memoirs*, **185**, 187–201.

Etheldred Benett (1776–1845): the first woman geologist?

S. LAMING[1] & D. LAMING[2]

[1]Gloucestershire Geoconservation Trust

[2]Herrington Geoscience, Exeter

Abstract: Etheldred Benett was chosen as the emblem for the Conference 'The role of women in the history of Geology' held in November 2005 at the Geological Society of London. This short synopsis was submitted as a poster to the conference.

Etheldred Benett (Fig. 1) was a highly esteemed geologist who lived in Wiltshire in the United Kingdom during the early part of the 19th century and made a large collection of local fossils. Some of these fossils were sent to collectors in the United States. In 1987 one specimen, named 'Drepanites striatus', a Cretaceous Greensand ammonite, was examined by curators in the Academy of Natural Sciences in Philadelphia. The staff of the Devizes Museum were approached for local topographical information, and the senior author, then Curator of Natural Sciences, undertook an enquiry into its origin (Nash 1990).

This led to an assessment of Benett's life history and her contribution to local geological collections. Her dates suggest that she was the first woman to be a scientific geologist in the country, being born 23 years before Mary Anning, and beginning her study of fossils in 1810 (Burek 2001).

Benett's only major publication was *A Catalogue of the Organic Remains of the County of Wiltshire* (Benett 1831). She was mentioned in early volumes of the *Wiltshire Archaeological and Natural History Magazine*, indicating that she was considered a geologist of some importance at the time.

Benett lived most of her life at the family home, Norton House, at Norton Bavant (Fig. 2), 4 km east–southeast of Warminster. She never married, but had the financial means to pursue the then new study of fossils, and collected from a wide range of sites within the county, especially in the SW part. An important factor in her development as a scientist was the influence of her

Fig. 1. Etheldred Benett.

Fig. 2. Norton House, Norton Bavant, northeast elevation.

brother-in-law, Aylmer Bourke Lambert, a keen botanist and antiquary. He was also an enthusiastic collector of fossils, and a Fellow of the Geological Society. Benett began to study fossils in 1810, and by 1813 had assembled an extensive collection, which included mammoth teeth and numerous Chalk fossils from around Salisbury, and was particularly rich in Greensand fossils from around Warminster.

She corresponded with several eminent geologists, including William Smith and James Sowerby, and her studies were of sufficient scientific merit to warrant publication in the latter's *Mineral Conchology*. In particular, correspondence with Gideon Mantell (1790–1852) of Sussex, the discoverer of *Iguanadon*, led to the definition of the general fossil sequence of Greensand and Chalk strata in Wiltshire and Sussex, and her contributions were generously acknowledged by him (Delair 1994).

One amusing incident came about when a visiting geologist from St Petersburg remarked on the excellence of her collections. This was mentioned to the Czar, and a selection of her fossils was sent to him for display in the Imperial collection. The Czar made her an 'honorary general' in the Russian army, thinking the name Etheldred indicated a male person! The fossils are still in St Petersburg.

Although after her death much of her fossil material was dispersed to locations unknown, numerous Jurassic, Cretaceous and Eocene fossils were donated to the British Museum (Natural History), the Geological Society, Yorkshire Museum and the Academy of Natural Sciences in Philadelphia. The latter received many specimens via Thomas B. Wilson, who acquired them in 1847.

The specimen (shown in Fig. 3), that gave rise to the original enquiry from Philadelphia, came from Shute Farm, 5 km southwest of Warminster. The exact site was probably a field now known as Picket Field, where there is a slight hollow of 'blue-grey clay' which still yields many fossils (Fig. 4),

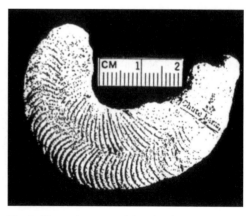

Fig. 3. '*Drepanites striatus*' found at Shute Farm and now at the Academy of Sciences, Philadelphia.

Fig. 4. There are still fossils to be found at Shute Farm.

although not of the quality which prevailed in the 19th century.

References

BENETT, E. 1831. *A Catalogue of the Organic Remains of the County of Wiltshire*. J. L. Vardy, Warminster.

BUREK, C. V. 2001. The first lady geologist or collector par excellence? *Geology Today*, **17**(5), 192–194.

DELAIR, J. B. 1994. Pioneer geologists of the Salisbury area. *Wiltshire Archaeological and Natural History Magazine*, **87**, 127–141.

NASH, S. E. 1990. The collections and life history of Etheldred Benett (1776–1845). *Wiltshire Archaeological and Natural History Magazine*, **83**, 163–169.

TORRENS, H. S. 1985. Women in geology. 2: Etheldred Benett. *Open Earth*, **21**, 12–13.

Grace Anne Milne (Lady Prestwich): more than an amanuensis?

J. D. MATHER[1] & I. CAMPBELL[2]

[1]Department of Geology, Royal Holloway, University of London, Egham, Surrey, TW20 0EX, UK

[2]Libraries, Culture and Community Learning, County Hall, St Anne's Crescent, Lewes, BN7 1SW, UK

Abstract: Grace Anne Milne, born in 1832, was the eldest child of James Milne of Findhorn in Morayshire and his wife Louisa Falconer, sister of the eminent botanist and palaeontologist Hugh Falconer. A marriage to George McCall in 1854 was short-lived. Widowed within 18 months and also losing her infant son, bereavement was followed by some years of depression. At this stage in her life an interest in geology was kindled by her uncle, who had been forced to give up his post in India because of the state of his health. From 1858 she travelled abroad with Falconer and subsequently resided with him in London until his death in 1865. During this period Grace both organized his household and acted as his secretary and companion. His regular and frank letters to her, about his scientific ideas, demonstrate that the relationship was more than one of domestic convenience. In 1870 Grace married Joseph Prestwich, a friend of Falconer and a London wine merchant who spent all his spare time 'geologizing'. They lived at Shoreham near Sevenoaks in Kent and, between 1874 and 1887, lived for part of the year in Oxford after Prestwich's appointment as professor of geology at the university. With Prestwich's encouragement, from 1874, she began to publish novels, travel articles and scientific papers, as well as helping him in the preparation of his own lectures and diagrams. Her six *Chapters on Geology* were published in 1880 as well as other articles on subjects such as Scottish scenery and the Channel Tunnel. Most were published in magazines, such as *Leisure Hour* and *Every Girl's Magazine*, designed for a readership from the prosperous middle classes. Her interest in geology meant that she was able to play a full part in Prestwich's life, helping him with his work and travelling with him on his geological adventures. She was a geologist in her own right and held in high regard by the Fellows of the Geological Society.

In the same year in which Grace Anne Milne (1832–99) was born in northeast Scotland there were 'grand disputes at the Geological Society about the propriety of admitting ladies' to Charles Lyell's lectures at King's College, London (Wilson 1972). The following year ladies were excluded 'because their presence diverted the attention of the young students' (Woodward 1908). Roderick Impey Murchison (1792–1871) was the President of the Geological Society: a man whose private journals were full of overtly sexist observations and who considered that women were of a lower status than men (Collie & Diemer 2004). At this time the role of women in science was largely that of assistant as illustrator, secretary or supportive wife, sister, daughter or niece.

During Grace's lifetime women took a more prominent role and by the time of her death, the work of women had been reported in numerous papers published in the *Quarterly Journal of the Geological Society*. However, even a sympathetic scientist, such as H. B. Woodward, still felt that geological mapping was unsuitable for women (Woodward 1895).

Grace Milne was associated with two prominent 19th-century scientists: firstly with her uncle, Hugh Falconer (1808–65), the botanist and palaeontologist, and secondly, after Falconer's death, with her husband, the geologist Joseph Prestwich (1812–96). She was not an invisible woman but left behind a body of published work, including two novels, travel articles, geological papers and a biography of her husband (G. A. Prestwich 1899). Her youngest sister, Louisa, edited some of her essays and wrote a brief memoir on her life (Milne 1901), and a collection of letters and manuscripts in the Falconer Museum in Forres is thought to have been assembled by Grace (Boylan 1977). The present paper reviews these various sources, together with some additional material, to provide an assessment of her own work and her contribution to the work of both Falconer and Prestwich.

Family background

Grace Anne Milne was born on 18 December 1832 in the village of Kinloss, 5 km northeast of the town of Forres, in Morayshire, close to the southern shore of the Moray Firth. Her father was James Milne (1798–1853) of Findhorn, a small port on the coast about 4 km northwest of Kinloss. In 1766 the estate on which Findhorn lies was acquired by

Sir Hector Munro of Novar House in Easter Ross on the northern side of the Moray Firth. An Act of Parliament gave him authority to create a harbour at Findhorn and to regulate shipping and collect harbour dues. In the 19th century his descendants, operating as Novar Estates, employed James Milne to carry out these tasks. Milne was also a shipowner in his own right and by 1842 he ran a fleet of eight ships out of Findhorn, including the 224-tonne *Grace Milne*, proudly named after his eldest child. At the same time he operated as a merchant and ship's chandler in Findhorn. The old chandlery building is now the Findhorn Village Hall, known as the James Milne Institute.

Grace's mother, Louisa Falconer (1802–78), was born in Forres, the daughter of a local merchant, David Falconer, and his wife Isabel McRae. Apart from Hugh, two more of Louisa's brothers, Alexander (1797–1857) and Charles (1804–79) were to play a significant role in Grace's life.

Grace was the eldest of five children, the rest of the family comprising twins Margaret (1834–1916) and Isabella (1834–1921), the only son Robert (1838–1919) and Louisa Elizabeth (1844–1920). The family home was at Abbeyside, adjoining the old abbey of Kinloss. Sometime between the death of her father in 1853 and the census of 1861 the family moved to a house in Forres and later still to Edinburgh, where her widowed mother and unmarried sisters were living in 1871. The family home in Kinloss still stands but is now called Kinloss House and appears to have been extensively altered by new owners in 1887. James Milne was a Justice of the Peace and an important and wealthy man. In consequence the family occupied a privileged position within the local community.

The young Grace is described as an attractive, bright child with an excellent memory and a gift for making portraits (Milne 1901). Brought up in an environment in which little was permitted on Sundays other than unlimited sermons, she could apparently write down from memory any sermon that she heard by the time she was 12 years old. At the age of 6 years she was placed as a boarder at school, perhaps in England, although that is unclear from her sister's memoir (Milne 1901).

Following her schooling, she returned to Abbeyside until, on 18 October 1854, she married George McCall (1816–56), a wine merchant and member of a large and prosperous Glasgow family. She became engaged to McCall on a visit to his widowed mother Janet and sister Ann in Glasgow (Milne 1901). It is not clear how Grace knew the family but she was some 18 months younger than Ann and it may be that they had become friends at school. A son, James, was born on 23 November 1855, but her happiness was short-lived. Her husband died of pleuro-pneumonia and haemoptesis in March 1856 after several months of illness and shortly afterwards, in May, her infant son died of hydrocephalus.

After these tragic events she returned to her mother's home at Abbeyside, where it is perhaps not surprising that she lapsed into a period of depression, writing to her mother-in-law that she felt 'so crushed'. A manuscript found in her papers, and dating from this period, refers to her lying 'half waking, half musing on the listless waste of my aimless life, thinking that I was a useless weed in the garden of well-growing plants and trees' (Milne 1901, p. 9). She found herself roles in nursing her uncle, Alexander, who had made his fortune as a merchant in Calcutta, through a long illness and death and helping her brother prepare for his military examinations to join the Indian Army. She also seems to have been sustained by her strong religious convictions, but it was to be her uncle Hugh who was to take charge of his niece and change the direction of her life.

Life with Hugh Falconer

Hugh Falconer (Fig. 1) was educated at the local grammar school and Aberdeen University, from where he went on to Edinburgh University, graduating as an MD in 1829. At Edinburgh he also attended botany classes, which were taught by Robert Graham, and geology classes, taught by Robert Jameson. Following short-term work in London, curating both botanical and fossil mammal collections, he went to India as a surgeon with the Bengal Establishment of the East India Company, arriving in Calcutta in September 1830. His work as a surgeon seems to have been short-lived and by 1832 he had become superintendent of the Saharanpur/Serempore Botanic Gardens (Moore 2004). Here he quickly established himself as an outstanding explorer and botanist, as well as a leading vertebrate palaeontologist. With P. T. Cautley, his investigations of the Tertiary fossil mammals of the Sewalik Hills earned him an international reputation as well as the Wollaston Medal of the Geological Society of London in 1837 at the age of 29 years.

He returned to the United Kingdom in 1842, as his health began to deteriorate, bringing with him about 5 tonnes of fossil bones, which he worked on over the next 5 years. He was elected a Fellow of the Royal Society in 1845 but in 1847 returned to India on his appointment as professor of botany at Calcutta Medical College and later as superintendent of the Royal Botanic Garden, where he undertook research associated with the introduction of tea into India (Moore 2004). However, he continued to

Fig. 1. Hugh Falconer, probably *c.* 1860. (From a photograph reproduced in G. A. Prestwich 1899.)

suffer from poor health and left India for the last time in 1855 to reside in London, first at 31 Sackville Street and then at 21 Park Crescent in Marylebone.

While back from India between 1842 and 1847 Falconer seems to have taken a particular interest in his intelligent young niece, an interest which continued when he returned to Calcutta. He wrote to her regularly and when, following a lecture she had attended, Grace had problems coming to terms with the scientific evidence that millions of years had elapsed between the first appearance of life on Earth and the present day, he wrote explaining the philosophical basis of geological evidence (letter from Hugh Falconer to Grace Milne, 19 January 1854; Milne 1901, p. 5). It seems that even at this stage in her life Grace had an interest in geology.

Following Falconer's return to London his health was still poor and he decided to spend the winter of 1858 in the western Mediterranean. He had already urged the young widow to travel abroad, writing to her: 'Till you had done so, I could not look upon you in another light than a demure little Puritan, whose strength lay in the ignorance of her own limited experience' (Milne 1901, p. 8). He finally persuaded her to accompany him with the demand that she took only two gowns and no band-boxes! Grace travelled with him as his secretary, carefully sketching fossils and geological sections (G. A. Prestwich 1895). During the trip she kept a journal which was used by her sister Louisa when preparing her memoir (Milne 1901) and by Grace herself when writing a number of travel articles later in her life.

They finally left England towards the end of October 1858, travelling leisurely through France. Their first stop was at Abbeville, the home of Jacques Boucher de Perthes, a visit described nearly 37 years later by Grace in an article in *Blackwood's Magazine* (G. A. Prestwich 1895). Falconer and his friend and colleague Joseph Prestwich were at that time the most active members of a Geological Society committee superintending excavations at Brixham Cave in Devon, recently commenced by William Pengelly. As early as 1846, Boucher de Perthes had published a description of the gravels of the Somme Valley, in northern France, in which he claimed to have found flint implements among fossil remains of extinct animals at the base of a bed of gravel underlying many metres of modern deposits. Abbeville was an out of the way place, very little visited, and his claims were ridiculed by French scientists, who met him only in Paris. With the Brixham excavations in mind, Falconer made a detour to Abbeville and he and Grace spent a day examining his collection. It was a cold November day and Grace complains that the gallery, in which the material was displayed, was like an ice-house, there was no fire and handling the flints was freezing work (G. A. Prestwich 1895). Falconer immediately alerted Prestwich to the probable authenticity of Boucher de Perthes' material and suggested that he visited him next time he was in France. Prestwich's visit and subsequent communications to the Royal Society (J. Prestwich 1859*a*) and to the French Académie des Sciences (J. Prestwich 1859*b*) caused a sensation and a pilgrimage of British and French scientists to the valley of the Somme. One can only imagine how this affected the young Grace McCall, now aged 26 years, who, from obscurity, suddenly found herself as one of the first people to examine material which was to lead to a whole new view of the antiquity of man.

From Abbeville the travellers went on to Paris in order to pick up an Italian maid whom Mary Mohl, the English wife of the philosopher Jules Mohl, had selected for Grace. Jules was a friend of Falconer and dined with him frequently when he visited England each autumn. Madame Mohl filled a prominent position in Paris society and Grace later described her visits to dinner parties at her salon in *Blackwood's Magazine* (G. A. Prestwich 1893).

From Paris they travelled via Lyons, Avignon and Montpellier to Marseilles and then on to Genoa, where they took a boat to Naples. Along the route Falconer visited museums, examining and comparing fossils, usually the bones and teeth of mammals, which Grace often figured for him. In Naples they visited Vesuvius, which at that time was in the middle of an eruption. The sight, in the moonlight, of a river of boiling lava, and the oppressive sulphurous atmosphere, greatly impressed Grace (Milne 1901). The problems they had in Naples, clearing through Customs a large wooden box containing casts of the head and bones of a rhinoceros, were described by Grace some 30 years later (G. A. Prestwich 1888).

They arrived at their destination, Sicily, on 1 February 1859, making Palermo their base for the next 2 months. Falconer spent his time in museums, looking at exhumed fossil material and exploring ossiferous caves including the 'Grotta di Maccagnone'. Here he found flint implements adhering to the roof matrix, mingled with the remains of hyaenas now extinct in Europe (Falconer 1860). Grace sketched fossils and artefacts for him (Fig. 2) while also drawing, tracing and colouring geological sections. They visited Mount Etna but, although she was impressed by the many fumeroles, found it was nowhere near as 'awful and striking' as Vesuvius (Milne 1901).

They returned north in March, where they were joined, in Naples, by Hugh's brother, Charles, recently back in Europe after 25 years in Australia, and Grace's sister Margaret. After spending Easter in Rome, the group moved on to Florence where in May they spent just over 2 weeks exploring the churches, palaces and pictures. While there, they visited their countrywoman, the distinguished mathematician Mary Somerville, who was an old friend of Hugh Falconer. Grace's recollections of this evening are found in an article written some 33 years later and published in *Leisure Hour* (G. A. Prestwich 1892). From Florence they continued northwards, at one time coming within 25 miles of the battle of Magenta which was being fought that day between the Austrian and Italian armies. After crossing the Alps they eventually arrived back in London on 29 June.

It is not clear whether or not Grace stayed in London or returned to her family in Scotland after her return. However, the winter of 1859/60 found her living in a villa in Torquay which had been rented by Falconer (Milne 1901), who at that time was busy with the exploration of the Brixham Cave. In the summer of 1860 she was with Falconer in North Wales where she drew a section of Cefn Cavern near St Asaph (Fig. 3). Together with her sister Margaret she was again abroad with her uncle during the winter of 1860/61. They halted at Amiens and Paris and then took a steamer down the Rhone from Lyons. Arriving in Nice on 3 November they remained until 1 April, when they started on a tour of northern Italy.

Travelling in northern Italy at this time could probably be regarded as foolhardy. War with Austria seemed likely and there were newly recruited troops everywhere. The party travelled from Nice to Genoa and then on to Turin. They spoke to Garibaldi in Turin and the sisters were present in the Italian chamber, on 18 April 1861, at Garibaldi's first appearance as a deputy. They then proceeded to Milan and the museum, where Grace acted as Falconer's secretary for 3 days, writing a description of a rhinoceros skull. Trips to Bologna and the Adriatic coast followed and ultimately they reached Venice, from whence they travelled to Innsbruck and then homewards via the Rhine (Milne 1901). During the later part of their trip they made a detour to Mantua and Verona, the two great Austrian military centres and the stronghold of Austrian power in Italy. Their experiences are detailed in an article written by Grace but not published until after her death (G. A. Prestwich 1901).

In the summer of 1861, Grace was with Falconer at Stouthall, on the Gower Peninsula in South Wales, the home of his friend Lieutenant-Colonel E. R. Wood, where Falconer explored more ossiferous caves (O'Connor 2006). She was able to write to her sister-in-law that 'I have come to England this year much stronger than for several years' (letter from Grace McCall to Annie McCall, 5 August 1861; Milne 1901, p. 50).

From this time onwards Grace made her home with her uncle at 21 Park Crescent on the edge of Regent's Park, but with frequent visits to her mother in Edinburgh. His letters to her when she was away suggest that she ran the household and that the servants eagerly awaited her return. Thus he writes 'Les domestiques of a morning speak daggers of impatient inquiry by their looks, but I make no sign ...' (letter from Hugh Falconer to Grace McCall, 14 October 1863; Milne 1901, p. 53). She spoke both French and Italian and acted as an accomplished hostess to his guests and scientific friends and as his companion when invitations were reciprocated. She also helped him geologically and, by this time, must have gained a wide general understanding of geology through both fieldwork and museum visits with him and interaction with his geological friends, such as Sir Charles Lyell and Joseph Prestwich. Between November 1860 and January 1863, Fellows of the

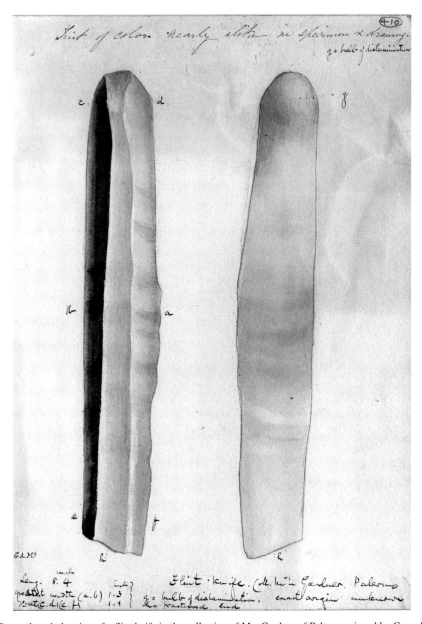

Fig. 2. Pen and wash drawing of a flint knife in the collection of Mrs Gardner of Palermo, signed by Grace McCall. The length of the knife is 21.3 cm; the greatest width (a–b) is 3.3 cm; the width at c–d and e–f is 2.8 cm; h is the fractured end and g is the bulb of dislamination. (The original is held by Moray Council Museums Service.)

Geological Society were invited to bring their wives, daughters or lady friends to evening meetings (Woodward 1908) and it is quite possible that Grace was one of the few ladies who attended.

In 1863 Lyell published his book *Geological Evidences of the Antiquity of Man* exploiting the discoveries of Falconer and Prestwich with little acknowledgment (Lyell 1863). This distressed Falconer and led to acrimonious exchanges. In the same year Jacques Boucher de Perthes announced that a workman from the gravel pit at Moulin Quignon near Abbeville had extracted a human jaw from a terrace some 33 m above the level of the River Somme. The authenticity of this jaw

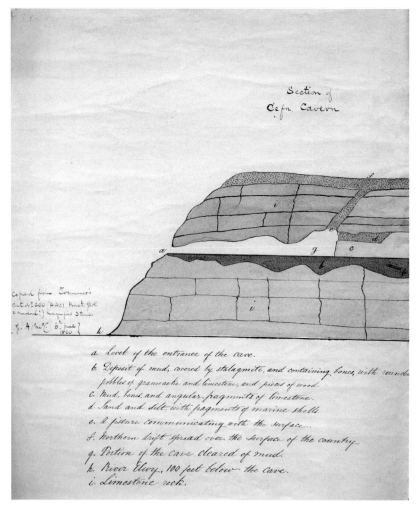

Fig. 3. Pen and watercolour drawing entitled 'Section of Cefn Cavern'. Copied from Trimmer's cut number 200 by Grace McCall on 6 June 1860. (The original is held by Moray Council Museums Service.)

was questioned and an Anglo-French Commission was set up to settle the argument. Falconer was a member of this Commission which met three times in Paris. While in France he wrote a series of revealing letters to Grace which showed his increasing concern about the affair. These letters and the story of the controversy have been described in detail by Boylan (1979). The letters demonstrate that the relationship which existed between Falconer and his niece was more than one of domestic convenience. Grace was also his scientific confidante to whom he reported on a day-to-day basis about the workings of the Commission, together with his ideas and fears.

The following year, Falconer, together with George Busk, visited Gibraltar as a consequence of the discovery of human and animal remains together with artefacts in caves within the Rock (Rose 2006). On the return journey a broken axle on the summit of the Sierra Morena in Spain meant two nights spent in the open air on the top of the coach. This resulted in a chill, which developed into rheumatic fever and ended fatally on 31 January 1865. The funeral on the 4 February at Kensal Green Cemetery must have been a sad affair for Grace as she was again deprived of someone she loved and who she described as 'a brilliant companion' (G. A. Prestwich 1899).

Following Falconer's death, Grace remained in London at 21 Park Crescent, which was now occupied by his brother Charles. Although nothing is known of her activities over the next 5 years it is assumed that she maintained contact with Falconer's geological friends as, in November 1869, she became engaged to one of them: his collaborator, the London wine merchant, Joseph Prestwich. With their marriage on 6 February 1870 at St Marylebone Church in London another chapter of Grace's life began.

Life with Joseph Prestwich

Joseph Prestwich was born in Clapham in south London, the eldest surviving son of Joseph Prestwich and his wife Catherine, daughter of Edward Blakeway, the squire of Broseley in Shropshire (G. A. Prestwich 1899). He went to school in both England and France becoming a fluent French speaker. In 1828, aged 16 years, he enrolled at University College, London, then recently opened. The family home in South Lambeth was 6.5 km away from the college so that, each day, he had a 13 km walk. He neglected other subjects in favour of chemistry and natural philosophy and 'all my spare time, spare pocket money and spare thoughts were spent on chemistry' (G. A. Prestwich 1899). As part of the chemistry course, Professor Edward Turner (1796–1837), a former pupil of Jameson at Edinburgh, gave three lectures on geology and mineralogy, which were delivered on Saturdays so that they did not interfere with the rest of the course (Kirk 1996). Prestwich found Turner 'a popular and excellent teacher' (G. A. Prestwich 1899) and it may have been his influence which drew him to geology.

In 1830, at the age of 18 years, he left University College and joined the family wine firm in Mark Lane in the City of London where, for the next 42 years, he worked a full week in the wine trade, taking control of the firm in 1842. Geological research was carried out in his own time, before and after work and over weekends. He drove himself hard, filling days which started at 6am and finished at 11pm with business, study and self-improvement (G. A. Prestwich 1899; Thackray 2004). His health was impaired and he suffered from stress-related illnesses for the rest of his life. He became an authority on the British Tertiary, a recognized expert on underground water supply and, with his friends Hugh Falconer and the archaeologist and geologist John Evans (1823–1908), a major figure in the debate on the antiquity of man.

For many years Joseph (Fig. 4) lived in a bachelor apartment in the City. However, the death of his father in 1856 led to his return to the family home where his youngest sister, Civil Mary, some 10 years his junior, had been left alone. According to Grace, 'She at once became his secretary and amanuensis, devoting her whole life to the furtherance

Fig. 4. Joseph Prestwich, probably c. 1850. (From a photograph reproduced in G. A. Prestwich 1899.)

of his scientific work, freeing his mind and time from all the wear and tear of petty distractions' (G. A. Prestwich 1899, p. 102). Although he now had an easier home life, stress levels were high and the two siblings decided to move out of London, building a house high on the chalk downland above the village of Shoreham in Kent, which they named Darent-Hulme. Unfortunately Civil did not live to occupy the house, dying after a short illness in December 1866. Prestwich finally moved into the Shoreham house in January 1869 and it was here (Fig. 5) that he brought his new bride just over a year later.

At the time of their marriage Joseph was almost 58 years old and Grace 20 years his junior (Fig. 6). Following a honeymoon trip to France and Italy they settled into a busy life at Darent-Hulme. Prestwich had assumed the presidency of the Geological Society a few days before his marriage. This role, together with 4 hours spent on the journey backwards and forwards to Mark Lane each day, meant that he had little time to write up his scientific work or study the material he had collected. His wife was clearly worried as, a year after their marriage, she wrote to John Evans asking for his help. Her letter, dated 8 February 1872 (Oxford University, Ashmoleum Museum, Department of Antiquities, personal papers of Sir John Evans,

Fig. 5. Darent-Hulme as it is today. (Photographed by J. D. Mather, April 2006.)

Fig. 6. Grace Prestwich in 1876, 2 years after she moved to Oxford. (From a pencil drawing by W. E. Miller, reproduced in Milne 1901.)

File JE/B/1/11) is significant as it demonstrates her role in his life and her concern for his welfare, which continued throughout their marriage. She wrote:

My dear Mr Evans,

I have been long thinking of writing to you concerning Joseph. He is working so much too hard that I cannot but feel anxious for him to get some comparative rest. Hard work at his age tells more upon a constitution than when younger. You are one of his best if not his very best friend and so I thought that I would consult you. My idea is for him to get some scientific appointment which would enable him to give up the city. The former would be a pleasing occupation; whilst the other is mere labour which combined with all his other pursuits becomes overwork.

I imagine that Sir Roderick's appointment on the Survey is likely to be given to someone on the staff – if given at all or let first. I thought that it would be for Mr Ramsay, but since then I have heard it stated very positively that he is not to have it and this by one who holds a prominent appointment himself and is likely to know.

Whilst there was a chance of Mr Ramsey's getting it, I know that Joseph would never have given it a thought, but should he not have it, it seems to me that it would be such a nice thing for Joseph – too good indeed almost to hope for. I am very ignorant concerning all these things but I feel sure that I may confide my desire to you and that your friendship for Joseph will be a significant excuse for my troubling you. I have long been anxious about him and am still more so now that he is married.

Trusting that Mrs Evans and all are well. Believe me very truly yours. G Prestwich.

Whoever her informant was he was wrong and the position went to Ramsay, a choice which was 'inevitable' according to Flett (1937).

Perhaps pressure from his wife eventually convinced Joseph as, on 1 August 1872, he sold the business and was able to devote himself entirely to his science. Two years later, Grace's wish that he got an academic appointment became reality when, although not a formal candidate, he was offered the chair of geology at Oxford, vacant after the death of John Phillips. From October 1974 until his retirement at the end of 1887 they spent part of the year in Oxford and part at Darent-Hulme.

Sometime after Grace's marriage, her mother and sisters moved from Edinburgh to live with her Uncle Charles at 21 Park Crescent. Here she and Joseph were regular visitors on their frequent trips to London. These visits continued after the death of her mother on 5 January 1878 and Charles on 13 December 1879. Both are buried, along with their brother Hugh, in Kensal Green Cemetery, beneath a pedestal of grey Aberdeen granite (Friends of Kensal Green Cemetery 1997).

Although Grace had stepped into the shoes of the deceased Civil, unlike her she did not devote herself entirely to the work of Prestwich. The comfortable lifestyle which they enjoyed at Shoreham and in Oxford enabled her to develop her own interests and over the next three decades she wrote a series of books and articles. Some of the articles are derived from the journal which she kept on her travels with Hugh Falconer, some are based on her childhood memories and others concern her geological excursions with Prestwich.

Her first published item is probably a letter dated 23 March 1872, which she wrote to *The Garden*, concerning the creeping myrtle (*Myrsiphyllum asparagoides*) which she had seen on her visit to Palermo in 1859. She advocated its cultivation in greenhouses in Britain on account of its long retention of green-ness and freshness, making it an ideal addition to the ballroom toilet. The letter is reproduced in her sister's memoir (Milne 1901, p. 25).

Her first novel – *The Harbour Bar: A Tale of Scottish Life* – was published in 1874 in two volumes (G. A. Prestwich 1874) and was reissued in a cheaper one-volume edition in 1876 (G. A. Prestwich 1876). A second novel – *Enga* – again in two volumes, was published in 1880 (G. A. Prestwich 1880). Both books were issued by Macmillan, one of seven publishers who dominated the book-selling market at that time. The novels were published anonymously and are now difficult to find, partly because of this anonymous attribution.

This was the period of the 'three-decker' novel, bought almost exclusively by wealthy rich women or by the circulating libraries, of which Mudie's was the best known, who then loaned them for a subscription to middle-class women. Although the Education Act had been passed in 1870, its impact had not yet been felt on the working classes. Therefore the type of novel that would be greeted most favourably among the small readership – a best-seller could expect to sell 10 000 copies – would be about a respectable heroine, who would have to deal with emotional, religious or moral dilemmas, or a combination of these, and would either live happily ever after – if she had been the victim of a miscarriage of justice, or would die in tragic circumstances, having heroically restored a damaged reputation (Campbell 1980).

The Harbour Bar was published a year after Hardy's *Far from the Madding Crowd* (Seymour-Smith 1980) Both are set in bucolic corners of the United Kingdom, but there the resemblance ends. The former is based around the small fishing community of Innan, 8 km from the ancient burgh of Inchmerle, on the northeast coast of Scotland. Innan is clearly a pseudonym for Findhorn and Inchmerle for Forres. It contrasts the lives of a fisherman, Walter Storm, and his extended family with those of the middle-class Captain Main and Doctor Munro. Main's son runs off and marries Munro's niece, but circumstances leave their daughter Mona an orphan and we find her adopted by the fisherman Storm. Using Mona's growing up as a catalyst we learn about the dangerous and hard lives of fishermen and their womenfolk and their relationships with their aristocratic employers and people like the captain and the doctor. The action shows that ethical behaviour, love and honour transcend all classes and characters are drawn sympathetically and allowed to develop their own morality and answers to life's pitfalls. Unfortunately, the novel is written in a densely plotted, descriptive and objective style, using an overview technique that does not allow bonding with the characters. The extensive use of Scottish dialect is a serious barrier, although it would have been less so for its 19th-century readers and was used by other writers, notably Sir Walter Scott. These factors lead to it being relatively inaccessible and, while the story line is of interest and the feel for life in a remote part of Scotland enlightening, it cannot be compared to contemporary books, such as Eliot's *Middlemarch* or Trollope's *The Way We Live Now* (Seymour-Smith 1980). A reviewer in *The Times* of 27 May 1875 thought that *The Harbour Bar* was 'chiefly to be commended for the fresh, breezy pictures, on almost every page, of the wild bleak coast, with its pale-blue sky and indistinct outline of sandy whinbanks'.

Enga was published in two volumes around the same time as Tolstoy's *Anna Karenina* and

Dostoyevsky's *Brothers Karamazov*, two of the greatest works of literature of their time. Grace's writing style has developed, as has her plotting, but it is the improvement in characterization that is most obvious. Marion Comyn, the sister of a poor but decent man, falls in love with and secretly marries the local laird and has a child. The laird, Hugh Innes, dies soon afterwards and the child is taken from Marion by Hugh's evil cousin, the new laird. Many years later, the gentle born Enga Malcolmson falls in love with orphan Alister Angus. He discovers that there is some secret about his ancestry and leaves England after betrothing himself to Enga, in order to seek his fortune and pay back his adoptive family. However, Laird Innes is also in love with Enga, and using his power over her bankrupt father, he tries to force Enga to marry him. But Enga remains true and, despite vicissitudes and obstacles, not least an unexplained complete silence from her fiancé, her integrity and morality are not breached, good triumphs over evil and Enga and Alister, Marion's lost son, find happiness. The novel ends with a catastrophic river flood which appears to be based on the Great Flood of Moray of August 1829, when the area around Forres was badly flooded and the fishermen of Findhorn sailed inland, plucking whole families from the beams of their wrecked cottages (Willing & Munro 2002).

The panoramic vision of the novel and its conventional plot are typical of the era and would have appealed to the potential readership (Campbell 1980). It is well written and paints a sympathetic if melodramatic picture of town and country life in Scotland and covers the range of social classes and shows how they interacted. There would have been a certain exotic flavour to the Scottish setting, although, as with *The Harbour Bar*, the Scottish dialect makes some sections of the story impenetrable. Unlike the great novels of the time, it lacks subtlety and humour and the morality and ethics under discussion are rather earnest. However, it is a major improvement on *The Harbour Bar* – which was so successful that it ran to a second edition – and holds its own as a readable and entertaining social comment on life at the time and the choices people had to make.

During her time at Oxford, Grace also wrote a series of articles for *Every Girl's Magazine* and *Good Words* on geological themes. Both these magazines were directed towards the leisured middle class, with *Good Words* having a significant percentage of religious articles. In the autumn of 1878 she accompanied Prestwich on a visit to Scotland to see the parallel roads of Lochaber. On their return they both wrote an account of their visit. Joseph's was published in the *Transactions of the Royal Society* (J. Prestwich 1879) and Grace's in *Good Words* (G. A. Prestwich 1879). Although Grace's article is more a travelogue than a scientific account it does include a discussion of the various theories for the origin of the parallel roads. She later wrote another article based on the same trip, describing their journey by coach through the Pass of Glencoe which she found 'inconceivably grand' (G. A. Prestwich 1885).

On 9 December 1873 Joseph read a paper at the Institution of Civil Engineers on the geological conditions affecting the construction of a tunnel between England and France, which was published the following year (J. Prestwich 1874). Grace used this paper to prepare her own account in *Good Words*. This has a greater geological content than previous articles, explaining to her readers the historic ideas for the construction of a bridge or tunnel across the Channel, outlining the geology of the rocks below the sea floor and discussing the choice of strata through which the tunnel might be driven (G. A. Prestwich 1881).

The geological emphasis of her work continued with the publication of *Chapters on Geology* in *Every Girl's Magazine* during 1880 and 1881 and was consolidated in *Every Girl's Annual* for 1882 (G. A. Prestwich 1882). This is a stratigraphical account and, following an introduction (Chapter 1), successive chapters deal with the Palaeozoic (Chapters 2 and 3), the Mesozoic (Chapters 4 and 5) and the Cainozoic (Chapter 6). Over the previous 25 years, Grace had seen many of the rocks she describes in the field and discussed geology with most of the famous geologists of the day. Her account, therefore, is not just based on the textbooks which were currently available but is enhanced by her own observations and with research with which she had been personally involved. According to her sister, it had been her intention to reprint these chapters with additions in book form but other commitments meant that this was never achieved (Milne 1901).

The way in which she reconciled her religious beliefs with geological timescales is well set out at the end of the first of her *Chapters on Geology* (G. A. Prestwich 1882, pp. 16–17):

Our world of to-day flourishes not upon a single tomb of the past, but upon a succession of tombs superimposed one above another, each representing a period of time too vast to be counted by finite years. For we must remember that before Him who made the earth and the heavens one day is as 'a thousand years, and a thousand years as one day'. One of the great difficulties in dealing with geology has been caused by the attempt (well meant, no doubt) to solve its problems, and to measure the vast operations of nature through cycles of ages by our small and finite standard of time. But God's handiwork is not to be strained and bent down to fit into the limits of fleeting days and years! What philosopher can tell the time needful to make a single grain of sand, or for the formation of one member of the Sedimentary Series? Or who can compute the number of ages required for the accumulation or consolidation of one vertical mile of strata? The answer is beyond our comprehension; we are impressed with awe and wonder. The sum and figures are beyond the range of human arithmetic.

Her *Chapters on Geology* were complemented by a long article, published in three parts in *Good Words*, on the forces which had shaped the present face of the Earth. This is an excellent review, designed for her audience of upper- and middle-class women, which discusses the origin of mountain chains, deep ocean basins, coral islands and lakes. It rehearses the theories which had influenced geological thought over the previous 50 years, presenting a well-balanced and well-written account (G. A. Prestwich 1883*a*).

Prestwich's life at Oxford was a busy one. He was elected a member of Christ Church and a number of dining clubs including the Ashmolean. He had lecture courses to prepare and he had to make himself familiar with the geological collections. In all his activities Grace played a major role. She was an able hostess at dinners for his friends and colleagues; she helped in the preparation of his scientific papers as well as his geological lectures and diagrams and accompanied him on annual geological excursions. Apparently, she was also a constant attendant at his lectures in the Museum (Milne 1901).

Towards the end of his time at Oxford, Prestwich produced a two-volume textbook (J. Prestwich 1886–88), which he wrote as an antidote to the rigid uniformitarian views that were the legacy of successive editions of Lyell's *Principles* and that he thought were harming geology (Thackray 2004). According to her sister (Milne 1901), some of the illustrations were done by Grace. However, there is no acknowledgement to her in the text, although there is one to Gertrude Woodward, daughter of Henry, who produced the lithographs of fossils in volume two, and to Mr J. D. Cooper, who produced most of the woodcuts.

As well as her support for Prestwich, Grace had a life of her own. She was active in the establishment of the Sarah Acland Home for Nurses, a regular visitor at the Oxford Infirmary and a member of the Council of Somerville Hall. Oxford was somewhat behind Cambridge in providing facilities for women, and it was not until 1877 that an association was formed at Oxford to provide courses of lectures specifically adapted for women. In 1879 two halls of residence for young women opened, one of which was Somerville Hall. In *Every Girl's Magazine* for April 1882 (reprinted in *Every Girl's Annual* for 1883), Grace wrote an article publicizing the purpose of Somerville Hall, the last two pages of which strongly advocate the value of higher education for women. Her argument is well summed up in the final sentences:

We firmly believe that by higher and wider education the lives of women will be enobled. Years, which might have been wasted in folly and frivolity, will be redeemed by earnest thought and work; and lots in life will be enriched and coloured by entrance into, and contact with, a wider world of thought (G. A. Prestwich 1883*b*, p. 321).

Following Prestwich's resignation in 1887, Darent-Hulme became their only home, although they spent some months each year in London with Grace's sisters. All three sisters still resided at 21 Park Crescent. Her brother Robert's career is something of a mystery. According to *The East-India Register and Army List* he joined the Indian army in 1857 with the rank of ensign. Promoted to lieutenant in July 1860, he is listed until January 1867 when he is marked as 'on furlough'. He would have been entitled to 2 years leave at the end of 10 years service but does not seem to have returned as there is no subsequent record for him. His gravestone at Shoreham refers to him as Lieutenant Robert Milne, Indian Army.

Although suffering from periodic illness, Prestwich was appointed President of the 4th International Geological Congress, which was held in London in September 1888. He was still actively engaged in fieldwork into his early 80s, visiting sites of raised beaches and tramping the downs with Benjamin Harrison, the village shopkeeper turned geologist from Ightham about 19 km from Shoreham. His health finally failed completely in November 1895 and he died on 23 June 1896. A knighthood had been conferred on him in the New Years honours list but he was never well enough to enjoy the honour.

Once they returned to Darent-Hulme, much of Grace's time seems to have been devoted to her husband. However, it was at this stage in her life that she returned to the journals that she had kept during her travels with Falconer, publishing four articles, in *Blackwood's Magazine* and *Leisure Hour*, between 1888 and 1895 (G. A. Prestwich 1888, 1892, 1893, 1895).

Following Joseph's death, her three sisters moved from Park Crescent and made Darent-Hulme their home. Unfortunately her husband's long illness and death affected her, accelerating the heart disease from which she suffered. She spent much of the time as an invalid on the sofa, writing a biography of her husband, which was published in June 1899 (G. A. Prestwich 1899). The book was well received and a review in *The Times* of 1 August 1899 describes the biography as:

lovingly and adequately told by his widow in this volume, which will delight all who knew the man and attract all who share his enthusiasm for the science of which he was so great a master.

Grace did not live long after the completion of her tribute to Prestwich, dying on the morning of 31 August 1899. She was only 66 years old. She is buried alongside her husband in the churchyard of the Parish Church of St Peter and St Paul in Shoreham. She is also commemorated inside the church, along with her husband and his sister Civil, by a stained-glass window from the workshop of William Morris, completed in 1903 (James c. 2001). Remarkably this window, in the south wall and emblematic of Joy, Creation and Love, escaped the blast from bombs which fell in the vicarage garden and churchyard during World War II. Her sisters may have remained for a time at Darent-Hulme but are not recorded as living anywhere in England in the 1901 census. However, the siblings were reunited in death, as the sisters, together with their brother Robert, are buried next to Grace and Joseph in Shoreham churchyard.

Subsequent to her death Grace received the rare distinction for a non-fellow, and a woman at that, of an obituary notice in the anniversary address of the President of the Geological Society, William Whitaker. He commented (Whitaker 1900, p. lxi):

I may fairly conclude these obituary notices with a reference to a person never enrolled among our Fellows, prevented from being so, indeed, by sex alone. The niece of Falconer, the wife of Prestwich, lived little more than long enough to see the production of the labour of her latest years, the Life of her husband. This work of love is one of the most notable of the geological books of last year, and it gives an account of the life and labours of one of our greatest geologists, illustrating the way in which he was led to take up those various lines of research wherein he so highly distinguished himself.

Discussion and conclusions

Grace Milne began life as the eldest daughter of a prosperous Scottish family. But for the death of her first husband and infant son when she was only 24 years of age, she would probably have had a conventional life amongst the leisured middle-class women of Victorian Glasgow. Following her bereavement, the cajoling of a persistent uncle took her first around Europe and then to London, where she became associated with an elite group of scientists centred on the Royal Society and the Geological Society of London. Her circle of contacts included Charles Lyell, at whose home she was a regular dinner guest, and Roderick Murchison, who was a guest at her wedding to Prestwich.

Her uncle, Hugh Falconer, was unmarried and had spent many of the previous 25 years of his life in India. He comes across as a gregarious extrovert but his letters to Grace also suggest a lonely man who valued her companionship and support. His 'boyish mirth' and 'laughter provoking sallies' (G. A. Prestwich 1899) contrasted with her own rather serious and earnest nature, which is apparent from her articles and novels. Her relationship with him was more than that of a housekeeper, secretary or amanuensis. She also acted as his scientific confidante, against whom he could bounce his ideas and vent his frustrations. However, during her time with Falconer she was largely invisible to those outside his circle of friends and colleagues. She is not mentioned in Charles Murchison's biographical sketch of Falconer (Murchison 1868) and, although she kept a journal, published nothing herself during the 10 years that she lived at 21 Park Crescent with first Hugh and then his brother Charles.

In contrast to Falconer, Prestwich was described as a quiet shy boy at school. In business he was conscientious and efficient, planning his life in considerable detail so that he could make time to pursue his passion for geology. He was serious and earnest and, in many ways, his personality probably mirrored that of his wife. In addition, both had partnerships with a close relative which had been broken by a premature death; they had moved within the same social circle and, by an interesting coincidence, Grace's first husband had also been a wine merchant, so she may even have known something about his business. Her interest in geology meant that she was able to play a full part in his life. She was not isolated at home but helped him with his work and travelled with him on his geological adventures. That she was also a geologist in her own right is demonstrated by the high regard in which she was held by Fellows of the Geological Society (Whitaker 1900).

Prestwich's relationship with his sister Civil was one in which she became his housekeeper and secretary, devoting her life to supporting him. After their marriage, Grace took over Civil's role but her relationship with Prestwich was much more of a partnership. Thus Grace was able to develop both as a novelist and a geologist. In order to do this in Victorian England she must have had the support and encouragement of Prestwich. Both of them were interested in improving the position of women within the scientific community. Grace was active in the establishment and management of Somerville Hall at Oxford, and Prestwich strongly supported the admission of women as Fellows of the Geological Society (Woodward 1908).

Both her novels and many of her articles were published anonymously or, subsequent to her first novel, 'By the author of The Harbour Bar'. Even her biography of her husband is 'written and edited by his wife' and does not identify Grace by name. In Victorian England ladies in her position did not commonly write fiction under their own names and many of the best-known women authors, such as the Bronte sisters earlier in the century, wrote their novels under pseudonyms. However, many of her geological papers identify her as the author. Thus her *Chapters on Geology*, dated 1882, are authored by Grace A. Prestwich, whereas her article on Somerville Hall, dated 1883, is 'By the author of the Harbour Bar'. In both cases the articles are published in *Every Girl's Annual*. It is interesting that she was prepared to bow to such conventions within the publishing world.

Grace Anne Milne became a successful novelist and author of factual articles for leisure magazines of the day, as well as being a respected member of the geological community. As Grace Prestwich she was the wife of an Oxford professor and active in promoting higher education for women. She was an intelligent and perceptive woman but very much a member of her class who, when travelling, would refuse a dinner invitation if she had no suitable gown (G. A. Prestwich 1885). She contributed significantly to the work of the two men with whom her name is associated but her role was much more than that of an amanuensis and she became their geological companion and confidante.

We would like to thank the staff of the British Library and the libraries at the Geological Society and the University of Exeter for locating published material. Although many of Grace Prestwich's publications have been referenced, it is possible that others have slipped through the net. The two novels were published anonymously and some of the articles are headed 'By the author of *The Harbour Bar*' with no other assignation. Also there is no author index to some of the 'leisure' magazines to which she contributed. JDM would also like to thank L. Munro of Findhorn for his help in obtaining information about J. Milne of Findhorn and the Milne family, Susanne Bangert of the Department of Antiquities at the Ashmolean Museum in Oxford for supplying copies of letters from the personal papers of Sir John Evans, and A. Joyce, Senior Museums Officer at the Falconer Museum in Forres, for supplying images from their collections. In addition, JDM would like to thank Mrs S. R. Hofmann of Darenth House, Shoreham, for showing him around her beautiful home.

References

BOYLAN, P. J. 1977. *The Falconer Papers, Forres*. Leicestershire Museums, Art Galleries and Records Service, Leicester.

BOYLAN, P. J. 1979. The controversy of the Moulin-Quignon jaw: the role of Hugh Falconer. *In*: JORDANOVA, L. J. & PORTER, R. S. (eds) *Images of the Earth: Essays in the History of the Environmental Sciences*. British Society for the History of Science, Monographs, **1**, 171–199.

CAMPBELL, I. 1980. *Victorian publishing, the novel and some aspects of its promotion*. MA Thesis, University College, London.

COLLIE, M. & DIEMER, J. (eds) 2004. *Murchison's Wanderings in Russia*. British Geological Survey, Occasional Publication, **2**, 474.

FALCONER, H. 1860. On the ossiferous Grotta di Maccagnone, near Palermo. *Quarterly Journal of the Geological Society, London*, **16**, 99–106.

FLETT, J. S. 1937. *The First Hundred Years of the Geological Survey of Great Britain*. HMSO, London.

FRIENDS OF KENSAL GREEN CEMETERY. 1997. *Paths of Glory or a select alphabetical and biographical list, illustrated with line drawings of their monuments, of persons of note commemorated at the Cemetery of All Souls at Kensal Green*. Friends of Kensal Green Cemetery, London.

JAMES, E. c. 2001. Sir Joseph Prestwich. *The Self-taught Professor Who Taught the World the Antiquity of Man*. Shoreham and District Historical Society, Shoreham, Kent.

KIRK, W. 1996. Thomas Webster (1772–1844): first Professor of Geology at University College London. *Archives of Natural History*, **23**, 309–326.

LYELL, C. 1863. *The Geological Evidences of the Antiquity of Man*. John Murray, London.

MILNE, L. E. 1901. Memoir. *In*: PRESTWICH, G. A. *Essays Descriptive and Biographical*. 1–70. Blackwood, Edinburgh and London.

MOORE, D. T. 2004. Falconer, Hugh (1808–1865). *In*: MATTHEWS, H. C. G. & HARRISON, B. (eds) *Oxford Dictionary of National Biography*, Vol. 18. Oxford University Press, Oxford, 967–968.

MURCHISON, C. 1868. *Palaeontological Memoirs and Notes of the Late Hugh Falconer, A. M., M. D. with a Biographical Sketch of the Author*. 2 vols. Robert Hardwicke, London.

O'CONNOR, A. 2006. Hugh Falconer, Joseph Prestwich and the Gower Caves. *Studies in Speleology*, **14**, 75–79.

PRESTWICH, G. A. 1874. *The Harbour Bar. A Tale of Scottish Life*. 2 vols. MacMillan, London.

PRESTWICH, G. A. 1876. *The Harbour Bar. A Tale of Scottish Life*. New and cheaper Edition. MacMillan, London.

PRESTWICH, G. A. 1879. Lochaber and the Parallel Roads. *Good Words*, **1879**, 119–123.

PRESTWICH, G. A. 1880. *Enga*. 2 vols. MacMillan, London.

PRESTWICH, G. A. 1881. Channel Tunnel and Channel bridges. *Good Words*, **1881**, 210–216.

PRESTWICH, G. A. 1882. Chapters on geology. *Routledge's Every Girl's Annual*, **1882**, 12–17; 116–122; 230–237; 306–313; 411–418; 492–500.

PRESTWICH, G. A. 1883a. The face of the globe. Short studies in physiography. *Good Words*, **1883**, 208–212; 438–441; 641–643.

PRESTWICH, G. A. 1883b. Somerville Hall, Oxford. *Routledge's Every Girl's Annual*, **1883**, 317–321.

PRESTWICH, G. A. 1885. Through the Pass of Glencoe. *Every Girl's Magazine*, **80**, 516–520.

PRESTWICH, G. A. 1888. Our white deal box, and the trouble it gave; or how the Custom-House at Naples dealt with some casts of fossil animals. *Leisure Hour*, **435**, 183–186.

PRESTWICH, G. A. 1892. An evening with Mrs Somerville. *Leisure Hour*, **491**, 38–42.

PRESTWICH, G. A. 1893. Evenings with Madame Mohl: reminiscences of a paris salon. *Blackwood's Magazine*, **154**, 41–48.

PRESTWICH, G. A. 1895. Recollections of M. Boucher de Perthes; being some account of the history of the discovery of flint implements. *Blackwood's Magazine*, **157**, 939–948.

PRESTWICH, G. A. 1899. *Life and Letters of Sir Joseph Prestwich M.A., D.C.L., F.R.S.* Blackwood, Edinburgh and London.

PRESTWICH, G. A. 1901. *Essays Descriptive and Biographical*. Blackwood, Edinburgh & London.

PRESTWICH, J. 1859a. On the occurrence of flint implements, associated with the remains of animals of extinct species in beds of a late geological period, in France at Amiens and Abbeville, and in England at Hoxne. *Proceedings of the Royal Society*, **10**, 50–59.

PRESTWICH, J. 1859b. Sur la découverte d'instruments en silex associés à des restes de mammifères d'espèces perdues dans des couches non remaniées d'une formation géologique récente. *Comptes Rendus Hebdomadaires des Séances de l'Académie des Sciences*, **49**, 634–636, 859.

PRESTWICH, J. 1874. On the geological conditions affecting the construction of a tunnel between England and France. *Proceedings of the Institution of Civil Engineers*, **37**, 110–145.

PRESTWICH, J. 1879. On the origin of the parallel roads of Lochaber, and their bearing on other phenomena of the glacial period. *Proceedings of the Royal Society*, **29**, 6–21.

PRESTWICH, J. 1886–88. *Geology: Chemical, Physical and Stratigraphical.* 2 vols. Clarendon Press, Oxford.

ROSE, E. P. F. 2006. Military geological cave exploration on Gibraltar 1862–68: the start of a saga. *Studies in Speleology*, **14**, 51–54.

SEYMOUR-SMITH, M. (ed.) 1980. *Novels and Novelists: A Guide to the World of Fiction.* Windward Press, London.

THACKRAY, J. C. 2004. Prestwich, Sir Joseph (1812–1896). *In*: MATTHEWS, H. C. G. & HARRISON, B. (eds) Vol. 45. *Oxford Dictionary of National Biography*, Oxford University Press, Oxford, 276–278.

WHITAKER, W. 1900. Anniversary address of the President for 1899. *Quarterly Journal of the Geological Society, London*, **56**, lxi.

WILLING, D. & MUNRO, M. 2002. *Findhorn. A Scottish Village.* Queen's Golden Jubilee and Tricentenary Edition, Findhorn Residents' Association, Findhorn, Moray.

WILSON, L. G. 1972. Charles Lyell. *The Years to 1841: The Revolution in Geology.* Yale University Press, New Haven and London.

WOODWARD, H. B. 1895. Geology in the field and in the study. *Proceedings of the Geologists Association*, **13**, 247–273.

WOODWARD, H. B. 1908. *The History of the Geological Society of London.* Longmans, Green and Co., London.

Anne Phillips: John Phillips's geological companion

N. MORGAN

Rose Cottage, East End, Chadlington, OXON OX73LX, UK
(e-mail: ninamorgan@lineone.net)

Abstract: In the late summer of 1842, Anne Phillips, working under her brother's instructions, found the crucial piece of evidence – known thereafter as Miss Phillips' conglomerate – that disproved Murchison's theories about the intrusive origin of the Malvern Hills. Later she travelled with her brother to examine the volcanics of the Auvergne. But these were not her first, or her only, geological achievements. From 1829 until her death in 1862, Anne served as housekeeper to her bachelor older brother. Orphaned at an early age, both John and Anne were taken in by their uncle, William Smith. Smith arranged for John's schooling and introduced John to the science of geology as a teenager. However, little evidence exists about Anne's upbringing and education. A series of 234 letters written by John to Anne between 1829 and 1841, and preserved in the Phillips archive at the Oxford University Museum of Natural History, reveals that Anne was clearly well educated and provided her brother with valuable scientific back-up as well as essential domestic and emotional support during their 33 years together.

Behind every successful man stands a strong and competent woman, and in the case of John Phillips that woman seems to have been his younger sister Anne (Fig. 1). The story of John Phillips is really one of rags to riches. Orphaned at the age of 7 years and without a university education, he worked his way up the scientific ranks to become the first professor of geology at Oxford, a post he

Fig. 1. Anne Phillips at the age of 57 years (photograph reproduced with permission from The Royal Institution London/Bridgeman Art Library) and John Phillips at the age of 60 years (reproduced with permission from the Oxford University Museum of Natural History). Both photographs date from 1860, and their family resemblance is striking! The photograph of Anne is the only known image of her and was discovered in the archives of Michael Faraday by Frank James of the Royal Institution.

A popular version of this paper, entitled Miss Phillips and the mystery of the Malverns, was published in Morgan (2006).

From: BUREK, C. V. & HIGGS, B. (eds) *The Role of Women in the History of Geology.* Geological Society, London, Special Publications, **281**, 265–275. DOI: 10.1144/SP281.16
0305-8719/07/$15.00 © The Geological Society of London 2007.

Fig. 2. The Oxford University Museum of Natural History shortly after its completion in 1861 (reproduced with permission from the Oxford University Museum of Natural History).

held from 1860 until his death in 1874. He was also keeper of the Ashmolean Museum in Oxford. In addition he helped to plan and arrange what is now the Oxford University Museum of Natural History (Fig. 2), and became the first keeper there in 1857.

Along the way he played important roles in the Yorkshire Philosophical Society, the British Association for the Advancement of Science and the new Geological Survey. He was a fluent lecturer and a prolific author and researcher, who did much to advance and popularize the new science of geology – and, incidentally, to open up scientific meetings to women.

By contemporary accounts he was well liked and well respected. For example, 150 members of the University accompanied his coffin to the railway station in Oxford, and his funeral in York brought the city to a standstill (Morrell 2005, p. 1; *Jackson's Oxford Journal*, 25 April & 2 May 1874). The huge number of his surviving letters reveals that he seemed to be in touch with everyone who was anyone in 19th-century science.

On the basis of his achievements one might expect Phillips to have been a supremely confident man, but a collection of 234 letters written by John to his sister Anne between 1829 and 1841, and housed in the Hope Library at the Oxford University Museum of Natural History shows how vulnerable he often felt, and how much he depended on Anne for emotional and practical support in both his domestic and scientific life.

Childhood

John Phillips was born on Christmas Day, 1800. His sister, Anne, was born in 1803. They had a younger brother, Jenkin, who was born in 1807. Their mother was the sister of William Smith (Fig. 3), author of the first geological map of England, whose pioneering worked formed the basis for the science of stratigraphy.

When both parents died in 1808, William Smith, along with his brothers, John and Daniel, took over responsibility for their orphaned nephews and niece. Jenkin seems to have remained in the care of his uncle Daniel in the Smith family's home village of Churchill, near Chipping Norton in Oxfordshire. John and Anne were taken by William Smith to stay with his brother, John Smith, at Broadfield Farm, near Hinton Charterhouse in Somerset (Morrell 2005, p. 11). And that is where the trail of evidence about Anne's upbringing runs cold.

Contemporary letters and diaries confirm that William Smith paid for John's education at David Thomson Arnot's school at Holt in Wiltshire, where John seems to have spent 5 years soaking up knowledge of all kinds (Phillips 1817–72; Morrell 2005, p. 13). In 1815 he left school and went to stay for a year with the Reverend Benjamin Richardson of Farleigh Hungerford, near Bath.

Richardson was a keen naturalist and long-time friend and admirer of William Smith and his work.

Fig. 3. William Smith at the age of 67 years (reproduced with permission from the Oxford Museum of Natural History) and the memorial to William Smith erected by Lord Ducie in Churchill, Oxfordshire, in 1891. The inscription reads: 'In memory of William Smith "the founder of British Geology" born at Churchill March 23 1769 died at Northampton August 28 1839' (photograph courtesy of Philip Powell).

On 21 November that year John Phillips joined his uncle in London and began working as William Smith's scientific assistant, sorting and cataloguing Smith's fossil collection. Together they travelled throughout England to carry out mineral surveys and John became, in effect, the first 'apprenticed' geologist (Edmonds 1981).

In 1824 John began what became a very successful career as a lecturer, first as a double act with his uncle, and later on his own. In 1826, he settled in York as the first keeper of the Yorkshire Philosophical Society Museum, but continued to supplement his income by lecturing. Both John and Anne stayed in close touch with their uncle until his death in 1839, and Anne often stayed with Smith when John was away from home.

A partnership formed

While there are many clues in letters and diaries about John's education, Anne's upbringing and education remains a mystery. For reasons not clear, Anne was living with Monsieur and Madam Goussaerts in Chateau Kockelberg near Brussels in 1829 (Phillips 1829–41, letters to Anne Phillips dated August & September 1829). John, who was enjoying his first trip to the Continent, geologizing with his friends the Reverend William Taylor and Jonathan Gray, arranged to meet Anne, first near Brussels and later in Paris (Phillips 1829). This was only the second time the pair had met for 14 years (Morrell pers. comm.). He invited Anne to return to York with him as his housekeeper. Neither married, and the two lived together and supported each other for the next 33 years, until her death in 1862.

There is nothing unusual about an unmarried sister keeping house for a brother – particularly in the 19th century. Her side of the correspondence appears to have been lost, but the letters John wrote to Anne reveal that the emotional ties between the two were particularly strong. In his letters John often made his feelings clear. For example, in 1831, (Phillips 1829–41, letter to Anne Phillips dated 18 October 1831) he wrote:

According to my wishes & your wishes, as well as my promise and your expectation I am composing a letter to my love.

And his letters are sometimes much more passionate than that, as for example, in 1838, when he begged her never to think of deserting him (Phillips 1829–41, letter to Anne Phillips dated 1 July 1838):

Through many years your counsel has been my guide & your gentle & true affection my reward, & though, at times, my rough nature must have hurt your feelings, you know I love you & am capable in your cause of bearing & doing much, which for none other I would bear or do. Whatever I possess is as much yours as mine, for without you I should not have won it: my home has

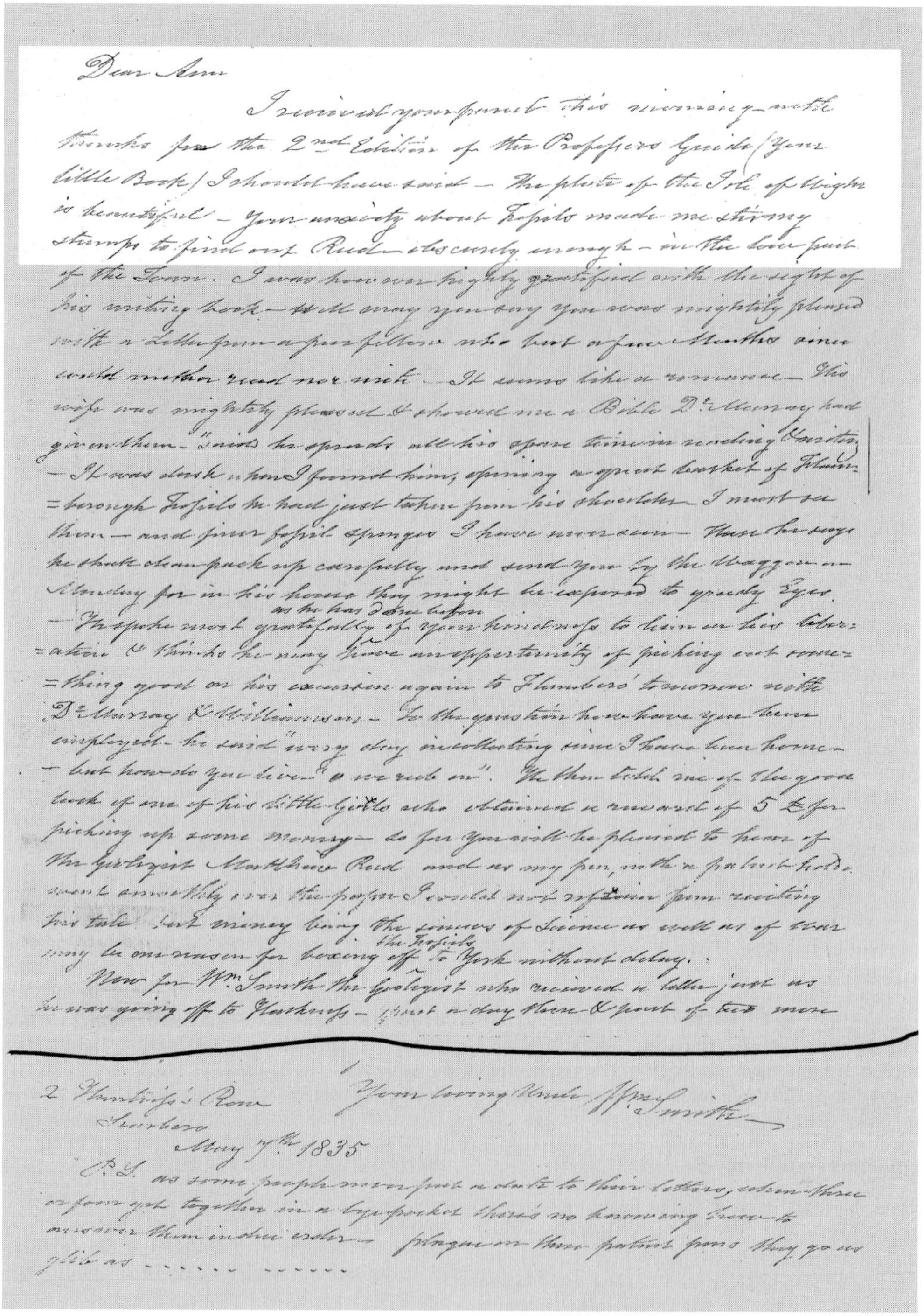

Dear Ann

I received your parcel this morning with thanks for the 2nd Edition of the Professors Guide (Your little Book) I should have said — The plate of the Isle of Wight is beautiful — Your anxiety about Fossils made me stirring stumps to find out Reed — obscurely enough — in the lower part of the Town. I was however highly gratified with the sight of his writing book — well may you say you was mightily pleased with a Letter from a poor fellow who but a few Months since could neither read nor write — It seems like a romance — His wife was mightily pleased & showed me a Bible Dr Murray had given them — "since he spends all his spare time in reading & writing" — It was dark when I found him, opening a great basket of Flam= =borough Fossils he had just taken from his shoulder — I must see them — and finer fossil sponges I have never seen — these he says he shall clean pack up carefully and send you by the Waggon on Monday for in his house they might be exposed to greedy Eyes as he has done before — He spoke most gratefully of your kindness to him on his liber= =ation & thinks he may have an opportunity of picking out some= =thing good on his excursion again to Flambro' tomorrow with Dr Murray & Williamson — To the question how have you been employed he said "every day in collecting since I have been home — but how do you live? o we rub on". He then told me of the good luck of one of his little Girls who obtained a reward of 5 £ for picking up some money — so far you will be pleased to hear of the Geologist Mark Lane Reed and as my pen, with a patent holder went smoothly over the paper I could not refrain from writing his tale but money being the sinews of Science as well as of War may be one reason for boxing off the Fossils to York without delay.

Now for Wm Smith the Geologist who received a letter just as he was going off to Hackness — spent a day there & part of two more

2 Huntriss's Row
Scarboro
May 7th 1835

Your loving Uncle Wm Smith

P.S. as some people never put a date to their letters, when three or four get together in a bag pocket there's no knowing how to answer them in due order — plague on these patent pens they go as glib as ....

Fig. 4. An extract from a letter written by William Smith to Anne Phillips, hinting at her strong involvement in the production of the second edition of John Phillips's *A Guide to Geology* (Phillips 1835 reproduced with permission from the Oxford Museum of Natural History).

been made happy: & my heart satisfied: You have been to me a blessing, for which I have thanked God not grudgingly.

Though in many ways the letters read more like letters between a husband and wife than between brother and sister, their relationship is most likely to have been platonic, and Anne seems to have been well accepted as a hostess and companion by John's scientific friends.

Field assistant

Practically from the moment they were reunited in Paris on 19 September 1829, Anne became involved in John's geological work. In his travel journal (Phillips 1829) John recorded that, on September 29:

We walked to day to the West of Havre and carefully examined the Geology of the Coast. The following observations written on the spot were published in the Philosophical Magazine March 1830, at the request of Mr De la Beche.

His letters to her include numerous requests for practical geological help – everything from selecting and sending boxes of fossils to him in London to replace those stolen on the way to a lecturing engagement – (an event he describes as 'What a malheur!') – to organizing the colouring of maps for his books (See, e.g., Phillips 1829–41, letters to Anne Phillips dated 15, 17 & 21 May 1837).

William Smith also often writes to Anne in a geological vein and refers to her work on her brother's books. For example, in a letter to Anne dated 7 May 1835 (Fig. 4), Smith (1819–39) writes:

I received your present this morning with thanks for the 2nd Edition of the Professor's Guide (Your Little Book) I should have said – The plate of the Isle of Wight is beautiful. Your anxiety about Fossils made me stir my stumps to find out Reed – obscurely enough – in the lower part of the Town.

Smith often referred to John Phillips affectionately as 'the Professor' or 'the Philosopher', and the book mentioned in this letter is the second edition of Phillips's very popular *Guide to Geology* (Phillips 1835; Fig. 5). Reed was, presumably, the local carrier.

But it is John's letters that give the greatest clues about Anne's intellectual achievements. A number of the letters are written in playful French (e.g., Phillips 1829–41, letter to Anne

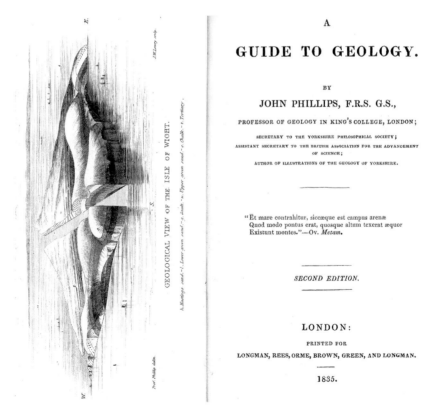

Fig. 5. The frontispiece and title page from the second edition of John Phillips's *A Guide to Geology* (Phillips 1835).

Phillips dated 23 January 1830), and sometimes he includes Latin or even Greek verses (e.g., Phillips 1829–41, letter to Anne Phillips dated 2 May 1831). He also writes to her in some detail about geology, complete with geological drawings. For example, a letter dated 22 November 1841, sent from Fishguard and written when John was engaged in fieldwork for the Geological Survey in Southwest Wales, includes a sketch of the folded beds in the cliffs (Fig. 6). along with the explanation:

The Strata very contorted. A few fossils in Slaty & sandstone rocks, belonging to what Sedgwick calls 'Cambrian' System.

Later that year he rented a house in Tenby so that Anne could join him to look after his instruments

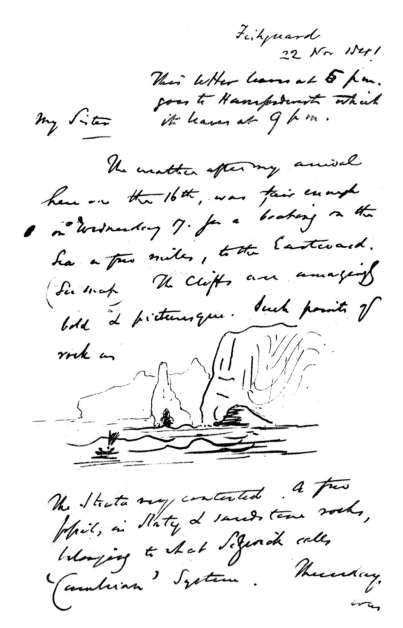

Fig. 6. Part of a letter written by John Phillips to his sister Anne, describing the geology on the coast near Fishguard, Pembrokeshire, South Wales (reproduced with permission from the Oxford University Museum of Natural History).

and books, and help to sort and catalogue the fossils and minerals he collected. She travelled to South Wales bringing her maid and their dog, Cholo. Cholo, incidently, was the second of two dogs the pair jointly owned. The first was called Prince, and both dogs were, apparently, partial to gingerbread (e.g., Phillips 1829–41, letters to Anne Phillips dated 1 & 2 July 1838 & 21 March 1840).

Miss Phillips's conglomerate

Anne also stayed with her brother in Malvern for 9 months in 1842, when he was working for the Geological Survey in the Malvern Hills (Morrell 2005, p. 180). This period provides the best-documented example, both of the extent of her geological insight and her brother's pride in, and respect for, her talents.

In general terms (see Fig. 7), the Malvern Hills are currently considered to consist of central core of Precambrian (Malvernian) igneous intrusives – often lumped together by Phillips as Trap or 'sienitic' (syenitic) rock – with Triassic sandstones and marls on the east side, and Cambrian and Silurian sediments on the west. The Triassic rocks are separated from the Precambrian intrusives by a north–south-trending fault. In contrast, the Cambrian and Silurian rocks on the western side of the hills lap unconformably onto the intrusives.

Even today, all the details of the origin of the Malverns are still not fully understood, but in the 19th century the subject was a very controversial one. On one side was Sir Roderick Murchison – author of a definitive contemporary geological guide, *The Silurian System* (Murchison 1839) – who believed that the Malvern ridge was intruded as a hot body after the Silurian sediments were laid down. On the other was John Phillips who, along with his colleagues in the Geological Survey, was coming to the conclusion that the ridge was already elevated and cooled before the Silurian rocks were laid down. Phillips based his case on the fact that, although the Silurian sediments are in some places in contact with the intrusive rocks, they are not affected by metamorphism. To clinch his argument, Phillips reasoned that, as the Silurian seas lapped up against the ridge, fragments could have been eroded off and incorporated to form a conglomerate at the base of the Silurian.

While Phillips was busy in late July and early August 1842, showing Murchison and other geologists around his field area to explain his theories, Anne went out in the field and, on 1 August, found broken blocks of the conglomerate.

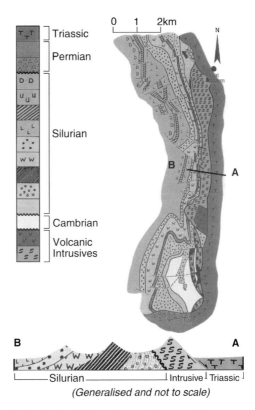

Fig. 7. Geological sketch map and cross-section across the Malvern Hills (adapted, with the permission of English Nature, from the geological map in *Malvern Hills: A Student's Guide to the Geology of the Malverns* by Bullard & Morris 1989).

This caused excitement all around, and within a week she and John, accompanied by John's boss, Sir Henry De la Beche, managed to locate a place where the conglomerate could be seen in contact with the intrusive rocks. In order to establish the Survey's priority, Phillips published a report of the find in October 1842 in the *London, Edinburgh and Dublin Philosophical Magazine and Journal of Science* (Phillips 1842), in which he gave full credit for the discovery to his sister. In this article Phillips notes that, after having discussed the origin of the intrusive rocks in the field with Murchison and Count Keyserling in the northern end of the Malverns:

Professor Sedgwick accompanied me on a leisurely survey of this and other points further south. On the day (August 1) while I was enjoying the advantage of his experience in examining the facts thus briefly adverted to, a discovery was made which threw a new and concentrated light on the phaenomena we were discussing.

My Sister, knowing the interest I felt in tracing out the history of the stratification visible in these trap hills, sought diligently for organic remains in the midst of and on the western flanks of the sienitic masses of the North hill and Sugar-loaf hill. In this most unpromising search she was entirely successful, and collected from the midst of heaps of fallen stones, which seemed to be all trap, several masses richly charged with organic remains, and full of felspar, quartz, and hornblende, in grains and large lumps ... It was, in fact, certainly and evidently a conglomerate full of Silurian shells, and pebbles and fragments of the sienitic, felspatho-quartzose and other rock-masses of the Malvern hills.

The same account, slightly abbreviated, also appears in his 1848 Geological Survey Memoir (Phillips 1848), along with a sketch of the contact (Fig. 8).

The rock became known as Miss Phillips's conglomerate – and samples of it soon became highly collectable items on geology field trips. A report of a Geologists' Association excursion to Malvern in 1873 (Lobley 1873), also published in the *Malvern Advertiser* of 9 August 1873 notes that:

Though the shades of evening were rapidly darkening, the enthusiasm of the geologists led them to the conglomerate of Miss Phillips, at West Malvern, specimens of which, by the aid of their well-used hammers they quickly obtained.

Anne's success in finding the conglomerate demonstrates that she must have had considerable geological knowledge and been very familiar with the local rock types. The contact between the conglomerate and the intrusive rocks is not very obvious to the casual observer, and it requires a very sharp eye to recognize the conglomerate in the field (Figs 9 & 10). Samples collected from the area near Phillips's original location contain rounded pebbles of intrusive rock less than a centimetre in length and width, in a well-cemented, sandstone matrix containing sparse fossil fragments

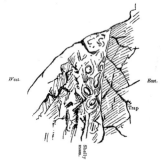

Fig. 8. Pages from the 1848 Geological Survey Memoir (Phillips 1848) describing the discovery of Miss Phillips's conglomerate with a sketch by John Phillips showing the contact between the volcanic intrusive and the Silurian conglomerate.

Fig. 9. A field trip to west Malvern undertaken in October 2005 by three Oxford geologists accompanied by two members of the Hereford and Worcestershire Earth Heritage Trust, one of whom had been researching for clues to the location of the site of the contact that Phillips described, revealed that the most likely site of the contact illustrated in Figure 8, near Lower Dingle Quarry, is now seriously overgrown (Payne 2007).

Fig. 10. A horizontal contact between the intrusive (lower bed) and the conglomerate less than 1 km from Lower Dingle Quarry illustrates how difficult it is to distinguish between the two rock types in the field.

Fig. 11. A hand specimen of Miss Phillips's conglomerate collected at the location shown in Figure 9. The arrow on the right points to a small rounded pebble of volcanic intrusive in the Silurian sandstone matrix. The arrow on the left points to a crinoid ossicle.

of a similar size (Fig. 11). These are visible under a hand lens, but are more clearly seen under the microscope (Fig. 12).

Conclusions

The 234 letters which John Phillips wrote to his sister Anne and which are housed in the Hope

Fig. 12. A photomicrograph of a bryozoan fossil and a small pebble of the intrusive in a hand specimen of Miss Phillips's conglomerate (photograph courtesy of Derek Siveter).

From the North.

The Malvern and Ledbury Hills, as seen from a part of the Abberley range.

Fig. 13. It is perhaps fanciful to think that this illustration, drawn by John Phillips and published in his 1848 *Geological Survey Memoir* (Phillips 1848), show him (in white) with Anne (in black) exploring the geology of the Malverns together. But it would be nice!

Library at the Oxford University Museum of Natural History, reveal a vivid portrait of many aspects of life and times – and science – in the 19th century. A full analysis of their content is still to be completed and will form the basis of future publications. Meanwhile, this paper highlights the many clues that they provide to Anne's geological awareness and knowledge suggest that she often served as her brother's 'geological assistant' (Fig. 13). Transcripts of letters written to Anne by her uncle, William Smith, also hint at Anne's interest in, and knowledge of, geology. However, more specific information about Anne's background and interests has not, so far, been found. Further sources are being sought.

I thank P. Powell and S. Brecknell of the Oxford Museum of Natural History for drawing my attention to the letters; the director of the Oxford University Museum of Natural History for allowing me access to museum archives; Derek Siveter of the Oxford University Museum of Natural History for sharing his knowledge of Silurian geology, both in the field and in the laboratory; R. Darwell-Smith, archivist at Magdalen College, Oxford, for providing access to John Phillips's travel journals; J. Morrell, whose book, *John Phillips and The Business of Victorian Science*, provided me with important background information about Anne Phillips for this paper, for sharing unpublished research; E. Robinson, Visiting Professor at University College, London, for drawing the *Malvern Advertiser* article to my attention; J. Payne and Moira Jenkins of the Hereford and Worcestershire Earth Heritage Trust for showing me around Miss Phillips's conglomerate locations in the Malverns; and M. Tomlinson for help in preparing the illustrations.

References

Anon. 1873. *Malvern Advertiser*, 9 August.
Anon. 1874. *Jackson's Oxford Journal*, 25 April, 2 May.
BULLARD, D. W. & MORRIS, L. 1989. *Malvern Hills: A Student's Guide to the Geology of the Malverns*. Nature Conservancy Council, Peterborough.
EDMONDS, J. M. 1981. The first 'apprenticed' geologist. *Wiltshire Archaeological and Natural History Magazine*, **76**, 141–154.
LOBLEY, J. L. 1873. Excursion to Malvern, July 21st 1873 and five following days. *Proceedings of the Geologists Association*, **3**, 269–280.
MORGAN, N. 2006. Miss Phillips and the mystery of the Malverns. *Geoscientist*, (7), 6–7, 12–15.
MORRELL, J. 2005. *John Phillips and The Business of Victorian Science*. Ashgate Publishing Ltd, Aldershot.
MURCHISON, R. I. 1839. *The Silurian System, Founded on Geological Researchers in the Counties of Salop, Hereford, Radnor, Montgomery, Caermarthen, Brecon, Pembroke, Monmouth, Gloucester, Worcester and Stafford*. John Murray, London.
PAYNE, M. J. P. 2007. The Miss Phillips's conglomerate of the Malvern Hills – where is Phillips's original site? *Proceedings of the Cotteswold Naturalist's Field Club*, **XLIV**(1), 67–76.
PHILLIPS, J. 1817–72. 'Notebooks and journals.' Held at the Hope Library, Oxford University Museum of Natural History.
PHILLIPS, J. 1829. 'Foreign travel journal.' Held at Magdalen College Library, Oxford.
PHILLIPS, J. 1829–41. 'Letters to Anne Phillips.' Held at the Hope Library, Oxford University Museum of Natural History.
PHILLIPS, J. 1835. *A Guide to Geology*. 2nd Edn. Longman, Rees, Orme, Brown, Green, and Longman, London.
PHILLIPS, J. 1842. On the occurrence of shells and corals in a conglomerate bed, adherent to the face of the trap rocks of the Malvern Hills, and full of rounded and angular fragments of those rocks, *London, Edinburgh and Dublin Philosophical Magazine and Journal of Science*, **21** (138), 288–293.
PHILLIPS, J. 1848. *The Malvern Hills Compared with the Palaeozoic Districts of Abberley, Woolhope, May Hill, Tortworth and Usk*. Geological Society of Great Britain: Memoirs, **2**.
SMITH, W. 1819–39. 'Letters to John Phillips and Anne Phillips.' Held at the Hope Library, Oxford University Museum of Natural History.

Keeping it in the family: the extraordinary case of Cuvier's daughters

M. ORR

School of Humanities (Modern Languages), University of Southampton, Avenue Campus, Southampton, SO17 1BJ, UK (e-mail: M.M.Orr@soton.ac.uk)

Abstract: In the history of French geology, Georges Cuvier (1769–1832) is largely figured as the antagonist to Etienne Geoffroy St Hilaire, and as the arch-conservative who blocked the path of more 'Darwinian' ideas about the evolution of species. If historians mention him as a key player in Catholic French secondary education reform, despite being a Protestant, it is his extraordinarily enlightened encouragement and employment of his daughter, and perhaps even more importantly his step-daughter, in his collaborative projects which this paper investigates. Trained up in the latest scientific ideas of the epoch (because Cuvier lived with his family at the Jardin des Plantes), and as his 'research assistants', theirs is a story within women's history of science that has never been told. While to label them 'women scientists' or 'geologists' is a misnomer for the period in France, the direct contributions of these daughters to cross-Channel, geological endeavour of the period can at last have a place in the history of geology.

In the foreword to Joan Chandler Smith's seminal annotated bibliography of Cuvier's published works (Smith 1993), Stephen Jay Gould sums up the stigma and reputation that Cuvier (1769–1832) enjoys:

We regard him today as having been wrong in his major scientific beliefs, particularly in two errors (one for each of the fields that he founded): his anti-evolutionism in comparative anatomy, and his catastrophism in palaeontology ... Nothing could be more arrogant, more conducive to the misunderstanding of history, and more unfair than our tendency to judge brilliant thinkers of the past according to our current ideas of truth value for their major beliefs. (p. x)

For Londa Schiebinger (1999, p. 17), Cuvier's reputation as a scientist rests on the double crime of racial and gender blindness:

In 1815, France's premier comparative anatomist performed his now infamous dissection of a South African woman known as Sarah Bartmann. The very name Cuvier gave this woman – Vénus Hottentote – emphasized her sexuality. In his memoir he made it very clear that Africans were not included among those who could do science: 'No race of Negro produced that celebrated people who gave birth to the civilization of ancient Egypt, and from whom we may say that the whole world inherited the principles of its laws, sciences, and perhaps also religion'. Like other women, Sarah Bartmann did not fit comfortably in 19th-century racial hierarchies where primarily men were studied for their comparative superiority. Like other Africans, she did not fit European gender ideals. Elite European naturalists who set such store by their own mothers, wives and sisters did not include African women in their new definitions of femininity.

A reinvestigation of these 'errors' in order to reinstate Cuvier as one of the geniuses of 19th-century French history of ideas including palaeontology, or to recast his woeful political incorrectness, is not the primary concern here.

And although a fuller reappraisal of the gender of Cuvier's science following the feminist methodologies of Keller (1985; Keller & Longino 1996), Jordanova (1989) or Harding (1993, 1998; Harding & O'Barr 1987) has yet to be undertaken, this essay offers some initial pointers and thereby ways to counter Schiebinger's overly negative view of Cuvier. Rather, these quotations frame current, trans-national, questions about who is occluded and excluded from the history of scientific progress. Even though women are more visible in 20th-century science and academe, Gould's foreword will still count more in citation indexes than Joan Chandler Smith's rich research tool. Such major blocks to proper recognition of the roles and place of *women* in the history of scientific endeavour will be shown in this essay to have particular legacies in early 19th-century Europe, especially France. There are in historical and national coverage telling gaps, even in important biographical dictionaries of women in science, such as Ogilvie (1988). Poirier (2002), which is French-specific, lists only two pre-19th-century women under 'geologists and mineralogists', while women 'naturalists' are in the main Enlightenment figures.

If one of Cuvier's most invaluable modern biographers, the historian Dorinda Outram, has rightly pinpointed how few scraps of personal information are available to understand his life (mainly because male historians largely discount aspects of the private in the making of dazzling public careers), it follows that information scraps on any

From: BUREK, C. V. & HIGGS, B. (eds) *The Role of Women in the History of Geology.* Geological Society, London, Special Publications, **281**, 277–286. DOI: 10.1144/SP281.17
0305-8719/07/$15.00 © The Geological Society of London 2007.

19th-century woman protagonist in this history of great scientists will be at best decorative; hers is the humanizing afterthought so as not to detract from his 'scientificness', as Outram (1984, p. 57) inelegantly calls it. Personal diaries and letters thus rarely survive or make their way into science archives. However, it is in the interstices of the personal details mentioned *en passant* by Cuvier's women historian-biographers, Joan Chandler Smith (1993), Dorinda Outram (1984, 1989) and his first, 19th-century English biographer, Sarah Lee (1833), that this essay is able to map for the first time the extraordinary place of Cuvier's daughters – Clémentine his biological daughter, and Sophie Duvaucel his step-daughter – in the collaborative projects that constitute all his major scientific works.

If I use the word 'map' as a geological metaphor to describe the inductive method of reconstructing some of the underlying strata in the larger project of women's history of science in early 19th-century France, this method is quintessentially Cuvier's inductive art of science. His famous public reconstructions of extinct forms from the evidence of one fossil fragment (a tooth, the bone of a creature's foot) were possible thanks both to his extensive knowledge of comparative anatomy and to a logic of complex, rather than directly linear or surface,

Fig. 2. Fountain erected in honour of Georges Cuvier.

relationships. It is interesting to note that Lee's *Memoirs of Baron Cuvier* (1833) devotes a whole final chapter to his family and is undoubtedly the model and source for the similar arrangement in Outram (1984). Lee's biography was interestingly published the same year in French, translated by M. T. Lacordaire. (All subsequent page references are from the French version). Mrs Sarah Lee's own detailed scientific achievements, particularly her contributions to the classification of new species of fish, are discussed in Orr (2007*a*).

The early 19th-century French context

The women's history project of digging up previously marginalized earlier period women scientists (or even geologists) to recuperate them retrospectively as independent 'professionals' in their own right is central to the work of feminist historians of science, such as Hubbard (1988). This approach is, however, both anachronistic and unthinkable regarding women in the France of the early 19th century. Prior to the Revolution, an intelligent aristocratic woman could enjoy the roles of *femme savante* and glittering salon hostess, whereby she could engage with the cultural

Fig. 1. Georges Cuvier.

and scientific minds of her period, or even participate in scientific pursuits herself. The cost of liberty, equality and fraternity for men bought by the populace ending Absolutist Monarchy in 1789, meant the total exclusion of women from public roles in this new democracy, especially intelligent and educated woman who, for these attributes, were also deemed unmarriageable by their own sex (Fraisse 1995).

The impact was more than the designation of women's place as the private, domestic and maternal sphere. When Napoleon Bonaparte set himself up as Emperor of France (1799–1814), his far-reaching reforms of all aspects of civic and public life codified and institutionalized a hierarchy of the sexes. His Code Civil of 1804, which still in large part informs France's property laws, gave a man full legislative power over his wife, children, goods, chattels and money, so that all inheritance followed a strictly male line (Articles 212, 213, 214). Women were truly second-class citizens, on the same legal footing as children and the insane; they were deemed unfit to vote, to sign a cheque or to have any right over property brought to the marriage as a dowry. Although the Revolution also abolished absolutist connections between Church and Throne, Napoleon's Concordat with the Pope in 1801 effectively reinstated a Catholic education for women through Catholic family values, even if girls could marry legally at the age of 15 years. As the primary goal for women was marriage and the task of motherhood, to produce sons for the Empire, girls received a convent education: rudimentary and not much beyond primary level for the working classes, it constituted a curriculum of refined ignorance for girls in higher classes.

By contrast, young men had a more secularized education at all levels (for a British comparison see Burek 2007). Napoleon's establishment of the University of Paris in 1806, which he deemed essential in the international power struggle, especially with England, for colonies and knowledge, offered a more scientific curriculum (i.e., non-theological). Women by law were not permitted to step inside the University, lest their nefarious influence disturb the concentration of the male students and staff who, like their Catholic precursors, were required to remain celibate. An intelligent daughter of middle rank, then, was utterly at the mercy of a very limited education, the dowry provided by a good father, and his permission to seal a suitable marriage. Her own free will and intelligence were excluded from all these equations since she had no rights to her person.

In the light of this constricting material, intellectual and educational context, which was specific to women in early 19th-century France, the term 'woman scientist' is now clearly an unthinkable concept. Doing the unthinkable, however, was precisely what both of Cuvier's daughters did, enabled indirectly by circumstance and directly by the 'unnatural' habitat and highly unconventional educational hothouse of their upbringing. In a nutshell, this was the tacit, but actively Protestant milieu of free enquiry that the Lutheran Cuvier established, whether in his many public and administrative roles in education and natural science at the University of Paris, the Collège de France, and briefly as Director of the Jardin des Plantes, or within his private household. Here, there was no demarcation of the public and the private. The education of his two sons and two daughters, the day-to-day work in the scientific collections and the Saturday salons hosted by his wife Anne-Marie, to which the most famous scientists and artists of the day were invited, were all seamlessly interconnected collaborations of people and ideas.

Cuvier's impartiality, religious and otherwise, his faith and science have elicited much heated debate (Rudwick 1997). When nomination to all posts of learning and public service was determined by a Catholic elite, Cuvier's life-long place as a Protestant in the highest echelons of French science, administration and public life are noteworthy but under-researched, as too is the 'protestantism' of his contributions to secondary education reform.

His wife Anne-Marie's intellect and personal qualities are only footnoted by Lee (1833, pp. 31–32, n. 3) and Outram (1984, pp. 213, n. 36). Anne-Marie's Catholicism and conversion prior to her (second) marriage to Cuvier in 1804 at the Swedish Embassy Church by Protestant rite is a detail in Duvernoy (1939, p. 30).

This is the local, multi-faceted context, within the wider early 19th-century French frame, by which to assess the contributions of Cuvier's daughters to the burgeoning disciplines of geology and palaeontology within the natural sciences.

Clémentine Cuvier (1805–27)

Probably because of the broad educational upbringing and curriculum encouraged by an extraordinary mother, Anne-Clémentine Chatel (Lee 1833, p. 11), Cuvier seemed to draw no quintessential gender or curriculum divisions in the education of his two boys and two girls. Clémentine Cuvier may inadvertently have been given special educational attention by her father because, by the age of 7 years, she was his sole surviving child. Be that as it may, it is her precocious intelligence and powers of reasoning that speak everywhere for themselves, even if these attributes will later be mentioned as only second to Clémentine's devout Protestantism (Wilks 1838), good works and angelic temperament, which are all heavily underscored in the treacly orations pronounced at her funeral service (Boissard n.d., p. 6). In the scanty accounts of Clémentine's life by Lee (1833, p. 44) or Outram (1984), they too focus on her sweet nature and charitable zeal, but not her intelligence and desire to bring education to other young girls of restricted means.

Clémentine participated fully and from an early age with her older half-sister Sophie in the famous Saturday salons hosted by her father and mother, which Cuvier suspended to mourn her death. Saturdays always saw a cosmopolitan guest list from the scientific world – including Charles Lyell in 1823 – and the arts. Without an informed knowledge and understanding of the latest scientific discoveries and debates, a young woman's conversation at such gatherings would plainly have been an embarrassment, whereas the opposite is recorded: guests and habitués were drawn as much by the scintillating conversation and intelligence of Cuvier's daughters as by their illustrious host (Buffetaut 2002, p. 17).

In fact, it was their prowess in English – Cuvier understood but did not speak it – that facilitated exchanges with English-speaking scientific visitors to Paris (Lee 1833, p. 283), and was probably of most use when Cuvier travelled to England, which he did twice, and on both occasions *en famille*. Clémentine was only 13 years old on Cuvier's first visit in 1818, surely a highly unusual involvement of a young girl by her father were she not altogether able to participate, and she was included in many of the scientific meetings on the itinerary. Cuvier probably met with Leach in London to see the collections at the British Museum and with Buckland in Oxford to see the fossil dinosaurs discovered at Stonesfield with Clémentine; the whole family also spent an evening with William Hershell (who showed them his giant telescope) and a day at Spring Grove, home of Sir Joseph Banks (Lee 1833, pp. 38–41).

The second visit to England was in 1830; then the July Revolution erupted, leaving Cuvier, accompanied by Sophie Duvaucel, on one side of the Channel and his wife on the other, in Paris. Dr J. B. C. Quoy, famous for his voyages of discovery on the *Astrolabe*, which furnished the Museum of the Jardin des Plantes with rich new collections, gave protection to Mme Cuvier during the 3 days of unrest.

By the age of 22 years, however, shortly before the date set for her marriage to Charles Duparquet, Clémentine became ill from tuberculosis, from which she died. Charles Duparquet was not a naturalist but a young colleague of Cuvier's in the 'Section of the interior at the Conseil d'Etate' [sic] (Outram 1984, p. 104). Interestingly, in an epoch where (bourgeois Catholic) daughters had their spouse vetted by their fathers, Clémentine's choice of husband was entirely her own – Duparquet shared her deeply held religious faith – but altogether supported by Cuvier. In the France into which she was born, marriage curtailed any degree of pre-marital intellectual (scientific) freedom that a spirited woman may have gained, although increasing numbers of married, bourgeois women 'botanized' as a leisure pursuit and attended Cuvier's public lectures. Had Clémentine survived, her married status even in what promised to be an enlightened union, would only have further channelled her many and well-documented public and charitable duties within the Protestant communities of Paris, especially the establishment of schools for working-class Protestant girls, as the exterior outlet for her prodigious intellect (Boissard n.d., notes 5–8). Clémentine Cuvier did, however, make significant contributions to her father's work *inside* the Jardin des Plantes. Our evidence is elicited from intriguing, stray but direct, references in two different kinds of published work to which I will now turn. The first is the *Eloge funèbre* (funerary oration), a form normally reserved to laud great men, such as Cuvier himself; the second is a work translated by Clémentine before she died and published posthumously in 1829.

Eloges funèbres

That various *Eloges funèbres* to Clémentine were published in French and English is due more to her own name and reputation than her father's. Although they deal principally with her *public* declarations of her Protestant faith through good works, there is one exception: the oration delivered at her tomb in the Père Lachaise Cemetery by M. de Salvandy, who was not a cleric but a '*Maître*' at the University of Paris. The litany of Clémentine's pious female qualities is developed by Salvandy into an unusually full account of her scientific mind. Prefaced by blandishments – the range of her intelligence, her seriousness in study – comes the following:

... her natural inclination involved her in his works [of her beloved father], without having an inkling as yet of what place these works occupied in the world's esteem. Soon she could understand them, but without being puffed up about it ... her charity therefore had both zeal and enlightenment at the same time. That was her genius ['son génie'].

However, you all know, Gentlemen, that she could have laid claim to something greater in her father's heritage. Providence had gifted her with **extraordinary** intellectual faculties [*facultés extraordinaires*]. There was no object of study so serious that did not delight her, so profound that she did not master it [*si forte qu'elle ne le dominât*]. Her broad and rigorous mind [*esprit étendu et ferme*] worked unceasingly to search out truth as her conscience did the same for goodness and justice. Easily recognisable in her was the generous and exalted sense of the mission of the human race on the earth and of the progress awaiting it ... There was always such deep pleasure for both mind and heart on hearing her speak ... when she would grasp, as it were *en passant*, the noble ideas in discussion before her, and in which she would always uncover hidden aspects, new premises and in an order of consideration which was more elevated than our own debates. This young hand seemed to have grasped the chain which links the earth to heaven and to have counted out its links. What foreigner ... would not have been surprised to discover such elevated and mature thoughts [*des pensées si hautes et si mûres*] in such youth ... Never was so benevolent a character combined with such intellectual riches [*richesses de l'esprit*]. (Salvandy in Boissard n.d., p. 7–10. The translation and emphasis in bold type are mine. The word 'foreigner' in the penultimate sentence is gendered masculine in the French.)

The superlatives and moral import of the *éloge funèbre* genre aside, this is an unprecedented account in the France of the time of any mind [*esprit*], let alone a female one. Without Salvandy's clear terms, Clémentine's research and enquiry could be falsely consigned to purely religious topics and questions since *esprit* in French covers both intellect and religious spirit (See the anonymous 'Notice Sur Clémentine Cuvier' [1928, p. 5], which lists only the pious works which inspired her and her 'victorious' defeat of illness and death). The female brain was smaller and less heavy than the male brain, thus the inferiority of the female brain was taken as scientific 'fact' and hence the reason why women rightfully had secondary legal, educational and political status. By retaining Salvandy's original French in square brackets with respect to the various qualities of Clémentine's intellect, what are usually male mental characteristics – *étendu et ferme, hautes et mûres* – appear more clearly, but in direct relation on both counts to the precocity of her mind; such attributes would normally be apportioned to the wisdom and intellectual experience of age rather than youth. Moreover, this enumeration is appositional to the word 'genius', a superlative term, usually exclusive of women and limited to men of science and poets. A fuller discussion of 'genius' as an overtly sexed and gendered idea in early 19th-century France lies outside this essay. In her biography of Clémence Royer, who translated Darwin's *Origin of Species* in 1862, Geneviève Fraisse (2002, p. 10) glosses a quotation by Ernest Renan about Royer:

She was a woman, 'almost a man of genius' ... and it is the derisory 'almost' which stands out, more than the strangeness of transforming a woman into a man (my translation).

But Salvandy's encomium is also unprecedented in its rhetoric of inclusion; of Clémentine and of informed, non-clerical, male high opinion of her. By recalling her known, respected and accepted intellectual participation in their male circle, this accolade is also richly suggestive of the many intellectual functions Clémentine fulfilled as Cuvier's *aide naturaliste* (research assistant). The term overlaps with the remit of a 'research assistant', but would in Cuvier's day have included someone with extensive knowledge of, and collaborative publication in, specialist scientific research. From this it can be inferred that she read key works to supplement Cuvier's current projects and discussed ideas, even to the point of challenging the thinking of his male collaborators and probably also of foreign scientific visitors. The evidence here is more overwhelming in the light of a later *Éloge*, to Cuvier himself by his successor Pierre Flourens (1869, pp. xlviii–lx), to crown the collected Éloges given by Cuvier of Great Men and Scientists (including Priestley). True to form, Flourens leaves family details to the very last moments in his *Éloge* of Cuvier, but it is here that Clémentine's *esprit* is mentioned (even if she is nameless) and comes on the heels of Flourens's encomium to Cuvier's eloquent history of science as in fact a history of *l'esprit humain*.

Translation as expression of personal experience

Unrecorded by biographers and critics hitherto is a final piece of inductive evidence regarding Clémentine's personal knowledge about being a woman in geology and natural science of the period. She left no personal diaries, only a translation of the prolific Mrs Barbara Hofland's novel-length work for children, *The Daughter of a Genius: A Tale for Youth* (Hofland 1823), with the title *La fille d'une femme de génie* (C. Cuvier 1829). Hofland's novel has elicited surprisingly little critical discussion. The recent work by Orr (2007b) is the first to address in any depth the issues of the history of women in science that underpin the themes and characterization of the work. Clémentine's overt gendering of the genius in question, a Mrs Albany, and the omission of the original subtitle, are not the only interesting features of the translation. How Clémentine came on this work is not known, but in the light of her own educational background and commitments to the education of girls as the intellectual equals of boys, the story has direct appeal to its translator's own educational and religious values, intellectual virtues and scientific interests and offers striking parallels. It tells how a Mrs Albany, an English woman of genius (and also a fluent French-speaker), on the death of her husband, sets up a boarding school for bright but impoverished young women with a curriculum which includes both French and science. She fails totally in the practical running of her school because she is in love, not with another man, but with her research in mineralogy and conchology, passions which also inspire her geological poetry. The opening lines of her poem to Derbyshire Blue John are recreated in the work (Hofland 1823, p. 101). It is only through the good auspices of her wise daughter that their financial plight is eventually resolved, so that the daughter may have a dowry to make a suitable marriage. This 'happy-ever-after' for the fictional daughter of a woman geologist and good mother was not, ironically, to be matched in Clémentine's own life.

Clémentine Cuvier's blank space in the biographies of her father and in her own right as a woman in early 19th-century French science has now been filled in. The occlusion, however, is also in part due to her much better-known, older stepsister, Sophie Duvaucel, friend and correspondent of the novelist Stendhal and model for the English portrait painter Thomas Lawrence. I now want to argue Clémentine's case further by analogy (another favoured principle in Cuvier's anatomy of criticism) with the many roles the arguably more 'brilliant' Sophie played in the life of her stepfather, if not as a 'scientist', at least as a leading *aide-naturaliste*. Certainly in the anonymous *Georges Cuvier: de son temps au notre*, the only mention of Sophie describes her using the adjective, 'brilliant' but qualifies it disparagingly thus: Completely *au courant* with Cuvier's work, she acted as his secretary and reader when he had retired for the evening (Anon. 1970, p. 56, my translation). Outram similarly paints Sophie as 'his faithful assistant and hostess' (Outram 1984, p. 57), but also fails to qualify either role.

Sophie Duvaucel (1789–1867)

Cuvier's stepdaughter enjoyed a life spanning various political régimes in France, but all of them inimical to the intelligent, accomplished and free-thinking Catholic that she was. Although many documents and letters concerning her are lost or destroyed, three accounts published in French, by Dorothy Gunnell (1925, ch. 8), Louis Royer (1930) and her only official biographer, Dr M. Duvernoy (1939), when read together, signal her extraordinary functions alongside her stepfather and help flesh out the singularity of her position within the male domains of the collections, cabinets and laboratories on 'the other side of the wall' from Cuvier's rambling apartments.

Sophie would have been 14 years old when her mother remarried and moved to Cuvier's house in the Jardin des Plantes. Her education would undoubtedly have been the Catholic curriculum for well-to-do girls, including music and drawing, but was enlarged by the intelligence and example of her mother and by conversations with her stepfather at dinner and in 'the Tent' (a curious room presided over by her mother) about the latest scientific discoveries and expeditions, as well as about the latest books and paintings (Gunnell 1925, p. 232). Although the 'love interest' is Gunnell's primary focus in an account of the life and French friends of Sutton Sharpe – he meets Sophie, who rapidly becomes more then 'just a friend', probably through Stendhal who regularly frequented the Cuvier's Saturday salons – lengthy descriptions of Sophie's superlative intelligence for a woman extend far beyond necessary character portraiture:

Considerably older than her step-sister, Sophie played a more visible role in the household. She was beautiful, with a graceful and touching beauty. She was also extremely witty and intelligent. A woman friend describes her conversation as 'the most seductive and the most brilliant that has ever fallen from the lips of a woman'; and other people, when speaking of the Saturday receptions, have admired the facility and grace of her conversations with all the invited guests, as at ease with her step-father's eminent scientific colleagues as she was with their purely society friends. (Gunnell 1925, p. 232–3, my translation.) (The woman friend in question is Sarah Lee.)

Sharpe was to attend Clémentine's marriage (he was engaged to Sophie, and their marriage and a move for the couple to London was planned), which turned, as we know from the above, into a funeral. If Gunnell then focuses on Sophie's moral battle between duty – to stay in Paris with a sick, bereft mother – and personal happiness with Sharpe in England, what is more revealing in the context of Cuvier's many public and scientific duties at the time is the portrait of Sophie's constancy and support for her *stepfather* at the time of Clémentine's death, even though she herself was under huge emotional strain and had unremitting insomnia. (Duvernoy 1939, p. 6).

Duvernoy is inclined to (over) emphasize Sophie's nerves and many illnesses and the similarities of mother and daughter (in line with the many clichés of the period of ailing, hysterical women), whereas it can equally be argued that Sophie put them second in her role as a tower of strength to her inconsolable mother and stepfather. Gunnell (1925, p. 236) notes how Sophie becomes Cuvier's constant and then indispensable companion, how she accompanies him to his public lectures, acts as his secretary, and reads novels aloud to him in the evening to refresh his tired mind in readiness for the following day. The word 'secretary' belies the responsibility of this function, since it entailed Cuvier's international correspondence with his many international science contacts and was, as Outram (1984, pp. 172–173) notes, the role that another longstanding aide, the Irish naturalist Joseph Barclay Pentland, fulfilled, alongside tasks such as preparing plaster casts of fossils and arranging the contents of the comparative anatomy cabinets. Outram (1984, pp. 172 & 188) alludes to Sophie's part with Pentland's on the unofficial payroll of co-workers for Cuvier, but not the unprecedented situation that this entailed for her as a woman of equal scientific merit. After Clémentine's death, Cuvier relinquished responsibility for the sorting of materials for his *Histoire naturelle des poissons* and the illustrations for it, not only to his long-time collaborator and fellow Montbéliardais, Laurillaud, but also, and more importantly, to Sophie. If critics note the collaborative nature of Cuvier's large-scale productions, it is this volume, which includes a woman collaborator for the first time, albeit invisibly unless we look at the illustrations and imagine the necessary knowledge and work entailed in setting up the specimens. Sophie's artistic talent clearly measured up to the meticulous drawings and exacting work Laurillard and Cuvier himself produced. Lee (1833, pp. 279–281) richly documents Cuvier's deep interest in anatomical drawing, from which I infer the exacting standards he would have set his stepdaughter. Lee also states the overtly scientific reasons for Cuvier's second visit to England accompanied by Sophie (Lee 1833, pp. 318, 322–23): research for his work on fish (notes and drawings for his *ossemens fossiles* and his ichthyologie). After lunch he spent time with Sophie, visiting art collections and exhibitions.

If Sharpe finally broke off the engagement a couple of years later (and never married), it was his friend, the writer Stendhal, who has perhaps rescued Sophie from undeserved oblivion through their correspondence, to which we will now turn, and as the probable inspiration for one of his feistiest and most intelligent women characters, Mathilde de la Mole in *The Red and the Black*. Although Royer (1930, p. 23) notes that Mme Cuvier was the likely inspiration for Mme de Serpierre in Stendhal's *Lucien Leuwen* (1894) and, in the same novel, Théodelinde as possibly a version of Sophie, a comparison of Sophie to Mathilde is valid because Stendhal sent Sophie a copy of *Le Rouge et le Noir* when it was published (1830), and sought her comment. Royer quotes only Stendhal's critical comment to Mareste, 'Je vois les figures que l'on fait chez Mammouth'. (The 'mammoth' was Cuvier's nickname, '*chez*' could encompass Sophie in this antediluvian household.) It seems to me that much of Mathilde's free spirit, intelligence, witty conversation and education in her father's library owe much to Sophie Duvaucel, and that Stendhal may have been wanted to elicit her direct response to his fictionalization of her.

Athough only a handful of letters remain (mostly from the years 1829 and 1830, which coincide with the period of the aftermath of Clémentine's death), Stendhal was indebted to the social connections of the Cuviers, both father and stepdaughter, both French and international. Sophie furnished him with an important introduction to the Countess Abrizzi ('The "Madame de Staël" of Venice according to Byron' (Royer 1930, p. 46) and again inductive reasoning allows us to conclude that Sophie could not have had contact with this echelon of the European female intelligentsia without belonging to it herself. Interestingly, Royer (1930) does not mention Sophie's intelligence once. That she was as fully cognizant of the arts as the sciences may indeed have been a rich stimulation for her father, whose voraciousness as a reader of history and fiction (imbued by his mother) never diminished in his lifetime. Unlike Clémentine, Sophie moved effortlessly amid those in the worldly and literary circles of Paris: writers who congregated at the Jardin des Plantes but also in other notable literary salons. Space does not permit a comparative study of the other important Paris salons in the 1820s, notably Charles Nodier's salon at the Arsenal (see Laisney 2002) which was attended by scientists, artists and writers (including women such as Louise Colet, George Sand, Delphine Gay and Marceline Desbordes-Valmore). Being female and conversant with the sciences and arts in equal measure was clearly singular and quintessential to Sophie's inestimable animation of the Cuviers' Saturday salon, which only she could persuade her father and mother to reopen in early August 1828 (Duvernoy 1939, p. 11).

Most important of the three biographies of Sophie for our purposes is Duvernoy's account of her correspondence (largely with his wife from the age of 15 years, but also with him), where glimpses of her scientific judgement and contribution appear. In a letter-travel note from the first visit to England in 1818, Sophie reports the following:

We have seen some very beautiful specimens of natural history, especially fossils. Their [the British] collections are however not to be compared to ours. The College of Surgeons houses some excellent preparation put together by Hunter and some skeletons besides; but the English anatomists, no more than their naturalists, are not strong in zoology and comparative anatomy, with the result that they do not even know the value of their treasures. In general, the scientific establishments in England are almost unspeakable [*nuls*], the government being interested only in favouring the art of earning money, an art carried to its perfection in this country. (Duvernoy 1939, p. 6, my translation.)

It would appear from this that Sophie accompanied her father (and Laurillard) when he went to see Dr Leach at the British Museum and William Buckland in Oxford, because she spoke English, but also because she was already acquainted with many British natural scientists who had graced the

Saturday salon, for example the geologists Roderick Murchison and Richard Owen.

But it is the year 1829/30 that is most significant for Sophie's position as Cuvier's *aide-naturaliste* and for which all these earlier encounters were the preface. Cuvier's sole alleviation of grief for Clémentine was the *Histoire naturelle des poissons*, to which he devoted all his energies, and this was largely thanks to Sophie. In January 1829, Cuvier sent Laurillard to the Mediterranean to collect various samples, and another of his collaborators, Valenciennes, to England and Germany to visit museums and take notes. This left Sophie as the replacement in Paris of both these men. The indubitable evidence is a letter from Sophie to Duvernoy, explaining her many tasks on the project:

I had to do the hundred and one things that they would have done. I spent the mornings mounting and drawing the fish for the 'magnum opus' and molluscs for the 'animal kingdom'. I made copies, classified the books. I read aloud every evening to allow the dear patron to rest a little from the over fatigue of the work; and over and above it all I slept badly myself as is my wont ... At the moment we are overwhelmed by the 85 crates full of natural history which the *Astrolabe* has brought back from its voyage around the world. All our men of learning are in raptures amid these old bones, miserable pelts and richly coloured birds, and M. Cuvier has been better served than anyone since M. Quoy has been particularly diligent concerning molluscs and fish. (Duvernoy 1939, p. 13, my translation.)

Sophie's place is now firmly established in the male, scientific space on the 'other side of the wall' of the Jardin des Plantes, but she testifies also to her place in the wider public forum of her father's lectures on the history of science, which he decided to recommence at the Collège de France in the Autumn of 1829. It is she who reports to Duvernoy her stepfather's clarity of argument and exposition, but with finer quality than she recalls 12 years previously (Duvernoy 1939, p. 13). These lectures continued until cholera struck Paris in February 1832, to which Cuvier himself succumbed and of which he died, leaving his widow and his remarkable stepdaughter, Sophie Duvaucel. In 1834 she met and then married 57-year-old widower, the Admiral Ducrest de Villeneuve, whose three children she adopted. Her enlightened marriage included frequent periods in Paris with her mother at the Jardin des Plantes (where the Ducrests moved in 1842 and remained until Anne-Marie's death of influenza in 1849), and visits to London to see her longstanding friend and scientific 'correspondante', Mrs Sarah Lee. The death of her stepfather, however, inevitably marks the end of Sophie's 'career' as the first woman *aide-naturaliste* at the Jardin des Plantes.

'Extraordinary' daughters of a father of genius

Stephen Jay Gould and Dorinda Outram, among Cuvier's modern-day advocates, are also his leading critics, respectively for the two 'errors' in his science with which we began and for his failure to professionalize science at the Muséum because he perpetuated a system of patronage and family connections (Outram 1984, p. 188). However, these 'faults' can now be assessed in the very different light of Cuvier's life on the 'private' side of the wall of the Jardin des Plantes. In terms of his attitudes to parenting and the education of his children, as evidenced particularly by this study with respect to his daughters, Cuvier is far from the arch-conservative when it came to educational opportunity and curriculum. It is only because he kept the appointment of *aides naturalistes* within the family that it was possible for both his daughters to engage with the rich developments within the natural sciences and palaeontology of the period and to contribute actively to this progress by being no less able cogs in the great collaborative works that Cuvier produced. From the viewpoint of modern feminist history, one might then even revive Cuvier 'the Mammoth' as an exemplary encourager of the female mind as neither sexed nor gendered, nor inferior to the male brain. His unprejudiced attitudes to his daughters' intelligence, like his commitment to public lectures to bring science to the general public, including women, are clear examples of this.

It is, however, more shocking to understand Cuvier's untold legacy regarding inclusion of French women in science within the historical context of his life. His instinctive and natural integration of Clémentine and Sophie into the daily life of the Muséum on both sides of its walls, and the active role of his wife and daughters in the vibrant scientific and cultural 'networking' gatherings that were the Saturday salons, speak of a man who put freedom of enquiry beyond any respect of persons, or what we would now call gender or racial bias. If having a patron was the only door that could open a future for a gifted man in early 19th-century France, as Cuvier had himself discovered, it was even more essential for those women of genius of the period in the more 'female' pursuits of art or writing, let alone in the sciences. For women of genius in the natural sciences in England, such as Cuvier's biographer Sarah Lee, marriage was one of the very few points of entry into this man's world. In the contrastingly unenlightened 19th-century French educational context, Cuvier's daughters are an almost more extraordinary case of active scientific endeavour. By keeping science within his family, Cuvier was both the literal and figurative father of the natural sciences for Clémentine and Sophie. Finally, if I have merely borrowed the superlative epithet 'extraordinary' from Salvandy's oration for Clémentine to describe the nature of their story in terms which fit its 19th-century French frames, I hope to have reconstructed à la Cuvier from this single 'bone' the fleshed-out implications of its significance for the collaborative contribution of Cuvier's daughters to the history of 19th-century French women in science and geology.

I would like to thank M. Rudwick for his interventions after I gave the conference version of this paper and for his editorial comments on the manuscript.

References

ANON. 1928. (no author) 'Notice sur Clémentine Cuvier'. Extrait des Archives du Christianisme au dix-neuvième siècle. Henry Servier Libraire. Paris, 3–9.

ANON. 1970. *Georges Cuvier: de son temps au notre*. Expansion Editeur, Langres.

BOISSARD, M. n. d. Discours prononcé dans le Temple de la Rue des Billettes. In: BOISSARD, M. *Clémentine: 28 septembre 1827*. Imprimerie de Marchand de Breuil, Paris, **80**, 3–8.

BUFFETAUT, E. 2002. *Cuvier: le découvreur de mondes disparus*. Belin, Paris.

CUVIER, C. 1829. *La Fille d'une femme de génie, traduit de l'anglais de Mme Hofland par Clémentine Cuvier*. Barbezant, Paris.

BUREK, C. V. 2007. The role of women in geological higher education – Bedford College, London (Catherine Raisin) and Newham College, Cambridge, UK. In: BUREK, C. V. & HIGGS, B. (eds) *The Role of Women in the History of Geology*. Geology Society, London, Special Publications, **281**, 9–38.

DUVERNOY, M. 1939. *Sophie Duvaucel d'après des correspondences inédites*. Société Anonyme d'Imprimerie Montbéliardaise, Montbeliard. Extrait des Mémoires de la Société d'Emulation.

FLOURENS, P. 1869. *Georges Cuvier: Eloges Historiques précédés de l'éloge de l'auteur*. Paul Ducrocq, Paris.

FRAISSE, G. 1995. *Muse de Raison: Démocratie et Exclusion des Femmes en France*. Gallimard, Paris.

FRAISSE, G. 2002. *Clémence Royer: Philosophe et Femme de Sciences*. La Découverte, Paris.

GOULD, S. J. 1993. Foreword. In: SMITH, J. C. *Cuvier: an Annotated Bibliography of His Published Works*. Smithsonian Institution Press, Washington and London, x–xi.

GUNNELL, D. 1925. *Sutton Sharpe et ses amis français*. Librairie ancienne Honoré Champion, Paris.

HARDING, S. 1993. *The 'Racial' Economy of Science: Toward a Democratic Future*. Indiana University Press, Bloomington.

HARDING, S. 1998. *Is Science Multicultural? Postcolonialisms, Feminisms and Epistemologies*. Indiana University Press, Bloomington.

HARDING, S. & O'BARR, J. F. (eds) 1987. *Sex and Scientific Enquiry*. University of Chicago Press, Chicago.

HOFLAND, B. 1823. *The Daughter of a Genius: A Tale for Youth*. John Harris & Son, London.

HUBBARD, R. 1988. Some thoughts about the masculinity of the natural sciences. In: MCCANNEY, G. M. (ed.) *Feminist Thought and the Structure of Knowledge*. New York University Press, New York & London, 1–15.

JORDANOVA, L. 1989. *Sexual Visions: Images of Gender in Science and Medicine between the Eighteenth and Twentieth Centuries*. Harvester Wheatsheaf, Hemel Hempstead.

KELLER, E. F. 1985. *Reflections on Gender and Science*. Yale University Press, New Haven.

KELLER, E. F. & LONGINO, H. E. (eds) 1996. *Feminism and Science*. Oxford University Press, Oxford and New York.

LAISNEY, V. 2002. *L'Arsenal romantique: le salon de Charles Nodier (1824–1834)*. Honoré Champion, Paris.

LEE, S. 1833. *Memoirs of Baron Cuvier*. Longman, Rees, Orme, Brown, Green and Longman, London. Translated by Théodore Lacordaire as *Mémoires du Bon Georges Cuvier*. H. Fournier Libraire. Paris [also 1833].

MATHIOT, C. & DUVERNOY, M. (eds) 1940. *Lettres inédites de Charles Laurillard à Georges Louis Duvernoy: introduction et notes*. Société Anonyme d'Imprimerie Montbéliardaise, Montbéliard.

OGILVIE, M. B. 1988. *Women in Science: Antiquity through to the Nineteenth Century. A Biographical Dictionary with Annotated Bibliography*. MIT Press, Cambridge, Massachusetts.

ORR, M. 2007a. Pursuing proper protocol: Sarah Bowdich's purview of the sciences of exploration. In: *Fourth Conference of the North American Victorian Studies Association, Purdue University, West Lafayette, 31 August–3 September 2006. Victorian Studies*, (Special Issue) **49**, 277–285.

ORR, M. 2007b. Women and daughters of genius: Mrs. Barbara Hofland and Mlle Clémentine Cuvier. In: DOW, G. (ed.) *Translators, Interpreters, Mediators: Women Writers, 1700–1900*. Peter Lang, Bern.

OUTRAM, D. 1984. *Georges Cuvier: Vocation, Science and Authenticy in Post-Revolutionary France*. Manchester University Press, Manchester.

OUTRAM, D. 1989. Before objectivity: wives, patronage and cultural reproduction in early nineteenth-century France. In: PNINA, G. A.-A. & OUTRAM, D. (eds) *Uneasy Careers and Intimate Lives: Women in Science 1789–1979*. Rudgers University Press, New Brunswick & London, 19–30.

POIRIER, J.-P. 2002. *Histoire des Femmes de Science en France: du Moyen Age à la Révolution*. Pygmalion, Paris.

ROYER, L. (ed.) 1930. *Stendhal au Jardin du Roi: Lettres inédites à Sophie Duvaucel publiées avec des notes par Louis Royer*. Arthaud, successeur des Editions J. Rey, Grenoble.

RUDWICK, M. 1997. *Georges Cuvier, Fossil Bones and Geological Catastrophes: New Translations and Interpretations of the Primary Texts.* University of Chicago Press, Chicago and London.

SALVANDY, M. de. n. d. Discours prononcé sur la Tombe. *In*: BOISSARD, M. n. d. Clémentine: 28 Septembre 1827. Disours prononcé dars le Temple de la Rue des Billottes Imprimeries de Marchand de Breuil, Paris, **80**, 1–18.

SCHIEBINGER, L. 1999. *Has Feminism changed Science?* Harvard University Press, Cambridge, Massachusetts.

SMITH, J. C. 1993. *Cuvier: an Annotated Bibliography of his Published Works.* Smithsonian Institution Press, Washington & London.

STENDHAL, H. B. 1830. *Le Rouge et le Noir.* 1972 Edn. Gallimard, Paris.

STENDHAL, H. B. 1894. *Lucien Leuwen.* 2 vols. 1973 Edn. Gallimard, Paris.

WILKS, M. 1838. *The Flower Faded: A Short Memoir of Clémentine Cuvier by the Rev. M. Wilks ... With Reflections and An Appendix by John Angell Adams.* Hamilton Adams & Co., London.

The influential Muriel Arber: a personal reflection

E. ROBINSON

Riverside Farm, Witchall, Watchet, Somerset TA23 OBB, UK

Abstract: Muriel Agnes Arber (1913–2004) was a female geologist of our time. However, she drew her energies from traditional sources, her family and their contribution to academic science, and the Geologists' Association. She benefited from the town and gown life of Cambridge and the influence of frequent visits to Lyme Regis, where family and science were drawn together, and where she met people to whom she could relate. This paper explore-these aspects of her life from a personal point of view.

In the competitive thrust of academia, influence can be sought and acquired for many reasons and in many different ways. Success may seem to depend upon power presentation and active canvassing. In complete contrast, however, 'influence' can come quite naturally and be recognized in those who act with a simple attention to a sense of duty, in whom recognized influence can seem a matter of indifference. To me, this would epitomize the life of Muriel Arber, a teacher to the core (Fig. 1).

Influence of academic family members

As an assistant lecturer in the Geology Department at University College London (UCL) in the 1950s, I first met Muriel when she was renewing her acquaintance with the Botany Department of Professor Pearsall in order to appreciate the work of her mother, Agnes Robertson, who had been a research student under Professor Oliver between 1889 and 1900. Women had long had access to UCL as a matter of policy, from it's foundation as 'The Godless Place on Gower Street', and Agnes, a first-rate systematic botanist, was just one product of this openness to talent. Her enquiries were equally focused upon the department of Professor T. C. Bonney, who himself tutored the formidable Dr Catherine Raisin, one of the first women to acquire a DSc from the University of London (Burek 2007). Muriel's father, E. A. Newell Arber, a palaeobotanist before he became an authority on the cliff geomorphology of the Devon coast, had worked in the Geology Department across the yard from the Botany Department.

When Newell Arber was appointed Demonstrator in Palaeobotany at Cambridge (1898), and the opportunity arose for her mother to move to the School of Botany, everything moved towards their marriage. This was solemnized by their friend Professor Bonney, who, in addition to being Yates-Goldsmid Professor of Geology at UCL, was an Honorary Canon of Manchester Cathedral and so capable of performing the ceremony. Bonney had just given way to Edmund Garwood, leaving him to cope with Marie Stopes, as a student, insisting on taking double honours in geology and botany. At his age, Bonney was entitled to certain eccentricities, one of which came out in the ceremony. Muriel always claimed that the marriage was doubly blessed as, when the vows were made, to their surprise, Bonney took them through a second time.

Demonstrators were poorly paid, but Newell Arber played an important part in the transfer of geological sciences into the newly opened Sedgwick Museum on Downing Street, while still actively researching in palaeobotany and making useful contributions to the study of Coal Measure plant fossils. Muriel was born in Cambridge in 1913. By this time her father had developed a keen interest in the Devon coast, and the geomorphology of its imposing cliffs,

Fig. 1. Portrait of Muriel Arber about 1936 (From *Geologwick Curator*, with Thanks to Patrick Wyse Jackson).

publishing the book by which he is best known: *The Coast Scenery of North Devon* (Arber 1911). The book was to be a prime influence on Muriel's later researches in south Devon and Dorset. Her interest developed, as Muriel herself said in her virtual autobiography *Lyme: Landscape with Figures* (Arber 1988), when, after the death of her father in 1918, her mother began the habit of spending long summer holidays in Lyme Regis with Muriel. Schooling and childhood were spent in Cambridge, and Muriel went up to Newnham College to read English in 1933. However, with all those family connections with the natural sciences, it was inevitable that she switched studies to geology, a change cemented when she was permitted to go into the field as a guest of the Sedgwick Club. On their excursion to Torquay she met her life-long friend T. C. Nicholas, already established as a Palaeozoic specialist through his paper on the Hell's Mouth Grits of St Tudwalls, North Wales. Another colleague was W. D. V. Jones, who worked on the Llandovery rocks of mid-Wales. In spite of these influences, on graduation, Muriel began her research in the attic floors of the Sedgwick Museum on the strophomenid brachiopods, under the supervision of O. M. B. Bulman, research that reached publication. Through this period of her life (Arber 2003), Muriel drew strength from the association with other women on those upper floors, who in turn were encouraged by the presence of Gertrude Elles, still building upon the work on graptolites which she and Miss Wood had done for the Palaeontographical Society monograph (Elles & Wood 1901–1918). Indeed, the Sedgwick Museum, which her father had played an important role in establishing, produced a team of researchers who must be given credit for improving the role of women in geology in the years immediate prior to World War II. Muriel herself was involved in the safety measure of sending much of the museum collections into safe storage when the threat of war became pressing. At the same time, she prepared for the evacuation of the staff and students of Bedford College from London to a stay in Cambridge, winning new friends in the process. She herself, however, was about to take up a teaching post at King's School, Ely, which she held until her retirement in 1973.

Just how important Muriel's family were to her I discovered quite early in our friendship. I was working as a volunteer geologist in the recording of gravestones in Kensal Green Cemetery on the Harrow Road when I came upon a substantial granite slab in a very prestigious position in the Dissenters' sector of the total space (about 22 hectares!). It was an Arber ancestor: Thomas Christian Newell Arber of Bryanston Square, who died 'on Sunday evening Dec. 6th, 1865'. Muriel was delighted and came to read the full and meaningful inscription (Fig. 2). What we recorded was as follows. 'His motto through life – Whatsoever thy hand findeth to do, do it with thy might for there is no work, nor device, nor knowledge, nor wisdom in the grave whither thou goest'. Added to that his "Hope in death" was the first verse of the hymn 'Just as I am, without one plea'.

It was one of those occasions when Muriel let slip some of her own feelings, including one of her chuckles over the Victorian piety of the lines, but, on reflection, I would say that what was termed his 'motto through life' could well have been her own dedication.

When her mother died in 1960 she donated her books to the Hunt Institute for Botanical Documentation in the Carnegie Mellon University in Pittsburg, Pennsylvania. At the time it is recorded that she was thanked by Mrs Hunt herself who said: 'I have used your mother's books over many years and to have this collection from her daughter means more to me than I can say at the moment . . . . I am so grateful to you for letting us have these' (Todd 2004). When Muriel died in 2004 her

Fig. 2. Kensal Green Cemetery. The gravestone of Thomas Christian Arber, who died 1865.

executives were instructed to present a further four boxes of her mother's botanical material to the same institute. It is gratifying to see a personal library being treated in this way and shows the esteem that her mother received from both her daughter and her peers.

The Geologists' Association

Muriel's years at Ely overlap with what we might see as the third claim to her energies: membership and promotion of the ideas behind the Geologists' Association (GA) as the society in geology devoting itself to the needs of the amateur. What is more, from its foundation in 1858, it had been a society in which women had equal rights with men. To quote from the Laws of the GA, 'unanimously adopted at a General Meeting held on the 17th December, 1858':

Law 3 Ladies shall be eligible for election as Members, and shall enjoy all the rights and privileges of either Town or Country Members, as the case may be.

Muriel joined the GA in 1935 while an undergraduate, and fully exploited Law 3 through membership of Council and, eventually, Presidency in 1972–74. In a GA context, I remember Muriel at the first annual Reunion, which I attended on the ground floor of Chelsea College in Manresa Road. She looked very much the headmistress figure as she struggled to pin up a panorama of photographs of the landslipping cliffs of Lyme Regis on hard wooden boards. This display was updated at almost all of the 18 Reunions that I was involved in running, later in the more spacious cloisters of UCL. Displaying the photo prints always seemed to create a problem. However, Muriel's record of the dynamic geomorphology of those cliffs fulfilled everything which the GA had been urging members to do: log and record temporary sections for posterity. Sadly, we never pressed Muriel to deposit those photographs with the GA.

Muriel lived up to the spirit of 1858 and the early years of the GA. Her drive through the GA was to present earth history where you were, in the most succinct way, in terms that the general public could understand. In GA publications she would always quote another early law that advised authors to 'avoid hard words!' This was her approach to the GA Council, where her contributions were measured and firm, but at the same time diplomatic. During her years as President of the GA, there were the usual financial crises. More worryingly there was a national outcry when the famous Norber erratic in Crummackdale was daubed with paint in the initials of a well-known university. Geological field parties immediately became open to accusation of mindless vandalism, including the unsightly results of hammering of crags and rock surfaces. At the time, Alan Stubbs was a GA member as well as second geologist to George Black in the Nature Conservancy Council (later English Nature). He brought to Council the need for a code of conduct for geological fieldwork that, at least, would make a point that geologists valued access to the classic localities in the United Kingdom, and were prepared to sign up to rules governing behaviour to retain the goodwill of landowners and the general public. A committee was formed to prepare such a leaflet and, as is the way with committees, produced the usual over-formal rubric full of stilted phrases, sounding much too stern and condemnatory. It was dire. Fortunately, the Council had the experience of Stanley Holmes to call upon, but it was Muriel, the King's School teacher, with her simple and direct style, which transformed the text. A list of 'Don'ts' were turned on their head to become 'Dos' and the whole tone changed. Together, Muriel and I went to British Petroleum (BP) in Britannic House to seek funding. They agreed to fund the printing of 100 000 copies of a simple folded leaflet, which was the first of what were soon to be other codes. Thanks to Muriel's guidance as President, it was and remains the neatest and most succinct of codes.

Another initiative in which Muriel had a hand was the approval of the newly formed Health and Safety at Work Executive, set up to prevent accidents in the workplace. A visit to their office on Millbank by Muriel and gave the Executive the chance to frame words that would cover those visits to quarries that are so essential a part of fieldwork. The code needed to reassure managers and companies that our behaviour on visits would recognize all the controls binding upon quarry workers. In due course, the code went to the Federations and the Union to become a guide-line document which had the widest acceptance possible. Later the GA Code became modified by 5 years of experience, but the original simple, folded, green-paper leaflet (Fig. 3) was a creditable achievement for the GA and it was thanks largely to the astute guidance of Muriel Arber.

Presidencies can be brief periods of influence, but in Muriel's case her sound advice continued for many more years after 1974. In this period, right up to the time when her blindness became a real handicap, she helped me as editor of the six GA Circulars, which went out to all members. It was a period when what had been simply the announcement of the programme of lectures and field meetings became an opportunity to highlight issues of conservation and matters of comment on museum charges and English Nature policies. On the museum policies, I had responded to the heartfelt feeling of Bev Halstead and John Evans

A Code for Geological Field Work

ISSUED BY THE GEOLOGISTS' ASSOCIATION

Benham and Company Limited, Colchester

The Scope of Geology

Geology embraces the scientific study of the history of the earth, and the past evolution of life upon it. The technical expertise derived from this study also enables our modern civilisation to gain its essential raw materials, for example coal, oil and gas, nuclear fuels, metallic ores, building materials, chemicals for agriculture, and water supplies. It also enables us to develop engineering skills to best advantage, as in the construction of tunnels, dams, roads, and sea defences.

Accessible field sites in good condition are essential for the training of Geologists and for the research needed to make the best use of the natural resources on which we all depend.

It is hoped that the widest possible publicity will be given to this Code, and that authors and teachers will quote it whenever possible.

Further copies are available from the Librarian at the above address.

This Code, initiated by the Geologists' Association, has the support of the following organisations:

The Geological Society of London
The Edinburgh Geological Society
The Geological Society of Glasgow
The Manchester Geological Association
The Liverpool Geological Society
The Yorkshire Geological Society
The East Midlands Geological Society
The Geological Society of Norfolk
The Palaeontological Association
The Geographical Association
The British Geomorphological Research Group
The National Museum of Wales, Cardiff
The Nature Conservancy Council

Note for Landowners

Landowners may wish to ensure that visiting Geologists are familiar with this Code. In the event of its abuse, they may choose to take the name and address of the offenders and the Institution or Society to which they belong.

Enquiries may be addressed to The Librarian, Geologists' Association, c/o Geology Department, University College London, Gower Street, London WC1E 6BT.

1975

A Code for Geological Field Work

A GEOLOGICAL 'CODE OF CONDUCT' has become essential if opportunities for field work in the future are to be preserved. The rapid increase in field studies in recent years has tended to concentrate attention upon a limited number of localities, so that sheer collecting pressure is destroying the scientific value of irreplaceable sites. At the same time the volume of field work is causing concern to many site owners. Geologists must be seen to use the countryside with responsibility; **to achieve this, the following general points should be observed.**

1. Obey the Country Code, and observe local byelaws. Remember to shut gates and leave no litter.
2. Always seek prior permission before entering private land.
3. Don't interfere with machinery.
4. Don't litter fields or roads with rock fragments which might cause injury to livestock, or be a hazard to pedestrians or vehicles.
5. Avoid undue disturbance to wildlife. Plants and animals may inadvertently be displaced or destroyed by careless actions.
6. On coastal sections, be sure you know the local tide conditions.
7. When working in mountainous or remote areas, follow the advice given in the pamphlet 'Mountain Safety', issued by the Central Council for Physical Education, and, in particular, inform someone of your intended route.
8. When exploring underground, be sure you **have the proper equipment**, and the necessary experience. **Never go alone.** Report to someone your departure, location, estimated time underground, and your actual return.
9. Don't take risks on insecure cliffs or rock faces. Take care not to dislodge rock, since other people may be below.
10. Be considerate. By your actions in collecting, do not render an exposure untidy or dangerous for those who follow you.

Collecting and Field Parties

1. Students, should be encouraged to observe and record but not to hammer indiscriminately.
2. **Keep collecting to a minimum.** Avoid removing *in situ* fossils, rocks or minerals unless they are genuinely needed for serious study.
3. For teaching, the use of replicas is commended. The collecting of actual specimens should be restricted to those localities where there is a plentiful supply, or to scree, fallen blocks and waste tips.
4. Never collect from walls or buildings. Take care not to undermine fences, walls, bridges or other structures.
5. The leader of a field party is asked to ensure that the spirit of this Code is fulfilled, and to remind his party of the need for care and consideration at all times. He should remember that his supervisory role is of prime importance. He must be supported by adequate assistance in the field. This is particularly important on coastal sections, or over difficult terrain, where there might be a tendency for parties to become dispersed.

Visiting Quarries

1. An individual, or the leader of a party, should have obtained **prior** permission to visit.
2. The leader of a party should have made himself familiar with the **current state** of the quarry. He should have consulted with the Manager as to where visitors may go, and what local hazards should be avoided.
3. On each visit, both arrival and departure must be reported.
4. In the quarry, the wearing of safety hats and stout boots is recommended.
5. Keep clear of vehicles and machinery.
6. Be sure that blast warning procedures are understood.
7. **Beware of rock falls.** Quarry faces may be highly dangerous and liable to collapse without warning.
8. Beware of sludge lagoons.

Research Workers

1. No research worker has the special right to 'dig out' any site.
2. Excavations should be back-filled where necessary to avoid hazard to men and animals and to protect vulnerable outcrops from casual collecting.
3. Don't disfigure rock surfaces with numbers or symbols in brightly coloured paint.
4. Ensure that your research material and note-books eventually become available for others by depositing them with an appropriate institution.
5. Take care that the publication of details does not lead to the destruction of vulnerable exposures. In these cases, do not give the precise location of such sites, unless this is essential to scientific argument. The details of such localities could be deposited in a national data centre for Geology.

Societies, Schools and Universities

1. Foster an interest in geological sites and their wise conservation. Remember that much may be done by collective effort to help clean up overgrown sites (with permission of the owner, and in consultation with the Nature Conservancy Council).
2. Create working groups for those amateurs who wish to do field work and collect, providing leadership to direct their studies.
3. Make contact with your local County Naturalists' Trust, Field Studies Centre, or Natural History Society, to ensure that there is coordination in attempts to conserve geological sites and retain access to them.

Fig. 3. *A Code for Geological Field Work*, leaflet issued by the Geologists' Association.

who, as GA Presidents, were entitled to set policy views on issues that seemed to touch upon those aims and objectives so ably framed in the laws of 1858. At times, the Circular could become controversial in ways which worried some. I had to excuse myself on some such occasions by saying that often the copy had been floated before Muriel for comment. With Muriel, if these things were alright, there would be helpful words, sometimes effusive praise (for the topic!); sometimes, there could be one of those chuckles which seemed to shake her frame. At other times, there was a silence, which I recognized as disapproval. This was 'influence' for which the GA should be grateful!

Muriel continued to attend the annual Reunions at UCL, producing her yearly record of landslips in Dorset; she also attended meetings through the winter months, but her increasing blindness saw her turning some of her support to activities in and around Cambridge. She had done much to raise the Cambridge Geological Society to affiliate status with the GA. So she returned to the strengths she could draw from Cambridge life in all of its facets. This I recognized through visits to her flat in Sherlock Close, which was crowded with books and papers relating to her many connections with geology and people, and a work place conveniently placed for town and university events. On one visit, following up my persistent habit of graveyard geology, after walking with her in the historic Histon Road cemetery, we ended up in the Ascension Cemetery off the Huntingdon Road. Here were buried academics whom she could identify, reeling off their relationships and claims to fame as Darwin's sons, Horace and Francis Eddington, Cockcroft, Wittgenstein? I wish that I had had an unobtrusive tape recorder to have taken down some of the flow, if only to supplement the record of the stones. One I do remember was a type of Celtic design with a coloured mosaic border. It was to Woods, the author of that textbook of palaeontology which we all had to use until better texts came along. 'Celtic', she volunteered because he was married to a daughter of Skeat the Anglo-Saxon scholar. And so it went on. It brought home to me how deeply soaked Muriel was in the life of the University and its characters, and how that grew in her later life. At concerts and functions she was often accompanied by T. C. Nicholas, to whom she always felt indebted for that early introduction to the Sedgwick Club and geology. When he was approaching his 100th year, Muriel forbade him to ride his bicycle out of concern for friends.

Family, the University, the GA – in many ways these strands are interwoven in Arber's life through to her last days (Arber 2003; Friend 2005; Robinson 2005).

Lyme Regis

As we know from her own account (Arber 1988), she fell in love with Lyme and all that went on there through family holidays with her mother as early as 1918. Visits had included outings on the shore with W. D. Lang, so fossils came into her life at an early stage. Over the years it was the slipping cliffs that caught her attention and, with her camera, she started a personal record which would have been valuable to conserve. She also published scientific papers on this subject as well as on landslides further afield in North Devon (Arber 1940, 1946, 1949, 1960, 1973, 1977) There was, however, in addition, her devotion to the town museum, named in honour of the Misses Philpot, collectors of Lyme Geology. This was a somewhat old-fashioned place, suffering from periods of neglect, Muriel rose to the task and helped as best she could (Arber 1982). She took heart when Liz-Anne Bawden retired to Lyme from the Slade School and took an interest in the museum. The latter extended our interest in the town by recalling that it was the home of Eleanor Coade, with her contribution to decorative architecture (and geology) with her Coade Stone.

Those beginnings were built on by the late John Fowles, who took up the curatorship of the museum and soon picked up the spirit of the place. He acknowledged that Muriel gave him the background context to the slipping cliffs, which allowed him to paint in the geology within his novel, *The French Lieutenants Woman*. He also acknowledged that she advised when he was drawn inevitably into the perpetual debate over collecting fossils and the role of professional collectors who save so many prized specimens as latter-day Mary Annings. It seems that Fowles was never the 'withdrawn individual' the critics often call him when Muriel Arber was around. In his introduction to her autobiographical *Lyme: Landscape with Figures*, Fowles named her 'genius of the place'. The local press could be more direct when they announced 'the arrival of our Geologist' when her seasonal visit came around. Possibly the honour which she valued, along with her Foulerton Award from the GA and the Worth Prize from the Geological Society, was again from the local press as 'Lyme's oldest tourist', a comment which she took in good part.

In truth, the Lyme that figured in her two presidential addresses to the GA (Arber 1973, 1974) seemed to bring together those interests that figured in her life. Her dedication to those principles we read on that Arber gravestone in Kensal Green Cemetery and her selfless work, won her the influence that so many people acknowledge when her name is mentioned.

References

ARBER, E. A. N. 1911. *The Coast Scenery of North Devon.* Dent, London, 1969 facsimile, Kingsmead Reprints, Bath.

ARBER, M. A. 1940. Coastal landslips in south east Devon. *Proceedings of the Geologists' Association,* **51**, 257–271.

ARBER, M. A. 1946. The valley system of Lyme Regis. *Proceedings of the Geologists' Association,* **57**(1), 8–15.

ARBER, M. A. 1949. Cliff profiles of Devon and Cornwall. *Geographical Journal, London,* **114**, pp. 191–197.

ARBER, M. A. 1960. Pleistocene sea-levels in north Devon. *Proceedings of the Geologists' Association,* **71**(2), 169–176.

ARBER, M. A. 1973. Landslips near Lyme Regis. *Proceedings of the Geologists' Association,* **84**, 121–133.

ARBER, M. A. 1974. The cliffs of North Devon. *Proceedings of the Geologists' Association,* **85**, 147–157.

ARBER, M. A. 1977. A brickfield yielding elephant remains at Barnstaple, north Devon. *Quaternary Newsletter,* **21**, 19–21.

ARBER, M. A. 1982. The Bideford Museum geological collection of Inkerman Rogers (1866–1959). *Geological Curator* **3**(4), 199–208.

ARBER, M. A. 1988. *Lyme: Landscape with Figures.* Dorset Books, Exeter, 72 pp.

ARBER, M. A. 2003. Early memories of the Sedgwick. *GeoCam,* **7**, 5–7.

BUREK, C. V. 2007. The role of women in geological higher education – Bedford College, London (Catherine Raisin) and Newnham College, Cambridge, UK. *In*: BUREK, C. V. & HIGGS, B. (eds) *The Role of Women in the History of Geology.* Geological Society, London, Special Publications, **281**, 9–38.

ELLIS, G. L. & WOOD, E. M. R. 1901–1918. A monograph of British graptolites. Palaeontographical Society, London, a–m, i–clxxi, pp. 1–530.

FRIEND, P. 2005. Obituary. Muriel Agnes Arber, 1913–2004. http://www.geolsoc.org.uk/gsl/null/lang/en/page2562.html.

ROBINSON, E. 2005. Obituary Muriel Agnes Arber, 1913–2004. *Proceedings of the Geologists' Association,* **116**, 61–63.

TODD, A. L. 2004. News from the Archives. *Bulletin of the Hunt Institute for Botanical Documentation,* **16**(2), 8–9.

A knowledge unique: the life of the pioneering explorer and palaeontologist, Dorothea Bate (1878–1951)

KAROLYN SHINDLER

Freelance science writer
(e-mail: KarolynShindler@aol.com)

Abstract: Dorothea Bate is recognized as a pioneer of archaeozoology – the study of animal bones from archaeological sites. She also worked extensively on cave sites. Her research in Cyprus, Crete, Majorca and Menorca, as well as pre World War II Palestine, showed her holistic approach to the use of animal bones in order to deduce dates, climate and environment of the sites. She was the first woman to work as a scientist at the Natural History Museum, then known as the British Museum (Natural History), and her research reports on the fossil faunas are still being used today.

In 1951, on the death of Dorothea Bate, the archaeologist Anthony Arkell wrote a series of tributes to her, culminating in this: 'Her widely lamented death has robbed archaeologists and prehistorians of a palaeontologist whose co-operation was invaluable and knowledge unique' (Arkell 1953, p. 11). Arkell (1951) called her work comparative zoology. Today she is recognized as a pioneer of archaeozoology – the study of animal bones from archaeological sites.

Dorothea was an exceptional scientist (Fig. 1) – an ornithologist, palaeontologist, zoologist and archaeologist. She was the first woman ever to work on the science staff at the British Museum (Natural History), or BM (NH), now known as the Natural History Museum. It became part of the Dorothea Bate legend that, if sufficient quantities of bony fragments were sent to her, she would not only identify the fauna, but would suggest the climate and environment as well (Edwards 1951*b*, p. lvii). Her knowledge ranged from the fossil fauna of British caves to that of East Africa, from the Tertiary period to Recent. Her courageous explorations of Mediterranean islands as a young woman in the early 1900s revealed for the first time the Pleistocene fauna of Cyprus, Crete and the Balearics. As for her work in the Near East 70 years ago, perhaps the greatest tribute is that her analyses and reports on the fossil fauna are still in use today.

Where does Dorothea fit in to the canon of women and their role in the history of geology? She was not the wife, child or sibling, or indeed any relative that I have discovered, of a geologist. That does not make her unique; look for example at Barbara Yelverton, the Marchioness of Hastings, who in the mid-19th century amassed a superb collection of Tertiary fossils which she subsequently sold to the British Museum (Hastings 1851–55; British Museum (Natural History) 1904, p. 296). The Marchioness's passion for fossilizing seems to have sprung from living near the fossil-filled cliffs of the Hampshire coast, rather than family influence. Even the most cursory glance through the history of the collections of the Geology Department of the British Museum in the 19th century reveals a significant number of women who were collectors in their own right (British Museum (Natural History) 1904). Some are well known – Lady Hester Stanhope, for example, in 1817 presented the Museum with her collection of Cretaceous fish from the Lebanon – but many of the others are not. They are women who collected quietly and diligently, but never wrote about their finds. They are remembered today on labels in the Natural History Museum's collections.

During her lifetime, Dorothea Bate had an international reputation as an innovative palaeontologist and fearless explorer. After her death, she slipped swiftly into obscurity. Yet the collections, libraries and archives of the Natural History Museum hold testament to her life and extraordinary achievements. She published more than 80 papers and wrote many more that exist only in manuscript form. Yet she was almost entirely self-educated. She used to laugh that her education was only briefly interrupted by school (Edwards 1951*b*, p. lvi), but somewhere along the way, she accumulated vast reserves of knowledge. The Curator of Quaternary Mammals at the Natural History Museum, Andrew Currant (who describes himself as Dorothea Bate's linear successor), told me that he had no idea how she had found the time to learn as much as she did (Currant, pers. comm. 2000).

Dorothea's childhood

Dorothea Bate was born in Carmarthen in South Wales in 1878, the second of three children. Her father, Henry, had left the Army with the rank of

Fig. 1. Dorothea Bate *c*. 1906, by her sister Leila Luddington. It was probably painted during the 5 years when her parents refused to allow her to travel abroad. (Copyright Natural History Museum, London.)

major when he was 38 years old, and was appointed Superintendent of Police in Carmarthenshire. Her mother, Elizabeth, was an accomplished musician. They were not wealthy. Henry's father had been a coastguard in Ireland, while Elizabeth's family were manufacturers in Scotland (Shindler 2005). There is nothing, on the face of it, in Dorothea's background that obviously accounts for her subsequent career, except, perhaps, the happy circumstance of where she lived.

When Dorothea was 10 years old, the family moved to Cenarth, near Newcastle Emlyn in South Wales. Their house stood in the hills above the valley of the River Teifi, near the famous Cenarth Falls. It was an idyllic place to play, and to feed the curiosity of an intelligent child. She was surrounded by opportunities to acquire knowledge, and with or without teachers or governesses, she did. She was an instinctive natural historian, and natural history was of course the pastime that obsessed so many in the 19th century – not just observing, but collecting. Everything natural seems to have been collectable: insects, stones, fossils, ferns, flowers, birds' eggs – all were desirable. Birds and mammals were also collected, or shot so their skins and skeletons could be examined,

and the finest examples were stuffed and displayed. Few were squeamish about the process of killing, skinning or dissecting creatures; why should they be? It was little different to the everyday business that went on in many kitchens. The advances in microscope technology in the 19th century were of great importance, even to amateurs (see Higgs & Wyse Jackson 2007), while a magnifying glass was *de rigueur* for every collector. Dorothea's father appears to have encouraged her (Bate 1900), pointing out particular birds and plants, and teaching her to shoot and fish. She learned how to skin and dissect birds and small mammals. But her passion then was ornithology, and it was that which led her to the BM (NH). Her arrival there is the stuff of legend.

Dorothea and the British Museum (Natural History)

According to her friend and colleague Wilfred Edwards, who was Keeper of Geology at the time of her death, in 1898 Dorothea marched into the BM (NH) and demanded a job (Edwards 1951*a*). She was taken to the Bird Room, which was ruled over by the brilliant but irascible Dr Richard Bowdler Sharpe, the Curator of Birds. Initially he told her to go away, he could not be bothered with anyone. Not only was she just 19 years old, but no women *at all* were then employed in a scientific capacity by the BM (NH). Yet such was Dorothea's charm and persistence that, within a few minutes, she was sorting bird skins into species with assurance and skill. It was the beginning of her association with the BM (NH) that was to last for more than 50 years. That achievement cannot be overstated. The museum was a wholly male preserve, a place of black frock coats and grave formality. Keepers would communicate through speaking tubes, and donned their silk hats even to cross the corridor to visit each other's offices (Riley 1964).

Dorothea – cave explorer and palaeontologist

How Dorothea became interested in palaeontology has to be a matter of surmise; there is no convenient letter or diary that reveals all. But in 1898, her family moved to Gloucestershire, near Symonds Yat in the Wye Valley, an area rich with limestone caves. It was there that Dorothea made her first palaeontological discoveries (Bate 1901). Many of these caves had been found by local iron-ore miners, and one of them showed her a cave that had not previously been excavated. It is now known as Merlin's Cave. It can only be reached by a climb of 45 metres up a steep wooded slope, and then, with the help of ladder, another 4.5 metres up a limestone cliff. It is almost hidden from view by shrubs and bushes. Palaeontologists today consider it to be particularly dangerous to reach. Dorothea must have had assistance from a miner; it would have been impossible for her to have reached the cave alone.

In her report (Bate 1901) she describes the cave vividly, with its two chambers going back into the depths of the cliff. The first was about 27 metres deep; the second, inner chamber had no natural light – the only light was from her candle lantern, and the roof was so low that she had to crawl on hands and knees. Embedded in the walls and floor, were numerous fossilized bones of small mammals. She describes how she had to teach herself how to extricate these tiny fragile fossil bones as she worked. At her first attempts, they fractured and broke. But then, learning on the job, she discovered that, if she chipped away at the surrounding matrix, 'similar bones were certain to be found loose in any soft or crumbly places' (Bate 1901, p. 102). These were the fossilized bones of small Pleistocene animals, including voles, lemmings and lagomorphs, dating back around 10 000 years. She brought her collection to the BM (NH) and, with the help of vertebrate palaeontologists in the Geology Department, she identified her finds. The quality of her report on her finds was such that it was published in the *Geological Magazine* with the title: 'A short account of a bone cave in the Carboniferous limestone of the Wye Valley' (Bate 1901). Dorothea was just 22 years old. The journal's editor, Dr Henry Woodward, who was Keeper of Geology, was renowned for his encouragement of young geologists (Anon. 1921, p. 482).

Dorothea and Mediterranean island cave research

Cyprus

In 1901, the year her Wye valley cave report was published, Dorothea embarked on the first of her pioneering explorations of the Mediterranean islands, when she was invited to stay with family friends in Cyprus, then under British administration. One of Dorothea's characteristics was to seize every opportunity that presented itself, and she turned what might have been a purely social occasion into an event that established her reputation as an explorer and palaeontologist. She became the first palaeontologist to search systematically the limestone caves of Cyprus and discover its extinct fossil fauna.

Crete

Crete followed in 1904; this was the time when the treasures of the Minoan civilization were being unearthed by Sir Arthur Evans and other teams of British, American and Italian archaeologists. Her work diaries for these early years (Bate 1901/2, 1904, 1910, 1911) record with wonderful directness the places she visited, the wildlife of the islands, details of her finds and the appalling difficulties she experienced in discovering them. There are tantalizing glimpses of the life she led and the people she met, but almost nothing of her private thoughts or emotions. These were work diaries and she could record nothing too personal; others might need to refer to them.

What she found in Cyprus and Crete included the fossilized remains of extinct tiny elephants and hippopotami, dwarf species that would have been no more than a metre or so high (Fig. 2). Both species of dwarf elephant, *Elephas cypriotes* Bate and *Elephas creticus* Bate, were new to science (Bate 1903, 1907). In Crete she also found the remains of an extinct full-sized elephant, possibly the ancestor of the dwarf species, deer of various sizes, and an extinct dormouse as large as a squirrel. This phenomenon of dwarfism and giganticism on islands was a subject Dorothea was to pursue throughout her life. She was one of the first to realize that this phenomenon occurred in parallel in a number of islands.

In spite of her growing reputation, on her return from Crete, the marriage of her elder sister meant Dorothea's parents refused to allow her to pursue her career for the next 5 years. As the unmarried daughter, they wanted her at home as their companion. It was not until 1909, when she was 30 years old, that she finally persuaded them to allow her to travel again, this time to the Balearics.

Balearic Islands

In Majorca and Menorca she discovered the fossil remains of an extraordinary species, a bizarre goat-antelope with rat-like teeth, which she named *Myotragus balearicus* Bate (Bate 1909). To this day, it is still the subject of academic enquiry and argument. *Myotragus* is so specialized there is no common name for it (Fig. 3). It is referred to variously as a rat-like goat, a cave goat, an antelope gazelle, or simply *Myotragus*, which translates from the Greek literally as 'mouse-goat'. Evolving from a goat that had become isolated on Majorca and Menorca several million years ago (Alcover *et al.* 1981, 1999), the animal had developed a strange shortened and thickened shape to enable it to climb very steep rocks and crags; the legs were wide apart and it is unlikely to have been able to

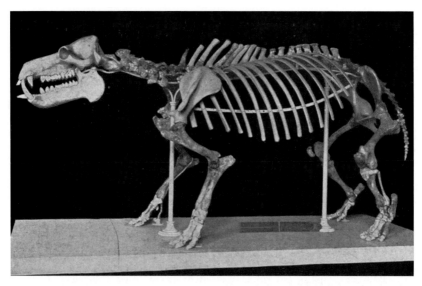

Fig. 2. *Hippopotamus minutus.* Dorothea had the dwarf hippo from Cyprus reconstructed from the fossil specimens she excavated and plaster. Completed in 1906, it was exhibited in the BM (NH) for many years. (Copyright Natural History Museum, London.)

run. What makes it most remarkable, however, are its teeth (Bate 1909; Andrews 1915). Instead of the usual six lower incisors, characteristic of ruminants, there are just two, enormously enlarged and rat-like, and continuously growing from open roots. It was, Dorothea concluded, 'an ungulate [hoofed mammal] which appears to be without parallel' (Bate 1909, p. 385).

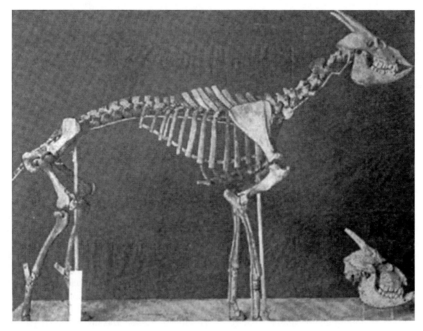

Fig. 3. *Myotragus balearicus* Bate. This reconstruction was exhibited in the BM (NH) galleries, together with the type skull and jaw, in 1915. (Copyright Natural History Museum, London.)

Fig. 4. Dorothea Bate's photograph of the Cuevas de los Colombs in northeast Majorca, one of the cave sites where she discovered fossil remains of *Myotragus*.

In all these islands, Dorothea established for the first time their extinct Pleistocene fauna. It was a remarkable feat for anyone, let alone a young woman working on her own. She also collected live species on all her expeditions. In Cyprus alone, she collected more than 200 specimens of live species of birds, mammals, butterflies, beetles, mosquitoes and other insects; the range of her interests and observations is astounding.

One of the most remarkable things about Dorothea is her persistence and determination in her search for the extinct fauna of islands, continuing her hunt for fossils when most people would have given up. That she found anything at all is extraordinary in itself. The Mediterranean islands in the early 20th century were relatively unknown; tourism was for the hardy few because accommodation and travel were considerably less than basic. In Crete for example, there were only 20 kilometres of paved road. In Majorca, because of the threat of pirates, towns were built inland with just the roughest of tracks to the coast. Even Cyprus under British rule had no harbour deep enough to accommodate even a moderate-sized ship (Green 1914, p. 5). Passengers would be taken as close to shore as possible and then had to be carried to dry land by porters. This, the Cyprus High Commissioner Sir William Haynes Smith

(1902), informed the Colonial Office, occasioned 'great irritation ... and female passengers regard having to be so handled as an outrage'.

Just finding bone caves presented her with enormous difficulties (Fig. 4). There was no convenient sign saying 'This way to the fossil bone deposit'. She had to work out for herself the most likely sites. Day after day she would ride or walk out in intense heat or drenching rain, covering great distances in her attempts to find bone sites. She climbed mountains and cliffs. In Crete, when the cliffs proved too dangerous to climb along, she even jumped into the sea and swam (presumably fully clothed) nearly 100 metres to a bone cave further down the coast (Bate 1904, 23 April). In Cyprus she did have some leads. For centuries, the Christian Greek population had worshipped what they believed to be the bones of saints, embedded in caves throughout the island. Often these bones would be ground to powder and drunk as medicine for all manner of ailments. Shrines to saints could still be found in Cyprus, with votive lights burning, well into the 20th century.

Most of these bones were in fact those of dwarf hippos, and Dorothea notes in her diaries that the men she hired to dig found it a source of great amusement that the so-called saints' bones were those of ancient beasts (Bate 1901/2, 6 May 1902). Many of these caves were in the most inaccessible places, in remote mountain passes, or were indescribably filthy and smelly because they were used as *mandras* (shelters for shepherds and their goats). It would often take her 4 or 5 hours simply to travel by foot or donkey across the most unforgiving terrain to find these bone caves. There then followed perhaps 3 or 4 hours of intensive digging, and then the same journey back. Often she stayed in flea-ridden hovels, or rented a room which might be clean but had no heating, even in the bitter cold of a Cretan winter. She contracted malaria in Cyprus, which recurred throughout all her expeditions, yet time and again, she would continue to work until she could no longer stand. Even then, she would spend just a few days in bed, and then insist on continuing her work, even with a fever.

She hired local men as guides and to do the heavy digging. When the rock matrix was too hard, she resorted to gunpowder. In Crete, in one particular cave (at Koutri), she used dynamite for the first time, which, unfortunately but not surprisingly, so damaged the deposit that later excavators found it 'unsuitable for scientific investigation' (Lax 1996, p. 4).

Money was a constant problem and she relied on grants and selling her finds to the BM (NH). She had to pay for her travel, for guides, for men to dig, for food and for lodging. Henry Woodward was impressed by her discovery of the dwarf hippo on Cyprus, and at her request he applied on her behalf to the Royal Society for a grant to enable her to continue her excavations. The thought of having to abandon her work for lack of money was unbearable to her, and she urged Woodward to send her a cable saying simply 'Bate, Cyprus Continue' if her application was successful (Bate 1902). She hoped for £20. In fact, the Royal Society awarded her £30 (Jowle 1902) – about £1700 in today's values.

Importance of Quaternary cave research published

Dorothea's papers on her Mediterranean discoveries were published, among others, by the Royal Society, the Zoological Society of London, the *Geological Magazine* and *Ibis*. Between 1903 and the outbreak of World War I in 1914, Dorothea wrote for publication more than 15 papers on this subject alone. In the early 1900s, Dorothea, as a woman, could not be elected a fellow of a learned society, nor present her own paper. Henry Woodward presented them for her.

Dorothea was more than a courageous and fearless explorer. Self taught as she was, she became a pioneering and innovative scientist. As a woman, she could not apply for a scientific staff post at the BM (NH) (women were not permitted to do that until 1928), but she could, and did, become one of the many temporary scientific workers, without whose dedication the work of the museum would surely have ground to a halt. Paid according to the number of fossils she prepared (British Museum (Natural History) 1902–37), during World War I Dorothea carried much of the main burden of preparation work in the Geology Department.

From the thousands of animal bones she examined, from millions of years old to a few thousand, Dorothea developed an encyclopaedic knowledge of anatomy and acquired skills that would lead to her pioneering work on the relationship between fauna, climate change and the environment. She was one of the first palaeontologists to realize how much faunal remains in human habitats could reveal about past societies, and how important that material was in the dating of sites. Her greatest work in this field was in collaboration with the Cambridge archaeologist and prehistorian, Dorothy Garrod, and her excavations at Mount Carmel in what was then Palestine.

Pre World War II Palestine

Dorothea's specimens and analyses of the Mount Carmel fossil fauna are still of critical importance today (Bate 1937).

Fig. 5. Dorothea Bate excavating in Bethlehem 1935. The ridged tooth of an extinct early elephant can be seen to the right of the workman. (Copyright Natural History Museum, London.)

Dorothea herself in the 1930s excavated an extraordinary site on the highest hill in Bethlehem (Fig. 5). Fossilized bones had been found which Dorothea identified as the remains of an ancient species of elephant (Bate 1934b). Funding came from the philanthropist Sir Henry Wellcome, and the scale of the excavation was astonishing: over three seasons, from 1935 to 1937, two vast pits were dug which revealed the remains of pre-Pleistocene animals, including elephants, rhinoceros, giant tortoises and *Hipparion*, an early species of horse.

Dorothea's notes and letters from the period, and the reports on the excavation (Bate 1934a, 1935–37; Wellcome Library 1934–38), vividly illustrate the excitement of the dig, and the underlying dangers of working in pre-World War II Palestine – a threat that became a reality in 1938 with the murder by Arab insurgents of Dorothea's colleague, the archaeologist James Starkey (Inge & Tufnell 1938).

Research recognition at last

By this stage, Dorothea, still an unofficial worker at the BM (NH) (Fig. 6), was a fellow of the Zoological Society of London, a fellow of the British Ornithological Union, and a fellow of the Royal Geographical Society. In 1940 she was awarded the prestigious Wollaston Fund of the Geological Society – and only after that was she elected a Fellow of this Society. Recognition by the BM (NH) did not come for another 8 years. In 1948, when she was nearly 70 – long after retirement age (then) for most government and quasi-government employees – Dorothea was given her first senior managerial role. She was appointed Officer-in-Charge of the BM (NH) offshoot in Tring: the Rothschild Zoological Museum (British Museum (Natural History) 1948; Woodisse 1948). The lack of a staff post had never bothered Dorothea, as she just believed in getting on with things. Despite her unofficial status, she had been Curator of Aves (fossil birds) and Pleistocene mammals since 1924, and it was her achievements, not any official title, that had created her reputation and international recognition. But she was now confronted with an overwhelming workload. After the war, collections were sent to her from all over the world. Archaeologists, anthropologists and prehistorians relied on her unique knowledge,

Fig. 6. This photograph of Dorothea Bate taken in 1938 after an illness is in marked contrast to the robust woman of the Bethlehem photo. (Copyright Natural History Museum, London.)

including Louis Leakey, Sir Max Mallowan (Agatha Christie's husband), John Desmond Clark and Anthony Arkell.

Arkell was just one of many archaeologists and palaeontologists who publicly recognized his debt to Dorothea. She challenged Arkell in the same way that she challenged young scientists in the BM (NH): to look at the implications of the finds, not just at the individual artefacts and faunal remains, but all relevant links across the region and beyond. They were to look from detail to pattern, from fragment to the entirety of the evidence and to use their scientific imagination.

Dorothea's last years were marred by recurring ill health. She died on 13 January 1951, working until just before her death. She is remembered in the museum as 'a great lady' (Currant pers. comm. 2000), an exceptional scientist and inspirational colleague.

Dorothea Bate's legacy

Most of her obituaries record the loss of a great and dear friend. In terms of the history of geology, her importance must lie in the many extinct species that were new to science that she so courageously discovered, and in her new and radical ways of interpreting the evidence. As well as that, for many, she was the spark that would ignite a project – encouraging and advising young scientists to pursue avenues of research she no longer had the time for herself. Dorothea Bate was in every respect a true pioneer, possessing in abundance that essential quality that characterizes our species: a ceaselessly enquiring mind.

References

ALCOVER, J. A., MOYÀ-SOLÀ, S. & PONS-MOYÀ, J. 1981. *Les Quimeres del Passat*. Editorial Moll, Mallorca.

ALCOVER, J. A., PEREZ-OBIOL, R., YLL, E.-I. & BOVER, P. 1999. The diet of *Myotragus balearicus* Bate 1909, (*Artiodactyla: Caprinae*), an extinct bovid from the Balearic islands: evidence from coprolites. *Biological Journal of the Linnean Society*, **66**(1), 57–74.

ANDREWS, C. W. 1915. A description of the skull and skeleton of a peculiarly modified rupicaprine antelope (*Myotragus balearicus* Bate) with a notice of a new variety, *M. balearicus* var. *major*. *Philosophical Transactions of the Royal Society of London, Series B*, **206**, 281–305.

ANON. 1921. Obituary Henry Woodward. *Geological Magazine*, **58**(11), 481–484.

ARKELL, A. J. 1951. Zoology and prehistoric archaeology: an opportunity. *Archaeological Newsletter*, **3**(11), 169–170.

ARKELL, A. J. 1953. *Shaheinab*. Oxford University Press, Oxford.

BATE, D. M. A. 1900. MSS notes etc on Syria and Palestine. Earth Sciences Library, Natural History Museum, London, **P MSS BAT**.

BATE, D. M. A. 1901. A short account of a bone cave in the Carboniferous limestone of the Wye Valley. *Geological Magazine*, **4**(8), 101–106.

BATE, D. M. A. 1901/2. 'Cyprus diary (3 vols)' Earth Sciences Library, Natural History Museum, London, **P MSS BAT**.

BATE, D. M. A. 1902. 'Letter to Dr Henry Woodward, 4 February'. Natural History Museum, London, Archives, **DF 100/160/9**.

BATE, D. M. A. 1903. Preliminary note on the discovery of a pigmy elephant in the Pleistocene of Cyprus. *Proceedings of the Royal Society of London*, **71**, 498–500.

BATE, D. M. A. 1904. 'Crete diary'. Earth Sciences Library, Natural History Museum, London, **P MSS BAT**.

BATE, D. M. A. 1907. On elephant remains from Crete, with description of *Elephas creticus* sp.n. *Proceedings of the Zoological Society of London*, 238–250.

BATE, D. M. A. 1909. Preliminary note on a new artiodactyle from Majorca, *Myotragus balearicus*, gen. et sp. nov. *Geological Magazine*, **5**(6), 385–388.

BATE, D. M. A. 1910. 'Diary of second trip to Majorca' Earth Sciences Library, Natural History Museum, London, **P MSS BAT**.

BATE, D. M. A. 1911. 'Diary of third trip to the Balearics'. Earth Sciences Library, Natural History Museum, London, **P MSS BAT**.

BATE, D. M. A. 1934*a*. 'Various letters re. Bethlehem excavations'. Natural History Museum, London, Archives, **DF 100/224**.

BATE, D. M. A. 1934*b*. Discovery of a fossil elephant in Palestine. *Nature*, **134**(1), 219.

BATE, D. M. A. 1935–37. 'Bethlehem notebooks' Earth Sciences Library, Natural History Museum, London, **P MSS BAT**.

BATE, D. M. A. 1937. Palaeontology: the fossil fauna of the Wady el-Mughara caves. *In*: GARROD, D. A. E. & BATE, D. M. A. *The Stone Age of Mount Carmel*, Vol. 1, Pt 2. Clarendon Press, Oxford, 135–240.

BRITISH MUSEUM (NATURAL HISTORY) 1904. *History of the Collections Contained in the Natural History Departments of the British Museum*, Vol. 1. BM (NH), London.

BRITISH MUSEUM (NATURAL HISTORY) 1902–37. 'Geology Department account book' Natural History Museum, London, Archives, **DF 102/9**.

BRITISH MUSEUM (NATURAL HISTORY) 1948. 'Standing Committee of Trustees, 24 July'. Natural History Museum, London, Archives, **DF 900/15**.

EDWARDS, W. N. 1951*a*. Obituary, Miss D. M. A. Bate. *The Times*, 23 January.

EDWARDS, W. N. 1951*b*. Dorothea Bate obituary. *Proceedings of the Geological Society of London*, **106**(3), lvi–lviii.

GREEN, COL. A. O. 1914. *Cyprus: A Short Account of Its History and Present State*. M. Graham Coltart, Kilmalcolm, Scotland.

HASTINGS, Barbara Yelverton, Marchioness of, 1851–55. 'Correspondence.' Natural History Museum, London, Archives, **DF 100/7**.

HAYNES SMITH, W. 1902. 'Colonial Office, Cyprus, Original Correspondence, Despatch 96, 15 May'. The National Archives (Public Record Office), **CO 67/131**.

HIGGS, B. & WYSE JACKSON, P. N. 2007. The role of women in the history of geological studies in Ireland. *In*: BUREK, C. V. & HIGGS, B. (eds) *The Role of Women in the History of Geology*, Geological Society, London, Special Publications, **281**, 137–153.

INGE, C. & TUFNELL, O. 1938. 'Extraordinary report II, 31 January, re murder of J.L. Starkey on 10 January 1938. Wellcome Archaeological Research Expedition to the Near East, Reports and correspondence'. Wellcome Library of the History of Medicine. **WA/HSW/AR/Lac/B13**.

JOWLE, F. A. 1902. 'Letter to Henry Woodward, 18 February' Natural History Museum, London, Archives, **DF 100/34/78**.

LAX, E. M. 1996. A gazeteer of Cretan palaeontological localities. *In*: REESE, S. R. (ed.) *Pleistocene and Holocene Fauna of Crete and Its First Settlers*. Prehistory Press, Madison, 1–32.

RILEY, N. D. 1964. *The Department of Entomology of the British Museum (Natural History) 1904–1964: A Brief Historical Sketch*. 12th International Congress of Entomology, London.

SHINDLER, K. 2005. *Discovering Dorothea: The Life of the Pioneering Fossil-hunter Dorothea Bate*. HarperCollins, London.

WELLCOME LIBRARY. WELLCOME ARCHAEOLOGICAL RESEARCH EXPEDITION TO THE NEAR EAST. 1934–38. 'Reports and correspondence'. Wellcome Library of the History of Medicine, **WA/HSW/AR/Lac**.

WOODDISSE, T. 1948. 'Memo to Dorothea Bate, 25 August'. Natural History Museum, London, Archives, **DF 1004/719/6**.

Maria Matilda Ogilvie Gordon (1864–1939): a Scottish researcher in the Alps

M. WACHTLER[1] & C. V. BUREK[2]

[1]*Rainer-Strasse 11, 39038 Innichen, Südtirol, Italy (e-mail: michael@wachtler.com)*
[2]*University of Chester, Parkgate Road, Chester, CH1 4BJ, UK*
(e-mail: c.burek@chester.ac.uk)

Abstract: Maria Ogilvie Gordon was one of the most prolific researchers of the later 19th century. Born and bred in Scotland she was the first woman to obtain a DSc from the University of London and a PhD from Munich University. Much of her research was in the Tyrol, in the high Alps between Austria and Italy. By 1900 she had published over 19 papers, many of them in German. However, it was not until later in life that she received recognition for her work. This paper explores her background, context and the work she undertook, and the contribution she made to the advancement of structural geology and palaeontology in the Alps.

Maria Ogilvie's childhood

Maria Matilda Ogilvie (Fig. 1) was born in Monymusk, Aberdeenshire, Scotland, on 30 April 1864 to Reverend Alexander Ogilvie LLD, headmaster of the Robert Gordon's Hospital (later Robert Gordon's College) in Aberdeen, Scotland, and his wife, Maria Matilda Nicol. She was one of eight children and the eldest daughter. Her uncles included Dr Robert Ogilvie, Chief Inspector of Schools; Dr Joseph Ogilvie, Rector of the Established Church Training College, Aberdeen; and Dr George Ogilvie, headmaster of George Watson's College, Edinburgh. Her five brothers, who all became eminent Victorian leaders, included a doctor, a minister of the Church, an engineer, a military lieutenant and a scientist, with her eldest brother, Francis, being knighted. Even her own sister, Emma, married the Inspector of Schools in Perthshire. Her youngest sister, Ida, died when she was an infant. Thus Maria was born and surrounded by eminent educationalists and high-flying siblings. It is not surprising that she had high aspirations and strove for acknowledgement of her research work in later years.

At the age of 9 years she was sent to one of the guild boarding schools – the Merchant Company Schools' Ladies College, Edinburgh – where she stayed for 9 years, becoming both head girl and the best academic pupil in the school. Later her sister Emma joined her at the college (1881 census). During the holidays at their country home in Ballater, near Balmoral Castle, she spent many hours exploring the landscapes of the Highlands in the company of her eldest brother Francis, who was also destined to become a geologist (Creese 1996).

Initially, Maria, or May as she was known, aspired to become a musician. At the age of 18 years, she went to London to the Royal Academy of Music, becoming a promising pianist. Within a year she had decided against this in favour of a science degree. After matriculating in London, she returned to Heriot-Watt University in Edinburgh, where her brother was principal. Here she pursued the first part of her BSc, returning to University College London (UCL) to finish it, specializing in geology, botany and zoology. She obtained her degree in 1890 (Burek 2005).

At this point in her life Maria could have chosen to go back to UCL or to Cambridge but she chose instead to go to Germany, hoping for Berlin University, but eventually settling for Munich.

Maria Ogilvie in Germany

Thus the year 1891 was to be a turning point in the life of the young Scotswoman Maria Matilda Ogilvie. Despite assiduous efforts by her friends, at the beginning of that year she was refused admission to lectures at Berlin University, as women were still not permitted to enrol for higher education in Germany. Even the famous Baron Ferdinand Freiherr von Richthofen (1833–1905) was unable to procure her a special permit. This was also in spite of the intercession of influential friends and relations. Disappointed, the Baron and his wife Irmgard personally accompanied the 27-year-old Maria Ogilvie, who was equipped with a grant from the Royal Society, to Munich. There the German palaeontologist Karl von Zittel (1839–1904) and the zoologist Richard von Hertwig (1850–1927) had agreed to let her carry out

Fig. 1. Maria Ogilvie, an early photo taken around 1900 (Michael Wachtler Archive).

research work in their institute privately, but not actually within the university. The precedent had already been set here by a Russian listening to von Zittel's lectures through an open door. To Maria's chagrin, at the very beginning, Paul Groth, the mineralogist, refused to allow the young woman to enter his laboratory (Creese 1996). Nevertheless, Maria Ogilvie came into contact with the world of corals through the very best contemporary scientists.

At the end of July 1891 Baron and Baroness von Richthofen invited the attractive young lady and nature enthusiast to join them on a 5-week trip to the Dolomites. Consequently this travel group met each other in St Ulrich (Ortisei), the main village of the Gröden Valley (Val Gardena). The Baron, who was almost 60 years old, was a famous scientist who had made history when, 30 years earlier, he was the first to discover that the Dolomites were nothing more than fossilized coral reefs, while she was the young British graduate with an immense thirst for knowledge who had been refused admission to doctorate courses both in the United Kingdom and in Germany. Maria Ogilvie was given the task of keeping Baroness Irmgard company and the Baroness had asked her not to open the curtains in the morning until she was called for. At last, around 9 o'clock, breakfast was served on the balcony of the small inn and Maria Ogilvie was quietly led outside:

> That was when all of a sudden I saw the Dolomites before me, a wonderful sight such as I had never experienced before. So captivating, it made an impression that stayed with me in later life like a sign from God. (Klebelsberg 1932)

First research work in the Dolomites

Maria Ogilvie was immensely impressed by the majestic wild and jagged mountain massifs of the Langkofel, the Plattkofel and the Sellastock. The Richthofens immediately entrusted her to the care of a good mountain guide, who introduced her to the techniques of rock climbing. Based in the Gröden valley she also travelled the entire Dolomite area and examined anew places which by then had become well known among geological researchers. In Predazzo she visited the 'Nave d'Oro' (*Golden Ship*) Inn and its famous (at least in geological and climbing circles) landlady, who by then was advanced in years and whose guest book bore the signatures of many great scientists, including Alexander von Humboldt, Roderick Impey Murchison and Leopold von Buch, to name but a few.

Thirty-three years earlier, the young Baron Ferdinand von Richthofen had stayed there before drawing up his groundbreaking interpretations on the evolution of the Dolomites. He had risen to become Vice President of the German and Austrian Alpine Association, which was gaining in reputation and status, and thus he wanted to return to those same mountains for the first time in three decades. He introduced the young Maria Ogilvie to numerous local guides, innkeepers and fossil collectors, which proved very useful to her future work.

The party went to Corvara and hiked to the Störeswiesen meadows which, for decades, had yielded a wealth of fossils that had increased geological knowledge of the area. It was the Baron who introduced her to alpine field geology. He advised her to become a geologist rather than a zoologist and urged her to start by mapping the area around San Cassiano. Since 1830 this area on the Austro-Italian border had produced a sensational yield of fossilized shellfish, corals and sponges discovered by researchers Georg Graf zu Münster, von Wissmann and August Klipstein.

Maria Ogilvie accepted this suggestion. The Richthofens then left and she started work, based at Corvara in Val Badia. Years later she recalls this time:

> When I began my field work, I was not under the eye of any Professor. There was no one to include me in his official round of visits among the young geologists in the field, and to subject my maps and sections to tough criticism on the ground. The lack

of supervision at the outset was undoubtedly a serious handicap. (Ogilvie Gordon 1932)

However, it did mean that she had no preconceived ideas and used only her own observations and reached her own conclusions.

Thus began a very long lonely research path which became her destiny. The terrain was very difficult and there were no proper inns or roads. However, she met Josef Kostner, who was later to form part of the famous climbing dynasty with his brother Franz. However, at the time he was just 20 years old and eager for knowledge and advancement. He was a gifted climber, teaching Maria how to climb safely and often accompanying her into the field. Franz and Josef Kostner, together with Gottfried Merzbacher, were later to explore the Caucasus and Tian Shan. After a hard field season Maria travelled back to Munich.

Maria Ogilvie returned to the Dolomites in the summer of 1892. In the company of Josef Kostner she went on hiking tours in the area around Störes (Fig. 2), then went to the Settsass, to Falzarego

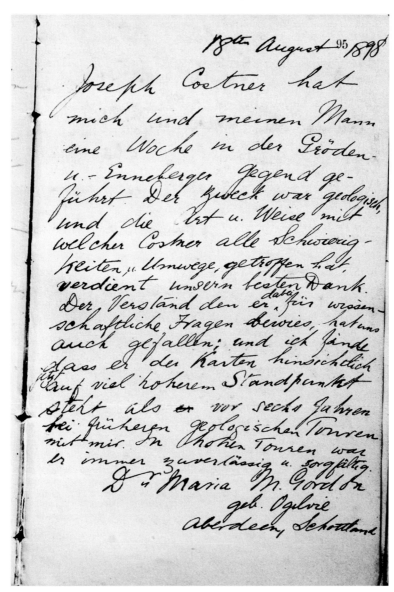

Fig. 2. Entry by Maria Ogilvie-Gordon in Josef Kostner's mountain guidebook, dated 18 August 1898 (Renata Pizzinini Archive, Hotel Cappella).

and on to Cortina d'Ampezzo. She was also active in the Prags Dolomites, as well as the Plätzwiese with its Alpine lake landscape. This latter area was becoming increasingly famous for the variety and excellent state of preservation of the Triassic corals and sponges it yielded. She instructed local collectors to keep accurate records of the location of the finds within the beds, and to particularly note the horizons. At the time, little was known of the evolutionary and environmental conditions prevalent in the worldwide development of the Alpine Triassic. This system was later to become important because of the evolution of organisms in the wake of the devastating Permian catastrophe. As stratigraphic research was in its infancy, she deemed it necessary to separate the various strata in order to understand the constantly changing and evolving animal and plant life.

As a result of her research Maria Ogilvie produced a 78-page article for the Geological Society's Quarterly Journal in February 1893, illustrated with plenty of drawings and entitled 'Contributions to the geology of the Wengen and St. Cassian Strata in southern Tyrol' (Ogilvie 1893). She had accomplished what other geologists before her had failed to do, namely to connect in an integrated manner the various stratigraphic horizons. She had also discovered sites which allowed her to draw conclusions about the development of corals and sponges and other marine life which took place 230 million years ago in this tropical world. She had accomplished the task in 1 year. For the first time she succeeded in separating the Wengen strata and fossil assemblage, which were deposited during the Triassic era in a period of volcanic activity, from the younger St Cassian strata with their infinite abundance of fossils. She had described 345 different faunal species, an impressive number. Today, 100 years later, over 1400 kinds of sponges, corals and crustaceans have been found, in some cases still preserved with their original colourings. However, Maria Ogilvie provided the baseline for this work. Her detailed research work is impressive and reveals how she frequently visited remote sites far apart in difficult terrain. The precision of the many detailed and accurate landscape descriptions became characteristic of all her work, and her conclusions are still relevant today.

The first woman DSc in the United Kingdom

For the first time British experts took notice of the young and ambitious female scientist. The research paper, submitted as a thesis entitled 'The geology of the Wengen and Saint Cassian Strata in southern Tyrol' earned her a DSc degree from the University of London, and she become the first female DSc in the United Kingdom (Kölbl-Ebert 2001). Although she was secretly hoping for recognition it came as a surprise for her and everyone else. 'No one could be worse prepared than I was at the beginning in 1891–93', she remarked later (Pia 1939).

Maria returned to the Dolomites to continue her research work in the summer of 1893. She became above all fascinated by two different questions: 'How had the Dolomites formed?' and 'How had the corals evolved in this mountainous terrain?'. However, difficult times lay ahead for Maria Ogilvie, psychologically as well as physically. She spent the winter in Munich and learned to speak German fluently. As soon as the weather permitted, she left Munich and headed straight for the Dolomites. In those days, working in the mountains was hardly a luxurious undertaking. There were few refuges or huts providing food or accommodation and she often found it necessary to set out from Corvara at 2 o'clock in the morning in the direction of the Boè-Spitze, Settsass or other sites, to return late in the evening completely exhausted, often alone and heavily burdened with all kinds of samples and fossils.

Being a single woman in this area she attracted attention, but she was ambitious and determined, as is shown in her perseverance to forward knowledge, even when she received no acknowledgement or recognition. This is something she highlighted in her reply to the President of the Geological Society years later, when she received her Lyell Medal (Ogilvie Gordon 1932), and once she had set her mind on something; nothing and nobody could distract her. The fact that she carried on her research in later years as a married woman showed the determination she had; she often took her husband and children with her into the high Alpine areas. In those days a professional woman still had enormous problems fitting into such a rural society. Indeed, she found a considerate helper in Rottonara from Corvara, the old landlady of the Post Inn, who advised and supported her, even to the point of sending farm hands up to her at the mountain huts with food, youths who, eager to help, would return out of breath with a heavy load of specimens.

However, in her scientific publications, she only described the 'Alpenhütten' and 'Kochhütten', to interested persons throughout the world, as places where one could spend a night alone in reasonable comfort. She did, however, hint at the uncomfortable conditions in which she was living, but did not go into further explanation (Ogilvie 1893). The mountains could be dangerous places and once, on the Störeswiesen pastures, a bolt of lightning struck and killed two cows a short distance away from her. Apparently, she lived thereafter in dread of thunderstorms.

Maria became a meticulous researching and passionate scientist, who was driven by her boundless curiosity to encroach on territory that until then had been only researched by men. Luckily, conditions of access improved in 1894, when a driving road was opened up from Brunico (Creese 1996) which would have eased the travelling to the research areas.

The same year, she had yet another article, entitled 'Coral in the Dolomites of south Tyrol', published in an English specialist periodical (Ogilvie 1894). Thus she started her next important project, which was to absorb her energy for numerous years: the environmental conditions under which corals, both modern and extinct, develop.

Marriage and family life

In autumn 1895 she returned to the city from where she had started out: Aberdeen, where a long-standing admirer, the physician Dr John Gordon was waiting patiently for her return. They married in 1895. They took a house in Rubislaw Terrace, a suitably distinguished address for a doctor and his future family (Morgan 2004).

Maria Ogilvie found in Dr John Gordon a sympathetic and, for the times, remarkably easy-going husband. Perhaps the age difference might account for his tolerance; he was 14 years her senior. He even accompanied her to the Dolomites in August 1898, where they went on hikes in the surrounding mountains in the company of Josef Kostner. Maria was full of praise for her former guide and how he had developed (Figs 2 & Fig. 3). She wrote the following in his mountain guide book in perfect German:

18th August 1898. Joseph Costner has guided me and my husband in the area of Gröden and Enneberg. The purpose was geological and the way Costner met all difficulties and impassable routes merits our heartfelt gratitude. We also appreciated the understanding he showed for scientific questions; and he has progressed considerably with regard to maps compared with the point he had reached six years ago when he joined me on earlier geological field trips. On high mountain hikes he always proved most reliable and solicitous. Dr Maria M. Gordon, née Ogilvie, Aberdeen, Scotland.

Her husband also found words to praise the mountain guide Kostner, who was in the process of preparing to take part in an expedition to the distant Caucasia to penetrate unknown territories:

18th Aug. 1898. I found Joseph Kostner during the week he acted as our guide, kind, considerate and most capable in every way. John Gordon Med. Aberdeen Scotland.

Maria Ogilvie gave birth to a daughter and a son in quick succession in 1897 and 1899 (1901 census) – the first two of her four children. She named her eldest Coral, a name which caused dismay among the social circles she frequented. Her second child was called John after his father, but was also called Ogilvie after his mother. He attended the Robert Gordon's College from 1907 to 1912. Here he wrote several articles for the school magazine, '*The Gordonian*', about family holidays to the Tyrol in Austria (Robert Gordon College archives). In spite of her commitments as a caring

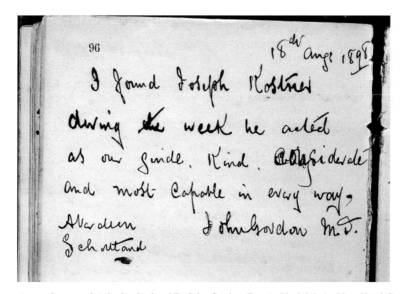

Fig. 3. The entry on the same date by her husband Dr John Gordon (Renata Pizzinini, Archive, Hotel Cappella).

mother and her responsibilities as a respected member of Aberdeen society, she never tired of writing and publishing. On the 1901 census return, for example, she is listed as student and investigator in science (1901 census).

Royal Society paper

In 1897 Maria Ogilvie Gordon had written a zoological paper entitled 'Microscopic and systematic study of madreporarian types of corals', which was so well thought out that Professor Edwin Ray Lankester and Sir Archibald Geikie put all their efforts into having it published in the exclusive Journal of the Royal Society of London, a very special honour and achievement. It was during this period that she continued seeking advice on publication from Sir Archibald Geikie:

Oct 21st 1895. I received on Saturday the 19th a printed intimation that the Royal Society Committee of Papers had directed mine to be published in the Philosophical Transactions. Allow me to thank you for this pleasant fulfilment of my hopes and of your kind promise to help the paper through. I have simply acknowledged the Secretary's formal notice and feel now rather in the dark as to my next step. (Robert Gordon's College archive)

It is obvious that Sir Archibald held her work in great esteem as he had helped the paper through the various editorial boards.

What was Maria Ogilvie Gordon's contribution to this unknown territory of research into corals? She explained it best:

... the genus can, as a rule, only be identified with certainty after a careful microscopic investigation. (Stacul 1969)

Researchers before her had tried to classify corals, whether living or fossilized, entirely by their outward appearance, a system which Maria Ogilvie Gordon showed to be defective on the basis of detailed examination. Building on the intricate internal structure of fossilized corals she decided to introduce some order into the chaotic system. For evidence she could only draw on her extensive collection of corals from the Dolomites, in the so-called Cassian strata from the Triassic period. Why the abundance and preservation of these corals is so great is not entirely clear. The causes could be the unfavourable ecological conditions prevailing at the time, relating to nutrients in the secluded tropical lagoons. In any case they provided excellent objects for study for this scientist who never tired of posing questions. An explosion of the most diverse types of corals from the Triassic period came to light.

Availing herself of the latest state-of-the-art equipment, such as improved microscopes, Maria Ogilvie Gordon succeeded in fundamentally improving knowledge about these organisms of such importance for marine life, in particular by realizing the necessity for examining all parts of modern corals, including their soft parts, polyps, as well as their fine skeletal structures. As a result, in 1897, she enthused scientific circles and consolidated her reputation as a meticulous researcher. However, in spite of having published her papers in German (Ogilvie Gordon 1897b), none of this sufficed to secure her a place on a DSc course at Munich University.

Women finally admitted to higher education in Germany

At the turn of the 20th century women were still excluded from higher education in Germany (Kölbl-Ebert 2001). In those days many people still held on to the belief about the inferiority of the female sex. Maria Ogilvie had been based mainly in Munich while carrying out research, although in reality she had only been tolerated at this university. Once her well-disposed supporter Karl von Zittel, head of the Palaeontology Institute had died, even the famous cartographer August Rothpletz (1853–1918), the well-known Alpine mapper, dissociated himself from her. Other celebrated university professors likewise opposed her and did everything in their power to prevent this woman with a thirst for knowledge from attending lectures even secretly. Finally, in 1900, women were admitted to higher education at Munich University, albeit half-heartedly. Initially female foreign students from abroad were allowed to enrol, from Russia, the United States and the United Kingdom, for the authorities were confident that they would return to their home countries after graduation. Likewise, the Scottish zoologist Agnes Kelly decided to take advantage of the situation and be the first to graduate.

Nevertheless, it was Maria Ogilvie-Gordon's two longstanding mentors, Karl von Zittel and Richard von Hertwig, who in their turn sided with her and ensured that she would be the first woman to qualify for a doctor's degree, on the basis of the long periods she had spent there earlier studying at the University of Munich, and of her research work and publications. In spite of being a mother, she spent several months in Munich in early summer 1900, preparing for the difficult examinations, which she passed with flying colours, becoming the first woman to obtain a PhD at Munich University. She thanked Karl von Zittel – by then old and ailing (he died in 1904) – in her own way: she painstakingly translated into English his work *Geschichte der Geologie und Palaeontologie* (*History of Geology and Palaeontology*), a monumental history of the progress of earth sciences. She even gained the praise and respect of the inveterate misogynist August Rothpletz, who otherwise continued to look down on women all his life. The reason for this was that, on his own field trips to the Dolomites, and in front of crowds of students, he was forced to confirm many of Maria Ogilvie Gordon's observations and conclusions. In fact, he even became one of her closest advisers and fatherly friend, who from then on directed her curiosity towards several unresolved questions in the spheres of plate tectonics, thrust faults and folds in the Dolomites, to the point of stimulating her to regard those mountains as the focus of her research work.

Inner turmoil

At the age of 36 years, Maria Ogilvie Gordon could look back on a wealth of achievements, although scant notice was taken of her, either in the United Kingdom or Germany. She had broken new ground and furthered the cause of equal rights for women in two countries, but in spite of her achievements she remained basically a housewife with three children and, although her husband was considerate and receptive, Maria still looked beyond domesticity. She remained ambitious, with ideas for work in various areas. She had also become engaged in civic and social issues and the newly founded 'National Council of Women', especially in matters relating to the advancement of women and the welfare of children. This placed demands on her time, which was already in short supply. Nevertheless her passion for geological research remained undiminished. Perhaps her husband, John Gordon, characterized her dedication best:

It is a lonely furrow you are ploughing, Mary; for your own sake I wish you had chosen some other interest for your hard work.

Between her geological research work and her involvement with social issues (which amounted to a second career), Maria found herself working day and night. She was plagued by doubts of all kinds, as she confided plaintively to the great geologist Julius Pia (Pia 1939) who later became a fatherly friend and curator of the Vienna Natural History Museum:

In my own country I never count at all. I am made to feel a complete outsider [1929]. (Pia 1939)

This melancholy was to grow year by year:

Here in this country no one ever refers to my work at all [1931].

She simply failed to attract the attention of other British researchers to her work:

Here it is obvious none of the British geologist have any interest in my work [1932].

Nevertheless she was swamped with work from other sources, as well as pointless activities:

I am not particularly happy this winter as I have had no time to do any scientific work [1934]. (Pia 1939)

As a result, inwardly, Maria Ogilvie Gordon became increasingly split and torn, although this did not prevent her from being prolific all those years. She was an amateur, working in a professional world in a subject far removed from the local British geologist in both language and location. Thus, despite appearances to the contrary (when doing her committee and social work, she seemed happy and contented) she was not.

I keep toiling hard at public work, meetings of all kind, while knowing inwardly that it is geology in my heart that I wish to do [1936].

From 1900 until the outbreak of World War I she continued to explore the Dolomites, as if drawn there by some irresistible force (Fig. 4). She researched the various valleys, from Val di Fassa to Gröden (Val Gardena), Val Badia, Cortina d'Ampezzo and Buchenstein (Ogilvie Gordon 1903). She had local mountain guides scale sheer, open rock faces for her and bring down rock samples, while they in turn became enthusiastic, trained collectors.

Fig. 4. Maria Ogilvie Gordon with the mountain guide Josef Kostner on a field trip around 1910 (Renata Pizzinini Archive, Hotel Cappella).

After numerous single publications on the subject of the Dolomites (Ogilvie Gordon 1903, 1906, 1909, 1910, 1911, 1913) she began formulating in her mind a lifetime's dream: to produce a comprehensive study of the area with special emphasis on geological features, the thrust faults and folds which she had already researched extensively. She wanted it to be a further milestone following Ferdinand von Richthofen's pioneering work *Geognostische Beschreibung der Umgegend von Predazzo, Sankt Cassian und der Seisser Alpe in Süd-Tyrol* (A geonostical description of the area of Predazzo, San Cassiano and the Seiser Alp in South Tyrol) published in 1860, in which he was the first to arrive at the correct conclusions with regard to the history of the origins of this area. That is, following the grandiose 1879 work by Edmund von Mojsisovic, one of the founders of the Austrian-German Alpine Association entitled *Die Dolomitriffe von Südtirol und Venetien* (The dolomite reefs of South Tyrol and the Veneto). She later went into even greater detail and published her own cartographic and deduced facts on this area (Ogilvie Gordon 1927).

Tragic losses of a lifetime's work

By 1913 Maria Ogilvie Gordon's geological surveys, countless analyses and sample collections had progressed to the point where the time had come to publish a general account of the geomorphological processes which had led to the formation of the Dolomites. She had produced hundreds of hand-written pages in English, although she intended to publish the work in German. She had personally produced a hand-coloured geological map in great detail and submitted everything to the capable hands of August Rothpletz, by then the head of the Palaeontological Institute of the University of Munich, the same man who had earlier opposed her, but with whom she had subsequently developed an affectionate and co-operative friendship. Rothpletz had found a student proficient in languages to translate the work. The first maps were already prepared on lithographic stone, ready to be printed, when World War I broke out in 1914. Germany and Austria were suddenly enemies of Britain. The work came to a halt. During the war, Maria Ogilvie Gordon was inundated with social responsibilities back home. She occupied herself with the distribution of food and clothing to the needy and supplies of food to hospitals, as well as doing her utmost for the wounded (Anon. 1939; Matheson 1939). There was very little room for geological work.

For Maria Ogilvie Gordon, two important people died in 1918 and 1919. Both had long-term consequences. In 1918, August Rothpletz died, and after the war neither her manuscript nor the name of the student who was translating it could be traced. This was a depressing blow for Maria Ogilvie Gordon.

A further even more fateful blow came in 1919, when her husband and understanding companion John Gordon died suddenly on 23 March. He was 68 years old. Figure 5 shows the Aberdeen granite tombstone and burial place for Dr John Gordon, and later Maria, in Allenvale Cemetery, by the banks of the River Dee in Aberdeen. It also shows that two of their four children are buried there.

By then Maria was 54 years old, an age at which many others in her position would have given up, but in spite of these disheartening setbacks she threw herself anew into her work, full of enthusiasm and courage. She returned to the Dolomites in 1922. The political landscape had changed and the area now belonged to Italy. She gathered the few old records and notes which remained, went into the field, compared, reconstructed and described. She found a friend in the young Julius Pia rather like she had earlier in Ferdinand von Richthofen. Julius was an enthusiastic palaeontologist and geologist who, during the war, had carried out research in the Prags Dolomites. He even visited her for several months in London during the winter of 1921/2 and together they planned innumerable hikes in the Dolomites between 1922 and 1925. Another famous Austrian geologist, Otto Ampferer, and his wife also accompanied her. Ampferer was to make history with his work on the origins of the Campanile Basso in the Brenta massif.

'Hunting for thrust-planes' is how she once described her methods in conversation with Julius Pia. Nothing could divert her from her conviction that all geological questions could be explained unequivocally through the power of one's own intellect. She could not accept that ultimately many questions must remain strewn with doubts. Still she treated her much younger colleagues as equals, remaining an unpretentious woman, able to fit in immediately with any kind of company in spite of her mature years and experience (Anon. 1923; Matheson 1939).

By then her children were reasonably independent. Her two eldest children, Coral and John, had decided to settle in India, while her third child, a daughter, married a medical doctor from London. She was now in a position to devote herself entirely to geological research.

Finally, in 1927, her major scientific work, voluminous, large-format and almost 400 pages long, was published as a treatise by the Geological Survey of Austria institute in Vienna, entitled '*The Gröden, Fassa and Enneberg areas in the South Tyrolean Dolomites*' with the subtitle '*Geological*

Fig. 5. The Aberdeen granite tombstone by the River Dee in Aberdeen marking the burial place of Dr John Gordon, his wife Maria Matilda Ogilvie Gordon, and two of their children: Mary Monica who died as an infant and their son, Dr John Ogilvie Gordon.

Descriptions with emphasis on overthrust fault phenomena' (Ogilvie Gordon 1927). She completed this work and a year later two geological guidebooks for amateurs: *Geological Tour Guide to the Western Dolomites* (Ogilvie Gordon 1928a) and *Guide for Geological Tours in the South Tyrolean Dolomites* (Ogilvie Gordon 1928b). Her knowledge and enthusiasm would, it was hoped, rub off on the increasing numbers of tourists who were visiting the region. The abstruse conclusions she had arrived at were to become common knowledge. She hoped that many people interested in the conclusions of her work on the history and formation of the Dolomites would travel there from the cities as tourists and hike to the important geological sites (Fig. 6). Her geological guide books were forerunners to the innumerable natural science guides which became popular much later and in this respect she

Fig. 6. The famous Störeswiesen pastures near Corvara, destination of many of Maria Ogilvie's field trips (Photo Martin Schönegger; Michael Wachtler Archive).

also broke new ground. She can be described as one of the first to recognize geotourism as an important activity.

Her works were based on her brilliant powers of observation, and her razor-sharp intellect, and they attested to the endurance and enormous amount of effort she had put into her work (Pia 1939). She noted the time required to walk or climb to the places indicated as well as the altitude above sea level. These details in a guidebook of this sort also broke new ground, even though one critic opined that fossils were nothing like as common as Maria Ogilvie Gordon had described in her exuberant enthusiasm.

She was now 63 years old, still tireless, still unrecognized and dedicated to a thorny niche domain.

Modest fame and a tough struggle for women's rights

At last, for the first time, scientific circles began to pay attention to and honour Maria Ogilvie Gordon. She became an honorary member of the Geological Survey of Austria Institute *(Geologische Bundesanstalt)* in Vienna (1931), of the University of Innsbruck and of the Natural History Museum of Trento. The Geological Society of London awarded her the much-coveted Lyell Medal in 1932. She was particularly thrilled by the recognition from her own people

All the more, Mr. President, because my work had to be done outside Great Britain, and was humanly of so isolated a character, I derive very special pleasure to-day in this recognition from the geologists of my homeland. (Ogilvie Gordon 1932)

In her address she mentions several key people who had helped her; Sir Archibald Geikie, William Topley and Professor Charles Lapworth. She also acknowledges the help of the two Scottish geologists Peach and Horne.

They accepted my work, so to say, on the bona fides of my thrust planes, which they said did precisely what thrust planes ought to do

... and from their work in the Highlands they should have been familiar with the problems Maria had encountered in the Alps. However, it was a breakthrough which would certainly not have been forthcoming without the support of her sympathetic German-speaking friends. She had to admit to her German-speaking colleagues:

I attribute its coming entirely to the previous generous treatment given me by your Geological Society.

She had found a new sphere of activity at the Geological Survey of Austria Institute, and it was from there that news of her reputation and audacious geological evaluations sped to the United Kingdom, where she was at long last taken seriously in scientific circles. Even her home country commemorated her in 1935 by awarding her an honorary doctorate from Edinburgh University. She had always regarded herself first and foremost as a Scotswoman and loved her homeland. She was invited to geological congresses in Canada and

Vienna and, in 1931, the Australian government even invited her to Perth, Melbourne and Adelaide.

However, in 1931, she received more bad news: she had to undergo major surgery. She had hardly recovered when she threw herself into her work once again. Ever since her student days she had experienced first hand how women everywhere were discriminated against and unfairly treated. Maria Ogilvie Gordon never became a militant campaigner but became first vice president of the International Council of Women, president of the National Council of Women and honorary president of the National Women's Citizen's Association (Anon. 1939). Her strength lay in constantly drawing attention to the unequal treatment of women. She supported the cause for working women to be protected during pregnancy and for men, women and children to be regarded as equal partners with equal rights. As early as 1908 she had aroused interest by publishing a *Handbook of Employment for Boys and Girls* (Ogilvie Gordon 1908), in which she expounded her view that children should grow up in a socially tranquil environment. She organised touring exhibitions to make adults aware of the importance of childcare. She drew attention to the importance of quality health insurance, general hygiene and safety measures in factories. Sometimes for weeks and even months she was away on selfless missions. Wherever she went she led from the front, met politicians, was on countless committees and sacrificed her time to create a better world.

Nothing could discourage her. She put her name forward as an MP candidate for the Liberal Party in seats with hardly any hope of success for the sole purpose of gaining votes for the movement, which she believed would campaign for sexual equality. Thus she was engaged in persuading more women to become involved in politics, to convince them not to fear expressing their opinions, to imbue them with a more active appreciation of democracy.

She herself was a Justice of the Peace and the first woman Chairman of the Marylebone Court of Justice. She served as honorary presidents of both the Associated Women's Friendly Society and the National Women's Citizens' Association and became president of the National Council of Women of Great Britain and Ireland in 1916 (and much later, in July 1936, first vice chair of the International Council of Women) and chair of the Mothercraft and Child Welfare Exhibitions Committee in 1919. She played a strong part in the negotiations following World War I at the Council for the Representation of Women in the League of Nations. For all this work she received a DBE in 1935 from King George V, thus becoming a Dame. The same year she received an honorary LL.B from Edinburgh University. By then she was 75 years old.

In 1935, she had travelled to the Langkofel massif in Val Gardena (Gröden) and undertaken field trips with Julius Pia to (*On the Geology of the Langkofel massif in the South Tyrolean Dolomites*) study the area's folds and thrust faults, followed by a trip to Vienna to go through the manuscript *Zur Geologie der Langkofelgruppe in den Südtiroler Dolomiten*. The intellectual analysis of the content visibly tired her out, and the food in local inns also took its toll on her:

I am oh! So very keen to have this work published soon.

Storm clouds gathered once again over Europe, a development which must have depressed Maria Ogilvie Gordon, given that she had spent such a large part of her life exploring and working in German-speaking regions and where she had large numbers of friends. Ailing, she returned to London.

She died on 24 June 1939 at Hanover Gate Mansions, NW London, but her ashes are buried in Aberdeen in the Scotland she loved.

It is a tribute to the high esteem in which she was held that, on the Continent, her colleague Julius Pia (1939) wrote a long obituary to her *after* the outbreak of World War II in 1939 in German in Vienna – surely a great tribute to a great lady.

What is Maria Ogilvie Gordon's legacy?

In a forceful letter to Julius Pia, dated 14 March 1937, Maria Ogilvie Gordon wrote:

But the work was a joy and I look back on the days of expecting discovery at every corner as my happiest time.

Despite all the other calls on her time, in her heart she always remained a geoscientist. All her life she was fascinated by Mother Nature, many of her other activities becoming mere irksome chores. She was worn down but at the same time spurred on to great achievements by her inner conflicts.

I am rather in despair about getting on with my work. [1925]

Following on from the classification of strata in the Dolomites, the nature and structure of corals and tectonics, she had wanted to break even more new ground in her geological research work. But time was running out for her. Certain Earth science enthusiasts soon began to call her the 'Dolomite Geologist'.

'Endless work to do in several of the women's Societies and have had no attempt at finding time for geology' [1927].

She had ventured into a field of activity which could yield no natural resources, and no powerful industries sponsored her in the hope of making vast profits – a fate shared by many other scientists. The fact that she had made inroads into new and uncharted territory was completely immaterial. She faded beneath a veil of oblivion.

Table 1. *The memorable highlights and achievements of Maria Ogilvie Gordon (after Burek 2004, 2005)*

Date	Achievements
1893	First woman to receive a DSc from University of London
1900	First woman to receive a PhD from Munich University19 published papers, many in German
1901	English translation of Zittel's German *History of Geology*
1908	Publishes *Handbook of Employment for Boys and Girls*
1916	President of the National Council of Women of Great Britain and Ireland
1919	Formed the Council for the representation of women in the League of Nations
1919	In first group of women to be elected to the Geological Society of London
1920	First JP and chairman of the Marylebone Court of Justice
1928	First geological guidebooks to the Dolomites
	Honorary membership of the University of Innsbruck
	Honorary correspondent of the Geological Survey of Austria
1931	First female honorary member of the Geological Survey of Austria Institute
1932	Received Lyell Medal from Geological Society of London
1935	DBE from King George V
	Honorary LL.B degree from Edinburgh University
1939	Died in London; ashes buried in Aberdeen with her husband
2000	A new fern genus named after Maria Gordon. The type specimen is called *Gordonopteris lorigae*; a tripinnate fern frond with aphlebia absent from the bases of the pinnae of the first and second order with small pinnules.

The movement for women's rights in which she had played a pivotal role in the early years progressed so agonisingly slowly that the early pioneers were soon forgotten. Today, when higher education for women is taken more for granted, nobody remembers the women who had to struggle even to be admitted to university courses because of their sex. Maria Ogilvie Gordon was a century ahead of her time. Her husband was right; she ploughed a lonely and arduous furrow all her adult days. She could certainly have chosen an easier path through life. However, Maria Matilda Ogilvie Gordon is an outstanding example of a highly intelligent woman who believed passionately in her research work but also had the foresight and sensitivity to fight for the rights of women in education, politics, peace and society at a time when many less determined women would have failed to rise to the challenge. She is considered one of the most prolific and original Scottish women of science during the 19th century with over 35 original scientific papers.

Honoured after 60 years

It is fitting that, at long last, Maria Matilda Ogilvie Gordon is honoured through palaeontological nomenclature, even if it is over 60 years after her death.

In the year 2000, the first author of this paper, Michael Wachtler, found an abundance of fossilized plants in the Prags Dolomites, among which was an unknown species of fern (Fig. 7). In 2005, Johanna van Konijnenburgh-van Cittert, Evelyn Kustatscher

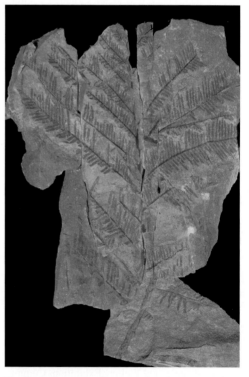

Fig. 7. *Gordonopteris lorigae*, a fern from the Triassic period found in the Prags Dolomites by Michael Wachtler. It was named '*Gordonopteris lorigae*' by Johanna van Konijnenburgh-van Cittert, Michael Wachtler and Evelyn Kustatscher in honour of the two female scientists and explorers Maria Ogilvie Gordon and Carmela Loriga Broglio (Michael Wachtler Archive).

and Michael Wachtler named it '*Gordonopteris lorigae*' in honour of the two women scientists, Maria Ogilvie Gordon and Carmela Loriga Broglio (Van-Konijnenburg-Van Cittert *et al.* 2006). The citation states: 'After Dr Marie Ogilvie Gordon who was one of the pioneers of Triassic palaeobotany in the Dolomites'. She would have been pleased and honoured.

Special thanks are due to the Department for Landscape Conservation of the Autonomous Province of South Tyrol; the descendants of J. Kostner's family, especially Renata Pizzinini; the Geological Survey of Austria Institute (*Geologische Bundesanstalt*) in Vienna; and T. O' Toole. Thanks also to Penny Hartley at Robert Gordon School; Captain M. Sweeney of Douneside House and the MacRobert Trust for access to their archives; Aberdeen and Edinburgh Universities; and finally to Anne Esson of Aberdeen for her inside knowledge and company. The paper has been greatly improved by comments from the referees. This paper is also dedicated to all those women who struggle to receive the recognition that they deserve but often do not get.

References

ANON. 1923. Value of organisation shown by improvement of woman's status. *Christian Science Monitor*, 23 April, 9.

ANON. 1939. Dame Maria Ogilvie Gordon – geologist and champion of women's causes, Obituary, *The Times*, 24 June.

BUREK, C. V. 2004. Gordon, Maria Matilda (née Ogilvie) (1864–1939) *In*: LIGHTMAN, B. (ed.) *Dictionary of Nineteenth Century British Scientists*, Vol. 2. Thoemmes Continuum Press, Bristol, 799–801.

BUREK, C. V. 2005. Who were they? The lives of geologists. 5. Dame Maria Matilda Ogilvie Gordon – A Britisher – and a woman at that (1864–1939). *Teaching Earth Science*, **30**(4), 42–44.

CREESE, M. R. S. 1996. Maria Ogilvie Gordon (1864–1939). *Earth Sciences History*, **15**(1), 68–75.

KLEBELSBERG, R. VON. 1932. Die Dolomitengeologin [The Dolomite Geologist]: Maria M. Ogilvie Gordon. *Der Schlern, Bozen*, **13**, 492–496.

KÖLBL-EBERT, M. 2001. On the origin of women geologists by means of social selection: German and British comparisons. *Episodes*, **24**(3), 182–193.

MATHESON, C. M. 1939. *Dame Maria Ogilvie Gordon*. Robert Gordon School, archives, newspaper article.

MORGAN, D. 2004. *Lost Aberdeen*, Birlinn, Edinburgh.

OGILVIE, M. M. 1893. Contributions to the geology of the Wengen and St. Cassian Strata in southern Tyrol. *Quarterly Journal of the Geological Society, London*, **49**(1), 4.

OGILVIE, M. M. 1894. Coral in the Dolomites of South Tyrol. *Geological Magazine*, **1**, 1–10, 49–60.

OGILVIE GORDON, M. M. 1903. The geological structure of Monzoni and Fassa. *Transactions of the Edinburgh Geological Society*, **8**, 1–180.

OGILVIE GORDON, M. M. 1897*a*. Microscopic and systematic study of madreporian types of corals. *Philosophical Transactions of the Royal Society*, **187 B**, 83–345.

OGILVIE GORDON, M. M. 1897*b*. Die Korallen der Stramberger Schichten. *Palaeontographica*, **7**, 73–282.

OGILVIE GORDON, M. M. 1906. Interference phenomena in the Alps. *Abstracts of the Proceedings of the Geological Society, 1905–6*, 118–120.

OGILVIE, GORDON. 1908. *Handbook of Employment for Girls and Boys*. National Council of Women, London.

OGILVIE GORDON, M. M. 1909. Die Überschiebungsmassen am Langkofel und im oberen Gröder Tal. *Verhandlungen der Kaiserlich-Königlichen Reichsanstalt, Wien*, 297.

OGILVIE GORDON, M. M. 1910. The thrust masses in the western district of the Dolomites, Tyrol. *Transactions of the Edinburgh Geological Society*, **9**, 1–91.

OGILVIE GORDON, M. M. 1911. Über Lavadiskordanzen und Konglomeratbildungen in den Dolomiten Südtirols. *Verhandlungen der Kaiserlich-Königlichen Reichsanstalt, Wien*, 212–222.

OGILVIE GORDON, M. M. 1913. Leithorizonte in der Eruptivserie des Fassa-Grödengebietes. *Verhandlungen der Kaiserllich-Königlichen Reichsanstalt, Wien*, 163–172.

OGILVIE GORDON, M. M. 1927. *Das Grödener-, Fassa- und Ennerberggebiet in den Südtiroler Dolomiten*, Geologische Beschreibung mit besonderer Berücksichtigung der Ueberschiebungscheinungen. Abhandlungen der Geologischen Bundesanstalt Wien, Vol. 24, fasc. 1&2.

OGILVIE GORDON, M. M. 1928*a*. *Geologisches Wanderbuch der Westlichen Dolomiten*. G. Freytag & Berndt, Wien.

OGILVIE GORDON, M. 1928*b*. *Guide for Geological Tours in the South Tyrolean Dolomites*. L. Waldmann, Erläut Exkurs. Deutsche Geol. Ges. Wien.

OGILVIE GORDON, M. M. 1932. Anniversary meeting: Lyell Medals. *Proceedings of the Geological Society, London*, **88**, 59–60.

PIA, J. 1939. Maria Matilda Ogilvie Gordon. *Mitteilungen des Alpenländischen geologischen Vereines*, Wien, **5**, 173–186.

STACUL, P. 1969. Die Dolomitengeologin [The Dolomite Geologist]: Maria M. Ogilvie-Gordon as Palaeontologist. *Der Schlern, Bozen*, **43**(12), 535–538.

VAN KONIJNENBURG-VAN CITTERT, J. H. A., KUSTATSCHER, E. & WACHTLER, M. 2006. Middle Triassic (Anisian) ferns from Kühwiesenkopf (Monte Prá Della Vacca), Dolomites, Northern Italy. *Palaeontology*, **49**(5), 943–968.

The role of Annie Greenly in the elucidation of the geology of Anglesey

T. P. T. WILLIAMS

Gwynedd & Môn RIGS (e-mail: iss006@bangor.ac.uk)

Abstract: Annie Greenly (née Barnard) is the great woman standing behind Edward Greenly. During his 25 years of surveying and producing the two volumes of the *Memoir of Anglesey*, she played a crucial role producing the index among other things. Her interaction with Greenly, first in the Highlands of Scotland and then on the Island of Anglesey, is detailed in this paper. Greenly outlived his wife by many years but he always acknowledged the debt he owed her.

The saying 'behind every great man there stands a great woman' has many exemplars in science. In the context of the history of geology in Wales, the case of Edward Greenly and his wife Annie (Fig. 1) stands out. Thanks to Greenly's taste for biography in later life (Greenly 1938), together with the fact that a significant amount of correspondence has survived, we know more about the role she played than that of most geologists' wives. This short paper attempts to briefly describe her role.

Edward Greenly was born on the 3 December 1861 at Bristol (Greenly 1938). His main claim to fame is his epic 25-year geological survey of Anglesey, from 1895 until 1920, usually said to have been carried out virtually unaided, except for the occasional supportive visit from one or other of his former colleagues from the Geological Survey. It is clear, however, that Annie Greenly's role was crucial at all times. There is little doubt that, without her support and assistance, Greenly would not have been able to complete this work, which culminated in the publication of a monumental two-volume memoir (Greenly 1919) and a new geological map of Anglesey, at a scale of 1 inch to the mile. The standard of this map was such that the island has not been resurveyed, and it remains standard issue after countless reprints, more recently to the modern 1:50000 scale. Even today, the only real criticism of this beautiful map is that the underlying Ordnance Survey details of roads and towns are now inevitably obsolete. However additional knowledge, and changes in nomenclature, will probably prompt a re-survey in the not too distant future.

Annie Greenly was born Ann Barnard on 8 June 1852 at No. 2 Cleveland Place, Bath. In later life she would say that her first introduction to geology came one night at the age of 11 years, when she was woken by a violent shaking of the house. Instead of panic, she recalled calmly informing her grandmother 'that's an earthquake!' It was indeed the Force 6 event of 6 October 1863, whose epicentre was calculated to be near Hereford.

She and Edward Greenly first met in 1875 at his family's house in Bristol. Fifty-eight years later he lovingly recollected how he found 'sitting in the firelight a young lady with Greek-like features, an unmistakably thoughtful expression and a wealth of golden hair'. During their first courtship Greenly was studying chemistry. He had a primitive laboratory, which Annie always referred to as the 'chemistry room'. Greenly writes: 'Here we spent many a happy hour at all manner of experiments, making oxygen, hydrogen and at analysis with test tube and precipitate.' Unfortunately, parental interference led to the breaking off of a first engagement with Annie in 1879 and they did not meet again for 11 years.

Greenly claims that, during these years, he was diverted from a career in law and towards geology by a walking holiday in Wales with friends. He had scaled Cader Idris and was profoundly impressed by its great sheets of microgranite exhibiting nearly vertical columnar jointing, lit by early sunlight.

'Marvelling what it might mean, we deplored our ignorance aloud. This was the crisis of my whole career. As soon as I returned to London, I began to read the subject up. That early morn on Cader Idris made me a geologist.'

Greenly (Fig. 2) attended University College London and learned petrology from Professor Bonney himself. As he put it 'owing to my chemical training I was drawn to petrology rather than to fossils'. He made the acquaintance of Professor A. H. Green, who introduced him to Archibald Geikie, Director of the Geological Survey. This led to Greenly joining the Survey in 1889 and being sent to help Benjamin Peach and John Horne in surveying the northwest Highlands of Scotland. His years from 1889 to 1895 as part of the Scottish Survey were his apprenticeship in the study of metamorphic rocks. Geikie continued being supportive, remarking when Greenly deplored his own lack of experience with these rocks, that this was

Fig. 1. Annie as a young woman (Greenly 1938, Vol. 1 frontispiece).

Fig. 2. Edward Greenly (Greenly 1938, Vol. 1 Plate xlviii, facing p. 273).

'all the better, you will have the less to unlearn'. This was the heroic period, during which Peach and Horne first unravelled the complexities of the Moine Thrust Zone and identified the importance of large-scale recumbent folding. Greenly was ever afterwards in thrall to this form of tectonics, constantly invoking it in explanation of phenomena observed in Anglesey, even when simpler mechanisms were available. This inevitably led him into error in determining the order of the succession he observed in the island, although it had no effect on the accuracy and quality of his survey.

Marriage to Edward Greenly

During all these years, Edward Greenly had retained fond memories of Annie Barnard and he tells us how, at the end of 1890, on leave in Bristol from Scottish Survey service, he finally felt impelled to write to her once more. As he would have us believe, she was none too pleased to receive his letter at first but, in the end, sent him a favourable reply, hoping that 'we might meet again in this world, if only once, yet in peace and reconciliation'. Finding their mutual feelings still to be strong, the couple were soon engaged and were married in August 1891. Geology was immediately an integral part of their relationship. Annie said: 'We have agreed that I am not to be ignorant of your subject... I can begin by reading. Put into our luggage some lucid elementary books'. We may gather from the resulting selection of Page and Lapworth's *Treatise*, along with Jukes Browne's *Historical Geology* and Green's *Physical Geology*, that 'elementary' was taken in a broad sense. Nor were these to suffice, for Greenly adds that he reserved Annie's tuition in metamorphism to himself (Greenly 1938).

Highlands of Scotland

Annie at once fell in with Greenly's routine of Survey life, accompanying him to Strathy Point between Cape Wrath and John O'Groats, where the couple put up at a local inn. It was to be their way of life for some years to come. Annie was not idle for, as Greenly puts it, 'she read geology and made my powders for the microscope.' He appears at this time to have considered writing a book on petrology, and was preparing a reference collection of some 124 slides of powdered mineral specimens. This 137-page unpublished manuscript work, written on foolscap paper, has survived and has been inspected by the present writer in the University of Wales, Bangor.

It was in 1891–1892 that Annie decisively influenced the development of her husband's mapping and surveying technique. As it is the quality of his mapping, embodying the famous 'green line method' for which Greenly is now principally held in regard, we must attach due weight to his statement that he:

had known [C. T.] Clough's maps for three years before it occurred to me to emulate them. It was my wife Annie who sowed in me the seeds of aspiration and steadily sustained me for thirty-five years in effort after improvement of my methods. (Greenly & Williams 1930)

In his autobiography (Greenly 1938) Greenly expands on this:

'It has been supposed that my inspiration was the wonderful mapping of Clough. Such however was not the case. True, when once I began to aspire [to a higher standard of geological mapping] it was the maps of Clough which became my models. *But the initial source of my inspiration was no other than Annie Greenly* [Greenly's italics].'

He goes on to describe how, while his early work showed a normal improvement from 1889 to 1891, his work in Sutherland that year showed 'a deplorable relapse, due ... to not having found out how to deal with the Gneissose Complex of the north.' While none of his seniors had criticized his work, he was well aware of the existence of a higher standard in Clough's work. However, he had heard Horne speak of it 'with a solemnity [which] seemed to indicate something unapproachable'.

In September 1891, Annie joined him in Sutherland, and came to see both her husband's and Clough's work. She immediately grasped the superiority of the latter, and urged Greenly to aim for the same high standard. But crucially, she pointed out the one weakness in Clough's quality, which was that, as Greenly says:

'where he could see his lines, nobody could find a fault. But where the rocks were covered up and he had to draw by inference, his lines have a certain rigidity ... he had, in fact, a certain lack of penetrative imagination, a faculty which my colleague was endowed with to intensity. (Greenly 1938)

As Annie said to him:

Let Clough be your model in precision, but do not follow him in style. Found your style on nature's curves. Watch these wherever you can and where you cannot see them, feel them. To be true, a map must be beautiful.

Thus was the green line method born, without which the Anglesey survey and mapping would have been incomparably inferior. Each exposure was delineated on the map with its own grass-green edging on the outer side. Thus it was always clear which lines on the map were actually visible on the ground (i.e., within a green line contour) or inferred.

Anglesey and the memoir

Throughout the Anglesey survey, Annie was continually involved in the logistics. She undertook the location of suitable lodgings with all the associated negotiations over rents and facilities. She often accompanied Greenly, usually as much more than a spectator. As he put it:

More than one geologist, Clough himself, alas, included, has been killed while examining railway cuttings. When the author was examining cuttings in Anglesey, Annie Greenly always came with him as train-watcher and with her at the top of the bank he was able to concentrate on the geology. (Greenly 1938).

It is important to realize that Annie was in no sense totally absorbed in her husband's work. Her independent musical interests, as a fine contralto singer and pianist, are well documented. She was also a long-standing committee member of the local branch of the North Wales RSPCA, at a time when that organization had far less power than it now possesses and relied much more on influence in high places, at which Annie was rather adept, being personally friendly with most of the local gentry wives, such as the great Lady Verney of Rhianfa on the Menai Strait.

The Greenlys were involved in the professional world of geology throughout the years of the Anglesey survey. As well as visits from colleagues, they attended scientific conferences and particularly the annual meetings of the Geological Section of the British Association for the Advancement of Science. In 1903, Annie was responsible for instituting a much-needed reform. The question of the Conference Dinner arose. No ladies had ever been allowed to attend this august gathering. As Greenly recalled:

Time and place being fixed, a gentle voice made all turn: 'Ah, very nice for you. What about us poor women, crouching by solitary fires? No dinner for us!' Who could resist that voice? 'Very well, why should not the ladies come?' said the Section President. 'Anyone against?' No. Then we invite them. Provided, that is, they will not object to *smoking*!'

As Greenly has it, Annie demurely glanced at Gertrude Elles and Lady Teall (both well-known 'modern' women) and replied: No, indeed. Smoke you shall, *to the top of your bent*! Such was gender politics in those distant, pre-politically correct days. (Greenly 1938)

The main survey phase was completed at Porth Eilian near Amlwch, on 8 October 1910. However, a great deal of additional fieldwork was necessary, including a week-long survey of the remote Skerries islands in July 1911. This punctuated the long process of composing the *The Geology of Anglesey*, a Geological Survey Memoir that was to accompany a new geological

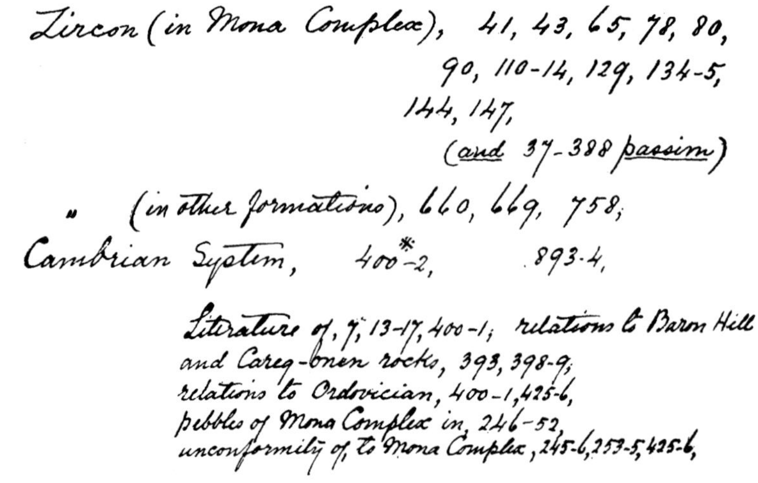

Zircon (in Mona Complex), 41, 43, 65, 78, 80, 90, 110–14, 129, 134–5, 144, 147, (and 37–388 *passim*)

" (in other formations), 660, 669, 758; Cambrian System, 400*–2, 893–4,

Literature of, 7, 13–17, 400–1; relations to Baron Hill and Careg-onen rocks, 393, 398–9, relations to Ordovician, 400–1, 425–6, pebbles of Mona Complex in, 246–52, unconformity of, to Mona Complex, 245–6, 253–5, 425–6,

Fig. 3. Annie's index (Greenly 1938, Vol. 2 Plate lxiii, facing p. 444).

map of the island. The two volumes of this epic work were eventually to run to 1020 pages.

On the outbreak of World War 1, Edward and Annie Greenly left their house *Achnashean* near Treborth, close to the Menai Bridge but on the mainland side, and took a succession of lodgings, first in Holyhead and later at Amlwch, Llanfairynghornwy and Llangefni, to enable Greenly to revisit various key sites and sections as his writing progressed. Annie worked alongside him, saying, 'Now I have no housework and no RSPCA, so I give myself wholly to ... this book'. She read each manuscript page, correcting in greater or lesser degree. According to Greenly, 'some sheets she pronounced to be hopeless and these, no less than 127 in number, she rewrote altogether ... we worked on opposite sides of the table, I writing, she *doctoring*.' It is clear that much of the final prose quality of the Anglesey memoir is due to Annie's work (Greenly 1938).

The Memoir was complete in manuscript by the end of 1915 and the couple decided to leave Anglesey for London, mainly to see the work through the press. It is of interest to note that, in transporting his manuscript survey maps by train to London, Greenly insured them for £1000 and found, on arriving at Euston to take delivery, that the London and North-Western had detailed a railway superintendent and a Scotland Yard detective to accompany the container to satisfy themselves as to its contents!

Annie undertook the onerous job of constructing the index to the Anglesey memoir (Fig. 3). It had to be written on individual slips of paper and threaded on individual lengths of string for each letter of the alphabet. In all there were 1879 subjects with 10 799 page entries. In *A Hand Through Time* Greenly describes how his friend Sir John Lloyd, the great historian of Wales, intervened after Annie's death to have the manuscript index preserved in the National Museum of Wales (Greenly 1938).

According to Greenly, Annie's taste was decisive in the final choice of colours used to print the 1-inch Geological Map of Anglesey, which was complete in manuscript form by 1911 but not printed until 1919 (Greenly 1938). The engraving was done by the Ordnance Survey draughtsmen at Southampton. The quality of their geological maps remains breathtaking, and it comes as no surprise to learn that a palette of no fewer than 763 tints was available, achieved by many different cross-hatchings over basic colours. Neither Greenly nor Sir Aubrey Strahan, Director of the Ordnance Survey, was able to decide on a satisfactory combination with which to represent the 58 different formations making up the solid geology of the island. So, as Greenly says:

I turned to my companion and soon wondered that I had not done so at the outset. She had a remarkable faculty for foreseeing the effects of colours even in quite small samples. Our problems were quickly solved. The one inch map is a thing of beauty, distinct from side to side, and has won general admiration. (Greenly 1938)

Annie as a geologist

Lest one suppose that Greenly was too fulsome in his praise of his wife's contribution, it is worth noting that he was not above a cool, detached criticism or two. He unhesitatingly relates to us her own

Fig. 4. Annie as an older woman (Greenly 1938, Vol. 2 Plate 1xi frontispiece).

verdict on herself 'I am not a scholar' (Greenly 1938) and comments 'she had little patience for the ponderous particulars far too often found in papers'. A section in *A Hand Through Time* entitled 'Gifts which she lacked' seems even slightly cruel and certainly inappropriate. However, it encourages us to accept his other assessment at face value, that, as is clear from almost the only sustained geological narrative of hers to survive (Greenly 1938), her judgment and depth of background knowledge were remarkable. A few extracts must suffice to give its general tenor:

The water of the earth was never wholly evaporated during the warm periods nor completely frozen during the ice ages, otherwise life would have perished. If we were emerging from the Pleistocene glaciation, we would have at least an interglacial period before us with more warmth, not less . . . .

As for the Permo-Carboniferous glaciation, its far-reaching extent, exceptional severity and duration make it so appalling that one wonders how the world could ever recover from it . . . . Most interesting, if only in its mystery.

American geologists are very sure that they know the ages of things! For instance, in describing the surface of the ancient rocks [it is said] 'there were no mountains, nor even high hills from which torrents could flow, or where mountain-glaciers could gather'. Is this not *purely hypothetical*?

. . . ideas of the earth subsisting on its internal heat, and an unbroken pall of cloud from the Permian to the Pliocene are surely rather wild? What of the mighty reptiles that ruled the world during the Mesozoic, with the rich flora needed [to sustain them] . . . *the writer could have had no sense of geological time and conditions.*

A final extract is remarkable for the wonderful vignette it affords us of the Greenlys' relationship and the intellectual quality of their shared life.

I do not wonder that Croll's theory [of the origin of glacial periods] was long a favourite with geologists. I remember how enchanted I was when Edward expounded it to me long years ago in the little

Fig. 5. The Greenly gravestone in the churchyard of Llangristiolus, Anglesey.

Highland croft where we were staying when he was on the Survey. Those days were very happy. It was a case of an Academy with one student ...

After the memoir

The years after the completion of the Anglesey survey saw the couple living in College Road, Bangor. Greenly continued his work on the mainland, in what is now the county of Gwynedd. In 1926 he undertook the writing of a short elementary textbook entitled *The Earth*. The publisher set very strict limits to the size of this little volume: no more than 54 pages. Greenly was defeated and turned once more to Annie (Fig. 4). She was firm.

This book is extremely difficult to write ... condensation is excessive ... [but] being for the general reader *it must be absolutely lucid* and ... of living interest, or [they] will not read it. I [shall] go through it ....

Go through it she did, but it was to be her last effort. She completed her meticulous reading on 21 February from what was to be her deathbed; only days later, on 1 March 1927, she died from heart failure. She was 75 years old.

Greenly survived her by nearly a quarter of a century, dying in 1951 at the great age of 90 years. They rest today in the churchyard of Llangristiolus, Anglesey, beneath a fine headstone of red 'Balmoral' granite (Fig. 5). Of all the letters of condolence Greenly received after Annie's death, that from Professor Shintaro Nakamura of Kyoto, who had been an officer of the Japanese Geological Survey and who had visited and corresponded with him, is perhaps most moving. Aware of his limited English but disdaining any intermediary, Nakamura wrote in his own hand:

Your loss surpasses any worldly substances in the gone-away of such a faithful colleague as your dearest, who did a strong devotion. I recollect always her tenderness and virtue. My memory on Bangor softens by your kindness and her quietness ... When we think of the true woman who devoted all her life to her faithful husband, we perceive that her life was very happy ... her life is living, though we cannot hear her voice; like that you and I are unable the discuss face to face the origin of the Menai Strait now. I told her virtuous womanhood to my wife and children, translating [the printed obituary] and all sympathised with you.

References

COLEMAN, A. P. 1926. *Ice Ages: Recent and Anicient*. MacMillan, New York.

GREENLY, E. 1919. *The Geology of Anglesey*. 2 vols. Memoirs of the Geological Survey of Great Britain, HMSO, London, 78, 980 pp.

GREENLY, E. 1938. *A Hand Through Time*. 2 vols. Murby & Co. London.

GREENLY, E. & WILLIAMS, H. 1930. *Methods in Geological Surveying*. Murby & Co. London.

Nancy Kirk: turning the world of graptolites upside down

A. R. WYATT

35 Livonia Road, Sidmouth, Devon, EX10 9JB, UK (e-mail: antony_wyatt@hotmail.com)

Abstract: Nancy Kirk was a protégée of O. T. Jones. Her early, unpublished, field studies in the Welsh Borders were an important foundation for some of the work on faunal distributions by Oxford workers in the 1960s and for regional studies by a group at Cambridge in the 1980s and 1990s. From the start, her graptolite work proved controversial but, for many people, enlivening. Interpretations (produced in conjunction with Denis Bates) based on detailed SEM examination, gave her an international reputation. As a teacher, she was noted for the care and effort that she put into stretching the minds of the brightest, while ensuring that even the weakest students had a chance. She was memorable for her dress sense (usually looking as if she had just come in from the field) and her sometimes colourful language.

Nancy Kirk (1916–2005) (Fig. 1) was born and brought up in Mansfield, Nottinghamshire, where she attended Queen Elizabeth's Girls' Grammar School. Her mother died when she was young, so she, and her somewhat older brother (who became a classics professor in the United States), were raised mainly by their father, who had worked his way from office boy to manager of a local factory. In 1935 she won an exhibition to go up to Newnham College, Cambridge, to read natural sciences.

As a student she never had much money, living a frugal lifestyle which continued for the rest of her life. By the end of each term she was usually so broke that she could not afford the train fare home, so would have to resort to her bicycle. She did, however, make use of the opportunities to develop her appreciation of art in the museums and libraries, and was willing to scrimp and save to be able to afford the odd object that caught her eye. At her death she still had a framed print of a van Gogh portrait that she bought as an undergraduate. It was at this stage of her life that she developed her political views, which, like many in Cambridge at the time, were distinctly left wing and pro-Russian.

Her original intention was to specialize in botany, a subject that fascinated her for the whole of her life, but she found the Cambridge teachers very dull. She was, however, inspired by some members (though by no means all) of the geology department, particularly Professor O. T. Jones, and two young workers, Brian Harland and Teddy Bullard. In 1939 she achieved first-class marks in the geology finals examination and was awarded the Bathurst Prize. As a female, however, she could not be awarded a degree.

The award of a Harkness Scholarship in 1939, followed by a Bathurst Research Studentship (1940–42), allowed her to start research on the geology of the faulted country between Pont Faen, Breconshire, and Presteigne, Radnorshire (both now in Powys), under the supervision of O. T. Jones. O. T. was a remarkable teacher, particularly in the field, where most colleagues marvelled at the speed at which he could appreciate a situation, unravel a problem and explain it clearly and simply. These abilities were clearly passed on to Nancy (Fig. 2). Not everyone appreciated O. T.'s abilities in the field. In one argument at the Geological Society he was accused of 'Alice in Wonderland' geology. Much the same would be said of some of Nancy's contributions.

Her studies were interrupted by World War 2. In 1942, following a period in the Land Army, she was sent to the Royal Ordnance Factory, where she remained until 1945. The experience was crucial in moulding her character, particularly in her attitude to authority, and in the factory floor language that she often used. On one occasion, when she was asked to investigate why so many bombs produced by the factory were being rejected, she discovered that a critical part in one of the machines had been fitted back to front. The only apparent outcome after she submitted her report was that she was banned from any further checking of machinery.

Following the war, Nancy returned to Cambridge on a 2-year Jenner Research Fellowship, after which she supported herself on money that she had saved. She was awarded her PhD in 1949 (Cambridge had by then changed the rules and allowed degrees to be awarded to females). In keeping with her views on public ceremonies she graduated by proxy. For the following 2 years Nancy worked at Birmingham University, before moving to live in her field area, where she continued her geological mapping, gradually extending the area covered. One inspiration for going it alone was Teddy Bullard, who, some years before, on a visit to her field area to try out some geophysical

Fig. 1. Nancy Kirk in the early 1980s, shortly before retirement (photo courtesy of Denis Bates).

from 1910 to 1919, Nancy was asked for interview at Aberystwyth. She travelled over on an old motorbike, fell in love with the countryside, and accepted the job as assistant lecturer. She was promoted to full lecturer in 1955, and continued working in the department long after her official retirement in 1983. She arrived in a department with a professor who was in his early 40s, and two senior lecturers, both in their 60s (Wyatt 2000). At first she had a heavy teaching load, mostly involving palaeontology, stratigraphy, map interpretation and fieldwork. With additional staff appointments, particularly during the 1960s, she was able to devote more time to research. She is remembered by generations of students for the care and effort that she put into teaching, particularly the help that she gave to people who were having difficulties.

Modern administrators would throw up their hands at some of the things she did. The culmination of her second-year geological maps course was an examination that started at 9 a.m. and could continue until late at night, depending upon the speed and stamina of the student. Students brought packed meals and were on their honour not to talk if they needed to go to the toilet. It was a long day, both for the students and for Nancy, but it certainly taught most of them how much information could be derived from detailed study of Geological Survey maps.

With expansion of staff numbers, clashes of temperament and personality began to arise. Dynamic young lecturers, who were often more impressed by style than substance, started to criticize her approach. Nancy found it particularly

equipment, asked her why she would ever want to leave such a beautiful area to try to work in the tense atmosphere of Cambridge.

In 1953, probably at the instigation of O. T. Jones, who had been the professor of geology

Fig. 2. O. T. Jones and Nancy Kirk in the field in the early 1960s (photo courtesy of Tony Jones).

irritating to find that one colleague told not only his student party, but also the warden of a Youth Hostel which she had used as a field centre for many years and recommended to the colleague, that he was going to be teaching real geology, not the soft and easy stuff that Nancy proffered. It was the same colleague who blithely assumed that the Geological Survey maps that Nancy used for second-year teaching were appropriate for Advanced-level examinations. It seems clear, given the personalities involved and other occurrences, that Nancy was being attacked for her academic background and experience rather than her sex, but there were times when she felt exposed as the only female on the staff.

Nancy's research work can be divided into two: the detailed fieldwork that she did prior to arriving at Aberystwyth, and her graptolite work that began after she had been on the Aberystwyth staff for many years. She was, of course, active in helping others in the intervening period; her down-to-earth approach and insistence on accurate and detailed mapping helped to prevent some research students and colleagues from following flights of fantasy too far.

A bibliography of her work is included as an appendix at the end of this paper. There is a separate list of references, which, to avoid duplication, excludes cited papers by Kirk.

Field studies

A report on part of Nancy's fieldwork can be read in her PhD thesis, stored in the university library at Cambridge. She always claimed that this was rather slapdash, as she had to finish it in a hurry in 1947 to comply with an unexpected deadline: the University was about to change its statutes to allow women full membership, including the right to be awarded degrees. Her thesis was officially approved at a meeting of the Board of Research Studies in June 1948, but a copy was not deposited in the University Library (required for the award of the degree) until February 1949. Following standard cataloguing rules (AACR2 1998), it is therefore listed as 1949 in the bibliography, despite previous citations in the literature as 1947.

The thesis contains a set of maps of her field area, which was later copied and used by a number of Oxford research students, who were working on the distribution of fossils in the Welsh Borders. Nancy's contribution to unravelling the geology of this part of Wales was also important to later workers from Cambridge. For example her thesis and the later Geological Society talks were referred to by Woodcock (1992) and several authors in Woodcock & Bassett (1993).

In the early 1950s she wrote up her more extensive fieldwork for presentation at three meetings of the Geological Society (Kirk 1951a & b, 1952). She always felt bitter that, at the conclusion of the third meeting, most of the assembled fellows went off to the Geological Society Dining Club (a private club that, at the time, did not allow females), leaving Nancy and the other speaker, Pam Robinson, at the door of Burlington House. Worse was to come when she submitted the work for publication. It included a large and detailed map, covering a greater area than that included in her PhD. The referees and editor originally proposed a much smaller, far less detailed map, at one stage suggesting removing most of the dip arrows that Nancy considered crucial for understanding her argument. Eventually she managed to persuade them that her map had to be in its original form, but then she was confronted with demands for drastic shortening of the text.

At the time the Geological Society had problems with paper rationing and was advertising the fact that they could not print any long papers. Years later Nancy admitted that she should have taken more note of this problem, but during the dispute she was convinced that the editor's approach was all part of a chauvinistic attitude to science, exemplified by the Dining Club. Unfortunately she did not feel able to consult anyone about the problem and withdrew her proposed work. The draft text of the three separate papers still exists, along with her fieldslips, but the original map seems to have disappeared.

Whatever Nancy may have thought, it is clear that there were some people in the Geological Society who appreciated her work. She was awarded the Murchison Fund in 1954.

Graptolites

There is no evidence that Nancy had more than a passing interest in graptolites until the late 1960s. She had been in the field with Gertie Elles during her Cambridge years, and would often tell of fossil hunting with her in Llanfawr Quarry, Llandrindod Wells, while it was still working, but graptolites were not their prime objective. O. M. B. Bulman had also been in Cambridge, but was not one of the staff that Nancy found inspiring. Following O. T.'s lead, she saw graptolites as tools for correlation purposes, best left to experts to identify. It was a chance mistake that started her off on graptolite research.

Nancy was painting her cottage, with almost all of her possessions piled up under dust sheets, when she realized that the notes for her first-year palaeontology course, which was about to start, were somewhere in the piles. Rather than cause havoc to the decorating she started to write out a new set of notes. With the extra maturity and experience of many years of teaching, she found herself questioning part of the received wisdom about some fossils, particularly the graptolites. For the moment the decorating was forgotten and she put her efforts into trying to put her thoughts on graptolites into a coherent picture. Years later, the possessions were still under the dust sheets.

The standard story was that floating graptolites evolved from dendroids attached to the sea floor. For years the assumption had been that the graptolites were attached to floating seaweed, although this had recently been discarded by Bulman (1964). The main difference between graptolites and dendroids was that the graptolites had only one type of theca, while in the dendroids the difference between autothecae and bithecae was attributed to sexual dimorphism.

Nancy thought that the sexual dimorphism story said more about the minds of the proponents than the reality. It seemed much more logical for an attached organism to have a mechanism for cleaning away bits of sediment stirred up from the sea bottom, and for the descendants, living much higher in the water column, to have no need for such cleaning individuals. After all, wherever you are in the water column, sex is just as important for survival of the race.

But it was her proposal for how the transition from attached to free living occurred, and the implications for lifestyle and evolutionary history, that were most controversial. She suggested that, as colonial animals, the individual zooids in a dendroid co-ordinated their feeding currents, causing a flow of water down into the inverted cone structure, bringing food for the colony. Of course such a flow would also bring in unwanted material that could plug the exit pores, so cleaning individuals were required to maintain efficiency. It was logical to infer that improved efficiency of the co-ordinated currents could eventually produce individuals that would break free from their attachment and begin the evolutionary path to graptolites.

Nancy saw the evolutionary history of the graptolites as a working out of ways of improving a colony that had evolved for life on the sea floor, but which was now living as a free-floating body. The cleaning individuals would soon be lost as they no longer had a function and were simply dead weight. She saw co-ordinated feeding currents as crucial in allowing the colony to rise through the water column while feeding and to sink back down during periods of relative relaxation. This concept of automobility, paralleling the movement that is seen today in much of the plankton, was, she thought, essential to ensure that the colony did not stay in the same water mass for long periods, as it would soon face starvation.

As part of her philosophy, Nancy always tried to keep to the lowest number of steps from one state to another. She reasoned that, because the co-ordinated feeding currents developed while the animals were at the evolutionary level of attachment, there was no need for the mechanism to change once they moved to life in the water column. One implication was that the experts had always thought of, and illustrated, graptolites upside down.

The orientation of graptolites became part of the Nancy mythology, and she often used her fingers to illustrate the difference between the conventional orientation of *Didymograptus* and her interpretation. Of course she knew that animals high up in the water column do not maintain a rigid, vertical orientation, and that movement is not necessarily simply up and down, but she never saw any need to modify her ideas about the mechanism of automobility and the consequent orientation of the colony.

Following an initial presentation to the Aberystwyth department, Nancy was persuaded to prepare a talk to be given at the Geological Society (Kirk 1969). At the time she thought that this would be the end of it all, and was expecting to return to her decorating. But the reception that she was given by some of the assembled experts upset her, and she thought that the Society editors were once again trying to stop her publishing her ideas. She felt that the graptolite world saw her as a middle-aged female interloper to be ignored. She was sure that, if she had been a young research student in a short skirt, her ideas would have been welcome.

In one sense there was a feeling among some graptolite workers that she was intruding. Most of them had devoted their working lives to taxonomy and correlation. They had spent years refining details and helping identify specimens for field mappers. They felt that, if anyone were to have original thoughts about graptolites, it should be one of them. But they had to accept that outsiders can, and do, rock the boat. Nancy did have difficulty to begin with in getting her ideas published, but her ideas could not be totally ignored, and there is evidence to show that her ideas were not suppressed. I remember being introduced to them in the early 1970s by Barrie Rickards, one of her great critics, as part of the Cambridge undergraduate course. My recollection is that, despite his known opposition, in teaching her ideas he was fairly neutral about their value, suggesting that we should come to our own conclusions.

A partial reprieve to the problem of finding outlets willing to publish her ideas came in 1972, when, to celebrate the centenary of the founding of the University College of Wales at Aberystwyth, the Geology Department started an occasional publication. Many of the numbers were papers by Nancy (Kirk 1972b, 1973a & b, 1974a & b, 1975b). It was a way of getting her ideas printed, but her critics claimed that it was just trying to get round proper refereeing.

One thing that the criticism ensured was that Nancy looked in far more detail at the available published information, and started to examine specimens under the scanning electron microscope (SEM), roping in her colleague Denis Bates, who rapidly developed skills in both the preparation of specimens and the production of photographic images. They soon accumulated thousands of hours of experience on the SEM and many thousands of images, meaning that they could no longer be accused of trying to fly before they could crawl. Her commitment to their work was such that she found delays intolerable. At one point there was a problem in funding a replacement SEM, so, to speed their work, Nancy produced most of the required money from her savings.

Their observations soon persuaded Nancy that the construction of the graptolite rhabdosome was far more complex than she had first envisaged, and she started to construct large-scale models to try to help her in understanding how things worked. She rapidly convinced herself that most of the detail could not have been produced by the standard model of zooids moving out of their thecae and licking on new layers. The only possible solution, as she saw it, was that most of the construction took place underneath a covering of extrathecal tissue (Kirk 1972b). This was another controversial suggestion to mainstream workers.

In the mid-1970's Nancy did try examining the sinking characteristics of her models in the college swimming pool. She was hoping to see forced rotation, and/or differences in sinking times when models were placed in the water upside down. Whether it was movement in the water caused by Denis Bates swimming down to retrieve the models, or imperfections in the models themselves, the results were neither good nor consistent. Certainly no clear picture emerged and the experiments were abandoned. Of course, the fact that the models could not include any of the extrathecal tissue that Nancy was sure had to be there meant that they may have born little resemblance to the real shape of the graptolites. Later work by Bates (1987) certainly suggested that there could have been a considerable volume of extrathecal tissue, particularly if it were required to give neutral buoyancy.

If nothing else, Nancy's contributions had produced controversy in an area of geology that, at the time, was seen by many as dry and dusty, if not downright dull. She found herself invited to give talks to student societies around the country, and her enthusiasm, impish sense of humour and slightly eccentric character certainly made a mark on many of her listeners. Whether they believed her ideas, or not, for many students it opened up the idea that palaeontology, and particularly graptolites, could be interesting.

But there was still reluctance to credit her views within the graptolite community. I remember sitting next to Nancy at a Palaeontological Association talk by a young research student, who was talking about how co-ordinated feeding activity could have allowed graptolites to move in the water. He was desperate to claim that his idea was not the same as Nancy's automobility, though, as a non-expert, I could not see the difference. After the talk I asked Nancy what she thought. 'No bloody difference', was the rather typical reply.

In later years, the quality of the descriptive work based on the material examined under the SEM began to produce a better result. To test the theories of extrathecal tissue and automobility, Nancy had decided to examine the retiolite graptolites (Figs 3 & 4). This resulted in the recognition that they had a more complex structure than the ordinary graptolites, with an additional skeletal structure: the outer ancora sleeve. Not only did this lead to a more detailed working out of current flow within the sleeve, but Nancy argued that the presence of the additional structure also makes the licking theory less convincing. Another result of this detailed work was the confirmation of the differences between the Ordovician archiretiolitids and the Silurian retiolitids that had originally been hinted at by Bulman (1970, p. 129), where, based on gross morphological differences and on time difference, he suggested that the group might be polyphyletic.

The proposed new edition of the graptolite volume of the *Treatise on Invertebrate Palaeontology* will draw heavily on the work of Nancy and Denis. The influx of new workers who had encountered Nancy's ideas before they started their own research also meant that she began to be seen as an old hand in the business. One outward sign was the dedication by Big G (the British and Irish Graptolite Group) of Palmer & Rickards (1990) to her. In typical fashion she made a bit of a fuss about the fact that they had managed to get a photograph of her for the frontispiece. She was always a little shy about being photographed.

Life outside geology

For many years the garden that Nancy had created on an isolated 3-acre site around her cottage near Llanafan in the Ystwyth Valley was her pride and

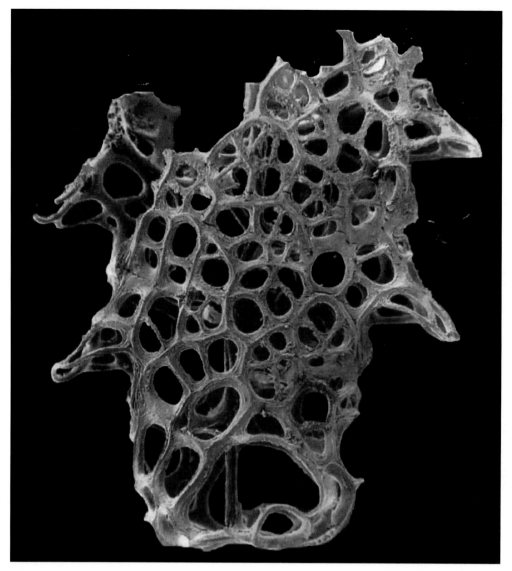

Fig. 3. An incomplete specimen of *Retiolites* as seen under the SEM. Examination of broken ends of skeletal elements was crucial for understanding growth. Specimen approximately 2 mm in length (photo courtesy of Denis Bates).

joy. She started with a set of formal rose gardens, but she was soon convinced that this was far too bourgeois and replaced it with a flowing mixture of lawns and sweeps of azaleas and rhododendrons. Most were grown from cuttings, usually taken with permission, but once the garden began to mature she grew new varieties from seed. In building and maintaining the garden she was helped by John Corfield, whom she first met when she was scrounging plants from the Aberystwyth Botany Department Gardens. Apart from assisting in her garden, John helped persuade her to join a botanical trip to Greece, which fired up her interest so that, for several years, the two of them would visit different parts of that country, looking partly at rocks, partly at the flora. For Nancy, who was otherwise used to a solitary existence, this gave her a platonic relationship with an enjoyable and knowledgeable companion.

In keeping with her interest in art, Nancy also developed into a talented potter. Her major pride was in the glazes that she produced, as she was almost never satisfied with the shape of the

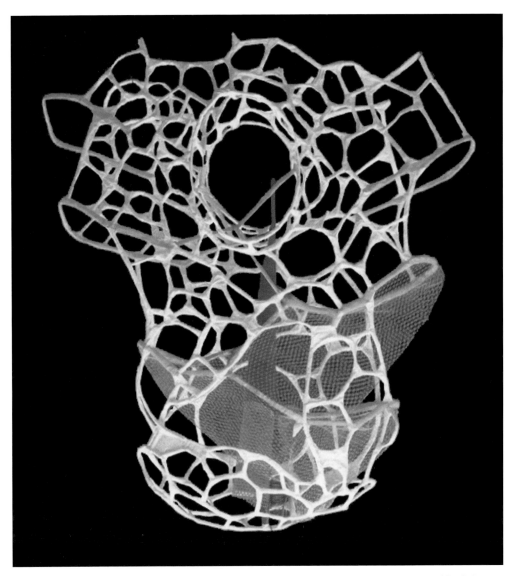

Fig. 4. Model of the early development of *Stomatograptus* produced by Nancy Kirk (photo courtesy of Denis Bates).

pots that she made. Some of her finer productions were collaborations with Bill Fitches, a colleague from the department, who provided her with the basic pots.

Nancy maintained her commitment to Marxism to the end of her life. She found much of modern consumer culture distasteful, was particularly concerned about the size and rapid growth of the world population, and had little time for most current politicians. She had wanted her body to be used for dissection, but bureaucracy got in the way and there had to be an autopsy, which ruled out donation. It was appropriate that the 'Red Flag' was played as her body was carried to its green burial.

Thanks are due to Denis Bates for his help in providing information, photographs and checking details, Tony Jones for Figure 2, and the referees who helped improve the text.

Appendix: Bibliography of Nancy Kirk

KIRK, N. H. 1949. *Geology of the anticlinal disturbance of Breconshire and Radnorshire: Pont Faen to Presteigne*. PhD thesis, University of Cambridge.

KIRK, N. H. 1951a. The Upper Llandovery and Lower Wenlock rocks of the area between Dolyhir and Presteigne, Radnorshire. *Proceedings of the Geological Society, London*, **1471**, 56–58.

KIRK, N. H. 1951b. The Silurian and Downtonian rocks of the anticlinal disturbance of Breconshire and Radnorshire: Pont Faen to Presteigne. *Proceedings of the Geological Society, London*, **1474**, 72–74.

KIRK, N. H. 1952. The tectonic structure of the 'Anticlinal Disturbance' of Breconshire and Radnorshire. *Proceedings of the Geological Society, London*, **1485**, 87–91.

KIRK, N. H. 1969. Some thoughts on the ecology, mode of life and evolution of the Graptolithina. *Proceedings of the Geological Society, London*, **1659**, 273–292.

KIRK, N. H. 1970. The structural framework around Cardigan Bay. *In*: DOBSON, M. R. (ed.) *Proceedings of the 4th Irish Sea Colloquium*. Geology Department, University College of Wales, Aberystwyth, 1–2.

KIRK, N. H. 1972a. More thoughts on the automobility of the graptolites. *Journal of the Geological Society, London*, **128**, 127–133.

KIRK, N. H. 1972b. *Some Thoughts on the Construction of the Rhabdosome in the Graptolithina, with Special Reference to Extrathecal Tissue and Its Bearing on the Theory of Automobility*. University College of Wales, Aberystwyth, Department of Geology, Publications, **1**, 31 pp.

KIRK, N. H. 1973a. *More Thoughts on Bithecae, Budding and Branching in the Graptolithina*. University College of Wales, Aberystwyth, Department of Geology, Publications, **2**, 16 pp.

KIRK, N. H. 1973b. *Some Thoughts on the Construction and Functioning of the Rhabdosome in the Retiolitidae*. University College of Wales, Aberystwyth, Department of Geology, Publications, **3**, 32 pp.

KIRK, N. H. ET AL. 1973. 'Unmutilated discussion of paper published by the Geological Society of London (Kirk 1972)'. Circulated with Kirk 1973b.

KIRK, N. H. 1974a. *Some Thoughts on Convergence and Divergence in the Graptolithina*. University College of Wales, Aberystwyth, Department of Geology, Publications, **5**, 33 pp.

KIRK, N. H. 1974b. *More Thoughts on the Construction of the Rhabdosome in the Dendroidea, in the Light of the Ultrastructure of the Dendroidea and of Mastigograptus* University College of Wales, Aberystwyth, Department of Geology, Publications, **6**, 11 pp.

KIRK, N. H. 1975a. Discussion of the function of extrathecal structures in graptoloids. *Geological Magazine*, **112**, 322–324.

KIRK, N. H. 1975b. *More Thoughts on the Construction and Functioning of the Rhabdosome in the Graptoloidea in the Light of Their Ultrastructure*. University College of Wales, Aberystwyth, Department of Geology, Publications, **7**, 21 pp.

BATES, D. E. B. & KIRK, N. H. 1978. Contrasting modes of construction of retiolite type rhabdosomes. *Acta Palaeontologica Polonica*, **23**, 427–448.

KIRK, N. H. 1978. Mode of life of graptolites. *Acta Palaeontologica Polonica*, **23**, 533–555.

KIRK, N. H. 1979. Thoughts on coloniality in the Graptolithina. *In*: LARWOOD, G. & ROSEN, B. R. (eds) *Biology and Systematics of Colonial Animals*. Systematics Association, Special Volume, **11**, 411–432.

KIRK, N. H. 1980. Controlling factors in the evolution of graptolites. *Geological Magazine*, **117**, 277–284.

BATES, D. E. B. & KIRK, N. H. 1984. Autecology of Silurian graptoloids. *Special Papers in Palaeontology*, **32**, 121–139.

BATES, D. E. B. & KIRK, N. H. 1985. The fine structure of graptolite periderm. *In*: BAIRATI, A. & GARRONE, R. (eds) *Biology of Invertebrate and Lower Vertebrate Collagens: Proceedings of NATO Advanced Research Workshop, Como, Italy, 24–28 June 1984*. Plenum Publishing Corporation, New York and London, 389–396.

BATES, D. E. B. & KIRK, N. H. 1986a. The mode of secretion of graptolite periderm, in normal and retiolite graptolites. *In*: HUGHES, C. P. & RICKARDS, R. B. (eds) *Palaeoecology and Biostratigraphy of Graptolites: Proceedings of 2nd International Conference of the Graptolite Working Group of the International Palaeontological Association held at Cambridge University, 1–15 September 1981*. Geological Society, London, Special Publications, **20**, 221–236.

BATES, D. E. B. & KIRK, N. H. 1986b. Graptolites, a fossil case-history of evolution from sessile, colonial animals to automobile superindividuals. *Proceedings of the Royal Society, London*, **B228**, 207–224.

BATES, D. E. B. & KIRK, N. H. 1987. The role of extrathecal tissue in the construction and functioning of some Ordovician and Silurian retiolitid graptoloids. *Bulletin of the Geological Society, Denmark*, **35**, 85–102.

BATES, D. E. B., KIRK, N. H., ZHAO, Y.-H. & LOYDELL, D. K. 1988. Ultrastructural studies on graptolites using scanning electron microscope. *Acta Palaeontologica Sinica*, **27**, 527–543.

KIRK, N. H. 1990. Thoughts on juvenile sessility, vertical automobility and lateral transport as factors in graptoloid evolution. *Modern Geology*, **14**, 153–187.

BATES, D. E. B. & KIRK, N. H. 1991. The ultrastructure, mode of construction and functioning of Ordovician retiolitid graptolites from the Viola Springs Limestone, Oklahoma. *Modern Geology*, **15**, 131–286.

KIRK, N. H. 1991. Construction, form and function in the Graptolithina: a review. *Modern Geology*, **15**, 287–311.

BATES, D. E. B. & KIRK, N. H. 1992. The ultrastructure, mode of construction and functioning of a number of Llandovery ancorate diplograptid and retiolitid graptolites. *Modern Geology*, **17**, 1–270.

KIRK, N. H. 1993. Thoughts on juvenile sessility, vertical automobility, and passive lateral transport as factors in graptoloid evolution. *In*: MU-XINAN (ed.) *Modern Theories and Hypotheses in the Study of Palaeontology*. Science Press, Beijing, 195–224. [Translation into chinese of Kirk 1990].

KIRK, N. H. 1994. Coordination of construction and functioning of the graptolite colony. *In*: CHEN XU ERDTMANN, B.-D. & NI, YU-NAN (eds) *Graptolite Research Today: Proceedings of the 4th International Graptolite Conference, Nanjing Institute of Geology and Paleontology, China, September 1990*. Nanjing University Press, Nanjing, 197–216.

BATES, D. E. B. & KIRK, N. H. 1997. *The Ultrastructure, Construction and Functioning of the Genera Stomatograptus and Retiolites, with an Appendix on The Incremental Construction of the Rhabdosome in Petalolithus, and its Comparison with that of the Thecal Framework in Retiolites and Stomatograptus.* Institute of Geography and Earth Sciences, University of Wales, Aberystwyth, Publications, **10**, 168 pp.

References

For references to Kirk see Appendix: Bibliography of Nancy Kirk.

AACR2 1998. *Anglo-American Cataloguing Rules, Prepared Under the Supervision of the Joint Steering Committee for the Revision of AACR.* 2nd edn. Library Association, London.

BATES, D. E. B. 1987. The density of graptoloid skeletal tissue, and its implication for the volume and density of the soft tissue. *Lethaia*, **20**, 149–156.

BULMAN, O. M. B. 1964. Lower palaeozoic plankton. *Quarterly Journal of the Geological Society, London*, **120**, 455–476.

BULMAN, O. M. B. 1970. Graptolithina (with sections on Enteropneusta and Pterobranchia). *In*: TEICHERT, C. (ed.) *Treatise on Invertebrate Paleontology*, **V**, part V. 2nd edn. Geological Society of America, Boulder, Colorado, and University of Kansas, Lawrence, Kansas.

PALMER, D. & RICKARDS, B. (eds) 1990. *Fossils Illustrated. Vol. 1: Graptolites: Writing in the Rocks.* Boydell Press, Woodbridge, Suffolk, 182 pp.

WOODCOCK, N. H. 1992. Dolyhir Quarries, Old Radnor (SO 245581). *In*: TREAGUS, J. E. (ed.) *Caledonian Structures in Britain South of the Midland Valley.* Geological Conservation Review Chapman and Hall, London, **3**, 154–158.

WOODCOCK, N. H. & BASSETT, M. G. (eds) 1993. *Geological Excursions in Powys, Central Wales.* University of Wales Press, National Museum of Wales, Cardiff, 366 pp.

WYATT, A. R. 2000. *A Turbid Tale: Geology at Aberystwyth.* Self-published, Berkeley, California, 159 pp.

Index

Note: Page numbers in *italic* denote figures. Page numbers in **bold** denote tables.

academia, careers for women, 20th century 89–90, *91*, *92*
Agassiz, Elizabeth C. Cary (1822–1907) *84*, 83
Ainsworth, F. Hilda (1891–1971) **74**
Alps, work of Maria Ogilvie Gordon 307–308
Anderson, Elizabeth *see* Gray, Elizabeth Anderson
Andrews, Mary K., Belfast Naturalists' Field Club 143
Apthorpe, Marjorie 179
Arber, Agnes (1879–1960) *54*
    Newnham College 55
    palaeobotany 54–56, 287
Arber, Edward A. Newell (1870–1918) *54*, 55, 287
Arber, Muriel Agnes (1913–2004) 2, *287*, 288–292
    Geologists' Association 289–292
    Lyme Regis 292
archaeozoology, work of Dorothea Bate 295–302
Archer, Jean Barbara, Geological Survey of Ireland 149
Armstrong College, Newcastle, Edith Bolton 66–67
Atkinson, Louisa (1834–72), New South Wales **168**, 172
Australia
    coal studies 194
    first universities 166, 172
    history of geology 196
    invertebrate palaeontology 184–186, 188–189
    mapping 192, 194
    micropalaeontology 180, 182–184
    mineralogy 194
    palaeobotany/palynology 190, 192
    petrology 192–193
    popularization of geology 196
    scientific societies 167
    vertebrate palaeontology 189–190
    women in geology 165–198, **195**
        19th century 167
        early 20th century 175–178
        first academic staff 175, 176
        late 20th century 178–180

Baguley, Pauline 18
Balfour Biological Laboratories 14, 55
Ball, Mary (1812–98) 140
Bancroft, Nellie **74**
Barnard, Annie *see* Greenly, Annie
Barton, Ethel Sarel (1864–1922), British Museum (Natural History) 104
basalt, origin debate 139
Bascom, Florence (1862–1945) 41, 123–134, *124*, *125*, *126*, *127*, *128*
    Bryn Mawr College 106, 123, 128, 130–131
    fieldwork 129–130, 132, *133*
    Geological Society of America 129
    Heidelberg 128–129
    Johns Hopkins University 127–128
    male mentors 124–129, 130
    Ohio State University 128
    PhD degree 128
    as role model 130–134
    University of Wisconsin 124, 125, 127
    USGS 129–130

Bascom, John (1827–1911) *125*, *126*
    influence on Florence Bascom 124–126
Basnett, Betty *173*, 194
Bassler, Ray Smith (1878–1961), National Museum of Natural History 103, *104*
Bate, Dorothea Minola Alice (1878–1951) *296*, *301*, *302*
    British Museum (Natural History) 107, 296, 297, 301–302
    cave exploration 296–300
    early life 295–296
    palaeontology 296–302
    Pleistocene fauna 86, 297–302, *298*
    as role model 86, 302
Beattie, Joan *see* Crockford, Joan
Bedford College, London 3, 11–12, 15–16, *16*
    Catherine Raisin 12, 23–32, 106
    destination of women geologists 32, *33*, **34**
    Emily Dix 69
    examinations and degrees 12
    fieldwork 119, *120*
    geology department 23–32
    laboratories 14–15, *15*, 23, 25–26
    Morton Sumner bequest 23, 24–25
    Natural History Society 26–27
    staff **23**, *24*, **24**, **25**, 26
Belfast Naturalists' Field Club, Sydney Mary Thompson 143
Benett, Etheldred (1776–1845) *247*
    fossil collection 105, 247–249
    problems of fieldwork 116–117
    professional difficulties 161, 162
Benson, Margaret Jane (1859–1936) *52*
    palaeobotany 51–54, 77
    Royal Holloway College 51
Berridge, Emily M. **74**, 75
Bitgood, Ellen Posey (d.1982), work on oilrigs 121
Blackburn, Kathleen (1892–1968), pollen analysis 89
Bliss, Eleanora *see* Knopf, Eleanora Bliss
'Boat Ladies' *see* 'Steamboat Ladies'
Bolton, Edith (1893–1974)
    palaeobotany 65–68
        Armstrong College Newcastle 66–67
        Bristol Museum and Art Gallery 66, 67–68
        coal measures 66–68
Brailsford, Beatrice 18
Brenchley, Winifred Elsie (1883–1953) **74**
Bristol Museum and Art Gallery, Edith Bolton 66
British Association for the Advancement of Science
    admission of women 140, 321
    first woman president 150
    lectures for women 158
British Geological Survey
    Audrey A. Jackson 150
    first female field geologist 3
    graptolites of Woods and Crosfield 210
    Memoir of Anglesey 319, 321–322

British Museum (Natural History) 3
   Dorothea Bate 3, 107, 295, 296, 301–302
   Marie Stopes 236
   role of women 99–100, 103–105, 107–108, 300
   women fossil collectors 295
Brown, Ida **168**, *173*, 176, 177, 178, *184*, 185, 192
Browne, Lady Isabel Mary Peyronnet (1881–1947) *61*
   palaeobotany 60–63, 77
Bryn Mawr College, Pennsylvania 40
   Florence Bascom 123, 128, 130–131
Buckland, Mary Morland (1797–1857), as assistant to husband 2, 98, 159, 161
Buckland, William (1784–1856), female assistants 159, 161
Buckley, Arabella (1840–1929)
   popular science books 2, 10, 97
   as secretary to Charles Lyell 2
Burek, Cynthia **5**
Burns, Veronica Conroy *109*, *144*
   Dublin Naturalists' Field Club 143, *144*
   Trinity College Dublin, geological museum 101–102, 143

calamites, work of Isabel Browne 62
Calder, Mary Gordon (?1906–92) *72*
   palaeobotany 69, 72–73
Cambridge University
   admission of women 11, 13
   degrees for women 12, 76, 105, 325
   laboratories for women 14
   mixed teaching 13, 17
   Tripos examinations 16–17
   see also Girton College; Newnham College
Carboniferous, palaeobotany, early 20th century 51–77
Carroll, Dorothy **168**, 194
Cary, Elizabeth C. *see* Agassiz, Elizabeth C. Cary
cave exploration
   Dorothea Bate 296–300
   Luella Owen 41–42
Cavendish Laboratories, admission of women 12, 14
cellulose peel technique 72–73
Central Association of Irish School Mistresses (CAISM) 145
Chandler, Marjorie E.J. 104
Chase, Mary Agnes (1869–1963) *109*
   National Herbarium, Smithsonian Institution 103
Cheeseman, Lucy Evelyn (1881–1969), British Museum (Natural History) 103–104
Cheltenham Ladies College 11
   Lucy Wills 63
Children, Anna 98
Cleve, Astrid (1874–1968), diatoms 43–44
Clwydian Range 221, *222*, *223*
   work of Woods and Crosfield 210–213, *211*, *214*–*215*, 216, *217*–*220*, 221
coal measures
   Northumberland and Durham, work of Edith Bolton 66–67
   South Wales, work of Emily Dix 68–69
coal studies, Australia 194
coal-ball palaeobotany 52, 54, 56
   work of Marie Stopes 58–59, 60
   work of Mary Calder 72, 73
Cohen, Fanny **168**, *172*, 176

Colditz, Joan *173*, 194
Cole, Blanche Vernon 98, 148
Connaught Coalfield, work of Patricia Yates 150
Cookson, Isabel Clifton (1893–1973) **168**, 176, 177, 190, *191*, 192
Cooksonia 190
cordaites
   work of Agnes Arber 55
   work of Margaret Benson 53
Coxon, Catherine, Trinity College, Dublin 150
Cranwell, Lucy 192
Crespin, Irene (1896–1980), micropalaeontology **168**, 177, 180, *181*, 182–183, *182*, 194, 196
Crete, cave research, Dorothea Bate 297, 300
Crockford, Joan (b.1919) **168**, *173*, 177, *184*, 185, 186, 194
Crosby, Dorothy 176
Crosfield, Margaret Chorley (1859–1952) 204, 205–206, *206*, 208
   collaboration with Ethel Skeat Woods 204, 209, 210–221
   early female geology student 17
   Geologists' Association *204*, 206, *207*, 209
   work with Gertrude Elles 209–210
Culey, Alma J. 193
curricula 6
cuticles, Carboniferous
   work of Edith Bolton 68
   work of Lucy Wills 63–65, *64*
Cuvier, Clémentine (1805–27) 279–282
Cuvier, Georges, Baron (1769–1832) 277–278, *278*
   daughters' assistance 2, 279–285
Cyprus, cave research, Dorothea Bate 297, 299, 300

David, Tannatt Edgeworth (1858–1934) *170*–*171*, *172*
   Sydney University, support for women students 172, 176
Davies, Dilys, first geology student Newnham College 17
De Fraine, Ethel **74**
Deer, Margaret 176
degrees for women 11, 12, 13, 76, 101, 105
   higher degrees 128, 166
Denmark, Inge Lehmann, seismology 47
diatoms
   Astrid Cleve 43–44
   Australia 194
dinosaurs, Mary Wade *189*, 190
discrimination, in academia 2–5
Dix, Emily (1904–72) *71*
   palaeobotany 68–69, *70*, 77
      Bedford College 32, 69
      fieldwork 119
Dolan, Eva M. 176
Dolomites, work of Maria Ogilvie Gordon 307–308
Donald, Mary Jane *see* Longstaff, Mary Jane
Dornwell, Edith Emily 172
Doyle, Ebhlin, Geological Survey of Ireland 149
Drepanites striatus 247, *248*
Drew, Helen 18
Drummond-Smith, Catherine **168**, *172*, 176
Drury, Susanna (c.1698–c.1770), artist 138–139
   Giant's Causeway *138*, 139
Dublin Naturalists' Field Club 143
   Veronica Burns 143, *144*

Dublin Society 138
Duigan, Suzanne Lawless (1924–93), palynology 89, **168**, 192
Duvaucel, Sophie (1789–1867), assistant to Georges Cuvier 282–285

education, girls
    17th and 18th centuries 9–10
    19th century 10
Elles, Gertrude Lilian (1872–1960) *109*, *206*
    degree from Trinity College, Dublin 13, 106, 146
    graptolites 21–22, 288
    Lower Palaeozoic stratigraphy 22
    Newnham College 13, 17, 21–23, *21*, 106–107, 203
    Sedgwick Club 20
    work with Ethel Skeat Woods and Margaret Crosfield 209–210
Equisetum, work of Isabel Browne 62
Evans, Gwyneth (1908–2003), as assistant to husband 150
examinations, admission of women 12, *13*, 16–17

Falconer, Hugh (1808–65) 251, 252–257, *253*
Farrell, Loreto Philomena Catherine, Geological Survey of Ireland 149
Fern Ledges 227, *229*, *234*, *235*, *239*
    early fossil collectors 228–230
    Geological Survey of Canada mapping project 230–231
    work of Marie Stopes 59–60, 231–243
ferns
    work of Margaret Benson 53
    work of Marie Stopes 59
fieldwork
    Australia 1920s–30s *173*, *174*
    Florence Bascom 129–130, 132, *133*
    Geologists' Association 116, *116*, *117*
    Code for Geological Fieldwork 289, 290–291
    Maria Ogilvie Gordon 118, 119, 306–308
    Nancy Kirk *326*, 327
    Tannatt Edgeworth David, Sydney University *170*, *171*
    and women scientists 4, 13, *14*
        chaperones 118–120, 128
        clothing 117, *124*, *170*, *171*, *173*, *174*
        travel 115–121
        'unseemly activity for women' 117–120
Finnegan, Susan, British Museum (Natural History) 108
foraminifera, Australia 182–183
fossil collectors 39–41, 295
    see also Benett, Etheldred; Burns, Veronica; Fern Ledges; Longstaff, Jane; Rosse, Mary, 3rd Countess of
France, early 19th century, role of women 278–279, 284–285
Fuji, Kenjiro, relationship with Marie Stopes 58, 59

Gardner, Julia Anna (1882–1960) 131
Garwood, Edmund Johnston (1869–1949), University College London, geological museum 101, *102*
Gatty, Margaret (1809–73), popular science books 97
Geikie, Professor Sir Archibald (1835–1924), support for women students 118, 205, 310
gender anonymity 130, 140, 300
geoconservation 223, *224*, 225

geodiversity 223, *224*, 225
geography, physical, women in academia 90, *91*, *92*
Geological Society of America
    attitude to women 129
    work of Florence Bascom 129
Geological Society of Australia, Nell Ludbrook 179
Geological Society of Dublin, exclusion of women 143
Geological Society of London
    attitude to women 3, 22, 83, 158–159, 251, 255, 327, 328
    early women fellows 22, 105, 106, 206
    Grace Milne obituary 262
    Lyell Fund 22, 28, 30, 107
    Lyell Medal 32, 107, 119, 308, 314
    Murchison Fund 106, 205, 327
    Murchison Medal 22, 107
    Prestwich Medal 87
    Wollaston Fund 301
Geological Survey of Canada
    Fern Ledges mapping project 230–231
        Memoir 41, 238–243
    work of Marie Stopes 232–243
Geological Survey of Ireland, role of women 148–149
geologists, female relatives
    as assistants 2, 98–99, 117, *118*, 158, 161, 167
        Anne Phillips 265–275
        Annie Greenly 319–324
        Civil Mary Prestwich 257
        Cuvier's daughters 277–285
        Grace Milne 251, 253–256, 261
        Gwyneth Evans 150
        Mary McKenny Hughes 86, *118*, 119
        Mary Morland Buckland 2, 98, 159, 161
    as chaperones 86, 118–119
Geologists' Association
    attitude to women 3, 83, 116, 289
    Code for Geological Fieldwork 289–291
    contribution of Margaret Crosfield *204*, 206, *207*, 209
    contribution of Muriel Arber 289–292
    field trips 116, *116*, *117*, 121, *204*, 206, *207*, 272
    Stopes Medal 86, 87
geomorphology, Marjorie Sweeting 86–87
Germany
    traditional *Hausfrau*, 19th century 156–157, *158*
    women admitted to higher education 310
    women in geology, early 19th century 155, 159, *160*, 162
GETSET initiative 4
Giant's Causeway, Susanna Drury *138*, 139
Gilbert-Tomlinson, Joyce (1916–81) *181*, 182
Girton College, Cambridge 11
Goldring, Winifred (1888–1971) *109*
    palaeontology 110
Goldschmidt, Victor Mordecai (1853–1933), influence on Florence Bascom *126*, 128–129, 130, *133*
Goodyear, Edith *109*
    University College London, geological museum 101, *102*
Gordon Cumming, Lady Eliza Maria (c.1798–1842), Old Red Sandstone, fossil fish 39, 40–41, 161–162
Gordon, Isabella (1901–88), British Museum (Natural History) 107

Gordon, Maria Matilda Ogilvie (1864–1939) *119*, 205, *306*, *311*
  fieldwork 118, 119
    Dolomites 306–308, 311–312, *314*
  first woman DSc 308–309
  marriage 309–310
  in Munich 119, 204, 305–306, 308, 310
  publications 312–314
  recognition 314, 316–317
  Royal Society paper on corals 310
  work for women's and children's rights 311, 312, 315
Gormley, Mary, Geological Survey of Ireland 149
Gortynskaia, Maria Vasilievna *see* Pavlova, Maria Vasilievna
graptolites
  Australia 184–185
  work of Gertrude Elles 21–22
  work of Nancy Kirk 327–329, *330*, *331*
  work of Woods and Crosfield 210–213, *215*, *216*
Gray, Elizabeth Anderson (1831–1924), Lower Palaeozoic fossils, Girvan 39, 41
Gray, Elizabeth 205
Greenly, Annie (1852–1927) 319, *320*, *323*
  as assistant to husband 320–322, 324
  as geologist 322–324
  work on Geological Survey Memoir of Anglesey 322
Greenly, Edward (1861–1951) *320*
  Geological Survey Memoir of Anglesey 319, 321–322
  Highlands of Scotland 320–321
Grey, Kath 179
Groom, Gillian E., Quaternary science 87
Grosvenor Museum, Chester, graptolites of Woods and Crosfield *208*, 210, *211*
Grove, Jean Mary Clark (1927–2001) 88
  Quaternary science 87–88
  The Little Ice Age 88

Halket, Ann C. **74**
Hastings, Anna Birchall (1902–77) *109*
  British Museum (Natural History) 108
Hastings, Barbara Yelverton, Marchioness of (1810–58), fossil collection 161, 295
Hertwig, Richard von (1850–1937), support for Maria Ogilvie Gordon 305, 310
Higgs, Bettie
  Geological Survey of Ireland 149
  University College, Cork 150
higher education
  attitudes to women 4–6
  17th and 18th centuries 9–10
  degrees for women 12, 13, 76, 101, 105
  geology, late 19th century 15–21
  institutions for women, 19th century 10–13
  role of women 9–36
  USA, 19th century 123, 124, 127
Hill, Dorothy (1907–97) 166, **168**, 176, 177, 178–179, 186, *187*, 194, 196
Hipparion 45
Hobson, Margaret, as assistant to husband 167, **168**
Hol, Jacoba (1886–1964), Quaternary science 84
Hoppin, Ruth 41–42
Hosking, Lucy (b.1919) **168**, 177, 185

illustrators, women 2, 98, 149, 167, 251, 254
  see also Cole, Blanche; Drury, Susanna; Leakey, Mary; Milne, Grace Anne; Ward, Mary; Wieser, Francisca
inclusivity, in academia 4
industry, female scientists 6, 179
Ireland
  education for women 139–140
  Mesolithic tools 137
  women in history of geology 137–151
  women in universities 4, 143–148
    staff 150–151
Irving, Roland Duer (1847–1880), influence on Florence Bascom 126, *126*

Jackson, Audrey A., Trinity College Dublin 150
Jebb, Louisa 20
Johns Hopkins University
  attitude to women students 127–128
  Florence Bascom 123, 127–128
Johnston, Joan 185
Johnston, Mary 205, 206
Jonas, Anna *see* Stose, Anna Jonas
Jones, Dilys Penelope Lindsey (b.1929), Geological Survey of Ireland 149
Joplin, Germaine (c.1910–89), petrology **168**, 176, 177, 192–193, *193*

King, Cuchlaine (b.1922), geomorphology 90
King, Georgina (1845–1932) 172
Kingsley, Mary Henrietta (1862–1900), British Museum (Natural History) 103
Kirk, Nancy (1916–2005) 325–331, *326*
  difficulties with Geological Society of London 327, 328
  fieldwork *326*, 327
  graptolites 327–329, *330*, *331*
  interest in botany 329–330
  University College of Wales, Aberystwyth 326–329
Knopf, Eleanora Bliss (1883–1974) 129, 130, 131, 132
Knowles, Matilda Cullen (1864–1933), Museum of Science and Art, Dublin 102–103

laboratories, for female students 14–15, 23
Lady Margaret Hall, Oxford 11
Lang, William Henry (1874–1960)
  Manchester University 59, 75
  support for women students 76
  work with Isabel Cookson 190
Lapworth, Professor Charles (1842–1920)
  fieldwork 119, *121*, 206, *207*
  Ordovician 203, 210
  support for female students 2
Lawless, Suzanne *see* Duigan, Suzanne Lawless
Leakey, Mary Nicol (1913–96), Quaternary science 88–89
Lee, Sarah (1791–1856) 278, 279, 282, 284
Lehmann, Inge (1888–1993), seismology 47
Lewarne, Gillian Clare, Trinity College, Dublin 150
Lexique stratigraphique international 194
Lindsey, Marjorie **74**

Linnean Society
    Margaret Benson 53
    women Fellows 106
Lister, Gulielma (1860–1949), British Museum (Natural
    History) 104
Little, Jessie Leonora (b.1867) 172
loess, work of Luella Owen 42
London University *see* University of London
Longstaff, Mary Jane Donald (1855–1935),
    gastropods 106
Lonsdale, Kathleen (1903–71), Bedford College 149–150
Lowe, Irene Helen, Bedford College 26, **27**, 30, **32**
Ludbrook, Nell Hooper (1907–95) 176–177, 179–180,
    180, 183, 186, 188
lycophytes
    work of Agnes Arber 55
    work of Margaret Benson 52–53
    work of Mary Calder 72

McBriar, E. Maude 194, 196
MacDonald, Jane 205
McInerny, Kathleen *see* Sherrard, Kathleen
McKenny Hughes, Mary Caroline (1862–1916)
    as assistant to husband 86, *118*, 119
McKenny Hughes, Professor Thomas (1832–1917) 22
    education of women in geology 2, 3, 13, *14*, 17, 18, *22*,
        106, 145, 206
Maclean, Constance 175
McMillan, Nora Fisher (1908–2003), Liverpool
    Museum 108
McWhae, Kathleen *see* Pike, Kathleen
Majorca, cave research, Dorothea Bate 297–298, *299*
Malvern Hills, work of Anne Phillips 271–272, *273*, 274
Manchester University
    Lang, W.H. 59, 75
    Marie Stopes 58–59
    Mary Calder 73
    see also Victoria University, Manchester
Manton, Irene (1904–88), British Museum (Natural
    History) 104–105
Manton, Sidnie (1902–79), British Museum (Natural
    History) 104–105
mapping, Australia 192
Marcet, Jane Haldimand (1769–1858), *Conversations* 139
Marker, Margaret, Quaternary science 87
Marr, Professor John Edward (1857–1933), education of
    women in geology 17, 18, 206
marriage, and academic careers 75, 130, 149
    Agnes Arber 55
    Australia 172, 175, 179, 184, 185, 186, 188, 192
    BMNH 3, 108
    British Geological Survey 3
    Ethel Wood 17
    Geological Survey of Ireland 149
    Gillian Lewarne 150
    Maria Gordon Ogilvie 309–310
    Museum of Science and Art Dublin 103
    USA 103, 130
Matheson, Bettie, *see* Higgs, Bettie
Matthew, George F. (1837–1923) 228, *230*, *233*
    work with Marie Stopes 233, 234, 235, 237–238
Maury, Carlotta Joaquina (1874–1938)
    petroleum geology and palaeontology 42–43
    fieldwork 120

Mediterranean, cave research, Dorothea Bate 297–301
Medwell, Lorna N. **168**, 190, 192
Melbourne University, first women students 172
Menorca, cave research, Dorothea Bate 297
micropalaeontology, Australia 180, 182–184
Miller, Barbara Thompson 143
Milne, Grace Anne (1832–99) 251–263, *258*
    as assistant to Joseph Prestwich 261
    as companion/secretary to Hugh Falconer 253–256
    illustration 254, *255*, *256*
    marriage to Joseph Prestwich 257
    novels 259–260
    popular geological writing 2, 260–261
mineralogy
    Australia 194
    Naima Sahlbom 44
mining
    Australia 166
    male attitude to women 3, 166
Moore, Kate, National University of Ireland 150
Morland, Mary *see* Buckland, Mary Morland
motherhood
    and geological careers 4, 75, 86, 87, 120, 150,
        309–310
    Australia 166, 179, 185, 186, 188, 192
Muir-Wood, Helen, (1895–1968) palaeobrachiopodology,
    British Museum (Natural History) **32**, 101,
    107, *108*
Munich University
    Ethel Woods 18, 204
    Evgenia Solomko 46
    Maria Gordon 119, 204, 305–306, 308, 310
    Marie Stopes 56, 58
    segregated lectures 46, 119, 306
    see also Zittel, Karl von
Munro, Madeline, Carboniferous bryozoans 101
Murchison, Charlotte (1788–1869) 98
Murchison, Roderick Impey (1792–1871), attitude
    towards women 251
Museum of Science and Art, Dublin 102–103
museums
    role of women
        early-mid 20th century 2, 66, 99–110
        volunteers 99–102, 103, 104

Nashar, Beryl *see* Scott, Beryl
National Museum of Natural History (Smithsonian
    Institution), Washington D.C. 103
natural history
    Irish societies 143
    a 'suitable activity' for women 97–98, 139
    women in museums
        academics 106–107
        early-mid 20th century 99–110
        national museums 102–105
        research scientists 107–108, *109*, 110
        researchers 105–106
        university collections 100–102
Natural History Museum *see* British Museum (Natural
    History)
Neptunists 139
Newnham College, Cambridge 3, 11, 75, 76
    Agnes Arber 55
    destinations of women geologists 32, *35*

Newnham College, Cambridge (*Continued*)
   Ethel Skeat 17, 204
   Ethel Wood 17–18
   freshers 1900 *19*
   geology 15–23
      first female students 17
   Igerna Sollas 99
   Lilian Elles 17, 106, 203
   Lucy Wills 63
   Margaret Benson 51
   Margaret Crosfield 17
   Nancy Kirk 325
   staff *18*
Noeggerathiales, work of Isabel Browne 62
North London Collegiate School 11, 27
   Marie Stopes 56

Ogilvie Gordon, Maria Matilda *see* Gordon, Maria Matilda Ogilvie
Ogilvie, Ida (1874–1963) 106, 131
Ohio State University, Florence Bascom 128
Old Red Sandstone, fossil fish, work of Lady Eliza Maria Gordon Cumming 40–41
Oliver, Francis W. (1864–1952), support for women students 51, 55, 56, 62, 73, **74**, 76
Open University, women scientists 5
Orton, Edward Francis Baxter (1829–1899), influence on Florence Bascom *126*, 128
Owen, Luella (1852–1932), cave exploration 41–42
Owen, Charlotte, as assistant to husband 98
Owen's College, Manchester 12
   see also Manchester University
Oxford University
   admission of women 11
   degrees for women 12, 105
   see also Lady Margaret Hall; Somerville College
Oxford University Museum of Natural History 266

palaeobotany
   Australia 190, 192
   Carboniferous, early 20th century 51–77
   male support for women students 76
      see also Lang, W.H.; Oliver, F.W.
palaeobrachiopodology, work of Helen Muir-Wood 101, 107, *108*
palaeontology
   Australia 180–190, **195**
   Carlotta Maury 42–43
   Dorothea Bate 296–302
   Evgenia Solomko 46
   invertebrate, Australia 184–186, 188–189
   Maria Pavlova 45–46
   Maria Tsvetaeva 46
Palaeozoic, Lower, work of Ethel Woods and Margaret Crosfield 203–225
Palestine, excavation of Dorothea Bate 300–301, *301*
palynology, Australia 190, 192
Pauss, Olga (1890/91?–1948) 175
Pavlova, Maria Vasilievna (1854–1938), Tertiary palaeontology 45–46
petroleum geology
   Carlotta Maury 42–43, 120
   discrimination against women 179, 183

   Ellen Bitgood 120
   micropalaeontology, Irene Crespin 182–183
petrology, Australia 192–193
Pettycur flora 52, 53
Phillips, Anne (1803–62) 265–275, *265*
   assistant to brother John 98, 269–275
   Malvern Hills conglomerate 271–272, *273*, *274*
Phillips, John (1800–74) 265–275, *265*
   Oxford University Museum of Natural History 266
Phillips, June R. (b.1931) 166, **168**, *178*, 179
Pike, Kathleen **168**, 190, *191*, 192
pleistocene fauna, Dorothea Bate 297–301
popular science books 19, 39, 97, 139, 260–261
Porter, Mary (1886–1980) 131
Powell, Dorothy K. **168**, *174*, 176
Prankerd, Theodora Lisle (1878–1939) **74**
Prendergast, Kathleen Laura (1913?–54) **168**, 177, 185
Prestwich, Lady Grace *see* Milne, Grace Anne
Prestwich, Joseph (1812–96) 251, 257–262, *257*
   chair of geology Oxford 259, 261
   marriage to Grace Milne 257
professors, female, UK 4, **5**, 75
pteridosperms
   work of Edith Bolton 67
   work of Margaret Benson 52, 53
   work of Marie Stopes 56, 59
   work of Mary Calder 72

quarries, 'unsuitable for women' 116, 117
Quaternary science
   role of women
      20th century **85**, 86–90, *91*, *92*
         Dorothea Bate 295–302
      21st century **85**, 90, *91*, *92*, 93
      pre-20th century 83–86, **85**
Queen's College, Cork 140
   admission of women 146–147
   women staff 147–148
Quisenberry, Adelaide Corinna (b.1881), National Museum of Natural History 103, *104*
Quodling, Florence Mabel (1901–85) **168**, 177, 194

Radcliffe College 83
Raisin, Catherine Alice (1855–1946) 17, *31*
   Bedford College 12, 23–32, 106
   fieldwork 119
   role model 27–32
Regionally Important Geological Sites (RIGS) 210, 223, *224*, 225
Reid, Eleanor Mary (1860–1953), British Museum (Natural History) 104
Reynolds, Doris Livesey (1899–1985)
   Bedford College 32, **32**
   work in Ireland 149
Richtohofen, Ferdinand, Freiherr von (1833–1905), friendship with Marie Ogilvie Gordon 118, 305, 306
Ripper, Elizabeth (Betty) (1909–2004) **168**, *173*, 184
role models 4–6, 34
   Catherine Raisin 27–32
   Dorothea Bate 86, 302
   Florence Bascom 130–134
   Gertrude Elles 21–23

Mary McKenny Hughes 86
Mary Somerville 86
Ross, June R. *see* Phillips, June R.
Rosse, Mary, 3rd Countess of (1813–85) *141*
   fossil collection 140–141, *141*, *142*
Rothschild, Miriam (1908–2005), British Museum (Natural History) 104
Royal Cork Institution 140
Royal Geographical Society, attitude to women 83, 85
Royal Holloway College, London 12, 75, 76
   Margaret Benson 51
   women professors 4, **5**
Royal Institution, London 140, 158
Royal Irish Academy 138
Royal Society
   attitude to women 10, 300
   first woman president 150
   paper by Maria Ogilvie Gordon 310
Russia
   Evgenia Solomko, palaeontology 46
   Maria Pavlova (1854–1938), Tertiary mammals 45–46
   Maria Tsvetaeva, palaeontology 46

Sahlbom, Naima (1871–1957), mineralogy 44
Sando, Margaret 194
Schreuder, Antje (1887–1952), Quaternary science 84
scientific societies
   Australia 167
   Ireland 138
   negative attitude to women 2–3, 10, 83, 140, 300
   see also under name of society
scientists, female, characteristics 2
Scott, Beryl (b.1923) 166, **168**, 177, *178*, 180, 194
Scott, Dukinfield Henry (1854–1934), support for women students 76
Scott, Henderina Victoria (Rina) **74**, 75
Sedgwick Club *20*, 288
   admission of women 18–21, **22**
Sedgwick Museum, women geologists 288
segregation, in universities 46, 119, 127, 306
seismology, Inge Lehmann 47
Shaen Wing, Bedford College 14
Shakespeare, Gilbert 20
Shakespear, Dame Ethel *see* Wood, Ethel
Sharpe, Hilda 18
Sherrard, Kathleen (1898–1975) **168**, 176, 177, 178, 184–185
Sidgwick, Eleanor Mildred (Nora) (1845–1936), on coeducation 145
Silurian
   Clwydian Range, work of Woods and Crosfield 210–221
   Malvern Hills, work of Anne Phillips 271–272, *273*, 274
'Silurian problem' 203, 210
Skeat, Ethel Gertrude *see* Woods, Ethel Gertrude Skeat
Slater, Ida 18, 20, 26
Smith, Annie Lorrain (1854–1937), British Museum (Natural History) 104
Smith, Catherine, Sydney University 167, 172
Smith, Edith Margaret (1913–67) 188–189, 194

Smithsonian Institution *see* National Museum of Natural History
society, attitudes to women 17th and 18th century 2–3, 9–10
Sollas, Hertha Beatrice Coryn (b.1875), assistant to William Johnson Sollas 99
Sollas, Igerna Brünhilda Johnson (1877–1965)
   at Newnham College 18, 99
   work with William Johnson Sollas 99, *100*
Solomko, Evgenia (1862–1898), palaeontology 46
Somerville College, Oxford 11, 139, 261
Somerville, Mary (Fairfax) (1780–1872) *85*
   indignation at role of *Hausfrau* 157
   Physical Geography 85, 139
   Quaternary science 85–86
South Wales Coalfield, work of Emily Dix 68–69
speleology *see* cave exploration
'Steamboat Ladies' 11, 12–13, 105, 106, 146, *147*
Stephens, Jane (b.1879), Museum of Science and Art, Dublin 103
Stopes, Marie Charlotte Carmichael (1880–1958) 57, 227, *228*, *232*
   palaeobotany 56–60
   Ancient Plants 59
   coal studies 60
   Fern Ledges 59–60, 231–243
   Manchester University 58–59
   relationship with Kenjiro Fuji 58, 59
Stose, Anna Jonas (1881–1974) 129, 130, 131
stratigraphy, Lower Palaeozoic, work of Gertrude Elles 22
Sullivan, Judy 179
Sweden
   Astrid Cleve (1874–1968), diatoms 43–44
   Naima Sahlbom (1871–1957), mineralogy 44
Sweet, Georgina (1875?–1946) 166, 167, 172
Sweet, Jessie M., British Museum (Natural History) 107–108
Sweeting, Marjorie (1920–94) *87*
   geomorphology 86–87
Sydney University
   Tannatt Edgeworth David *172*
   fieldwork *170*, *171*
   women in geology **175**, 176

Talbot, Mignon (1869–1950) 106, 131
Tappan, Helen, fieldwork 120
Tharp, Maria, mid-ocean ridges 47
Thompson, Sydney Mary (1847–1923), Belfast Naturalists' Field Club 143
Tomlinson, Mabel (1916–81) *89*, **168**
   Quaternary science 88, 89
travel, historical problems 115–121
Trewavas, Ethelwynn (1900–93), British Museum (Natural History) 107
Trinity College, Dublin
   admission of women 144–146
   degrees for women 11, 12–13, 146, *147*, 205
   exclusion of women 140
   geological collection 101, 138
   Veronica Conroy Burns 143
Tsvetaeva, Maria (b.1854), palaeontology 46
Tutin, Winifred Pennington, Lake Windermere 90

United States of America, early women geologists 41–43, 106, 123
United States Geological Survey
   work of Florence Bascom 123, 129–130
universities
   Australia 166, 172
   degrees for women 12, 13, 76, 101, 105
   Ireland 143–148
      female staff 150–151
   natural history collections, role of women 100–102
   UK, female professors 4, **5**, 6, 75, 90
   USA
      admission to women 123, 124, 127
      inequality 124
University College London
   Agnes Arber 55, 287
   degrees for women 76, 101
   geological museum 101
   Isabel Browne 62
   Margaret Benson 51
   Marie Stopes 56, 59–60
University College, Cork 150
University of London 11–12
   examinations and degrees for women 12, *13*
   see also Bedford College; Royal Holloway College
University of Wisconsin
   attitude to women students 124
   Florence Bascom 124, 126, 127

Van Hise, Charles Robert (1857–1919), influence on Florence Bascom 126, *126*, 131–132
Vernon, Blanche *see* Cole, Blanche Vernon
vertebrate palaeontology
   Australia 189–190
   Dorothea Bate 297–301
   Maria Pavlova 45–6
Victoria University
   Manchester 12
   see also Manchester University
Vincent, Adele V. (b.1896?) **168**, 177, 190
volunteers, museums 99–102, 103, 104
Vulcanists 139

Wade, Mary (1932–2005) **168**, 183–184, *184*, 188, 189–190, *189*
Ward, Mary (1827–69) 140, 141, 143
Westfield College, London 73, 75, 76
Wieser, Francisca, National Museum of Natural History 103
Wigglesworth, Grace **74**, 75
Wilkinson, Jean Margaret (b.1937), Geological Survey of Ireland 149
Williams, George Huntington (1856–94), influence on Florence Bascom *126*, 127–128
Wills, Lucy (1888–1964) 65
   palaeobotany 63–65
      cuticle studies 63–65, *64*
      Newnham College, Cambridge 63
   Royal Free Hospital 65
   work on folic acid 65
Women into Science and Engineering (WISE) initiative 4
Wood, Ethel (1871–1946) *17*, 106, 107
   at Newnham College 17–18
   work on graptolites 21
Woods, Ethel Gertrude Skeat (1865–1939) 203–205, *204*
   at Newnham College 17
   Chester 18, 204–205, *208*
   collaboration with Margaret Crosfield 204, 209, 210–221
   degree from Trinity College, Dublin 13, 205
   Munich 18, 204
   Murchison Fund 205
   Newnham College 204
   work with Gertrude Elles 209–210
Woods, Nell *see* Ludbrook, Nell Hooper
World War I, effect on employment of women 105
Wright, Mabel Crawford MacDowell 148

Yates, Patricia (1932–60), work in Connaught Coalfield 150

Zittel, Karl von (1839–1904)
   segregated lectures 46, 119, 306
   support for women students 18, 46, 204, 305–306, 310